MULTISCALE STOCHASTIC VOLATILITY FOR EQUITY, INTEREST RATE, AND CREDIT DERIVATIVES

Building upon the ideas introduced in their previous book, *Derivatives in Financial Markets with Stochastic Volatility*, the authors study the pricing and hedging of financial derivatives under stochastic volatility in equity, interest rate, and credit markets. They present and analyze multiscale stochastic volatility models and asymptotic approximations. These can be used in equity markets, for instance, to link the prices of path-dependent exotic instruments to market implied volatilities. The methods are also used for interest rate and credit derivatives. Other applications considered include variance-reduction techniques, portfolio optimization, forward-looking estimation of CAPM "beta," and the Heston model and generalizations of it.

"Off-the-shelf" formulas and calibration tools are provided to ease the transition for practitioners who adopt this new method. The attention to detail and explicit presentation make this also an excellent text for a graduate course in financial and applied mathematics.

JEAN-PIERRE FOUQUE studied at the University Pierre et Marie Curie in Paris. He held positions at the French CNRS and Ecole Polytechnique, and at North Carolina State University. Since 2006, he is Professor and Director of the Center for Research in Financial Mathematics and Statistics at the University of California Santa Barbara.

GEORGE PAPANICOLAOU was Professor of Mathematics at the Courant Institute before coming to Stanford University in 1993. He is now Robert Grimmett Professor in the Department of Mathematics at Stanford.

RONNIE SIRCAR is a Professor in the Operations Research and Financial Engineering department at Princeton University, and an affiliate member of the Bendheim Center for Finance and the Program in Applied and Computational Mathematics.

KNUT SØLNA is a Professor in the Department of Mathematics at the University of California at Irvine. He received his undergraduate and Master's degrees from the Norwegian University of Science and Technology, and his doctorate from Stanford University. He was an instructor at the Department of Mathematics, University of Utah before coming to Irvine.

MULTISCALE STOCHASTIC VOLATILITY FOR EQUITY, INTEREST RATE, AND CREDIT DERIVATIVES

JEAN-PIERRE FOUQUE
University of California, Santa Barbara

GEORGE PAPANICOLAOU
Stanford University

RONNIE SIRCAR
Princeton University

KNUT SØLNA
University of California, Irvine

CAMBRIDGE
UNIVERSITY PRESS

CAMBRIDGE
UNIVERSITY PRESS

Shaftesbury Road, Cambridge CB2 8EA, United Kingdom

One Liberty Plaza, 20th Floor, New York, NY 10006, USA

477 Williamstown Road, Port Melbourne, VIC 3207, Australia

314–321, 3rd Floor, Plot 3, Splendor Forum, Jasola District Centre, New Delhi – 110025, India

103 Penang Road, #05–06/07, Visioncrest Commercial, Singapore 238467

Cambridge University Press is part of Cambridge University Press & Assessment,
a department of the University of Cambridge.

We share the University's mission to contribute to society through the pursuit of
education, learning and research at the highest international levels of excellence.

www.cambridge.org
Information on this title: www.cambridge.org/9780521843584

© J.-P. Fouque, G. Papanicolaou, R. Sircar, and K. Sølna 2011

First published 2011

A catalogue record for this publication is available from the British Library

ISBN   978-0-521-84358-4   Hardback

To our families and students

Contents

Introduction   *page* xi

1   The Black–Scholes Theory of Derivative Pricing   1
   1.1   Market Model   1
   1.2   Derivative Contracts   10
   1.3   Replicating Strategies   13
   1.4   Risk-Neutral Pricing   21
   1.5   Risk-Neutral Expectations and Partial Differential
        Equations   27
   1.6   American Options and Free Boundary Problems   32
   1.7   Path-Dependent Derivatives   36
   1.8   First-Passage Structural Approach to Default   43
   1.9   Multidimensional Stochastic Calculus   46
   1.10   Complete Market   49

2   Introduction to Stochastic Volatility Models   51
   2.1   Implied Volatility Surface   52
   2.2   Local Volatility   57
   2.3   Stochastic Volatility Models   62
   2.4   Derivative Pricing   65
   2.5   General Results on Stochastic Volatility Models   74
   2.6   Summary and Conclusions   83

3   Volatility Time Scales   86
   3.1   A Simple Picture of Fast and Slow
        Time Scales   86
   3.2   Ergodicity and Mean-Reversion   88
   3.3   Examples of Mean-Reverting Processes   95
   3.4   Time Scales in Synthetic Returns Data   110
   3.5   Time Scales in Market Data   114
   3.6   Multiscale Models   118

4   **First-Order Perturbation Theory**   121
  4.1   Option Pricing under Multiscale Stochastic Volatility   121
  4.2   Formal Regular and Singular Perturbation Analysis   125
  4.3   Parameter Reduction   135
  4.4   First-Order Approximation: Summary and Discussion   137
  4.5   Accuracy of First-Order Approximation   138

5   **Implied Volatility Formulas and Calibration**   148
  5.1   Approximate Call Prices and Implied Volatilities   149
  5.2   Calibration Procedure   154
  5.3   Illustration with S&P 500 Data   155
  5.4   Maturity Cycles   163
  5.5   Higher-Order Corrections   174

6   **Application to Exotic Derivatives**   179
  6.1   European Binary Options   179
  6.2   Barrier Options   181
  6.3   Asian Options   185

7   **Application to American Derivatives**   189
  7.1   American Options Valuation under Stochastic
        Volatility   189
  7.2   Stochastic Volatility Correction for American Put   190
  7.3   Parameter Reduction   196
  7.4   Summary   197

8   **Hedging Strategies**   199
  8.1   Black–Scholes Delta Hedging   200
  8.2   The Strategy and its Cost   200
  8.3   Mean Self-Financing Hedging Strategy   206
  8.4   A Strategy with Frozen Parameters   209
  8.5   Strategies Based on Implied Volatilities   217
  8.6   Martingale Approach to Pricing   220
  8.7   Non-Markovian Models of Volatility   226

9   **Extensions**   232
  9.1   Dividends and Varying Interest Rates   232
  9.2   Probabilistic Representation of the Approximate Prices   237
  9.3   Second-Order Correction from Fast Scale   238
  9.4   Second-Order Corrections from Slow and Fast Scales   247
  9.5   Periodic Day Effect   249
  9.6   Markovian Jump Volatility Models   251
  9.7   Multidimensional Models   254

10 **Around the Heston Model**   259
  10.1 The Heston Model   260
  10.2 Approximations to the Heston Model   265

	10.3	A Fast Mean-Reverting Correction to the Heston Model	271
	10.4	Large Deviations and Short Maturity Asymptotics	276
11	**Other Applications**		**283**
	11.1	Application to Variance Reduction in Monte Carlo Computations	283
	11.2	Portfolio Optimization under Stochastic Volatility	287
	11.3	Application to CAPM Forward-Looking Beta Estimation	296
12	**Interest Rate Models**		**307**
	12.1	The Vasicek Model	307
	12.2	The Bond Price and its Expansion	315
	12.3	The Quadratic Model	327
	12.4	The CIR Model	330
	12.5	Options on Bonds	335
13	**Credit Risk I: Structural Models with Stochastic Volatility**		**342**
	13.1	Single-Name Credit Derivatives	342
	13.2	Multiname Credit Derivatives	353
14	**Credit Risk II: Multiscale Intensity-Based Models**		**377**
	14.1	Background on Stochastic Intensity Models	377
	14.2	Multiname Credit Derivatives	385
	14.3	Symmetric Vasicek Model	388
	14.4	Homogeneous Group Structure	402
15	**Epilogue**		**424**
	References		430
	Index		439

Introduction

This book is about pricing and hedging financial derivatives under stochastic volatility in equity, interest rate, and credit markets. We demonstrate that the introduction of two time scales in volatility, a fast and a slow, is needed and is efficient for capturing the main features of the observed term structures of implied volatility, yields, or credit spreads. The present book builds on and replaces our previous book, *Derivatives in Financial Markets with Stochastic Volatility*, published by Cambridge University Press in 2000.

We present an approach to derivatives valuation and hedging which consists of integrating singular and regular perturbation techniques in the context of stochastic volatility. The book has a dual purpose: to present "off-the-shelf" formulas and calibration tools, and to introduce, explain, and develop the mathematical framework to handle the multiscale asymptotics.

There are many books on financial mathematics (mostly for introductory courses at the level of the Black–Scholes model). Primarily, these books deal with the case of constant volatilities, be it for stock prices, interest rates, or default intensities. This book is about analyzing these models in the presence of stochastic volatility using the powerful tools of perturbation methods. The book can be used for a second-level graduate course in Financial and Applied Mathematics.

Our goal is to address the following fundamental problem in pricing and hedging derivatives: how can traded call and put options, quoted in terms of implied volatilities, be used to price and hedge more complicated contracts? Modeling the underlying asset usually involves the specification of a multifactor Markovian model under the risk-neutral pricing measure. Calibration of the parameters of that model to the observed implied volatilities, including the market prices of risk, is a challenging task because of the complex relation between option prices and model parameters (through a pricing partial differential equation, for instance). The main difficulty is to find models which will produce stable parameter estimates. We like to think of this problem as the "(K,T,t)-problem": for a given present time t and a fixed maturity T, it is usually easy with low-dimensional models to fit the skew with respect to strikes K. Getting a good fit of the term structure of implied volatility, that is when a range of observed maturities T is taken into account, is a much harder problem that can be handled with a sufficient number of parameters. The main problem remains: the stability with respect to t of these calibrated parameters. This is a crucial quality to have if one wants to use the model to compute no-arbitrage prices of more complex path-dependent derivatives, since in this case the distribution over time of the underlying is central.

Modeling directly the evolution of the implied volatility surface is a promising approach but involves some complicated issues. One has to make sure that the model is free of arbitrage or, in other words, that the surface is produced by some underlying under a risk-neutral measure. This is known to be a difficult task, and the choice of a model and its calibration is also an important issue in this approach. But most importantly, in order to use this modeling to price other path-dependent contracts, one has to identify a corresponding underlying which typically does not lead to a low-dimensional Markovian evolution.

Wouldn't it be nice to have a direct and simple connection between the observed implied volatilities and prices of more complex path-dependent contracts! Our objective is to provide such a linkage. This is done by using a combination of singular and regular perturbation techniques corresponding respectively to fast and slow time scales in volatility. We obtain a parametrization of the implied volatility surface in terms of Greeks, which involves four parameters at the first order of approximation. This procedure leads to parameters which are exactly those needed to price other contracts at this level of approximation. In our previous work presented in Fouque *et al.* (2000), we used only the fast volatility time scale combined with a statistical estimation of an effective constant volatility from historical data. The introduction of the slow volatility time scale enables us to capture more accurately the behavior of the term structure of implied volatility at long maturities. Yet, we preserve a parsimonious parametrization which effectively and robustly captures the main effects of time scale heterogeneity. Moreover, in the framework presented here, statistics of historical data are not needed for the calibration of these parameters.

Thus, in summary, we directly link the implied volatilities to prices of path-dependent contracts by exploiting volatility time scales. Furthermore, we extend this approach to interest rate and credit derivatives.

In Chapter 1 we review the basic ideas and methods of the Black–Scholes theory as well as the tools of stochastic calculus underpinning the models used. Chapter 2 provides a general introduction to stochastic volatility models. In Chapter 3, we identify time scales in financial data and introduce them in stochastic volatility models. In Chapter 4 we present the first-order perturbation theory in the context of European equity derivatives and identify the important parameters arising in this asymptotic analysis. This is the central chapter on the mathematical tools used in our multiscale modeling approach. In Chapter 5 we provide a calibration procedure for these parameters using observed implied volatilities. Indeed, these are the parameters that provide a parsimonious linkage between various contracts. We also show in this chapter how to extend the perturbation techniques to the case with time-dependent parameters needed for practical fitting of the presented S&P 500 data. The extensions to exotic and American claims are described in Chapters 6 and 7. It is also natural to exploit the presence of a skew of implied volatilities for designing hedging strategies of part of the volatility risk by trading the underlying. This is achieved in Chapter 8 by using the asymptotic analysis presented in the previous chapters combined with a martingale argument, which in turns can be used to derive asymptotics in the case of non-Markovian models of volatility. In Chapter 9 we present several extensions to the perturbation theory, including the cases with dividends and varying interest rates, and the derivation of the second-order corrections. Next, in Chapter 10, we discuss the Heston model, which is very popular for its computational tractability. We implement our perturbation theory on this particular model, we show how to generalize it while retaining its tractability, and we derive large deviation results in the regime of short maturities and fast mean-reverting volatility. Applications to variance-reduction techniques for Monte Carlo simulations, to portfolio optimization, and to estimation of CAPM Beta parameters are presented in Chapter 11. After introducing the basics of fixed income markets, we demonstrate in Chapter 12 that our perturbation approach is also effective for interest rates models with stochastic volatility. Then, we introduce the fundamental concepts used in credit risk modeling, and we apply our method to both single-name and multiname credit derivatives using structural models in Chapter 13 and intensity-based models in Chapter 14.

One cannot write a book in 2011 on financial mathematics without commenting on the recent financial crisis. We choose to do so in the Epilogue – Chapter 15 – since it involves judgement and behavior of the market players rather than mathematical modeling as presented in this book.

1

The Black–Scholes Theory of Derivative Pricing

The aim of this first chapter is to review the basic objects, ideas, and results of the classical Black–Scholes theory of derivative pricing. It is intended for readers who want to enter the subject or simply refresh their memory. This is not a complete treatment of this theory with detailed proofs but rather an intuitive but precise presentation including a few key calculations. Detailed presentations of the subject can be found in many books at various levels of mathematical rigor and generality, a few of which we list in the notes at the end of the chapter.

This book is about extending the Black–Scholes theory using perturbation methods in order to handle markets with stochastic volatility. The notation and many of the tools used in the constant volatility case will be used for the more complex markets throughout the book.

1.1 Market Model

In this simple model, suggested by Samuelson and used by Black and Scholes, there are two assets. One is a riskless asset (bond) with price β_t at time t described by the ordinary differential equation

$$d\beta_t = r\beta_t\, dt, \qquad (1.1)$$

where r, a non-negative constant, is the instantaneous interest rate for lending or borrowing money. Setting $\beta_0 = 1$, we have $\beta_t = e^{rt}$ for $t \geq 0$. The price X_t of the other asset, the risky stock or stock index, evolves according to the stochastic differential equation

$$dX_t = \mu X_t dt + \sigma X_t\, dW_t, \qquad (1.2)$$

where μ is a constant mean return rate, $\sigma > 0$ is a constant *volatility*, and $(W_t)_{t\geq 0}$ is a standard *Brownian motion*. This fundamental model and the intuitive content of equation (1.2) are presented in the following sections.

1.1.1 Brownian Motion

Brownian motion is a stochastic process whose definition, existence, properties, and applications were the subject of numerous studies during the twentieth century (and still are, in the twenty-first). Our goal here is to give a very intuitive and practical presentation.

A Brownian motion is a real-valued stochastic process with continuous trajectories that have independent and stationary increments. The trajectories are denoted by $t \to W_t$ and for the standard Brownian motion, we have that:

- $W_0 = 0$;
- for any $0 < t_1 < \cdots < t_n$, the random variables $(W_{t_1}, W_{t_2} - W_{t_1}, \ldots, W_{t_n} - W_{t_{n-1}})$ are independent;
- for any $0 \le s < t$, the increment $W_t - W_s$ is a centered (mean-zero) normal random variable with variance $\mathbb{E}\{(W_t - W_s)^2\} = t - s$. In particular, W_t is $\mathcal{N}(0,t)$-distributed.

Denote by $(\Omega, \mathscr{F}, \mathbb{P})$ the probability space where our Brownian motion is defined and the expectation $\mathbb{E}\{\cdot\}$ is computed. For example, it could be $\Omega = \mathscr{C}([0, +\infty) : \mathbb{R})$, the space of all continuous trajectories ω, with $W_t(\omega) = \omega(t)$. The σ-algebra $\mathscr{F}$ contains all sets of the form $\{\omega \in \Omega : |\omega(s)| < R, s \le t\}$; the Wiener measure, $\mathbb{P}$, is the probability distribution of the standard Brownian motion.

The increasing family of σ-algebras $\mathscr{F}_t$ generated by $(W_s)_{s \le t}$, the information on W up to time t, and all the sets of probability 0 in $\mathscr{F}$, is called the *natural filtration* of the Brownian motion. This *completion* by the null sets is important, in particular for the following reason. If two random variables X and Y are equal almost surely ($X = Y$ $\mathbb{P}$-a.s. means $\mathbb{P}\{X = Y\} = 1$) and if X is $\mathscr{F}_t$-measurable (meaning that any event $\{X_t \le x\}$ belongs to $\mathscr{F}_t$) then Y is also $\mathscr{F}_t$-measurable.

A stochastic process $(X_t)_{t \ge 0}$ is *adapted* to the filtration $(\mathscr{F}_t)_{t \ge 0}$ if the random variable X_t is $\mathscr{F}_t$-measurable for every t. We say that (X_t) is $(\mathscr{F}_t)$-adapted. If another process (Y_t) is such that $X_t = Y_t$ $\mathbb{P}$-a.s. for every t then it is also $(\mathscr{F}_t)$-adapted.

The independence of the increments of the Brownian motion and their normal distribution can be summarized using *conditional characteristic functions*. For $0 \le s < t$ and $u \in \mathbb{R}$

$$\mathbb{E}\left\{e^{iu(W_t - W_s)} \mid \mathscr{F}_s\right\} = e^{-\frac{u^2(t-s)}{2}}. \tag{1.3}$$

If W is a Brownian motion, by independence of the increment $W_t - W_s$ from the past $\mathscr{F}_s$, the left-hand side of (1.3) is simply $\mathbb{E}\left\{e^{iu(W_t - W_s)}\right\}$, which is the characteristic function of a centered normal random variable with variance $t - s$, and is equal to the right-hand side. Conversely, if (1.3) holds, then the continuous process (W_t) is a standard Brownian motion.

This independence of increments makes the Brownian motion an ideal candidate for defining a complete family of independent infinitesimal increments dW_t, which are centered, normally distributed with variance dt and which will serve as a model of (Gaussian white) noise. The drawback is that the trajectories of (W_t) cannot be "nice" in the sense that they are not of bounded variation, as the following simple computation suggests. Let $t_0 = 0 < t_1 < \cdots < t_n = t$ be a subdivision of $[0, t]$, which we may suppose evenly spaced so that $t_i - t_{i-1} = t/n$ for each interval. The quantity

$$\mathbb{E}\left\{\sum_{i=1}^n |W_{t_i} - W_{t_{i-1}}|\right\} = n\mathbb{E}\{|W_{\frac{t}{n}}|\} = n\sqrt{\frac{t}{n}}\mathbb{E}\{|W_1|\}$$

goes to $+\infty$ as $n \nearrow +\infty$, indicating that the integral with respect to dW_t cannot be defined in the usual way "trajectory by trajectory." We describe how such integrals can be defined in the next section.

1.1.2 Stochastic Integrals

For T a fixed finite time, let $(X_t)_{0 \leq t \leq T}$ be a continuous stochastic process adapted to $(\mathscr{F}_t)_{0 \leq t \leq T}$, the filtration of the Brownian motion up to time T, such that

$$\mathbb{E}\left\{\int_0^T X_t^2 dt\right\} < +\infty. \tag{1.4}$$

Using iterated conditional expectations and the independent increments property of Brownian motion, we note that with $t_0 < t_1 < \cdots < t_n = t$

$$\mathbb{E}\left\{\left(\sum_{i=1}^n X_{t_{i-1}}(W_{t_i} - W_{t_{i-1}})\right)^2\right\} = \mathbb{E}\left\{\sum_{i=1}^n (X_{t_{i-1}})^2 (t_i - t_{i-1})\right\},$$

for $t \leq T$, which is a basic calculation in the construction of stochastic integrals. Note also that the Brownian increments on the left are forward in time and that the sum on the right converges to $\mathbb{E}\{\int_0^t X_s^2 ds\}$, which is finite by (1.4).

The *stochastic integral* of (X_t) with respect to the Brownian motion (W_t) is defined as a limit in the mean-square sense $(L^2(\Omega))$

$$\int_0^t X_s dW_s = \lim_{n \nearrow +\infty} \sum_{i=1}^n X_{t_{i-1}} \left(W_{t_i} - W_{t_{i-1}} \right), \qquad (1.5)$$

as the mesh size of the subdivision goes to zero.

As a function of time t, this stochastic integral defines a continuous square integrable process such that

$$\mathbb{E}\left\{ \left(\int_0^t X_s dW_s \right)^2 \right\} = \mathbb{E}\left\{ \int_0^t X_s^2 ds \right\}, \qquad (1.6)$$

and has the *martingale property*

$$\mathbb{E}\left\{ \int_0^t X_u dW_u \mid \mathscr{F}_s \right\} = \int_0^s X_u dW_u \qquad \mathbb{P}\text{-a.s., for } s \leq t, \qquad (1.7)$$

as can easily be deduced from the definition (1.5). The *quadratic variation* $\langle Y \rangle_t$ of the stochastic integral $Y_t = \int_0^t X_u dW_u$ is

$$\langle Y \rangle_t = \lim_{n \nearrow +\infty} \sum_{i=1}^n (Y_{t_i} - Y_{t_{i-1}})^2 = \int_0^t X_s^2 ds \qquad (1.8)$$

in the mean-square sense.

Stochastic integrals are zero-mean, continuous, and square integrable martingales. It is interesting to note that the converse is also true: every zero-mean, continuous, and square integrable martingale is a Brownian stochastic integral. This representation result will be made precise and used in Section 1.4.

1.1.3 Risky Asset Price Model

The Black–Scholes model for the risky asset price corresponds to a continuous process (X_t) such that, in an infinitesimal amount of time dt, the infinitesimal return dX_t/X_t has mean μdt, proportional to dt, with a constant *rate of return* μ, and centered random fluctuations independent of the past up to time t. These fluctuations are modeled by σdW_t, where σ is a positive constant *volatility* which measures the strength of the noise, and dW_t the infinitesimal increments of the Brownian motion. The corresponding formula for the infinitesimal return is

$$\frac{dX_t}{X_t} = \mu dt + \sigma dW_t, \qquad (1.9)$$

which is the stochastic differential equation (1.2). The right-hand side has the natural financial interpretation of a return term plus a risk term. We are also assuming that there are no dividends paid in the time interval that we are considering. It is easy to incorporate a continuous dividend rate in all that follows, but for simplicity we shall omit this here.

In integral form, this equation is

$$X_t = X_0 + \mu \int_0^t X_s ds + \sigma \int_0^t X_s dW_s, \tag{1.10}$$

where the last integral is a stochastic integral as described in Section 1.1.2 and where X_0 is the initial value, which is assumed to be independent of the Brownian motion and square integrable.

Equation (1.10), or (1.2) in the differential form, is a particular case of a general class of stochastic differential equations driven by a Brownian motion:

$$dX_t = \mu(t, X_t) dt + \sigma(t, X_t) dW_t, \tag{1.11}$$

or in integral form

$$X_t = X_0 + \int_0^t \mu(s, X_s) ds + \int_0^t \sigma(s, X_s) dW_s. \tag{1.12}$$

In the Black–Scholes model, $\mu(t, x) = \mu x$ and $\sigma(t, x) = \sigma x$; these are independent of t, differentiable in x, and linearly growing at infinity (since they are linear). This is enough to ensure existence and uniqueness of a continuous adapted and square integrable solution (X_t). The proof of this result is based on simple estimates like

$$\mathbb{E}\{X_t^2\} = \mathbb{E}\left\{\left(X_0 + \mu \int_0^t X_s ds + \sigma \int_0^t X_s dW_s\right)^2\right\}$$
$$\leq 3\left(\mathbb{E}\{X_0^2\} + (\mu^2 T + \sigma^2) \int_0^t \mathbb{E}\{X_s^2\} ds\right),$$

where we used the inequality $(a + b + c)^2 \leq 3(a^2 + b^2 + c^2)$, the Cauchy–Schwarz inequality

$$\mathbb{E}\left(\int_0^t X_s ds\right)^2 \leq t \int_0^t \mathbb{E}\{X_s^2\} ds,$$

and (1.6). We deduce

$$0 \leq \mathbb{E}\{X_t^2\} \leq c_1 + c_2 \int_0^t \mathbb{E}\{X_s^2\} ds,$$

for $0 \leq t \leq T$ and constants c_1 and $c_2 \geq 0$. By a direct application of Gronwall's lemma, we deduce that the solution is *a priori* square integrable. The construction of a solution and the proof of uniqueness can be obtained by similar and slightly more complicated estimates that use the Kolmogorov–Doob inequality for martingales.

Looking at equation (1.9), it is very tempting to write X_t/X_0 explicitly as the exponential of $(\mu t + \sigma W_t)$. However, this is not correct because the usual chain rule is not valid for stochastic differentials. For instance W_t^2 is not equal to $2\int_0^t W_s dW_s$ as might be expected since, by the martingale property (1.7), this last integral has an expectation equal to zero but $\mathbb{E}\left\{W_t^2\right\} = t$.

This discrepancy is corrected by Itô's formula, which we explain now.

1.1.4 Itô's Formula

A function of the Brownian motion W_t defines a new stochastic process $g(W_t)$. We suppose in the following that the function g is twice continuously differentiable, bounded, and has bounded derivatives. The purpose of the chain rule is to compute the differential $dg(W_t)$, or equivalently its integral $g(W_t) - g(W_0)$. Using the subdivision $t_0 = 0 < t_1 < \cdots < t_n = t$ of $[0,t]$, we write

$$g(W_t) - g(W_0) = \sum_{i=1}^{n} (g(W_{t_i}) - g(W_{t_{i-1}})).$$

We then apply Taylor's formula to each term to obtain

$$g(W_t) - g(W_0) = \sum_{i=1}^{n} g'(W_{t_{i-1}})(W_{t_i} - W_{t_{i-1}})$$

$$+ \frac{1}{2}\sum_{i=1}^{n} g''(W_{t_{i-1}})(W_{t_i} - W_{t_{i-1}})^2 + R,$$

where R contains all the higher-order terms.

If (W_t) were differentiable only the first sum would contribute to the limit as the mesh size of the subdivision goes to zero, leading to the chain rule $dg(W_t) = g'(W_t)W_t' dt$ of classical calculus. In the Brownian case, (W_t) is not differentiable and, by (1.5), the first sum converges to the stochastic integral

$$\int_0^t g'(W_s) dW_s.$$

The correction comes from the second sum which, like (1.8), converges to

$$\frac{1}{2}\int_0^t g''(W_s) ds,$$

as can be seen by comparing it in L^2 with $\frac{1}{2}\sum_{i=1}^{n} g''(W_{t_{i-1}})(t_i - t_{i-1})$. The higher-order terms contained in R converge to zero and do not contribute to the limit, which is

$$g(W_t) - g(W_0) = \int_0^t g'(W_s)dW_s + \frac{1}{2}\int_0^t g''(W_s)ds. \qquad (1.13)$$

This is the simplest version of Itô's formula. It is often written in differential form:

$$dg(W_t) = g'(W_t)dW_t + \frac{1}{2}g''(W_t)dt. \qquad (1.14)$$

The next step is deriving a similar formula for $dg(X_t)$, where X_t is the solution of a stochastic differential equation like (1.11). We give here this general formula for a function g depending also on time t:

$$dg(t,X_t) = \frac{\partial g}{\partial t}(t,X_t)dt + \frac{\partial g}{\partial x}(t,X_t)dX_t + \frac{1}{2}\frac{\partial^2 g}{\partial x^2}(t,X_t)d\langle X\rangle_t, \qquad (1.15)$$

where dX_t is given by the stochastic differential equation (1.11) and

$$\langle X\rangle_t = \int_0^t \sigma^2(s,X_s)ds$$

is the quadratic variation of the martingale part of X_t: that is, of the stochastic integral on the right-hand side of (1.12). In terms of dt and dW_t the formula is

$$dg(t,X_t) =$$
$$\left(\frac{\partial g}{\partial t} + \mu(t,X_t)\frac{\partial g}{\partial x} + \frac{1}{2}\sigma^2(t,X_t)\frac{\partial^2 g}{\partial x^2}\right)dt + \sigma(t,X_t)\frac{\partial g}{\partial x}dW_t, \qquad (1.16)$$

where all the partial derivatives of g are evaluated at (t,X_t).

As an application we can compute the differential of the discounted price $g(t,X_t) = e^{-rt}X_t$:

$$d\left(e^{-rt}X_t\right) = -re^{-rt}X_t dt + e^{-rt}dX_t$$
$$= e^{-rt}\left(-rX_t + \mu(t,X_t)\right)dt + e^{-rt}\sigma(t,X_t)dW_t, \qquad (1.17)$$

since the second derivative of $g(t,x) = xe^{-rt}$ with respect to x is zero. In the particular case of the price X_t given by (1.2), $\mu(t,x) = \mu x$ and $\sigma(t,x) = \sigma x$ so we obtain

$$d\left(e^{-rt}X_t\right) = (\mu - r)\left(e^{-rt}X_t\right)dt + \sigma\left(e^{-rt}X_t\right)dW_t. \qquad (1.18)$$

The discounted price $\widetilde{X}_t = e^{-rt}X_t$ satisfies the same equation as X_t where the return μ has been replaced by $\mu - r$.

1.1.5 Lognormal Risky Asset Price

Coming back to the stochastic differential equation (1.9) for the evolution of
the stock price X_t, it is natural to suspect from the ordinary calculus formula
$\int dx/x = \log x$ that $\log X_t$ might satisfy an equation that we can integrate
explicitly. We compute the differential of $\log X_t$ by applying Itô's formula
(1.16) with $g(t,x) = \log x$, $\mu(t,x) = \mu x$, and $\sigma(t,x) = \sigma x$:

$$d\log X_t = \left(\mu - \frac{1}{2}\sigma^2\right)dt + \sigma dW_t.$$

The logarithm of the stock price is then given explicitly by

$$\log X_t = \log X_0 + \left(\mu - \frac{1}{2}\sigma^2\right)t + \sigma W_t,$$

which leads to the following formula for the stock price:

$$X_t = X_0\exp\left((\mu - \frac{1}{2}\sigma^2)t + \sigma W_t\right). \tag{1.19}$$

The return X_t/X_0 is *lognormal*: it is the exponential of a nonstandard Brownian motion which is normally distributed with mean $\left(\mu - \frac{1}{2}\sigma^2\right)t$ and
variance $\sigma^2 t$ at time t. The process (X_t) is also called *geometric* Brownian motion. The stock price given by (1.19) satisfies equation (1.9). It can
also be obtained as a diffusion limit of binomial tree models which arise
when Brownian motion is approximated by a random walk.

Notice that, if $X_0 = 0$, X_t stays at zero at all times thereafter. Thus in this
model, bankruptcy (zero stock price) is a permanent state. However, W_t is
finite at all times, and therefore, if $X_0 > 0$, X_t remains positive at all times.

In Figure 1.1, we show a sample path or realization of a geometric
Brownian motion (X_t) in which $\mu = 0.15$, $\sigma = 0.1$, and $X_0 = 95$. This
path exhibits the "average growth plus noise" behavior we expect from this
model of asset prices.

1.1.6 Ornstein–Uhlenbeck Process

Many financial quantities, volatility amongst them, are modeled as *mean-
reverting* processes, a term we shall explain in more detail in Chapters 2
and 3. The simplest example of a mean-reverting diffusion is the Ornstein–
Uhlenbeck process, defined as a solution of

$$dY_t = \alpha(m - Y_t)dt + \beta\,dW_t, \tag{1.20}$$

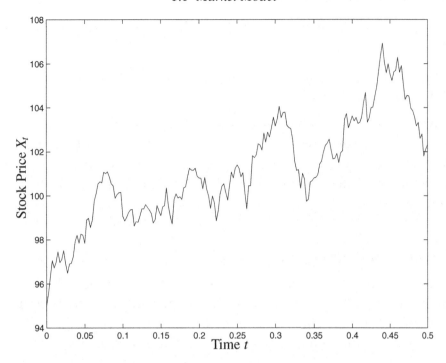

Figure 1.1   A sample path of a geometric Brownian motion defined by the stochastic differential equation (1.9), with $\mu = 0.15$, $\sigma = 0.1$, and $X_0 = 95$.

where α and β are positive constants. This is one of the few explicitly solvable stochastic differential equations, which we illustrate here as an application of Itô's formula.

First, we rearrange the terms to write

$$dY_t + \alpha Y_t\, dt = \alpha m\, dt + \beta\, dW_t.$$

Multiplying through by the "integrating factor" $e^{\alpha t}$ gives

$$d(e^{\alpha t} Y_t) = \alpha m e^{\alpha t}\, dt + \beta e^{\alpha t}\, dW_t,$$

where the left-hand exact integral is easily checked from Itô's formula (1.15). Integrating from zero to t and multiplying through by $e^{-\alpha t}$ gives

$$Y_t = m + (y - m)e^{-\alpha t} + \beta \int_0^t e^{-\alpha(t-s)}\, dW_s, \qquad (1.21)$$

where y is its (assumed known) starting value.

From this representation, it follows that Y is a Gaussian process and the distribution of Y_t is $\mathcal{N}\left(m + (y - m)e^{-\alpha t}, \frac{\beta^2}{2\alpha}(1 - e^{-2\alpha t})\right)$. Its long-run distribution, obtained as $t \to \infty$, is $\mathcal{N}(m, \beta^2/2\alpha)$, which does not depend on y.

The concept of long-run (invariant) distribution will be discussed in detail in Chapter 3.

1.2 Derivative Contracts

Derivatives are contracts based on the underlying asset price (X_t). They are also called *contingent claims*. We will be interested primarily in *options*, which can be European, American, path-independent, or path-dependent. The definition of the options discussed in this first chapter is given in the following sections.

1.2.1 European Call and Put Options

A *European call option* is a contract that gives its holder the right, but not the obligation, to buy one unit of an underlying asset for a predetermined *strike price* K on the *maturity* date T. If X_T is the price of the underlying asset at maturity time T, then the value of this contract at maturity, its *payoff*, is

$$h(X_T) = (X_T - K)^+ = \begin{cases} X_T - K & \text{if } X_T > K, \\ 0 & \text{if } X_T \le K, \end{cases} \qquad (1.22)$$

since in the first case the holder will exercise the option and make a profit $X_T - K$ by buying the stock for K and selling it immediately at the market price X_T. In the second case the option is not exercised, since the market price of the asset is less than the strike price.

Similarly, a *European put option* is a contract that gives its holder the right, but not the obligation, to sell a unit of the asset for a strike price K at the maturity date T. Its payoff is

$$h(X_T) = (K - X_T)^+ = \begin{cases} K - X_T & \text{if } X_T < K, \\ 0 & \text{if } X_T \ge K, \end{cases} \qquad (1.23)$$

since in the first case buying the stock at the market price and exercising the put option yields a profit of $K - X_T$, and in the second case the option is simply not exercised.

More generally, we will consider European derivatives defined by their maturity time T and their non-negative payoff function $h(x)$. This will be a contract which pays $h(X_T)$ at maturity time T when the stock price is X_T. The standard European-style derivatives are *path-independent* because the payoff $h(X_T)$ is only a function of the value of the stock price at maturity time T.

At time $t < T$ this contract has a value, known as the *derivative price*, which will vary with t and the observed stock price X_t. This option price at time t and for a stock price $X_t = x$ is denoted by $P(t,x)$: the problem of *derivative pricing* is determining this pricing function. The fact that this option price will depend only on the observed value at time t and not on the past values of the stock price is closely related to the *Markov property* shared by the solutions of stochastic differential equations like (1.11), by which we shall model the stock price. More details on this will be given in Section 1.5.

Perhaps the simplest way to price such a derivative is as the expected value of its discounted payoff. More precisely, if the stock price is the process (1.2) and the observed stock price $X_0 = x$, the option price at time $t = 0$ would be

$$P(0,x) = \mathbb{E}\left\{ e^{-rT} h(X_T) \right\}$$
$$= \mathbb{E}\left\{ e^{-rT} h\left(x e^{(\mu - \frac{\sigma^2}{2})T + \sigma W_T} \right) \right\}, \qquad (1.24)$$

where we have used the explicit formula (1.19) with $X_0 = x$ and time T. The expectation reduces to a Gaussian integral since W_T is $\mathcal{N}(0,T)$-distributed. In general (unless $\mu = r$) the option price given by formula (1.24) leads to an *arbitrage opportunity*, meaning that there will be a risk-free way to make a profit with strictly positive expectation by holding a particular portfolio. This is one of the key ideas in Section 1.3 that is used to determine the *fair option price*.

1.2.2 American Options

An *American option* is a contract in which the holder decides whether to exercise the option or not at any time of his choice before the option's expiration date T. The time τ at which the option is exercised is called the *exercise time*. Because the market cannot be anticipated, the holder of the option has to decide to exercise or not at time $t \leq T$ with information up to time t contained in the σ-algebra $\mathcal{F}_t$. In other words, τ is a random time such that the event $\{\tau \leq t\}$ (and its complement $\{\tau > t\}$) belongs to $\mathcal{F}_t$ for any $t \leq T$. Such a random time is called a *stopping time* with respect to the filtration $(\mathcal{F}_t)$. If the payoff function of the derivative is h, then its value at the exercise time τ is $h(X_\tau)$, where X_τ is the stock price at the stopping time τ.

For an *American call option* the payoff is $h(X_\tau) = (X_\tau - K)^+$ for a given strike price K and a stopping time $\tau \leq T$ chosen by the holder of the option. Note that even if τ is an exercise time, the option will be exercised only if

$X_\tau > K$ but, in any case, the contract is terminated at time τ. Similarly, the payoff of an *American put option* is $h(X_\tau) = (K - X_\tau)^+$ and the option is exercised only if $K > X_\tau$.

As in the case of European derivatives, an intuitive way to price an American derivative at time $t = 0$ is to maximize the expected value of the discounted payoff over all the stopping times $\tau \leq T$:

$$P(0,x) = \sup_{\tau \leq T} \mathbb{E}\left\{e^{-r\tau}h(X_\tau)\right\}. \tag{1.25}$$

Again, this price leads in general to an opportunity for arbitrage and therefore cannot be the fair price of the derivative.

1.2.3 Other Exotic Options

The term "exotic option" refers here to any option contract which is not a standard European or American option described in the previous sections. Our aim is not to write a catalogue of existing options but rather to give some examples of exotic options that we will use in the rest of the book.

Barrier options are path-dependent options whose payoff depends on whether or not the underlying asset price hits a specified value during the option's lifetime. For instance, a *down-and-out call option* becomes worthless, or *knocked out*, if, at any time t before the expiration date T, the stock price X_t falls below a predetermined level B. The payoff at expiration T is a function of the trajectory of the stock price

$$h = (X_T - K)^+ \mathbf{1}_{\{\inf_{t \leq T} X_t > B\}}. \tag{1.26}$$

Here $\mathbf{1}_A(x) = 1$ if $x \in A$ and $\mathbf{1}_A(x) = 0$ if x is not in A. It is the indicator function of the set A. This option is obviously less valuable than a standard European call option given by (1.22) with the same strike K and maturity T, and it will lead to a *knock-out discount*.

Lookback options are path-dependent options whose payoff functions depend on the minimum or maximum price of the underlying asset during the lifetimes of the options. In particular, a *standard lookback call option* has a payoff at maturity given by

$$h = \left(X_T - \inf_{t \leq T} X_t\right)^+ = X_T - \inf_{t \leq T} X_t, \tag{1.27}$$

where the lowest price plays the role of a *floating* strike price. Similarly, a *standard lookback put option* has a payoff given by

$$h = \left(\sup_{t \leq T} X_t - X_T \right)^+ = \sup_{t \leq T} X_t - X_T, \qquad (1.28)$$

where the highest price plays the role of the floating strike price. Note that these options are not genuine option contracts since they are almost always exercised, since $h > 0$ ($\mathbb{P}$-a.s.) as can be seen from (1.27) and (1.28).

A **forward-start** or **cliquet** option is like a call option for instance, where the strike price is set at a later time. If $t < T_1 < T$, then the payoff at maturity T is given by

$$h = (X_T - X_{T_1})^+, \qquad (1.29)$$

where the stock price at time T_1 becomes the strike price.

Compound options are options on options. For instance, a *call-on-call* is the right to buy a call option at a later time for a predetermined price. It is an option written on an underlying which is itself an option with longer maturity. If $t < T_1 < T$, then the payoff at maturity T_1 is given by

$$h = (C_{T_1}(K,T) - K_1)^+, \qquad (1.30)$$

where $C_{T_1}(K,T)$ is the price at time T_1 of a call option which pays $(X_T - K)^+$ at maturity time T.

Our last example is **Asian options** whose payoff depends on the average stock price during a specified period of time before maturity. They can be European or American with typical payoffs like

$$h = \left(X_T - \frac{1}{T} \int_0^T X_s ds \right)^+, \qquad (1.31)$$

for an *arithmetic-average strike call option* (European style), where the strike price is the average stock price.

1.3 Replicating Strategies

The Black–Scholes analysis of a European-style derivative yields an explicit trading strategy in the underlying risky asset and riskless bond whose terminal payoff is equal to the payoff $h(X_T)$ of the derivative at maturity, no matter what path the stock price takes. Thus, selling the derivative and holding a dynamically adjusted portfolio according to this strategy "covers" an investor against all risk of eventual loss, because a loss incurred at the final time from one part of this *portfolio* will be exactly compensated by a gain in the other part. This *replicating strategy*, as it is known, therefore provides an insurance policy against the risk of being short the derivative. It is called a *dynamic hedging strategy*, since it involves continuous trading, where to hedge means to eliminate risk. The essential step in the Black–Scholes methodology is the construction of this replicating strategy and arguing, based on *no-arbitrage*, that the value of the replicating portfolio at time t is the fair price of the derivative. We develop this argument in the following sections.

1.3.1 Replicating Self-Financing Portfolios

We consider a European-style derivative with payoff $h(X_T)$, a function of the underlying asset price at maturity time T. Assume that the stock price (X_t) follows the geometric Brownian motion model (1.19), solution of the stochastic differential equation (1.2). A *trading strategy* is a pair (a_t, b_t) of adapted processes specifying the number of units held at time t of the underlying asset and the riskless bond, respectively. We suppose that $\mathbb{E}\left\{\int_0^T a_t^2 dt\right\}$ and $\int_0^T |b_t| dt$ are finite so that the stochastic integral involving (a_t) and the usual integral involving (b_t) are well-defined.

Assuming, as in (1.1), that the price of the bond at time t is $\beta_t = e^{rt}$, the value at time t of this portfolio is $a_t X_t + b_t e^{rt}$. It will *replicate* the derivative at maturity if its value at time T is almost surely equal to the payoff:

$$a_T X_T + b_T e^{rT} = h(X_T). \tag{1.32}$$

In addition, this portfolio is to be *self-financing*, meaning that the variations of its value are due only to the variations of the market – that is, the variations of the stock and bond prices. No further funds are required after the initial investment so that if, for example, more of the asset is bought (a_t is increased), then money would have to be obtained by selling bonds (b_t decreased) to pay for it. This is expressed in differential form as

$$d\left(a_t X_t + b_t e^{rt}\right) = a_t dX_t + r b_t e^{rt} dt. \tag{1.33}$$

In integral form, the self-financing property is

$$a_t X_t + b_t e^{rt} = a_0 X_0 + b_0 + \int_0^t a_s dX_s + \int_0^t r b_s e^{rs} ds, \; 0 \le t \le T.$$

An intuitive way to understand this relation is to think in terms of discrete trading times $\{t_n, n = 0, 1, \ldots\}$. The portfolio consists of a_{t_n} and b_{t_n} of the stock and bond (respectively) at time t_n. When the prices change to $X_{t_{n+1}}$ and $e^{rt_{n+1}}$, we observe the change and *then* we adjust our holdings to $a_{t_{n+1}}$

and $b_{t_{n+1}}$. As no further cash input or output is allowed, the value of the portfolio after adjustment must equal the value before, so

$$a_{t_n} X_{t_{n+1}} + b_{t_n} e^{rt_{n+1}} = a_{t_{n+1}} X_{t_{n+1}} + b_{t_{n+1}} e^{rt_{n+1}}.$$

This says that

$$a_{t_{n+1}} X_{t_{n+1}} + b_{t_{n+1}} e^{rt_{n+1}} - \left(a_{t_n} X_{t_n} + b_n e^{rt_n} \right)$$
$$= a_{t_n} \left(X_{t_{n+1}} - X_{t_n} \right) + b_{t_n} \left(e^{rt_{n+1}} - e^{rt_n} \right),$$

which in continuous time becomes (1.33).

1.3.2 The Black–Scholes Partial Differential Equation

As in Section 1.2.1, the pricing function for a European-style contract with payoff $h(X_T)$ is denoted by $P(t,x)$. At this stage we do not know that we can find such a function relating the option price only to the present risky asset price and not to its history. Nevertheless, we shall assume that such a pricing function $P(t,x)$ exists and is regular enough to apply Itô's formula (1.16). Our goal is to construct a self-financing portfolio (a_t, b_t) that will replicate the derivative at maturity (1.32).

The **no-arbitrage** condition requires that

$$a_t X_t + b_t e^{rt} = P(t, X_t), \text{ for any } 0 \le t \le T. \tag{1.34}$$

For if at some time $t < T$ the left-hand side of (1.34) is (say) less than the right-hand side, an *arbitrage opportunity* exists by selling the over-priced derivative security immediately and investing in the under-priced asset–bond trading strategy, yielding an instant profit with no exposure to future loss since the terminal payoff of the trading strategy is equal to the payoff of the derivative.

Differentiating (1.34) and using the self-financing property (1.33) on the left-hand side, Itô's formula (1.16) on the right-hand side, and equation (1.2), we obtain

$$\left(a_t \mu X_t + b_t r e^{rt} \right) dt + a_t \sigma X_t dW_t$$
$$= \left(\frac{\partial P}{\partial t} + \mu X_t \frac{\partial P}{\partial x} + \frac{1}{2} \sigma^2 X_t^2 \frac{\partial^2 P}{\partial x^2} \right) dt + \sigma X_t \frac{\partial P}{\partial x} dW_t, \tag{1.35}$$

where all the partial derivatives of P are evaluated at (t, X_t). Equating the coefficients of the dW_t terms gives

$$a_t = \frac{\partial P}{\partial x}(t, X_t). \tag{1.36}$$

From (1.34) we get

$$b_t = (P(t,X_t) - a_t X_t) e^{-rt}. \tag{1.37}$$

Equating the dt terms in (1.35) gives

$$r\left(P - X_t \frac{\partial P}{\partial x}\right) = \frac{\partial P}{\partial t} + \frac{1}{2}\sigma^2 X_t^2 \frac{\partial^2 P}{\partial x^2}, \tag{1.38}$$

which is satisfied for any stock price X_t if $P(t,x)$ is the solution of the *Black–Scholes partial differential equation*

$$\mathscr{L}_{BS}(\sigma)P = 0, \tag{1.39}$$

where the Black–Scholes operator is defined by

$$\mathscr{L}_{BS}(\sigma) = \frac{\partial}{\partial t} + \frac{1}{2}\sigma^2 x^2 \frac{\partial^2}{\partial x^2} + r\left(x\frac{\partial}{\partial x} - \cdot\right). \tag{1.40}$$

Equation (1.39) holds in the domain $t \le T$ and $x > 0$, since in our model the stock price remains positive. It is to be solved *backward in time* with the final condition $P(T,x) = h(x)$, because at expiration the price of the derivative is simply its payoff.

The partial differential equation (1.39) with its final condition has a unique solution $P(t,x)$, which is the value of a self-financing replicating portfolio. Knowing P, the portfolio (a_t, b_t) is uniquely determined by (1.36) and (1.37).

Surprisingly, the rate of return μ does not enter at all in the valuation of this porfolio, owing to the cancellation of the μ terms in equation (1.35) after the determination (1.36) of a_t. This is a key fact in the setting of the fair price of the derivative as an expected payoff given in Section 1.4. It is a remarkable feature of the Black–Scholes theory that if two investors have different speculative views about the growth rate of the risky asset – meaning that they have different values of μ but agree that the (commonly estimated and stable) historical volatility σ will prevail – then they still agree on the no-arbitrage price of the derivative, since P does not depend on μ.

1.3.3 Pricing to Hedge

There is another way to derive the Black–Scholes partial differential equation that emphasizes risk elimination or hedging. It is a reinterpretation of the calculations in the previous section as follows.

Let $P_t = P(t, X_t)$ be the price of the option. If we sell N_t options and hold A_t units of the risky asset X_t, then the change in the value of this portfolio is $A_t dX_t - N_t dP_t$ because it is assumed to be self-financing. We now determine (A, N) so that this portfolio is riskless, which means that we set to zero the coefficient of dW_t. The change in the value of the portfolio should then equal that of a riskless asset, so

$$A_t dX_t - N_t dP_t = r(A_t X_t - N_t P_t) dt.$$

Using (1.9) and Itô's formula we have

$$A_t(\mu X_t dt + \sigma X_t dW_t) -$$
$$N_t \left\{ \left(\frac{\partial P}{\partial t} + \mu X_t \frac{\partial P}{\partial x} + \frac{1}{2}\sigma^2 X_t^2 \frac{\partial^2 P}{\partial x^2} \right) dt + \sigma X_t \frac{\partial P}{\partial x} dW_t \right\}$$
$$= r(A_t X_t - N_t P_t) dt.$$

Eliminating the dW_t terms gives

$$A_t = N_t \frac{\partial P}{\partial x}(t, X_t),$$

and then the terms involving μ cancel also. We are thus left with the Black–Scholes partial differential equation (1.39) for $P(t, x)$.

In this derivation of the Black–Scholes pricing equation, the role of hedging is clear. Selling the option and holding a dynamically adjusted amount of the risky asset so as to eliminate risk determines the price of the option P_t and the *hedge ratio* A_t/N_t. This is known as **Delta hedging** and the ratio $a_t = A_t/N_t$ given by (1.36) is called the Delta.

1.3.4 The Black–Scholes Formula

For European call options described in Section 1.2.1, the Black–Scholes partial differential equation (1.39) is solved with the final condition $h(x) = (x - K)^+$. Prices of European calls at time t and for an observed risky asset price $X_t = x$ will be denoted by $C_{BS}(t, x)$. In this particular case, there is a closed-form solution known as the *Black–Scholes formula*:

$$C_{BS}(t, x) = xN(d_1) - Ke^{-r(T-t)}N(d_2), \tag{1.41}$$

where

$$d_1 = \frac{\log(x/K) + \left(r + \frac{1}{2}\sigma^2\right)(T - t)}{\sigma\sqrt{T - t}}, \tag{1.42}$$

$$d_2 = d_1 - \sigma\sqrt{T - t}, \tag{1.43}$$

and

$$N(z) = \frac{1}{\sqrt{2\pi}} \int_{-\infty}^{z} e^{-y^2/2} dy, \qquad (1.44)$$

is the standard Gaussian cumulative distribution function. The probabilistic derivation of the Black–Scholes formula (1.41) is given in Section 1.4.4.

This convenient formula for the price of a call option – in terms of the current stock price x, the *time-to-maturity* $T - t$, the strike price K, the volatility σ, and the short rate r – explains the popularity of the model in the financial services industry since the mid-1970s. We will also denote C_{BS} by $C_{BS}(t,x;K,T;\sigma)$ to emphasize the dependence on K, T, and σ. Only the volatility σ, the standard deviation of the returns scaled by the square root of the time increment, needs to be estimated from data, assuming that the constant short rate r is known.

The fact that $C_{BS}(t,x)$ given by (1.41) satisfies equation (1.39) with the final condition $h(x) = (x - K)^+$ can easily be checked directly. A probabilistic representation of this solution is presented in the following section.

In Figure 1.2 the pricing function $C_{BS}(0,x;100,0.5;0.1)$ is plotted against the present $(t = 0)$ stock price x. Notice how it is a smoothed version of the "ramp" terminal payoff function.

The Delta hedging ratio a_t for a call is given by

$$\frac{\partial C_{BS}}{\partial x} = N(d_1).$$

There is a similar formula for European put options. Let $P_{BS}(t,x)$ be the price of a European put option (Section 1.2.1). We then have the *put–call parity* relation

$$C_{BS}(t,X_t) - P_{BS}(t,X_t) = X_t - Ke^{-r(T-t)}, \qquad (1.45)$$

between put and call options with the same maturity and strike price. This is a model-free relationship that follows from simple *no-arbitrage* arguments. If, for instance, the left-hand side is smaller than the right-hand side then buying a call and selling a put and one unit of the stock, and investing the difference in the bond, creates a profit no matter what the stock price does.

Under the lognormal model, this relationship can be checked directly since the difference $C_{BS} - P_{BS}$ satisfies the partial differential equation (1.39) with the final condition $h(x) = x - K$. This problem has the unique simple solution $x - Ke^{-r(T-t)}$. Using the Black–Scholes formula (1.41) for C_{BS} and the put–call parity relation (1.45), we deduce the following explicit formula for the price of a European put option:

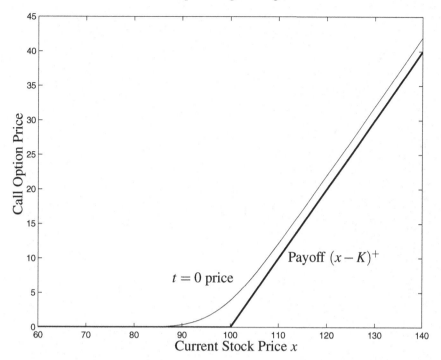

Figure 1.2   Black–Scholes call option pricing function at time $t = 0$, with $K = 100$, $T = 0.5$, $\sigma = 0.1$, and $r = 0.04$.

$$P_{BS}(t,x) = Ke^{-r(T-t)}N(-d_2) - xN(-d_1), \qquad (1.46)$$

where d_1, d_2, and N are as in (1.42), (1.43), and (1.44), respectively.

In Figure 1.3 the pricing function $P_{BS}(0,x; 100, 0.5; 0.1)$ is plotted against the present ($t = 0$) stock price x. Here we see that the pricing function crosses over its terminal payoff for some (small enough) x, which does not happen with the call option function in Figure 1.2. This observation be important when we look at American options in Section 1.6.

Other types of options do not lead in general to such explicit formulas. Determining their prices requires solving numerically the partial differential equation (1.39) with appropriate boundary conditions. Nevertheless, probabilistic representations can be obtained as explained in the following section. In particular, American options lead to *free boundary value problems* associated with equation (1.39).

1.3.5 The Greeks

The sensitivity of the price of an option to the variations of its parameters is measured by using partial derivatives. These quantities, known as Greeks of

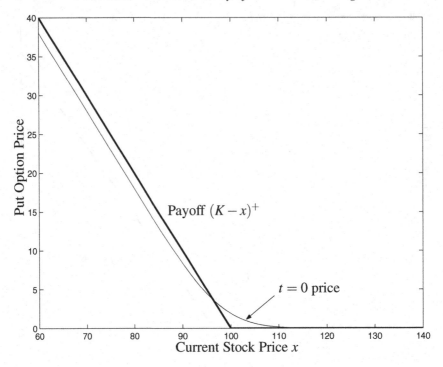

Figure 1.3   Black–Scholes put option pricing function at time $t = 0$, with $K = 100$, $T = 0.5$, $\sigma = 0.1$, and $r = 0.04$.

the option, play an important role in the derivative markets. For instance, we have already introduced in the previous section the *Delta* of a call option:

$$\Delta_{BS} = \frac{\partial C_{BS}}{\partial x} = N(d_1). \tag{1.47}$$

The second derivative with respect to the stock price is known as the *Gamma* which, in the case of a call option, is given by:

$$\Gamma_{BS} = \frac{\partial^2 C_{BS}}{\partial x^2} = \frac{\partial \Delta_{BS}}{\partial x} = \frac{e^{-d_1^2/2}}{x\sigma\sqrt{2\pi(T-t)}}. \tag{1.48}$$

The sensitivity to the volatility level, represented by the first derivative with respect to σ, is called the *Vega*:

$$\mathcal{V}_{BS} = \frac{\partial C_{BS}}{\partial \sigma} = \frac{xe^{-d_1^2/2}\sqrt{T-t}}{\sqrt{2\pi}}. \tag{1.49}$$

The sensitivities with respect to time to maturity $T - t$ and short rate r are respectively named the *Theta* and the *Rho*.

In the general case of a European derivative whose price satisfies the Black–Scholes partial differential equation (1.39) with a terminal condition $P(T,x) = h(x)$, there are simple and important relations between some of the Greeks. These are obtained by differentiating (1.39). For instance, differentiating with respect to σ leads to the following equation for the *Vega*:

$$\mathcal{L}_{BS}(\sigma)\mathcal{V} + \sigma x^2 \frac{\partial^2 P}{\partial x^2} = 0, \tag{1.50}$$

with a zero terminal condition. One can easily check that the Black–Scholes operator $\mathcal{L}_{BS}(\sigma)$ commutes with $x^2 \frac{\partial^2}{\partial x^2}$, and therefore that $(T-t)\sigma x^2 \frac{\partial^2 P}{\partial x^2}$ satisfies equation (1.50). If the second derivative with respect to x remains bounded as $t \to T$, this solution satisfies the zero terminal condition, and therefore we obtain the following relation between the *Vega* and the *Gamma*:

$$\frac{\partial P}{\partial \sigma} = (T-t)\sigma x^2 \frac{\partial^2 P}{\partial x^2}. \tag{1.51}$$

In the case of a call option this relation can be obtained directly from (1.48) and (1.49). Using the same argument, by differentiating the Black–Scholes equation with respect to r, one can obtain the relation

$$\frac{\partial P}{\partial r} = (T-t)\left(x\frac{\partial P}{\partial x} - P\right). \tag{1.52}$$

It is important to note that these relations may not be satisfied by more complex derivatives involving additional boundary conditions, such as barrier options for instance.

1.4 Risk-Neutral Pricing

We mentioned in Section 1.2.1 that, unless $\mu = r$, the expected value under the *real-world* probability $\mathbb{P}$ of the discounted payoff of a derivative (1.24) would lead to an opportunity for arbitrage. This is closely related to the fact that the discounted price $\tilde{X}_t = e^{-rt}X_t$ is not a martingale since, from (1.18),

$$d\tilde{X}_t = (\mu - r)\tilde{X}_t dt + \sigma \tilde{X}_t dW_t, \tag{1.53}$$

which contains a nonzero *drift* term if $\mu \neq r$.

The main result we want to build in this section is that there is a unique probability measure $\mathbb{P}^\star$ equivalent to $\mathbb{P}$ such that, under this probability, (i) the discounted price $\tilde{X}_t$ is a martingale and (ii) the expected value under $\mathbb{P}^\star$ of the discounted payoff of a derivative gives its no-arbitrage price. Such a probability measure describing a *risk-neutral* world is called an *equivalent martingale measure*.

1.4.1 Equivalent Martingale Measure

In order to find a probability measure under which the discounted price $\widetilde{X}_t$ is a martingale, we rewrite (1.53) in such a way that the drift term is "absorbed" in the martingale term:

$$d\widetilde{X}_t = \sigma \widetilde{X}_t \left[dW_t + \left(\frac{\mu - r}{\sigma} \right) dt \right].$$

We set

$$\theta = \frac{\mu - r}{\sigma}, \tag{1.54}$$

called the *market price of asset risk*, and we define

$$W_t^\star = W_t + \int_0^t \theta \, ds = W_t + \theta t, \tag{1.55}$$

so that

$$d\widetilde{X}_t = \sigma \widetilde{X}_t \, dW_t^\star. \tag{1.56}$$

Using the characterization (1.3), it is easy to check that the positive random variable ξ_T^θ defined by

$$\xi_T^\theta = \exp\left(-\theta W_T - \frac{1}{2} \theta^2 T \right) \tag{1.57}$$

has an expected value (with respect to $\mathbb{P}$) equal to 1 (the Cameron–Martin formula). More generally, it has a conditional expectation with respect to $\mathscr{F}_t$ given by

$$\mathbb{E}\{\xi_T^\theta \mid \mathscr{F}_t\} = \exp\left(-\theta W_t - \frac{1}{2} \theta^2 t \right) = \xi_t^\theta, \quad \text{for } 0 \le t \le T,$$

which defines a martingale denoted by $(\xi_t^\theta)_{0 \le t \le T}$.

We now introduce the probability measure $\mathbb{P}^\star$. It is the equivalent measure to $\mathbb{P}$, meaning that they have the same null sets (they agree on which events have zero probability), which has the density ξ_T^θ with respect to $\mathbb{P}$:

$$d\mathbb{P}^\star = \xi_T^\theta \, d\mathbb{P}. \tag{1.58}$$

Denoting by $\mathbb{E}^\star\{\cdot\}$ the expectation with respect to $\mathbb{P}^\star$, for any integrable random variable Z we have

$$\mathbb{E}^\star\{Z\} = \mathbb{E}\{\xi_T^\theta Z\},$$

and one can check that, for any adapted and integrable process (Z_t),

$$\mathbb{E}^\star\{Z_t \mid \mathscr{F}_s\} = \frac{1}{\xi_s^\theta}\mathbb{E}\{\xi_t^\theta Z_t \mid \mathscr{F}_s\}, \tag{1.59}$$

for any $0 \le s \le t \le T$. The process $(\xi_t^\theta)_{0 \le t \le T}$ is called the Radon–Nikodym process.

The main result of this section asserts that the process $(W_t^\star)$ given by (1.55) is a standard Brownian motion under the probability $\mathbb{P}^\star$. This result in its full generality (when θ is an adapted stochastic process) is known as *Girsanov's theorem*. In our simple case (θ constant), it is easily derived by using the characterization (1.3) and formula (1.59) as follows:

$$
\begin{aligned}
\mathbb{E}^\star\left\{e^{iu(W_t^\star - W_s^\star)} \mid \mathscr{F}_s\right\} &= \frac{1}{\xi_s^\theta}\mathbb{E}\left\{\xi_t^\theta\, e^{iu(W_t^\star - W_s^\star)} \mid \mathscr{F}_s\right\} \\
&= e^{\theta W_s + \frac{1}{2}\theta^2 s}\mathbb{E}\left\{e^{-\theta W_t - \frac{1}{2}\theta^2 t}\, e^{iu(W_t - W_s + \theta(t-s))} \mid \mathscr{F}_s\right\} \\
&= e^{\left(-\frac{1}{2}\theta^2 + iu\theta\right)(t-s)}\mathbb{E}\left\{e^{i(u+i\theta)(W_t - W_s)} \mid \mathscr{F}_s\right\} \\
&= e^{\left(-\frac{1}{2}\theta^2 + iu\theta\right)(t-s)}\, e^{-\frac{(u+i\theta)^2(t-s)}{2}} \\
&= e^{-\frac{u^2(t-s)}{2}}.
\end{aligned}
$$

1.4.2 Self-Financing Portfolios

As in Section 1.3.1, a portfolio comprises a_t units of stock and b_t in bonds; we denote by V_t its value at time t:

$$V_t = a_t X_t + b_t e^{rt}.$$

The self-financing property (1.33), namely $dV_t = a_t dX_t + rb_t e^{rt} dt$, implies that the discounted value of the portfolio, $\widetilde{V}_t = e^{-rt}V_t$, is a martingale under the risk-neutral probability $\mathbb{P}^\star$. This essential property of self-financing portfolios is obtained as follows:

$$
\begin{aligned}
d\widetilde{V}_t &= -re^{-rt}V_t dt + e^{-rt}dV_t \\
&= -re^{-rt}(a_t X_t + b_t e^{rt})dt + e^{-rt}(a_t dX_t + rb_t e^{rt} dt) \\
&= -re^{-rt}a_t X_t dt + e^{-rt}a_t dX_t \\
&= a_t d(e^{-rt}X_t) \\
&= a_t d\widetilde{X}_t \\
&= \sigma a_t \widetilde{X}_t dW_t^\star \quad \text{(by (1.56))}, \tag{1.60}
\end{aligned}
$$

which shows that $(\widetilde{V}_t)$ is a martingale under $\mathbb{P}^\star$ as a stochastic integral with respect to the Brownian motion $(W_t^\star)$. Indeed, the same computation shows that if a portfolio satisfies $d\widetilde{V}_t = a_t d\widetilde{X}_t$, then it is self-financing.

A simple calculation demonstrates the connection between martingales and no-arbitrage. Suppose that $(a_t, b_t)_{0 \le t \le T}$ is a self-financing arbitrage strategy; that is,

$$V_T \ge e^{rT} V_0 \ (\mathbb{P}\text{-a.s.}), \tag{1.61}$$

with

$$\mathbb{P}\{V_T > e^{rT} V_0\} > 0, \tag{1.62}$$

so that the strategy never makes less than money in the bank and there is some chance of making more. By the martingale property of self-financing strategies

$$\mathbb{E}^\star\{V_T\} = e^{rT} V_0, \tag{1.63}$$

so that (1.61) and (1.62) cannot both hold. This is because $\mathbb{P}$ and $\mathbb{P}^\star$ are equivalent and therefore, (1.61) and (1.62) also hold with $\mathbb{P}$ replaced by $\mathbb{P}^\star$. The inequality (1.61) (which holds $\mathbb{P}^\star$-a.s.) and (1.63) imply $V_T = e^{rT} V_0$ ($\mathbb{P}^\star$-a.s. and $\mathbb{P}$-a.s.), contradicting (1.62).

1.4.3 Risk-Neutral Valuation

Assume that (a_t, b_t) is a self-financing portfolio satisfying the same integrability conditions of Section 1.3.1 and replicating the European-style derivative with non-negative payoff H:

$$a_T X_T + b_T e^{rT} = H, \tag{1.64}$$

where we assume that H is a square integrable $\mathscr{F}_T$-adapted random variable. This includes European calls and puts or more general standard European derivatives for which $H = h(X_T)$, as well as other European-style exotic derivatives presented in Section 1.2.3.

On the one hand, a no-arbitrage argument shows that the price at time t of this derivative should be the value V_t of this portfolio. On the other hand, as shown in Section 1.4.2, the discounted values $(\widetilde{V}_t)$ of this portfolio form a martingale under the risk-neutral probability $\mathbb{P}^\star$ and consequently

$$\widetilde{V}_t = \mathbb{E}^\star\left\{\widetilde{V}_T \mid \mathscr{F}_t\right\},$$

which gives

$$V_t = \mathbb{E}^\star \left\{ e^{-r(T-t)} H \mid \mathscr{F}_t \right\}, \tag{1.65}$$

after reintroducing the discounting factor and using the replicating property (1.64).

Alternatively, given the **risk-neutral valuation formula** (1.65), we can find a self-financing replicating portfolio for the payoff H. The existence of such a portfolio is guaranteed by an application of the **martingale representation theorem**: for $0 \leq t \leq T$

$$M_t = \mathbb{E}^\star \left\{ e^{-rT} H \mid \mathscr{F}_t \right\},$$

defines a square integrable martingale under $\mathbb{P}^\star$ with respect to the filtration $(\mathscr{F}_t)$, which is also the natural filtration of the Brownian motion $W^\star$. This representation theorem says that any such martingale is a stochastic integral with respect to $W^\star$, so that

$$\mathbb{E}^\star \left\{ e^{-rT} H \mid \mathscr{F}_t \right\} = M_0 + \int_0^t \eta_s \, dW_s^\star,$$

where (η_t) is some adapted process with $\mathbb{E}^\star \left\{ \int_0^T \eta_t^2 dt \right\}$ finite. By defining $a_t = \eta_t / (\sigma \tilde{X}_t)$ and $b_t = M_t - a_t \tilde{X}_t$, we construct a portfolio (a_t, b_t), which is shown to be self-financing by checking that its discounted value is the martingale M_t and using the characterization (1.60) obtained in Section 1.4.2. Its value at time T is $e^{rT} M_T = H$, and therefore it is a replicating portfolio.

1.4.4 Using the Markov Property

If H is a function of the path of the stock price after time t – as, for instance, for a standard European derivative with payoff $H = h(X_T)$ – then the Markov property of (X_t) says that conditioning with respect to the past $\mathscr{F}_t$ is the same as conditioning with respect to X_t, the value at the current time; this gives

$$V_t = \mathbb{E}^\star \left\{ e^{-r(T-t)} h(X_T) \mid X_t \right\}.$$

We will come back to this property in the next section.

Denoting by $P(t,x)$, as in Section 1.3.2, the price of this derivative at time t for an observed stock price $X_t = x$, we obtain the **pricing formula**

$$P(t,x) = \mathbb{E}^\star \left\{ e^{-r(T-t)} h(X_T) \mid X_t = x \right\}. \tag{1.66}$$

If we compare this formula (at time $t = 0$) with (1.24), our first intuitive idea for pricing a standard European derivative in Section 1.2.1, we see that the essential step is to replace the "real-world" $\mathbb{P}$ by the "risk-neutral world" $\mathbb{P}^\star$ in order to obtain the fair no-arbitrage price.

Knowing that $X_t = x$, one can generalize the formula (1.19) obtained in Section 1.1.5 and obtain an explicit formula for X_T by solving the stochastic differential equation (1.2) from t to T starting from x:

$$X_T = x\exp\left((\mu - \frac{\sigma^2}{2})(T - t) + \sigma(W_T - W_t) \right). \qquad (1.67)$$

Using (1.55), this formula can be rewritten in terms of $(W_t^\star)$ as

$$X_T = x\exp\left((r - \frac{\sigma^2}{2})(T - t) + \sigma(W_T^\star - W_t^\star) \right).$$

Since $(W_t^\star)$ is a standard Brownian motion under the risk-neutral probability $\mathbb{P}^\star$, the increment $W_T^\star - W_t^\star$ is $\mathcal{N}(0, T - t)$-distributed, and (1.66) gives the Gaussian integral

$$P(t,x) = \frac{1}{\sqrt{2\pi(T-t)}} \int_{-\infty}^{+\infty} e^{-r(T-t)} h\left(xe^{(r-\frac{\sigma^2}{2})(T-t)+\sigma z} \right) e^{-\frac{z^2}{2(T-t)}} \, dz.$$

$$(1.68)$$

In the case of a European call option, $h(x) = (x - K)^+$, this integral reduces to the Black–Scholes formula (1.41) obtained in Section 1.3.4, as the following computation shows:

$$P(t,x) = \frac{x}{\sqrt{2\pi\tau}} \int_{z^\star}^{+\infty} e^{-\frac{(z-\sigma\tau)^2}{2\tau}} \, dz - \frac{Ke^{-r\tau}}{\sqrt{2\pi\tau}} \int_{z^\star}^{+\infty} e^{-\frac{z^2}{2\tau}} \, dz,$$

where $\tau = T - t$ and $z^\star$ is defined by $x\exp\left((r - \frac{1}{2}\sigma^2)\tau + \sigma z^\star\right) = K$. We then set

$$\frac{z^\star - \sigma\tau}{\sqrt{\tau}} = -d_1, \quad \frac{z^\star}{\sqrt{\tau}} = -d_2,$$

which coincide with the definitions (1.42) and (1.43) of d_1 and d_2. The Black–Scholes formula (1.41) follows by introducing the Gaussian cumulative distribution function N given by (1.44).

Another important example is given by the *binary* or *digital* options which, for instance, pay at time T a fixed amount (say one), if $X_T \geq K$, and nothing otherwise. The corresponding discontinuous payoff function is simply $h(x) = 1_{\{x \geq K\}}$. Its value at time t is given by (1.66), which, in this case, becomes

$$P_{\text{digital}}(t,x) = \frac{e^{-r\tau}}{\sqrt{2\pi\tau}} \int_{z^*}^{+\infty} e^{-\frac{z^2}{2\tau}} dz = e^{-r\tau} N(d_2). \qquad (1.69)$$

The two approaches developed in Sections 1.3 and 1.4 should give the same fair price to the same derivative. This is indeed the case, and is the content of the following section, where we explain that a formula like (1.66) is just a probabilistic representation of the solution of a partial differential equation like (1.39).

1.5 Risk-Neutral Expectations and Partial Differential Equations

In Section 1.4.4, we used the Markov property of the stock price (X_t) and, in order to compute X_T knowing that $X_t = x$ at time $t \leq T$, we solved the stochastic differential equation (1.2) between t and T. This was a particular case of the general situation where (X_t) is the unique solution of the stochastic differential equation (1.11). We denote by $(X_s^{t,x})_{s \geq t}$ the solution of that equation starting from x at time t:

$$X_s^{t,x} = x + \int_t^s \mu(u, X_u^{t,x}) \, du + \int_t^s \sigma(u, X_u^{t,x}) \, dW_u, \qquad (1.70)$$

and we assume enough regularity in the coefficients μ and σ for $(X_s^{t,x})$ to be jointly continuous in the three variables (t,x,s). The *flow property* for deterministic differential equations can be extended to stochastic differential equations like (1.70); it says that, in order to compute the solution at time $s > t$ starting at time 0 from point x, one can use

$$x \longrightarrow X_t^{0,x} \longrightarrow X_s^{t,X_t^{0,x}} = X_s^{0,x}. \qquad (1.71)$$

In other words, one can solve the equation from 0 to t, starting from x, to obtain $X_t^{0,x}$. Then we solve the equation from t to s, starting from $X_t^{0,x}$. This is the same as solving the equation from 0 to s, starting from x.

The **Markov property** is a consequence and can be stated as follows:

$$\mathbb{E}\{h(X_s) \mid \mathscr{F}_t\} = \mathbb{E}\{h(X_s^{t,x})\}|_{x=X_t}, \qquad (1.72)$$

which is what we have used with $s = T$ to derive (1.66). Observe that the discounting factor could be pulled out of the conditional expection since the interest rate is constant (not random). In the time-homogeneous case (μ and σ independent of time) we further have

$$\mathbb{E}\{h(X_s^{t,x})\} = \mathbb{E}\left\{h(X_{s-t}^{0,x})\right\},$$

which could have been used with $s = T$ to derive (1.68) since $W_{T-t}^\star$ is $\mathscr{N}(0, T-t)$-distributed.

1.5.1 Infinitesimal Generators and Associated Martingales

For simplicity, we first consider a time-homogeneous diffusion process (X_t), solution of the stochastic differential equation

$$dX_t = \mu(X_t)dt + \sigma(X_t)\,dW_t. \tag{1.73}$$

Let g be a twice continuously differentiable function of the variable x with bounded derivatives, and define the differential operator $\mathscr{L}$ acting on g according to

$$\mathscr{L}g(x) = \frac{1}{2}\sigma^2(x)g''(x) + \mu(x)g'(x). \tag{1.74}$$

In terms of $\mathscr{L}$, Itô's formula (1.16) gives

$$dg(X_t) = \mathscr{L}g(X_t)dt + g'(X_t)\sigma(X_t)\,dW_t,$$

which shows that

$$M_t = g(X_t) - \int_0^t \mathscr{L}g(X_s)ds \tag{1.75}$$

defines a martingale. Consequently, if $X_0 = x$, we obtain

$$\mathbb{E}\{g(X_t)\} = g(x) + \mathbb{E}\left\{\int_0^t \mathscr{L}g(X_s)ds\right\}.$$

Under the assumptions made on the coefficients μ and σ and on the function g, the Lebesgue dominated convergence theorem is applicable and gives

$$\frac{d}{dt}\mathbb{E}\{g(X_t)\}|_{t=0} = \lim_{t\downarrow 0}\frac{\mathbb{E}\{g(X_t)\} - g(x)}{t}$$

$$= \lim_{t\downarrow 0}\mathbb{E}\left\{\frac{1}{t}\int_0^t \mathscr{L}g(X_s)ds\right\} = \mathscr{L}g(x).$$

The differential operator $\mathscr{L}$ given by (1.74) is called the *infinitesimal generator* of the Markov process (X_t).

Considering now a nonhomogeneous diffusion $(\sigma(t,x), \mu(t,x))$ and functions $g(t,x)$ which depend also on time, (1.75) can be generalized by using the full Itô formula (1.16) to yield the martingale

$$M_t = g(t,X_t) - \int_0^t \left(\frac{\partial g}{\partial t} + \mathscr{L}_s g\right)(s,X_s)ds, \tag{1.76}$$

where the infinitesimal generator $\mathscr{L}_t$ is defined by

$$\mathscr{L}_t = \frac{1}{2}\sigma^2(t,x)\frac{\partial^2}{\partial x^2} + \mu(t,x)\frac{\partial}{\partial x}, \tag{1.77}$$

and g is any smooth and bounded function. Finally, it is possible to incorporate a discounting factor by using the integration-by-parts formula to compute the differential of $e^{-rt}g(t,X_t)$ and obtain the martingales

$$M_t = e^{-rt}g(t,X_t) - \int_0^t e^{-rs}\left(\frac{\partial g}{\partial t} + \mathcal{L}_s g - rg\right)(s,X_s)\,ds, \quad (1.78)$$

which introduces the *potential* term $-rg$. This can also be generalized to the case of a potential depending on t and x, e^{-rt} being replaced by the discounting factor $e^{-\int_0^t r(s,X_s)\,ds}$.

1.5.2 Conditional Expectations and Parabolic Partial Differential Equations

Suppose that $u(t,x)$ is a solution of the partial differential equation

$$\frac{\partial u}{\partial t} + \frac{1}{2}\sigma^2(t,x)\frac{\partial^2 u}{\partial x^2} + \mu(t,x)\frac{\partial u}{\partial x} - ru = 0, \quad (1.79)$$

with the final condition $u(T,x) = h(x)$ and assume that it is regular enough to apply Itô's formula (1.16). Using (1.78) we deduce that $M_t = e^{-rt}u(t,X_t)$ is a martingale when $\mathcal{L}_t$, given by (1.77), is the infinitesimal generator of the process (X_t) – in other words, when μ and σ are the drift and diffusion coefficients of (X_t).

The martingale property for times t and T reads $\mathbb{E}\{M_T \mid \mathcal{F}_t\} = M_t$, which can be rewritten as

$$u(t,X_t) = \mathbb{E}\left\{e^{-r(T-t)}h(X_T) \mid \mathcal{F}_t\right\},$$

since $u(T,X_T) = h(X_T)$ according to the final condition. Using the Markov property (1.72), we deduce the following probabilistic representation of the solution u:

$$u(t,x) = \mathbb{E}\left\{e^{-r(T-t)}h(X_T^{t,x})\right\}, \quad (1.80)$$

which may also be written as

$$u(t,x) = \mathbb{E}\left\{e^{-r(T-t)}h(X_T) \mid X_t = x\right\} \text{ or } u(t,x) = \mathbb{E}_{t,x}\left\{e^{-r(T-t)}h(X_T)\right\}.$$

If r depends on t and x, the discounting factor becomes $e^{-\int_t^T r(s,X_s)\,ds}$. The representation (1.80) is then called the *Feynman–Kac formula*.

1.5.3 Kolmogorov Equations

When $r = 0$, equation (1.79) reads

$$\frac{\partial u}{\partial t} + \mathscr{L}_t u = 0, \qquad u(T,x) = h(x), \tag{1.81}$$

with $\mathscr{L}_t$ given in (1.77). This equation is known as the *backward Kolmogorov equation* for the conditional expectation. Let $p(t,x;T,\xi)$ denote the transition probability density of X_T starting at time t at point x, and assume that it is sufficiently regular for the following calculations. Then

$$u(t,x) = \mathbb{E}\{h(X_T) \mid X_t = x\} = \int h(\xi')p(t,x;T,\xi')\,d\xi', \tag{1.82}$$

and by choosing $h(x) = \delta(x - \xi)$, the point mass or Dirac delta function at ξ, gives $u(t,x) = p(t,x;T,\xi)$, and therefore

$$\frac{\partial p}{\partial t} + \mathscr{L}_t p = 0, \qquad p(T,x;T,\xi) = \delta(x - \xi),$$

where the operator $\mathscr{L}_t$ acts on the variable x. This equation is known as the backward Kolmogorov equation for the transition probability density.

It is also useful to have an equation for p with respect to its forward variables (T,ξ). To this end, we define the *adjoint operator* $\mathscr{L}_T^*$ of $\mathscr{L}_T$ by

$$\int \psi(\xi)\mathscr{L}_T \phi(\xi)\,d\xi = \int \phi(\xi)\mathscr{L}_T^* \psi(\xi)\,d\xi, \tag{1.83}$$

for rapidly decaying smooth test functions ϕ and ψ. Integration by parts yields

$$\mathscr{L}_T^* = \frac{1}{2}\frac{\partial^2}{\partial \xi^2}\left(\sigma^2(T,\xi)\cdot\right) - \frac{\partial}{\partial \xi}\left(\mu(T,\xi)\cdot\right), \tag{1.84}$$

where the rapid decay of the test functions ensures that the boundary terms are zero. Let us define $\tilde{p}(T,\xi)$ as the solution of the equation

$$\frac{\partial \tilde{p}}{\partial T} = \mathscr{L}_T^* \tilde{p}, \tag{1.85}$$

with the initial condition $\tilde{p}(t,\xi) = \delta(\xi - x)$ at $T = t$. Now, for $t \leq s \leq T$, compute

$$\frac{d}{ds}\int \tilde{p}(s,\xi')u(s,\xi')\,d\xi'$$

$$= \int \left(u(s,\xi')\mathscr{L}_s^* \tilde{p}(s,\xi') - \tilde{p}(s,\xi')\mathscr{L}_s u(s,\xi')\right)d\xi' = 0,$$

where we have used the equations (1.81) and (1.85) for u and $\tilde{p}$, and the defining property (1.83) of the adjoint. Consequently,

$$\int \tilde{p}(t,\xi')u(t,\xi')d\xi' = \int \tilde{p}(T,\xi')u(T,\xi')d\xi',$$

and using the initial condition for $\tilde{p}$ and the terminal condition for u, we obtain

$$u(t,x) = \int \tilde{p}(T,\xi')h(\xi')d\xi',$$

for an arbitrary function h. By (1.82), we identify $\tilde{p}(T,\xi) = p(t,x;T,\xi)$. Therefore, the transition probability $p(t,x;T,\xi)$ satisfies the *forward Kolmogorov equation*, also known as the Fokker–Planck equation:

$$\frac{\partial p}{\partial T} = \mathscr{L}_T^* p, \quad T > t, \tag{1.86}$$
$$p(t,x;t,\xi) = \delta(\xi - x).$$

The rigorous derivation of these Kolmogorov equations can be found in the references given in the Notes at the end of the chapter.

1.5.4 Application to the Black–Scholes Partial Differential Equation

In the previous section, we have assumed the existence, uniqueness, and regularity of the solution of the partial differential equation (1.79) in order to apply Itô's formula. A sufficient condition for this is that the coefficients μ and σ are regular enough and that the operator $\mathscr{L}_t$ is *uniformly elliptic*, meaning (in this one-dimensional situation) that there exists a positive constant A such that

$$\sigma^2(t,x) \geq A > 0 \quad \text{for every } t \geq 0 \text{ and } x \in \mathscr{D}, \tag{1.87}$$

so that the diffusion coefficient $\sigma(t,x)$ cannot become too small. Here $\mathscr{D}$ is the domain of the process (X_t), which may be natural (e.g., $\mathscr{D} = \{x > 0\}$ for the geometric Brownian motion) or imposed externally from other modeling considerations.

When $\mu(t,x) = rx$ and $\sigma(t,x) = \sigma x$ in (1.79), we have the Black–Scholes partial differential equation (1.39) for the option price $P(t,x)$ on the domain $\{x > 0\}$, since

$$\mathscr{L}_{BS} = \frac{\partial}{\partial t} + \mathscr{L} - r,$$

where $\mathscr{L}$ is the infinitesimal generator of the geometric Brownian motion X. The ellipticity condition (1.87) is clearly not satisfied since the diffusion coefficient σx goes to zero as the state variable approaches zero. We get around this difficulty here (and also in more general situations) with the change of variable $P(t,x) = u(t,y = \log x)$, so that equation (1.39) becomes

$$\frac{\partial u}{\partial t} + \frac{1}{2}\sigma^2 \frac{\partial^2 u}{\partial y^2} + \left(r - \frac{1}{2}\sigma^2\right)\frac{\partial u}{\partial y} - ru = 0, \tag{1.88}$$

to be solved for $0 \le t \le T$, $y \in \mathbb{R}$, and with the final condition $u(T,y) = h(e^y)$. The operator

$$\mathscr{L} = \frac{1}{2}\sigma^2 \frac{\partial^2}{\partial y^2} + \left(r - \frac{1}{2}\sigma^2\right)\frac{\partial}{\partial y}$$

is the infinitesimal generator of the (nonstandard) Brownian motion defined by

$$dY_t = \left(r - \frac{1}{2}\sigma^2\right)dt + \sigma dW_t^\star,$$

where $(W_t^\star)$ is a standard Brownian motion under $\mathbb{P}^\star$. We use here the same notation as in the equivalent martingale measure context, but the only important fact is that $W^\star$ is a standard Brownian motion with respect to the probability used to compute the expectation in the Feynman–Kac formula (1.80). Applying this formula to Y_t yields

$$u(t,y) = \mathbb{E}^\star\left\{e^{-r(T-t)}h\left(e^{y + (r - \frac{1}{2}\sigma^2)(T-t) + \sigma(W_T^\star - W_t^\star)}\right) \mid Y_t = y\right\},$$

which is indeed the same as (1.68) by undoing the change of variable $e^y = x$.

1.6 American Options and Free Boundary Problems

We recall here basic facts about American options under constant volatility which will be used in Chapter 7 where we derive corrections due to stochastic volatility.

1.6.1 Optimal Stopping

Pricing American derivatives is mathematically more involved than the European case. Using the theory of *optimal stopping*, it can be shown that the price of an American derivative with payoff function h is obtained by maximizing over all the stopping times the expected value of the discounted payoff. As in the European case, the expectations have to be taken with respect to the risk-neutral probability to avoid arbitrage opportunities. In other words, the intuitive idea presented in Section 1.2.2 is correct when $\mathbb{E}$ is replaced by $\mathbb{E}^\star$:

$$P(0,x) = \sup_{\tau \le T} \mathbb{E}^\star \left\{ e^{-r\tau} h(X_\tau) \right\}$$

is the price of the derivative at time $t = 0$, when $X_0 = x$ and where the supremum is taken over all the possible stopping times less than the expiration date T. This formula can be generalized to get the price of American derivatives at any time t before expiration T:

$$P(t,x) = \sup_{t \le \tau \le T} \mathbb{E}^\star \left\{ e^{-r(\tau-t)} h\left(X_\tau^{t,x}\right) \right\}, \qquad (1.89)$$

where $(X_s^{t,x})_{s \ge t}$ is, as in Section 1.5, the stock price starting at time t from the observed price x.

By taking $\tau = t$ we deduce that $P(t,x) \ge h(x)$, which is natural since if $P(t,x) < h(x)$ then there would be an obvious instant arbitrage at time t. Moreover, by choosing $t = T$ we obtain $P(T,x) = h(x)$.

Because an American derivative gives its holder more rights than the corresponding European derivative, the price of the American is always greater than or equal to the price of the European derivative which has the same payoff function and the same expiration date. By taking $\tau = T$ in (1.89), we see that this is indeed the case.

Formula (1.89) gives the price of an American derivative. The supremum in (1.89) is reached at the *optimal stopping time*,

$$\tau^\star = \tau^\star(t) = \inf \left\{ t \le s \le T \mid P(s,X_s) = h(X_s) \right\}, \qquad (1.90)$$

the first time after t that the price of the derivative drops down to its payoff. In order to determine $\tau^\star$, one must first compute the price. In terms of partial differential equations, this leads to a so-called *free boundary value problem*. To illustrate, we consider the case of an American put option defined in Section 1.2.2.

It can be shown by a no-arbitrage argument that, for non-negative interest rates and no dividend paid, the price of an American call option is the same as its corresponding European option. The price of an American put option

$$P^a(t,x) = \sup_{t \le \tau \le T} \mathbb{E}^\star \left\{ e^{-r(\tau-t)} \left(K - X_\tau^{t,x}\right)^+ \right\},$$

is in general strictly higher than the price of the corresponding European put option which has been obtained in closed form (1.46). In fact, we saw in Figure 1.3 that the Black–Scholes European put option pricing function crosses below the payoff "ramp" function $(K - x)^+$ for small enough x. This violates $P(t,x) \ge h(x)$, so the European formula for a put cannot also give the price of the American contract, as is the case for call options. We therefore use a put option as our canonical example of an American security in Chapter 7.

1.6.2 Free Boundary Value Problems

Pricing functions for American derivatives satisfy partial differential *inequalities*. For the non-negative payoff function h, the price of the corresponding American derivative is the solution of the following *linear complementarity problem*:

$$
\begin{aligned}
P &\geq h, \\
\mathscr{L}_{BS}(\sigma)P &\leq 0, \\
(h - P)\mathscr{L}_{BS}(\sigma)P &= 0,
\end{aligned}
\tag{1.91}
$$

to be solved in $\{(t,x) : 0 \leq t \leq T, x > 0\}$ with the final condition $P(T,x) = h(x)$, and where $\mathscr{L}_{BS}(\sigma)$ is the Black–Scholes operator given by (1.40). The second inequality is linked to the supermartingale property of $e^{-rt}P(t,X_t)$ through (1.78) applied to $g = P$.

To see that the price (1.89) is the solution of the differential inequalities (1.91), with the optimal stopping time given by (1.90), we assume that we can apply Itô's formula to the solution P of (1.91). For any stopping time $t \leq \tau \leq T$ we have

$$
e^{-r\tau}P(\tau,X_\tau^{t,x}) = e^{-rt}P(t,x) + \int_t^\tau e^{-rs}\left(\frac{\partial}{\partial t} + \mathscr{L} - r\right)P(s,X_s^{t,x})ds
$$
$$
+ \int_t^\tau e^{-rs}\sigma X_s^{t,x}\frac{\partial P}{\partial x}(s,X_s^{t,x})dW_s^\star,
$$

where $\mathscr{L}$ is the infinitesimal generator of X. The integrand of the Riemann integral is nonpositive by (1.91) and, since τ is bounded, the expectation of the martingale term is zero by Doob's optional stopping theorem. This leads to

$$
\mathbb{E}^\star\left\{e^{-r(\tau-t)}P(\tau,X_\tau^{t,x})\right\} \leq P(t,x),
$$

and, using the first inequality in (1.91),

$$
\mathbb{E}^\star\left\{e^{-r(\tau-t)}h(X_\tau^{t,x})\right\} \leq P(t,x).
$$

It is easy to see now that if $\tau = \tau^\star$, the optimal stopping time defined in (1.90), then we have equalities throughout. This verifies that if (1.91) has a solution to which Itô's formula can be applied then it is the American derivative price (1.89).

In the case of the American put option there is an increasing function $x^\star(t)$ – the *free boundary* – such that, at time t,

$$
\begin{aligned}
P(t,x) &= K - x \quad \text{for } x < x^\star(t),\\
\mathscr{L}_{BS}(\sigma)P &= 0 \qquad\;\; \text{for } x > x^\star(t),
\end{aligned}
\tag{1.92}
$$

with

$$P(T,x) = (K - x)^+, \tag{1.93}$$

$$x^\star(T) = K. \tag{1.94}$$

In addition, P and $\frac{\partial P}{\partial x}$ are continuous across the boundary $x^\star(t)$, so that

$$P(t,x^\star(t)) = K - x^\star(t), \tag{1.95}$$

$$\frac{\partial P}{\partial x}(t,x^\star(t)) = -1. \tag{1.96}$$

The exercise boundary $x^\star(t)$ separates the *hold* region, where the option is not exercised, from the *exercise* region, where it is; this is illustrated in Figure 1.4. In the corresponding Figure 1.5, we show the trajectory of the stock price and the optimal exercise time $\tau^\star$.

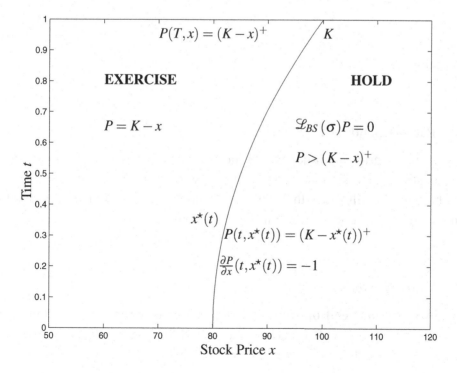

Figure 1.4   The American put problem for $P(t,x)$ and $x^\star(t)$, with $\mathscr{L}_{BS}(\sigma)$ defined in (1.40).

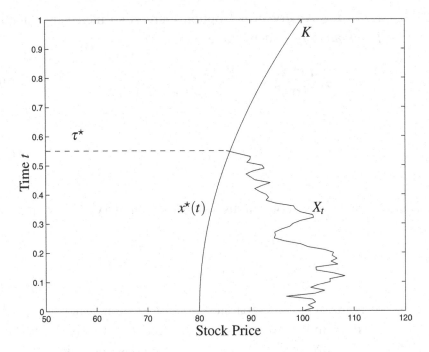

Figure 1.5   Optimal exercise time $\tau^\star$ for an American put option.

As in (1.88), the change of variable $y = \log x$ is convenient for analytical and numerical purposes. Notice that this is a system of equations and boundary conditions for $P(t,x)$ *and* the free boundary $x^\star(t)$.

1.7 Path-Dependent Derivatives

In order to price path-dependent derivatives, one has to compute the expectations of their discounted payoffs with respect to the risk-neutral probability. To illustrate this we give examples which will be developed in the context of stochastic volatility in Chapter 6.

1.7.1 Barrier Options

A **down-and-out call option** (European style) is an example of a barrier option that has a payoff function given by (1.26). Its no-arbitrage price at time $t = 0$ for a stock price equal to x is given by

$$P(0,x) = \mathbb{E}^\star \left\{ e^{-rT} (X_T - K)^+ \mathbf{1}_{\{\inf_{0 \leq t \leq T} X_t > B\}} \mid X_0 = x \right\}.$$

The price at time $t < T$ of this option is given by

$$P_t = \mathbb{E}^\star \left\{ e^{-r(T-t)} (X_T - K)^+ \mathbf{1}_{\{\inf_{0 \leq s \leq T} X_s > B\}} \mid \mathscr{F}_t \right\}$$

$$= \mathbf{1}_{\{\inf_{0 \leq s \leq t} X_s > B\}} \mathbb{E}^\star \left\{ e^{-r(T-t)} (X_T - K)^+ \mathbf{1}_{\{\inf_{t \leq s \leq T} X_s > B\}} \mid \mathscr{F}_t \right\}$$

$$= \mathbf{1}_{\{\inf_{0 \leq s \leq t} X_s > B\}} u(t, X_t),$$

where we have used the Markov property of (X_t), and $u(t,x)$ is defined by

$$u(t,x) = \mathbb{E}^\star \left\{ e^{-r(T-t)} (X_T - K)^+ \mathbf{1}_{\{\inf_{t \leq s \leq T} X_s > B\}} \mid X_t = x \right\}. \quad (1.97)$$

These expectations can be computed by using classical results on the joint probability distribution of the Brownian motion and its minimum, obtained by the *reflection principal*. This involves working with the nonstandard Brownian motion $\log X_t$, and then performing a change of measure, by Girsanov's theorem, in order to remove the drift (we refer to Etheridge (2002), for instance). Alternatively, the function $u(t,x)$ given by (1.97) satisfies the following boundary value problem on $\{x > B\}$:

$$\mathscr{L}_{BS}(\sigma)u = 0,$$
$$u(t,B) = 0,$$
$$u(T,x) = (x - K)^+.$$

We briefly explain here how the *method of images* leads to a formula for $u(t,x)$ in terms of the Black–Scholes price of the corresponding call option. As in the probabilistic approach mentioned above, we first use log-coordinates, $y = \log x$, by defining

$$v(t,y) = u(t, e^y),$$

which satisfies the following problem on $\{y > \log B\}$:

$$\frac{\partial v}{\partial t} + \frac{1}{2} \sigma^2 \frac{\partial^2 v}{\partial y^2} + \left(r - \frac{1}{2} \sigma^2 \right) \frac{\partial v}{\partial y} - rv = 0,$$
$$v(t, \log B) = 0,$$
$$v(T,y) = (e^y - K)^+.$$

We then reduce the problem to the *heat equation* by defining

$$k = 2r/\sigma^2, \quad (1.98)$$

and introducing

$$w(t,y) = e^{\left[\frac{1}{2}(k-1)y - \frac{\sigma^2}{8}(k+1)^2 t \right]} v(t,y),$$

which satisfies

$$\frac{\partial w}{\partial t} + \frac{1}{2}\sigma^2 \frac{\partial^2 w}{\partial y^2} = 0,$$

$$w(t, \log B) = 0,$$

$$w(T, y) = e^{\left[\frac{1}{2}(k-1)y - \frac{\sigma^2}{8}(k+1)^2 T\right]} (e^y - K)^+,$$

on the domain $\{y > \log B\}$. We now consider the solution $w_1(t, y)$ of the heat equation

$$\frac{\partial w_1}{\partial t} + \frac{1}{2}\sigma^2 \frac{\partial^2 w_1}{\partial y^2} = 0,$$

on the full domain $\{-\infty < y < +\infty\}$ with the terminal condition

$$w_1(T, y) = e^{\left[\frac{1}{2}(k-1)y - \frac{\sigma^2}{8}(k+1)^2 T\right]} (e^y - K)^+ \text{ if } y > \log B, \text{ and } 0 \text{ elsewhere.}$$

One can easily check that the image function $w_2(t, y) = w_1(t, 2 \log B - y)$ satisfies the same equation on the full space with the terminal condition

$$w_2(T, y) = \begin{cases} e^{\left[\frac{1}{2}(k-1)(2\log B - y) - \frac{\sigma^2}{8}(k+1)^2 T\right]} (e^{2\log B - y} - K)^+ & \text{if } y < \log B, \\ 0 & \text{elsewhere.} \end{cases}$$

Consequently, the function $w_1(t, y) - w_2(t, y)$ satisfies the heat equation, its value at the boundary $\{y = \log B\}$ is $w_1(t, \log B) - w_2(t, \log B) = 0$, and it has the same terminal condition as w on $\{y > \log B\}$. Therefore its restriction to the domain $\{y > \log B\}$ is equal to w, and one has

$$w(t, y) = w_1(t, y) - w_1(t, 2 \log B - y).$$

Undoing the previous changes of variable, we get

$$\begin{aligned} u(t, x) &= v(t, \log x) \\ &= e^{\left[\frac{1}{2}(k-1)\log x - \frac{\sigma^2}{8}(k+1)^2 t\right]} w(t, \log x) \\ &= e^{\left[\frac{1}{2}(k-1)\log x - \frac{\sigma^2}{8}(k+1)^2 t\right]} (w_1(t, \log x) - w_1(t, 2 \log B - \log x)) \\ &= u_{BS}(t, x) - \left(\frac{x}{B}\right)^{1-k} u_{BS}\left(t, \frac{B^2}{x}\right), \end{aligned} \tag{1.99}$$

where $u_{BS}(t, x)$ is the Black–Scholes price of the European derivative with payoff function $h(x) = (x - K)^+ \mathbf{1}_{\{x > B\}}$. In the case $B \leq K$, where the knock-out barrier is below the call strike, $u_{BS}(t, x)$ is simply the price $C_{BS}(t, x)$ of a call option given by the Black–Scholes formula (1.41).

1.7.2 Lookback Options

We consider for instance a *floating strike lookback put* which pays the difference of the realized maximum of the underlying asset during the option's life and the asset price itself at the expiration time T. Defining the running maximum $J_t = \sup_{0 \leq s \leq t} X_s$, its payoff is $J_T - X_T$, and its value at time $t = 0$, for $X_0 = x$, is given by

$$P(0,x) = \mathbb{E}^\star \left\{ e^{-rT} \left(J_T - X_T \right) \mid X_0 = x \right\}$$

$$= x e^{-rT} \mathbb{E}^\star \left\{ \sup_{0 \leq t \leq T} \left(e^{(r - \frac{1}{2}\sigma^2)t + \sigma W_t} \right) \right\} - x,$$

where we used the martingale property of the discounted stock price under the risk-neutral probability $\mathbb{P}^\star$, and the explicit form of X_t. By using log-variables and a change of measure, the computation is reduced to integrals involving the joint distribution of a driftless Brownian motion and its running maximum. Alternatively, we briefly present the partial differential equation approach which will be used in Chapter 6.

The price $P(t,x,J)$ of this option satisfies the problem

$$\mathscr{L}_{BS}(\sigma)P = 0 \quad \text{in } x < J \text{ and } t < T,$$

$$\frac{\partial P}{\partial J}(t,J,J) = 0,$$

$$P(T,x,J) = J - x.$$

The boundary condition at $J = x$ expresses the fact that the price of the lookback option for $X_t = J_t$ is insensitive to small changes in J_t because the realized maximum at time T is larger than the realized maximum at time t with probability one.

The problem of finding $P(t,x,J)$ can be reduced to a one (space)-dimensional boundary value problem with the following similarity reduction:

$$\xi = x/J, \quad \text{and} \quad P(t,x,J) = JQ(t,\xi).$$

We can express $Q(t,\xi)$ as the solution of

$$\mathscr{L}_{BS}(\sigma)Q = 0 \quad \text{for } \xi < 1 \text{ and } t < T,$$

$$\left(\frac{\partial Q}{\partial \xi} - Q \right)(t,1) = 0,$$

$$Q(T,\xi) = 1 - \xi,$$

where, in a slight abuse of notation, we redefine $\mathcal{L}_{BS}(\sigma)$ as the Black–Scholes operator with respect to the variable ξ. We now use the log-variable

$$\eta = \log \xi, \quad u(t,\eta) = Q(t,\xi),$$

and find that $u(t,\eta)$ satisfies the problem

$$\frac{\partial u}{\partial t} + \frac{1}{2}\sigma^2 \frac{\partial^2 u}{\partial \eta^2} + \left(r - \frac{1}{2}\sigma^2\right)\frac{\partial u}{\partial \eta} - ru = 0 \quad \text{in } \eta < 0 \text{ and } t < T,$$

$$\left(\frac{\partial u}{\partial \eta} - u\right)(t,0) = 0,$$

$$u(T,\eta) = 1 - e^\eta.$$

We first find $w(t,\eta) = \frac{\partial u}{\partial \eta}(t,\eta) - u(t,\eta)$ which solves the following (Dirichlet) boundary value problem:

$$\frac{\partial w}{\partial t} + \frac{1}{2}\sigma^2 \frac{\partial^2 w}{\partial \eta^2} + \left(r - \frac{1}{2}\sigma^2\right)\frac{\partial w}{\partial \eta} - rw = 0 \quad \text{in } \eta < 0 \text{ and } t < T,$$

with the conditions $w(t,0) = 0$ and $w(T,\eta) = -1$. The solution for $w(t,\eta)$ can be found via the method of images explained in Section 1.7.1:

$$w(t,\eta) = e^{-r(T-t)}\left[e^{(1-k)\eta}N\left(c_1(T-t)\right) - N\left(c_2(T-t)\right)\right],$$

where

$$c_1(\tau) = \frac{\eta}{\sigma\sqrt{\tau}} + \frac{1}{2}(1-k)\sigma\sqrt{\tau} \quad \text{and} \quad c_2(\tau) = \frac{-\eta}{\sigma\sqrt{\tau}} + \frac{1}{2}(1-k)\sigma\sqrt{\tau},$$

and k was defined in (1.98). Restoring all transformations and using the notation of Wilmott *et al.* (1996), we get

$$P(t,x,J) = -x + x\left(1 + \frac{\sigma^2}{2r}\right)N(d_7)$$

$$+ Je^{-r(T-t)}\left(N(d_5) - \frac{\sigma^2}{2r}\left(\frac{x}{J}\right)^{1-k}N(d_6)\right), \qquad (1.100)$$

where

$$d_5 = \frac{\log(J/x) - \left(r - \frac{1}{2}\sigma^2\right)(T-t)}{\sigma\sqrt{T-t}}, \quad d_6 = \frac{\log(x/J) - \left(r - \frac{1}{2}\sigma^2\right)(T-t)}{\sigma\sqrt{T-t}},$$

$$d_7 = \frac{\log(x/J) + \left(r + \frac{1}{2}\sigma^2\right)(T-t)}{\sigma\sqrt{T-t}}.$$

1.7.3 Forward-Start Options

Recall from Section 1.2.3 that a typical forward-start option is a call option maturing at time T such that the strike price is set equal to X_{T_1} at time $T_1 < T$. Its payoff at maturity T is given by $h = (X_T - X_{T_1})^+$. If $T_1 \leq t \leq T$, the contract is simply a call option with (known) strike price $K = X_{T_1}$; its price $P(t,x)$ is given by the Black–Scholes formula (1.41), where $X_t = x$. When $t < T_1 < T_2$, which is the case when the contract is initiated, its price at time t is given by $P(t, X_t)$, where

$$
\begin{aligned}
P(t,x) &= \mathbb{E}^\star \left\{ e^{-r(T-t)} (X_T - X_{T_1})^+ \mid X_t = x \right\} \\
&= \mathbb{E}^\star \left\{ e^{-r(T_1-t)} \mathbb{E}^\star \left\{ e^{-r(T-T_1)} (X_T - X_{T_1})^+ \mid \mathscr{F}_{T_1} \right\} \mid X_t = x \right\} \\
&= \mathbb{E}^\star \left\{ e^{-r(T_1-t)} C_{BS}(T_1, X_{T_1}; T, K = X_{T_1}) \mid X_t = x \right\} \\
&= \mathbb{E}^\star \left\{ e^{-r(T_1-t)} X_{T_1} \left(N(\bar{d}_1) - e^{-r(T-T_1)} N(\bar{d}_2) \right) \mid X_t = x \right\},
\end{aligned}
$$

where $\bar{d}_1$ and $\bar{d}_2$ are given here by

$$
\bar{d}_1 = \left(r + \frac{1}{2}\sigma^2 \right) \frac{\sqrt{T-T_1}}{\sigma} \ , \quad \bar{d}_2 = \left(r - \frac{1}{2}\sigma^2 \right) \frac{\sqrt{T-T_1}}{\sigma},
$$

because the underlying call option is computed at the money $K = X_{T_1}$. We then deduce

$$
\begin{aligned}
P(t,x) &= \left(N(\bar{d}_1) - e^{-r(T-T_1)} N(\bar{d}_2) \right) \mathbb{E}^\star \left\{ e^{-r(T_1-t)} X_{T_1} \mid X_t = x \right\} \\
&= x \left(N(\bar{d}_1) - e^{-r(T-T_1)} N(\bar{d}_2) \right),
\end{aligned} \tag{1.101}
$$

by using the martingale property of the discounted stock price under the risk-neutral probability $\mathbb{P}^\star$.

1.7.4 Compound Options

We consider the example of a call-on-call option introduced in Section 1.2.3. For $t < T_1 < T$, at time T_1, the maturity time of the option, the payoff is given by

$$
h(C_{BS}(T_1, X_{T_1}; K, T)) = (C_{BS}(T_1, X_{T_1}; K, T) - K_1)^+,
$$

where $C_{BS}(T_1, X_{T_1}; K, T)$ is the price at time T_1 of a call option with strike K and maturity T. The price at time t of this call-on-call is given by

$$P(t,x) = \mathbb{E}^\star \left\{ e^{-r(T_1-t)} \left(C_{BS}(T_1, X_{T_1}; K, T) - K_1 \right)^+ \mid X_t = x \right\}$$

$$= \mathbb{E}^\star \left\{ e^{-r(T_1-t)} \left(C_{BS}(T_1, X_{T_1}; K, T) - K_1 \right) \mathbf{1}_{\{X_{T_1} \geq x_1\}} \mid X_t = x \right\},$$

where x_1 is defined by $C_{BS}(T_1, x_1; K, T) = K_1$. We then write

$$P(t,x)$$

$$= \mathbb{E}^\star \left\{ e^{-r(T_1-t)} \left(\mathbb{E}^\star \left\{ e^{-r(T-T_1)} (X_t - K)^+ \mid X_{T_1} \right\} - K_1 \right) \mathbf{1}_{\{X_{T_1} \geq x_1\}} \mid X_t = x \right\}$$

$$= e^{-r(T-t)} \mathbb{E}^\star_{t,x} \left\{ X_T \mathbf{1}_{\{X_T \geq K\}} \mathbf{1}_{\{X_{T_1} \geq x_1\}} \right\}$$

$$- K e^{-r(T-t)} \mathbb{E}^\star_{t,x} \left\{ \mathbf{1}_{\{X_T \geq K\}} \mathbf{1}_{\{X_{T_1} \geq x_1\}} \right\} - K_1 e^{-r(T_1-t)} \mathbb{E}^\star_{t,x} \left\{ \mathbf{1}_{\{X_{T_1} \geq x_1\}} \right\}$$

$$= x N_2 \left(d_1^{(1)}, d_1; \rho \right) - K e^{-r(T-t)} N_2 \left(d_2^{(1)}, d_2; \rho \right)$$

$$- K_1 e^{-r(T_1-t)} N \left(d_2^{(1)} \right), \tag{1.102}$$

where we used the following transformations and notations:

$$X_{T_1} = x \exp \left((r - \tfrac{1}{2}\sigma^2)(T_1 - t) + \sigma \sqrt{T_1 - t} Z_1 \right), \quad Z_1 = \frac{W_{T_1} - W_t}{\sqrt{T_1 - t}}$$

$$X_T = x \exp \left((r - \tfrac{1}{2}\sigma^2)(T - t) + \sigma \sqrt{T - t} Z \right), \quad Z = \frac{W_T - W_t}{\sqrt{T - t}}$$

$$d_1^{(1)} = \frac{\log(x/x_1) + (r + \tfrac{1}{2}\sigma^2)(T_1 - t)}{\sigma \sqrt{T_1 - t}}, \quad d_2^{(1)} = d_1^{(1)} - \sigma \sqrt{T_1 - t}$$

$$d_1 = \frac{\log(x/K) + (r + \tfrac{1}{2}\sigma^2)(T - t)}{\sigma \sqrt{T - t}}, \quad d_2 = d_1 - \sigma \sqrt{T - t}$$

$$N(z) = \frac{1}{\sqrt{2\pi}} \int_{-\infty}^{z} e^{-y^2/2} dy, \quad \rho = \sqrt{T_1 - t}/\sqrt{T - t}$$

$$N_2(z_1, z; \rho) = \frac{1}{2\pi} \int_{-\infty}^{z_1} \int_{-\infty}^{z} e^{-(y_1^2 + 2\rho y_1 y + y^2)/2} dy_1 dy.$$

1.7.5 Asian Options

As an example we consider an **Asian (European-style) average-strike option** whose payoff is given by a function of the stock price at maturity and of the arithmetically averaged stock price before maturity like in an average strike call option (1.31). Without entering into further details, one can introduce the integral process

$$I_t = \int_0^t X_s \, ds,$$

and redo the replicating strategies analysis or the risk-neutral valuation argument for the pair of processes (X_t, I_t). Observe that (I_t) does not introduce new risk or, in other words, there is no new Brownian motion in the equation $dI_t = X_t \, dt$. Using a two-dimensional version of Itô's formula presented in the Section 1.9, one can deduce the partial differential equation

$$\frac{\partial P}{\partial t} + \frac{1}{2}\sigma^2 x^2 \frac{\partial^2 P}{\partial x^2} + r\left(x\frac{\partial P}{\partial x} - P\right) + x\frac{\partial P}{\partial I} = 0, \qquad (1.103)$$

to be solved, for instance, with the final condition $P(T,x,I) = (x - \frac{I}{T})^+$, in order to obtain the price $P(t, X_t, I_t)$ of an arithmetic-average strike call option at time t. This is solved numerically in most examples. Dimension reduction techniques have been proposed (see for instance Vecer (2002)).

1.8 First-Passage Structural Approach to Default

In this section, in the context of *credit risk*, we present Merton's approach to default as an illustration of the use of the objects introduced in this chapter. We consider here the problem of pricing a defaultable zero-coupon bond which pays a fixed amount (say \$1) at maturity T unless default occurs, in which case it is worth nothing. In other words, we consider the simple case of no recovery in case of default.

1.8.1 Merton's Approach

In Merton's approach, the underlying X_t follows a geometric Brownian motion, and default occurs if $X_T < B$ for some threshold value B. In this case the price at time t of the defaultable bond is simply the price of a European digital option which pays one if X_T exceeds the threshold and zero otherwise, as in (1.69). Assuming that the underlying is tradable and the risk-free interest rate r is constant, by the no-arbitrage argument, the price of this option is explicitly given by $u^d(t, X_t)$, where

$$
\begin{aligned}
u^d(t,x) &= \mathbb{E}^{\star}\left\{ e^{-r(T-t)} \mathbf{1}_{X_T > B} \mid X_t = x \right\} \\
&= e^{-r(T-t)}\mathbb{P}^{\star}\left\{ X_T > B \mid X_t = x \right\} \\
&= e^{-r(T-t)}\mathbb{P}^{\star}\left\{ \left(r - \frac{\sigma^2}{2}\right)(T-t) + \sigma(W_T^{\star} - W_t^{\star}) > \log\left(\frac{B}{x}\right) \right\} \\
&= e^{-r(T-t)}\mathbb{P}^{\star}\left\{ \frac{W_T^{\star} - W_t^{\star}}{\sqrt{T-t}} > -\frac{\log\left(\frac{x}{B}\right) + \left(r - \frac{\sigma^2}{2}\right)(T-t)}{\sigma\sqrt{T-t}} \right\} \\
&= e^{-r\tau}N(d_2(\tau)), \qquad (1.104)
\end{aligned}
$$

with the usual notation $\tau = T - t$ and the *distance to default*:

$$d_2(\tau) = \frac{\log\left(\frac{x}{B}\right) + \left(r - \frac{\sigma^2}{2}\right)\tau}{\sigma\sqrt{\tau}}. \tag{1.105}$$

1.8.2 First-Passage Model

In the first-passage structural approach, default occurs if X_t goes below B at some time before maturity. In this extended Merton, or Black and Cox model, the payoff is

$$h(X) = 1_{\{\inf_{0 \le s \le T} X_s > B\}}.$$

The defaultable bond can then be viewed as a path-dependent derivative. Its value at time $t \le T$, denoted by $P^B(t, T)$, is given by

$$P^B(t, T) = \mathbb{E}^\star\left\{e^{-r(T-t)} 1_{\{\inf_{0 \le s \le T} X_s > B\}} \mid \mathscr{F}_t\right\} \tag{1.106}$$

$$= 1_{\{\inf_{0 \le s \le t} X_s > B\}} e^{-r(T-t)} \mathbb{E}^\star\left\{1_{\{\inf_{t \le s \le T} X_s > B\}} \mid \mathscr{F}_t\right\}.$$

Indeed $P^B(t, T) = 0$ if the asset price has reached B before time t, which is reflected by the factor $1_{\{\inf_{0 \le s \le t} X_s > B\}}$. Introducing the *default time* τ_t defined by

$$\tau_t = \inf\{s \ge t, X_s \le B\},$$

one has

$$\mathbb{E}^\star\left\{1_{\{\inf_{t \le s \le T} X_s > B\}} \mid \mathscr{F}_t\right\} = \mathbb{P}^\star\{\tau_t > T \mid \mathscr{F}_t\},$$

which shows that the problem reduces to the characterization of the distribution of default times. Observe that τ_t is a stopping time (as defined in Section 1.2.2). Moreover, it is a *predictable* stopping time in the sense that there exists an increasing sequence of stopping times $\tau_t^{(n)} < \tau_t$ such that $\lim_{n \to \infty} \tau_t^{(n)} = \tau$ a.s. For instance one can consider the sequence $(\tau_t^{(n)})$ defined by $\tau_t^{(n)} = \inf\{s \ge t, X_s \le B + 1/n\}$. These stopping times are illustrated in Figure 1.6.

An alternative *intensity-based* approach to default, presented in Chapter 14, consists of introducing default times which are *not* predictable.

In the first passage model, a defaultable zero-coupon bond is in fact a *binary down-and-out barrier option* where the barrier level and the strike price coincide. As presented in Section 1.7.1, we have

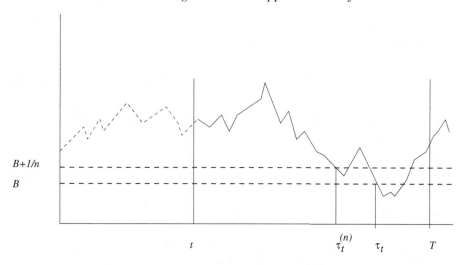

Figure 1.6   A sample trajectory of the geometric Brownian motion X_t, and the corresponding values of the first hitting times $\tau_t^{(n)}$ and τ_t after t of the levels $B+1/n$ and B.

$$\mathbb{E}^\star \left\{ \mathbf{1}_{\{\inf_{t \leq s \leq T} X_s > B\}} \mid \mathscr{F}_t \right\} =$$

$$\mathbb{P}^\star \left\{ \inf_{t \leq s \leq T} \left((r - \frac{\sigma^2}{2})(s-t) + \sigma(W_s^\star - W_t^\star) \right) > \log\left(\frac{B}{x}\right) \mid X_t = x \right\},$$

which can be computed by using the distribution of the minimum of a (nonstandard) Brownian motion. In this Markovian setting, we have

$$\mathbb{E}^\star \left\{ e^{-r(T-t)} \mathbf{1}_{\{\inf_{t \leq s \leq T} X_s > B\}} \mid \mathscr{F}_t \right\} = u(t, X_t),$$

where $u(t,x)$ is the solution of the following partial differential equations problem:

$$\mathscr{L}_{BS}(\sigma)u = 0 \text{ on } x > B, t < T, \tag{1.107}$$
$$u(t,B) = 0 \text{ for any } t \leq T,$$
$$u(T,x) = 1 \text{ for } x > B,$$

which is to be solved for $x > B$. This problem can be solved by introducing the corresponding European digital option which pays \$1 at maturity if $X_T > B$ and nothing otherwise. Its price at time $t < T$ is given by $u^d(t, X_t)$, where $u^d(t,x)$ is computed explicitly in (1.104). The function $u^d(t,x)$ is the solution to the partial differential equation

$$\mathscr{L}_{BS}(\sigma)u^d = 0 \text{ on } x > 0, t < T, \tag{1.108}$$
$$u^d(T,x) = 1 \text{ for } x > B, \text{ and } 0 \text{ otherwise.}$$

By using the method of images presented in Section 1.7.1, the solution $u(t,x)$ of the problem (1.107) can be written

$$u(t,x) = u^d(t,x) - \left(\frac{x}{B}\right)^{1-k} u^d\left(t, \frac{B^2}{x}\right),\qquad(1.109)$$

where k was defined in (1.98). Combining the expression (1.104) for $u^d(t,x)$ with (1.109), we get

$$u(t,x) = e^{-r(T-t)}\left(N(d_2^+(T-t)) - \left(\frac{x}{B}\right)^{1-k} N(d_2^-(T-t))\right),(1.110)$$

where we denote

$$d_2^\pm(\tau) = \frac{\pm\log\left(\frac{x}{B}\right) + \left(r - \frac{\sigma^2}{2}\right)\tau}{\sigma\sqrt{\tau}}.\qquad(1.111)$$

This simple model of default will be discussed further and generalized in Chapter 13.

1.9 Multidimensional Stochastic Calculus

In the following chapters we consider models where volatility is driven by additional processes (or factors). In order to handle these situations, we will need multidimensional versions of the tools introduced in the previous sections. We briefly summarize here the multidimensional Itô formula, Girsanov's theorem, and the Feynman–Kac formula.

1.9.1 Multidimensional Itô's Formula

We consider the generalization of the stochastic differential equations (1.11) to the case of systems of such equations:

$$dX_t^i = \mu_i(t,X_t)dt + \sum_{j=1}^d \sigma_{i,j}(t,X_t)dW_t^j,\ i=1,\ldots,d,\qquad(1.112)$$

where W_t^j, $j=1,\ldots,d$, are d independent standard Brownian motions, and

$$X_t = (X_t^1,\ldots,X_t^d)$$

is a d-dimensional process. We assume that the functions $\mu_i(t,x)$ and $\sigma_{i,j}(t,x)$ are smooth and at most linearly growing at infinity, so that this system has a unique solution adapted to the filtration $(\mathscr{F}_t)$ generated by the Brownian motions (W_t^j). We now consider real processes of the form $f(t,X_t)$, where the real function $f(t,x)$ is smooth on $\mathbb{R}_+ \times \mathbb{R}^d$ (for instance continuously differentiable with respect to t, and twice continuously differentiable in the x-variable). The d-dimensional Itô formula can then be written:

$$df(t,X_t) = \frac{\partial f}{\partial t}(t,X_t)dt + \sum_{i=1}^{d} \frac{\partial f}{\partial x^i}(t,X_t)dX_t^i$$

$$+ \frac{1}{2} \sum_{i,j=1}^{d} \frac{\partial^2 f}{\partial x^i \partial x^j}(t,X_t)d\langle X^i, X^j \rangle_t, \qquad (1.113)$$

where

$$d\langle X^i, X^j \rangle_t = \sum_{k=1}^{d} \sigma_{ik}(t,X_t)\sigma_{jk}(t,X_t)dt$$

$$= (\sigma\sigma^T)_{i,j}(t,X_t)dt \qquad (1.114)$$

follows from the cross-variation rules $d\langle t, W_t^j \rangle = d\langle W_t^j, t \rangle = 0, d\langle W_t^i, W_t^j \rangle = d\langle W_t^j, W_t^i \rangle = 0$ for $i \neq j$, and $d\langle W_t^i, W_t^i \rangle = dt$. Formula (1.113) can then be rewritten:

$$df(t,X_t) = \left(\frac{\partial f}{\partial t} + \sum_{i=1}^{d} \mu_i \frac{\partial f}{\partial x^i} + \frac{1}{2} \sum_{i,j=1}^{d} (\sigma\sigma^T)_{i,j} \frac{\partial^2 f}{\partial x^i \partial x^j} \right) dt$$

$$+ \sum_{i=1}^{d} \frac{\partial f}{\partial x^i} \left(\sum_{j=1}^{d} \sigma_{i,j} dW_t^j \right),$$

where the partial derivatives of f and the coefficients μ and σ are evaluated at (t,X_t).

1.9.2 Girsanov's Theorem

In Section 1.4.1 we have used a change of probability measure so that the one-dimensional process $W_t^\star = W_t + \theta t$ becomes a standard Brownian motion under the new probability $\mathbb{P}^\star$. We now give a multidimensional version of this result in the case where θ may also be a stochastic process. To simplify the presentation we assume that the d-dimensional process (θ_t) is of the form $(\theta_j(X_t), j = 1, \ldots, d)$, where the functions $\theta_j(x)$ are bounded (for less restrictive conditions, such as the Novikov condition, we refer to Karatzas and Shreve (1991)). Generalizing (1.57), we define the real process $(\xi_t^\theta)_{0 \leq t \leq T}$ by:

$$\xi_t^\theta = \exp\left(-\sum_{j=i}^{d} \left(\int_0^t \theta_j(X_s)dW_s^j + \frac{1}{2} \int_0^t \theta_j^2(X_s)ds \right) \right), \quad (1.115)$$

which satisfies

$$d\xi_t^\theta = -\xi_t^\theta \sum_{j=1}^d \theta_j(X_t)dW_t^j.$$

and, therefore, is a martingale. We then define, on $\mathscr{F}_T$, the probability $\mathbb{P}^\star$ by $d\mathbb{P}^\star = \xi_T^\theta d\mathbb{P}$ as in (1.58). Girsanov's theorem states that the processes $(W_t^{j\star})_{0\leq t\leq T}, j = 1,\dots,d$, defined by

$$W_t^{j\star} = W_t^j + \int_0^t \theta_j(X_s)ds, \quad j = 1,\dots,d, \tag{1.116}$$

are independent standard Brownian motions under $\mathbb{P}^\star$. Using norm and dot product in $\mathbb{R}^d$, the following formal computation, for all $u \in \mathbb{R}^d$, explains this result:

$$
\begin{aligned}
\mathbb{E}^\star &\left\{ e^{iu\cdot(W_t^\star - W_s^\star)} \mid \mathscr{F}_s \right\} \\
&= \mathbb{E}\left\{ (\xi_s^\theta)^{-1}\xi_t^\theta e^{iu\cdot(W_t^\star - W_s^\star)} \mid \mathscr{F}_s \right\} \\
&= \mathbb{E}\left\{ e^{-\int_s^t \theta_v\cdot dW_v - \frac{1}{2}\int_s^t \|\theta_v\|^2 dv} e^{iu\cdot\left(W_t - W_s + \int_s^t \theta_v dv\right)} \mid \mathscr{F}_s \right\} \\
&= e^{-\frac{1}{2}(t-s)\|u\|^2}\mathbb{E}\left\{ e^{-\int_s^t (\theta_v - iu)\cdot dW_v - \frac{1}{2}\int_s^t \|\theta_v - iu\|^2 dv} \mid \mathscr{F}_s \right\} \\
&= e^{-\frac{1}{2}(t-s)\|u\|^2}\mathbb{E}\left\{ (\xi_s^{\theta - iu})^{-1}\xi_t^{\theta - iu} \mid \mathscr{F}_s \right\} \\
&= e^{-\frac{1}{2}(t-s)\|u\|^2},
\end{aligned}
$$

where we used the martingale property of $(\xi_t^{\theta-iu})$, and the characterization of independent standard Brownian motions by conditional characteristic functions.

1.9.3 The Feynman–Kac Formula

The infinitesimal generator of the (possibly nonhomogeneous) Markovian process $X = (X^1,\dots,X^d)$, introduced in (1.112), is given by

$$\mathscr{L}_t = \sum_{i=1}^d \mu_i(t,x)\frac{\partial}{\partial x^i} + \frac{1}{2}\sum_{i,j=1}^d (\sigma\sigma^T)_{i,j}(t,x)\frac{\partial^2}{\partial x^i\partial x^j}.$$

If $r(t,x)$ is a function on $\mathbb{R}_+ \times \mathbb{R}^d$ (for instance bounded), then the function $u(t,x)$ defined by

$$u(t,x) = \mathbb{E}\left\{ e^{-\int_t^T r(s,X_s)ds}h(X_T) \mid X_t = x \right\}$$

satisfies the partial differential equation

$$\frac{\partial u}{\partial t} + \mathcal{L}_t u - ru = 0,$$

with the terminal condition $u(T,x) = h(x)$. Here h does not need to be smooth, and, in particular, can be the payoff of a call, put, or digital option.

Such parabolic partial differential equations with an additional *source* will be important in our perturbation theory: if the function $g(t,x)$ is, for instance, bounded, then the backward problem

$$\frac{\partial v}{\partial t} + \mathcal{L}_t v - rv + g = 0,$$
$$v(T,x) = h(x),$$

admits the solution

$$v(t,x) = \mathbb{E}\left\{ e^{-\int_t^T r(s,X_s)ds} h(X_T) + \int_t^T e^{-\int_t^s r(u,X_u)du} g(s,X_s)ds \mid X_t = x \right\}.$$

1.10 Complete Market

The model we have analyzed in this chapter is an example of a *complete market* model. The simplest definition of a complete market is one in which every contingent claim can be replicated by a self-financing trading strategy in the stock and bond.

In Section 1.3.2, we constructed such a strategy for any European derivative with payoff $h(X_T)$, using the Markov property and Itô's lemma, and outlined the arguments for American and other exotic contracts. In fact, in this model, *any* security whose payoff H is known on date T (H is any $\mathscr{F}_T$-measurable random variable with $\mathbb{E}\{H^2\} < \infty$) can be replicated by some unique self-financing trading strategy, as the martingale representation theorem of Section 1.4.3 tells us. Equivalently, such a claim can be *perfectly* hedged (without overshooting) by trading in the underlying stock and bond. If there is no early exercise feature to the contract (nothing is paid out except on date T and possibly dividends on *fixed* dates, or at a fixed rate), then the risk-neutral valuation formula (1.65) prices the derivative.

Finally, we mention that another characterization of an arbitrage-free complete market is that there is a *unique* equivalent martingale measure $\mathbb{P}^\star$ under which the discounted prices of traded securities are martingales. When looking at stochastic volatility market models in the next chapter, we shall see that the market is *incomplete*: there is a whole family of equivalent martingale measures and derivatives securities cannot be perfectly hedged with just the stock and bond.

Notes

The original derivation of the *no-arbitrage* price of a European call option under the lognormal model appeared in Black and Scholes (1973), with related results in Merton (1973). The geometric Brownian motion model for the risky asset and many other issues regarding the pricing of options prior to the Black–Scholes theory are discussed in Samuelson (1973).

For further details about the material outlined in Section 1.1, namely Brownian motion, stochastic integrals, stochastic differential equations, and Itô's formulas, we recommend Oksendal (2007) or Mikosch (1999). These books also cover Girsanov's and martingale representation theorems discussed in Section 1.4.1 and 1.4.3 as well as the optimal stopping theory introduced in Section 1.6 for the American option pricing problem, and an extensive list of further references on the subject. For the relation with partial differential equations, Kolmogorov's equations, and the Feynman–Kac formula, we refer for instance to Freidlin (1985) or Friedman (2006).

There are many books discussing the finance topics we have summarized in this chapter. Among them are Duffie (2001), Hull (2008), Björk (2004), Lamberton and Lapeyre (1996), Musiela and Rutkowski (2002), Etheridge (2002), Shreve (2004), Platen and Heath (2006), and Jeanblanc *et al.* (2009).

A reference for the method of images approach to pricing barrier options mentioned in Section 1.7 is Wilmott *et al.* (1996). Details about the linear complementarity and partial differential inequality formulations of the American pricing problem are found here also. The other examples of exotic options can also be found in the above-mentioned references, and in Lipton (2001). We refer to the original work of Goldman *et al.* (1979) for lookback options, and to Geske (1979) for compound options.

For an introduction to *Credit Risk* we refer for instance to the survey article by Giesecke (2004), and the books by Duffie and Singleton (2003), Lando (2004), and Schönbucher (2003).

2

Introduction to Stochastic Volatility Models

The Black–Scholes model makes a number of simplifying assumptions. Among these are that the asset price has independent Gaussian returns and constant volatility. We shall focus here on relaxation of these assumptions by allowing volatility to be randomly varying, for the following reason: a well-known discrepancy between Black–Scholes-predicted European option prices and market-traded options prices, the **implied volatility skew**, can be accounted for by stochastic volatility models. That is, this modification of the Black–Scholes theory has *a posteriori* success in one area where the classical model fails.

In fact, modeling volatility as a stochastic process is motivated *a priori* by empirical studies of stock price returns in which estimated volatility is observed to exhibit "random" characteristics. As we will see, stochastic volatility has the effect of thickening the tails of returns distributions compared with the normal distribution, and therefore modeling more extreme stock price movements. Stochastic volatility modeling is a powerful modification of the Black–Scholes model that describes a much more complex market.

In Chapter 1, we introduced the notation and tools for pricing and hedging derivative securities under a constant volatility lognormal model (1.2). This is the simplest continuous-time example of pricing in a complete market, while pricing in a market with stochastic volatility is an incomplete markets problem, a distinction we shall explain further below, and one which has far-reaching consequences, particularly for the hedging problem and the problem of parameter estimation. It is the latter *inverse* problem that is the biggest mathematical and practical challenge introduced by such models, and also perhaps the one that benefits most from the asymptotic methods of Chapter 4.

After a brief review of *local volatility* models and in particular the Dupire fomula, we turn to stochastic volatility models. The incompleteness of such markets leads to a family of risk-neutral probability measures which we describe by means of Girsanov's transform and associated pricing partial differential equations in Markovian models. We start with one-factor models and then we generalize to multidimensional models motivated by the multi-factor stochastic volatility models considered in the rest of the book.

2.1 Implied Volatility Surface

Implied volatility is a convenient quantity frequently used to express discrepancies between European options prices predicted by the Black–Scholes theory, and traded options prices.

Given an observed European call option price C^{obs} for a contract with strike price K and expiration date T, the **implied volatility** I is defined to be the value of the volatility parameter that must go into the Black–Scholes formula (1.41) to match this price:

$$C_{BS}(t,x;K,T;I) = C^{\text{obs}}. \qquad (2.1)$$

Remarks

(i) A non-negative implied volatility $I > 0$ can be found uniquely provided

$$C^{\text{obs}} > C_{BS}(t,x;K,T;0)$$

because of the monotonicity of the Black–Scholes formula in the volatility parameter:

$$\frac{\partial C_{BS}}{\partial \sigma} = \frac{xe^{-d_1^2/2}\sqrt{T-t}}{\sqrt{2\pi}} > 0. \qquad (2.2)$$

(ii) The implied volatilities from put and call options of the same strike price and time-to-maturity are the same because of put–call parity (1.45).

Many formulas, for example the asymptotic ones of Chapter 4, are most neatly expressed in terms of implied volatility. It is common practice for traders to quote derivative prices in terms of I, the conversion to price being through the Black–Scholes formula appropriate for the contract.

In general, $I = I(t,x;K,T)$, but if the observed options prices equalled the Black–Scholes prices, $C^{\text{obs}} = C_{BS}(t,x;K,T;\sigma)$, then $I = \sigma$, the historical volatility. That is, if observations matched Black–Scholes exactly, I would be the same constant for all derivative contracts.

The most quoted phenomenon testifying to the limitations of the standard Black–Scholes model is the *implied volatility skew or smile*: for options with the same time-to-maturity, but different strike prices, the implied volatilities computed from their market prices are not constant. Before the 1987 crash, the graph of $I(K)$ against K for fixed t, x, T obtained from market options prices was often observed to be U-shaped with minimum at- or near-the-money ($K_{\min} \approx x$). This is called the smile curve, and is illustrated in Figure 2.1.

More typically since then, it is downward sloping at- and around-the-money, and is referred to as the *skew*. An example from six-month S&P 500 index options is shown in Figure 2.2. The variation of implied volatilities with respect to maturity is known as the *term structure*, and implied volatility as a function of strike K and maturity T is called the *implied volatility surface*. Typical qualitative features of implied volatilities from stock index options are that they are higher than historical volatility of the index, and the skews are steeper for shorter maturities. An illustration is given in Figure 2.3, where each strand corresponds to a given maturity as a function of *moneyness*, that is the ratio of the strike to the current underlying price K/x.

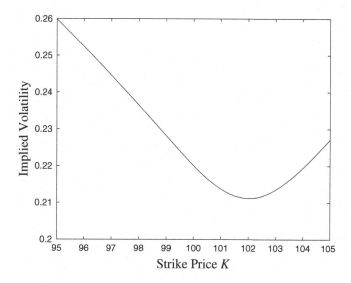

Figure 2.1   Illustrative smile curve of implied volatilities from European options with the same time-to-expiration. The current stock price is $x = 100$, which is close to the minimum point. In fact, the minimum is at $xe^{r(T-t)} \approx 102$.

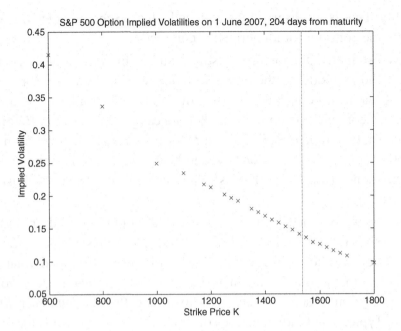

Figure 2.2   S&P 500 implied volatilities approximately six months from maturity as a function of strike on June 1, 2007. The vertical line shows at-the-money.

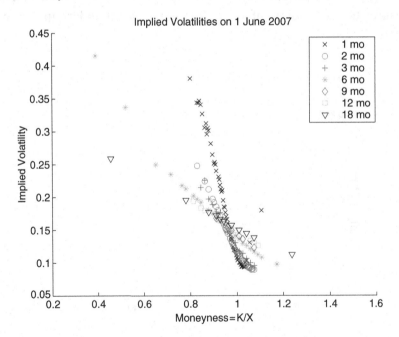

Figure 2.3   S&P 500 implied volatilities as a function of moneyness from S&P 500 index options on June 1, 2007.

2.1.1 Interpretation of the Skew

An implied volatility skew tells us that there is a premium charged for out-of-the-money put options (low K relative to x) above their Black–Scholes price computed with at-the-money implied volatility. The market prices as though the lognormal model fails to capture probabilities of large downward stock price movements and supplements the Black–Scholes prices to account for this. This higher premium on out-of-the-money puts is often known as *crashophobia*.

One can obtain some broad bounds on the permissible slope of the implied volatility curve $I(K)$ by noting that call prices must be decreasing in the strike price K (or else there is an arbitrage opportunity). Differentiating (2.1) with respect to K gives

$$\frac{\partial C^{\mathrm{obs}}}{\partial K} = \frac{\partial C_{BS}}{\partial K} + \frac{\partial C_{BS}}{\partial \sigma}\frac{\partial I}{\partial K} \leq 0,$$

from which we conclude that

$$\frac{\partial I}{\partial K} \leq -\frac{\partial C_{BS}/\partial K}{\partial C_{BS}/\partial \sigma}.$$

Similarly, put prices must be increasing in K, and since prices of puts and calls with the same K and T must have the same implied volatility I, we also get

$$\frac{\partial I}{\partial K} \geq -\frac{\partial P_{BS}/\partial K}{\partial P_{BS}/\partial \sigma}.$$

Using the explicit Black–Scholes formulas (1.41), we get the bounds

$$-\frac{\sqrt{2\pi}}{x\sqrt{T-t}}(1-N(d_2))e^{-r(T-t)+\frac{1}{2}d_1^2} \leq \frac{\partial I}{\partial K} \leq \frac{\sqrt{2\pi}}{x\sqrt{T-t}}N(d_2)e^{-r(T-t)+\frac{1}{2}d_1^2},$$

where d_1 and d_2 are defined in (1.42)–(1.43) with σ replaced by I. In other words, the slope cannot be too positive or too negative. Observe that this result is model-free. Other bounds for the asymptotic behavior of implied volatilities for large and small strikes have been studied, and we refer to Benaim *et al.* (2009) for a survey.

This type of calculation, where information about the geometry of the implied volatility curve is obtained from the Black–Scholes formula and its derivatives will come up often, for example to show that stochastic volatility models predict the smile (Section 2.5.2), and in the asymptotic calculations of Chapter 4.

At this stage, it is clear that the specification of a constant volatility is significantly inaccurate, first from statistical studies of stock price history that indicate the random character of volatility (measured as the normalized standard deviation of returns), and second from the nonflat implied volatility surface $I(K,T)$ that tells us how the market is actually pricing derivatives contrary to the Black–Scholes theory.

We also note that we use implied volatility simply as a convenient unit to express market and model options prices. It is *not* the fundamental object we wish to model. As discussed in Figlewski (2010), "the smile is stable enough over short time intervals that traders use the BS model anyway, by inputting different volatilities for different options according to their moneyness. This jury-rigged procedure, known as 'practitioner Black–Scholes,' is an understandable strategy for traders, who need some way to impose pricing consistency across a broad range of related financial instruments, and don't care particularly about theoretical consistency with academic models. This has led to extensive analysis of the shape and dynamic behavior of volatility smiles, even though it is odd to begin with a model that is visibly inconsistent with the empirical data and hope to improve it by modeling the behavior of the inconsistency."

2.1.2 What Data to Use

When it comes to estimation, there is a radical departure from the spirit of the Black–Scholes model both in practice and in the literature. Of course it is tempting to use implied volatility as a proxy for volatility. In other words, options are priced with the Black–Scholes formula, but using implied volatilities instead of historical volatilities.

In particular, out-of-the-money puts $(K < x)$ are typically priced with a higher volatility coming, for example, from yesterday's smile curve. The smile curve is thus translated forward in time, and other than the current stock price, no historical stock price data is used at all. Today's options are priced using yesterday's *options* data.

The main reason for this procedure, which is of course mathematically inconsistent since volatility is assumed constant (independent of the derivative contract) in the derivation of the Black–Scholes formula, is the belief that implied volatilities are better predictors of future realized volatility. Empirical studies disagree as to whether this is borne out by the data.

In tests of new models in the academic literature, calibration of parameters is also usually from derivative data (traded call option prices, for example), ignoring the underlying time series. This is because the latter, so-called cross-sectional fitting, is easier to do compared with econometric methods for time series, especially when there is a formula for call option prices to fit to. Once the parameters have been estimated, the model can be used to price *other* derivatives in a consistent (no-arbitrage) manner, for example by simulation.

Thus some derivative data, most often at-the-money European option prices, are part of the basic data. The market in at-the-money calls (and puts) is liquid enough for this to be a valid procedure (we can trust the data).

When we look at stochastic volatility models, we shall see that derivative data contains some information that we need for pricing that is not contained in historical data, so we shall always have to use it in part. We address this further in Section 2.4.3.

2.2 Local Volatility

One popular way to modify the lognormal model is to suppose that volatility is a deterministic positive function of time and stock price: $\sigma = \sigma(t, X_t)$. The stochastic differential equation modeling the stock price is

$$dX_t = \mu X_t dt + \sigma(t, X_t) X_t dW_t,$$

and the function $P(t, x)$ giving the no-arbitrage price of a European derivative security at time t when the asset price $X_t = x$ then satisfies the generalized Black–Scholes partial differential equation

$$\frac{\partial P}{\partial t} + \frac{1}{2}\sigma^2(t, x)x^2\frac{\partial^2 P}{\partial x^2} + r\left(x\frac{\partial P}{\partial x} - P\right) = 0, \tag{2.3}$$

the derivation being identical to that given in Section 1.3.2 for the constant σ case. The coefficient σ becomes $\sigma(t, X_t)$ in equations (1.35) and (1.38), and $\sigma(t, x)$ in (1.39). The terminal condition is the payoff function: $P(T, x) = h(x)$.

The hedging ratio is given by the delta of the solution to this partial differential equation problem, $\partial P/\partial x$, and a perfect hedge is achieved by holding this amount of stock. This can be seen by repeating the argument of Section 1.3.2 replacing the constant σ coefficient by $\sigma(t, X_t)$, provided the function $\sigma(t, x)$ is sufficiently regular.

The market is still complete as the randomness of the volatility was introduced as a function of the existing randomness of the lognormal model. There is a unique risk-neutral measure $\mathbb{P}^\star$ under which the stock price is a geometric Brownian motion with drift rate r and the same volatility $\sigma(t, X_t)$:

$$dX_t = rX_t\, dt + \sigma(t, X_t)X_t\, dW_t^\star, \qquad (2.4)$$

with $(W_t^\star)$ a $\mathbb{P}^\star$-Brownian motion.

2.2.1 Time-Dependent Volatility

In the special case $\sigma(t,x) = \sigma(t)$, a deterministic function of time, we can solve the stochastic differential equation

$$dX_t = rX_t dt + \sigma(t)X_t dW_t^\star,$$

analogously to the calculation of Section 1.1.5, by the logarithmic transformation to obtain

$$X_T = X_t \exp\left(r(T-t) - \frac{1}{2}\int_t^T \sigma^2(s)ds + \int_t^T \sigma(s)dW_s^\star \right),$$

so that $\log\left(X_T/X_t\right)$ is $\mathscr{N}\left((r - \frac{1}{2}\overline{\sigma^2})(T-t), \overline{\sigma^2}(T-t) \right)$-distributed, where

$$\overline{\sigma^2} = \frac{1}{T-t}\int_t^T \sigma^2(s)ds.$$

When we compute the call option price from

$$C = \mathbb{E}^\star\left\{ e^{-r(T-t)}(X_T - K)^+ \mid \mathscr{F}_t \right\},$$

the answer will just be the Black–Scholes formula with volatility parameter $\sqrt{\overline{\sigma^2}}$, the root-mean-square volatility. For fixed t, T, all call options are priced with the Black–Scholes formula using the same (time-averaged) volatility, so there is no smile across strike prices. There is, however, a term structure of implied volatility, as $\overline{\sigma^2}$ is different for different maturities. The implied volatility I produced in this model is a function only of t and T defined by:

$$(T-t)I(t,T)^2 = \int_t^T \sigma^2(s)\,ds.$$

Suppose the data actually had implied volatilities that were flat across strikes, but with a term structure. Then, for two given maturities $T_1 < T_2$, we could back-out the integral of σ^2 between the maturities from the implied volatilities:

$$\int_{T_1}^{T_2} \sigma^2(s)\,ds = (T_2 - t)I(t, T_2)^2 - (T_1 - t)I(t, T_1)^2.$$

For the model to be consistent with data, the right-hand side of this expression would have to be independent of t, but this is not generally the case with market implied volatilities. A common practice, called *recalibration*, is to change the function σ on subsequent days, as the data changes. But this is inconsistent with the model's specification, and the way it is used to value other options.

2.2.2 Level-Dependent Volatility and Dupire's Formula

To have a skew across strike prices, we need σ to depend on x as well as t in this framework: $\sigma(t, X_t)$. One disadvantage is that volatility and stock price changes are now perfectly correlated (positively or negatively). Empirical studies suggest that when volatility goes up, prices tend to go down and *vice versa*, but not that there is a perfect (-1) inverse correlation.

The volatility function $\sigma(t, x)$ can either be given a parametric specification, such as in the CEV (constant elasticity of variance) model where

$$\sigma(t, x) = \kappa x^{-\gamma},$$

for parameters κ and γ to be estimated from data. For $\gamma > 0$, volatility goes down when the stock price goes up. There are reasonably tractable formulas for call and put option prices under this model, and we provide references in the Notes. To match steep short-term implied volatility skews, the parameter γ has to be around 3–4, to supply sufficient crashophobia to match the data. This implies that instantaneous volatility would drop 25–32% for a 10% increase in the underlying, which appears to be overly stabilizing.

An alternative approach is to estimate the local volatility surface $\sigma(t, x)$ from options data in a nonparametric way. This is an inverse problem for the backward partial differential equation (2.3), except that the data on a fixed date t when the stock price is x is parametrized by the strikes K and the maturities T in the terminal condition. It turns out that it is possible to take advantage of the "hockey stick" structure of the call and put option payoffs to derive a forward equation in the variables K and T. For fixed (t, x), we denote by $C(T, K)$ the price of the corresponding call option

$$C(T, K) = \mathbb{E}^{\star}\{e^{-r(T-t)}(X_T - K)^+ \mid X_t = x\}$$
$$= e^{-r(T-t)} \int_0^{\infty} (\xi - K)^+ p(t, x; T, \xi) \, d\xi,$$

where $p(t, x; T, \xi)$ is the transition density of X, as introduced in Section 1.5.3. Differentiating twice with respect to K gives

$$\frac{\partial C}{\partial K}(T,K) = -e^{-r(T-t)} \int_0^\infty \mathbf{1}_{\{\xi>K\}} p(t,x;T,\xi)\,d\xi, \qquad (2.5)$$

$$\frac{\partial^2 C}{\partial K^2}(T,K) = e^{-r(T-t)} \int_0^\infty \delta(\xi-K) p(t,x;T,\xi)\,d\xi$$

$$= e^{-r(T-t)} p(t,x;T,K), \qquad (2.6)$$

where δ is the Dirac delta function. Formula (2.6) rewritten as $p(t,x;T,K) = e^{r(T-t)} \frac{\partial^2 C}{\partial K^2}(T,K)$ suggests a way of finding the risk-neutral transition density in terms of the curvature of option prices with respect to strike. So far, we did not use a particular model for X, and this model-free result is known as the Breeden–Litzenberger formula. However, these marginal distributions of X are not enough to price path-dependent options, so we return to the framework (2.4) of local volatility models to proceed further.

Differentiating $C(T,K)$ with respect to T gives

$$\frac{\partial C}{\partial T}(T,K) = e^{-r(T-t)} \int_0^\infty (\xi-K)^+ \frac{\partial p}{\partial T}(t,x;T,\xi)\,d\xi$$

$$- re^{-r(T-t)} \int_0^\infty (\xi-K)^+ p(t,x;T,\xi)\,d\xi$$

$$= e^{-r(T-t)} \int_0^\infty (\xi-K)^+ \mathscr{L}_T^* p(t,x;T,\xi)\,d\xi$$

$$- re^{-r(T-t)} \int_0^\infty (\xi-K)^+ p(t,x;T,\xi)\,d\xi,$$

where we have used the forward Kolmogorov equation (1.86) for p, where under $\mathbb{P}^\star$, we have

$$\mathscr{L}_T^* = \frac{1}{2}\frac{\partial^2}{\partial \xi^2}\left(\sigma^2(T,\xi)\xi^2\cdot\right) - \frac{\partial}{\partial \xi}(r\xi\cdot).$$

Then we compute

$$\frac{\partial C}{\partial T}(T,K)$$

$$= e^{-r(T-t)} \int_0^\infty p(t,x;T,\xi)\mathscr{L}_T(\xi-K)^+\,d\xi$$

$$- re^{-r(T-t)} \int_0^\infty (\xi-K)^+ p(t,x;T,\xi)\,d\xi,$$

$$= e^{-r(T-t)} \int_0^\infty p(t,x;T,\xi)\left(\frac{1}{2}\sigma^2(T,\xi)\xi^2\delta(\xi-K) + r\xi\mathbf{1}_{\{\xi>K\}}\right)d\xi$$

$$- re^{-r(T-t)} \int_0^\infty (\xi-K)\mathbf{1}_{\{\xi>K\}} p(t,x;T,\xi)\,d\xi,$$

$$= \frac{1}{2}\sigma^2(T,K)K^2 e^{-r(T-t)} p(t,x;T,K)$$

$$+ rKe^{-r(T-t)} \int_0^\infty \mathbf{1}_{\{\xi > K\}} p(t,x;T,\xi)\,d\xi$$

$$= \frac{1}{2}\sigma^2(T,K)K^2 \frac{\partial^2 C}{\partial K^2}(T,K) - rK\frac{\partial C}{\partial K}(T,K),$$

where we have used (2.5) and (2.6) in the last line. In this calculation, there are no boundary terms from the integrations by parts because the call payoff is zero for $\xi \le K$, and the decay of p at infinity.

We have derived the forward partial differential equation, known as Dupire's equation,

$$\frac{\partial C}{\partial T} = \frac{1}{2}\sigma^2(T,K)K^2\frac{\partial^2 C}{\partial K^2} - rK\frac{\partial C}{\partial K}, \quad T > t, \tag{2.7}$$

with the initial condition at $T = t$ given by $C(t,K) = (x-K)^+$, where x is the price of the underlying at time t.

For discrete data, meaning options prices at a finite set of strikes and maturities, the calibration of the local volatility surface is a classical inverse problem for the forward equation (2.7). We give a reference for the numerical aspects of this approach in the Notes. However, if we assume we can observe option prices for a continuum of strikes and maturities, then the local volatility surface is given explicitly by the Dupire formula obtained by inverting (2.7):

$$\sigma^2(T,K) = \frac{\frac{\partial C}{\partial T}(T,K) + rK\frac{\partial C}{\partial K}(T,K)}{\frac{1}{2}K^2\frac{\partial^2 C}{\partial K^2}(T,K)}.$$

In practice, the use of this formula requires interpolation of option prices across strikes and maturities to estimate the partial derivatives. This is a delicate issue, in particular when the (K,T) grid of traded options is sparse.

There are a number of competing ways to estimate a local volatility surface $\sigma(t,x)$ from traded option prices, some parametric and some nonparametric, as we have seen. It has the advantage that we still have a complete market model. Where the problem lies is in the stability of the fits over time: with new data, a week later for instance, the fits are often completely different, even though, given the large degree of freedom, they fit present option prices very well. References for empirical evidence of this are given at the end of the chapter.

2.3 Stochastic Volatility Models

In "pure" stochastic volatility models (as opposed to local volatility), the asset price $(X_t)_{t\geq 0}$ satisfies the stochastic differential equation

$$dX_t = \mu_t X_t dt + \sigma_t X_t dW_t^{(0)}, \tag{2.8}$$

where $W^{(0)}$ is a standard Brownian motion, $(\mu_t)_{t\geq 0}$ is an adapted returns process, and $(\sigma_t)_{t\geq 0}$ is called the volatility process. It must satisfy some regularity conditions for the model to be well-defined, but it does not have to be an Itô process: it can be a jump process, a Markov chain, among other examples. In order for it to be a volatility, it should be positive. Unlike the local volatility models, the volatility process is *not* perfectly correlated with the Brownian motion $W^{(0)}$. Therefore, volatility is modeled to have an independent random component of its own. We shall see in a number of ways that this leads to an incomplete market and there is no *unique* equivalent martingale measure. Stochastic volatility models can be seen as continuous time versions of ARCH-type models, which have been introduced by R. Engle. The importance of volatility modeling is reflected by the fact that R. Engle was awarded the 2003 Nobel Prize for Economics, shared with C. Granger.

In the following, we will specialize to *one-factor* stochastic volatility models, and we discuss multifactor models in Section 2.5.4.

2.3.1 One-Factor Stochastic Volatility Models

We consider here models where volatility is a function of a one-dimensional Itô's process satisfying a stochastic differential equation driven by a second Brownian motion. This is the easiest way to incorporate correlation with stock price changes by correlating the respective driving Brownian motions. We want a model in which volatility is positive, and some choices are lognormal or square-root (CIR) processes described below. Within this framework, we have finite-difference or tree methods available for computation.

Let us denote $\sigma_t = f(Y_t)$ where f is some smooth, positive, increasing function and, for now, Y is a one-dimensional process defined by the stochastic differential equation

$$dY_t = \alpha(Y_t)\,dt + \beta(Y_t)\,dW_t^{(1)},$$

where $\alpha : \mathbb{R} \to \mathbb{R}$, $\beta : \mathbb{R} \to \mathbb{R}_+$, and $W^{(1)}$ is a standard Brownian motion. Writing volatility as a function of a driving factor makes it convenient to introduce other factors, as will be needed later, and allows us easily to enforce positivity by choice of the function f.

The Brownian motion $W^{(1)}$ is typically correlated with the Brownian motion $W^{(0)}$ driving the asset price equation (2.8). We denote by $\rho \in [-1, 1]$ the instantaneous correlation coefficient defined by

$$d\langle W^{(0)}, W^{(1)} \rangle_t = \rho \, dt,$$

using the notation of Section 1.9.1. The parameter ρ can also be viewed as the correlation between stock price and volatility shocks as follows. By Itô's formula applied to $f(Y_t)$, we see that the martingale part of $d\sigma_t$ is given by $f'(Y_t)\beta(Y_t)\, dW_t^{(1)}$. Therefore, we have the covariance

$$\mathrm{cov}\,(d\sigma_t, dX_t) = \rho f'(Y_t)\beta(Y_t) f(Y_t) X_t \, dt,$$

and hence the correlation

$$\mathrm{corr}\,(d\sigma_t, dX_t) = \rho.$$

It is often found from financial stock price data that $\mathrm{corr}\,(d\sigma_t, dX_t) < 0$, and there are economic arguments for a negative correlation or *leverage effect* between stock price and volatility shocks. From common experience and empirical studies, when volatility goes up, stock prices tend to go down. In general, the correlation may depend on time $\rho(t) \in [-1, 1]$, but we shall assume it a constant from now on for notational simplicity and because, in most practical situations, it is taken to be such.

One desirable feature is *mean-reversion*, which loosely refers to an increased probability to drop when it is high and to rise when it is low, relative to a long-run mean level. This is made precise in Chapter 3. From a financial modeling perspective, mean-reverting refers to a linear pull-back term in the drift of volatility driving factor Y. Then, mean-reverting stochastic volatility means that the stochastic differential equation for (Y_t) looks like

$$dY_t = \alpha(m - Y_t)dt + \beta(Y_t)\, dW_t^{(1)}.$$

Here $\alpha > 0$ is called the *rate of mean-reversion*, m is the long-run mean level of Y, and $\beta(Y_t)$ is some diffusion coefficient. The drift term pulls Y towards m.

Some common driving processes (Y_t) are

LN lognormal

$$dY_t = \alpha Y_t dt + \beta Y_t \, dW_t^{(1)},$$

OU Ornstein–Uhlenbeck

$$dY_t = \alpha(m - Y_t)dt + \beta \, dW_t^{(1)},$$

CIR Feller, square-root, or Cox–Ingersoll–Ross

$$dY_t = \alpha(m - Y_t)dt + \beta \sqrt{Y_t} \, dW_t^{(1)}.$$

Note that the lognormal model is *not* mean-reverting, but OU and CIR are. An explicit expression for the lognormal process was derived in Section 1.1.5, and for the OU process in Section 1.1.6.

Some stochastic volatility models studied in the literature are

Authors	Correlation	$f(y)$	Y Process
Hull–White	$\rho = 0$	$f(y) = \sqrt{y}$	Lognormal
Scott	$\rho = 0$	$f(y) = e^y$	OU
Stein–Stein	$\rho = 0$	$f(y) = |y|$	OU
Ball–Roma	$\rho = 0$	$f(y) = \sqrt{y}$	CIR
Heston	$\rho \neq 0$	$f(y) = \sqrt{y}$	CIR

These particular one-factor models are chosen for their nice properties (for example, positivity and mean-reversion) and analytical tractability, rather than for deeper financial reasons (the Heston model is revisited in detail in Chapter 10). Empirical evidence points in the direction of at least two factors driving stochastic volatility, which we shall discuss in more detail in Chapter 3. From the point of view of this book, we will try to develop a framework for stochastic volatility that does not depend upon specific modeling of the volatility process. One requirement for the asymptotic theory developed in this book is that one of the stochastic volatility factors is an ergodic process, explained in Chapter 3, the OU or CIR processes being perfect examples.

2.3.2 Stock Price Distribution under Stochastic Volatility

What effect does stochastic volatility have on the probability density function of the stock price? To answer this question qualitatively, we plot in Figure 2.4 the density function (estimated from simulation) of the expOU stochastic volatility model in which $f(y) = e^y$ and (Y_t) is an OU process.

Notice the thicker tails because of the random volatility. In particular, the negative correlation causes the tails to be asymmetric: the left tail is thicker.

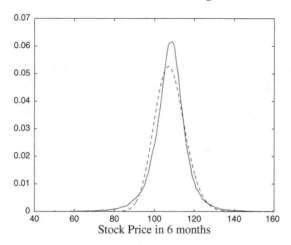

Figure 2.4   Density functions for the stock price (under the subjective measure) in six months when the present value is 100. The solid line is estimated from simulation of an expOU stochastic volatility model with $\alpha = 1, \beta = \sqrt{2}$, long-run average volatility $\bar{\sigma} = 0.1$, and negative correlation $\rho = -0.2$. The dotted line is the corresponding Black–Scholes lognormal density function with volatility $\bar{\sigma}$. The mean growth rate of the stock is $\mu = 0.15$.

2.4 Derivative Pricing

We consider next the problem of pricing a European derivative with payoff $h(X_T)$ when the volatility is a diffusion process. We will find a pricing function of the form $P(t, X_t, Y_t)$ from no-arbitrage arguments. The function $P(t,x,y)$ satisfies a partial differential equation with *two* "space dimensions" (x and y), with terminal condition $P(T,x,y) = h(x)$. The price of the derivative depends on the value of the process y which is not directly observable. We return to this issue later.

We will derive the pricing partial differential equation assuming the following *Markovian* stochastic volatility model:

$$
\begin{aligned}
dX_t &= \mu(Y_t)X_t\,dt + \sigma_t X_t\,dW_t^{(0)}, \\
\sigma_t &= f(Y_t), \\
dY_t &= \alpha(Y_t)\,dt + \beta(Y_t)\,dW_t^{(1)},
\end{aligned} \tag{2.9}
$$

where for simplicity, we have restricted the stochastic return process to be a function $\mu(Y_t)$ of Y only. It is also convenient to decompose $W^{(1)}$ in terms of $W^{(0)}$ and an *independent* standard Brownian motion denoted by $W^\perp$:

$$W_t^{(1)} = \rho W_t^{(0)} + \sqrt{1 - \rho^2}\, W_t^{\perp}. \tag{2.10}$$

We continue to denote the underlying probability space by $(\Omega, \mathscr{F}, \mathbb{P})$, where now we can take $\Omega = \mathscr{C}([0, \infty) : \mathbb{R}^2)$, the space of all continuous trajectories $(W_t^{(0)}(\omega), W_t^{\perp}(\omega)) = \omega(t)$ in $\mathbb{R}^2$. The filtration $(\mathscr{F}_t)_{t \geq 0}$ represents the information on the two Brownian motions, so for example, $\mathscr{F}_t$ is the σ-algebra generated by sets of the form $\{\omega \in \Omega : |W_s^{(0)}| < R_1, |W_s^{\perp}| < R_2, s \leq t\}$ completed by the null sets.

2.4.1 Hedging Argument

We look for a pricing function $P(t, x, y)$ by trying to construct a hedged portfolio of assets which can be priced by the no-arbitrage principle. Unlike in the Black–Scholes case, it is not sufficient to hedge solely with the underlying asset, since the $dW_t^{(0)}$ term can be balanced, but the $dW_t^{\perp}$ term cannot. Thus we try to hedge with the underlying asset *and* another option which has a different expiration date, and which we also trade continuously.

Let $P^{(1)}(t, x, y)$ be the price of a European derivative with expiration date T_1, and $P^{(2)}(t, x, y)$ be the price of another European derivative with expiration date $T_2 > T_1$. We try to find processes $\{N_t, A_t, \Sigma_t\}$ such that the self-financing portfolio

$$\Pi_t = N_t P^{(1)}(t, X_t, Y_t) - A_t X_t - \Sigma_t P^{(2)}(t, X_t, Y_t) \tag{2.11}$$

is instantaneously riskless (or hedged) at any time $t < T$.

We shall use the two-dimensional version of Itô's formula (1.115) to compute $dg(t, X_t, Y_t)$ for smooth functions g:

$$dg = \left(\frac{\partial g}{\partial t} + \mathscr{L}_{(X,Y)} g \right) dt + \left(f(Y_t) X_t \frac{\partial g}{\partial x} + \rho \beta(Y_t) \frac{\partial g}{\partial y} \right) dW_t^{(0)}$$
$$+ \sqrt{1 - \rho^2}\, \beta(Y_t) \frac{\partial g}{\partial y} dW_t^{\perp}, \tag{2.12}$$

where $\mathscr{L}_{(X,Y)}$ is the infinitesimal generator of (X, Y) given by

$$\mathscr{L}_{(X,Y)} = \frac{1}{2} f^2(y) x^2 \frac{\partial^2}{\partial x^2} + \rho \beta(y) f(y) x \frac{\partial^2}{\partial x \partial y} + \frac{1}{2} \beta^2(y) \frac{\partial^2}{\partial y^2}$$
$$+ \mu(y) x \frac{\partial}{\partial x} + \alpha(y) \frac{\partial}{\partial y}.$$

Since the portfolio Π_t is to be self-financing, we have

$$d\Pi_t = N_t dP^{(1)}(t,X_t,Y_t) - A_t\,dX_t - \Sigma_t dP^{(2)}(t,X_t,Y_t),$$

and applying (2.12) to $P^{(1)}$ and $P^{(2)}$ yields

$$d\Pi_t = \left(N_t\left[\frac{\partial}{\partial t}+\mathscr{L}_{(X,Y)}\right]P^{(1)} - A_t\mu(Y_t)X_t - \Sigma_t\left[\frac{\partial}{\partial t}+\mathscr{L}_{(X,Y)}\right]P^{(2)}\right)dt$$
$$+\left(X_t f(Y_t)\left[N_t\frac{\partial P^{(1)}}{\partial x}-\Sigma_t\frac{\partial P^{(2)}}{\partial x}-A_t\right]\right.$$
$$+\rho\beta(Y_t)\left[N_t\frac{\partial P^{(1)}}{\partial y}-\Sigma_t\frac{\partial P^{(2)}}{\partial y}\right]\right)dW_t^{(0)}$$
$$+\sqrt{1-\rho^2}\,\beta(Y_t)\left[N_t\frac{\partial P^{(1)}}{\partial y}-\Sigma_t\frac{\partial P^{(2)}}{\partial y}\right]dW_t^{\perp}.$$

To hedge the risk from $W^{\perp}$, we choose

$$\Sigma_t = N_t\left(\frac{\partial P^{(2)}}{\partial y}\right)^{-1}\left(\frac{\partial P^{(1)}}{\partial y}\right), \tag{2.13}$$

where we assume that the second derivative is sensitive to volatility so that $\frac{\partial P^{(2)}}{\partial y}\neq 0$. To hedge the remaining risk from $W^{(0)}$, we choose

$$A_t = N_t\frac{\partial P^{(1)}}{\partial x}-\Sigma_t\frac{\partial P^{(2)}}{\partial x}. \tag{2.14}$$

For there to be no arbitrage, the now riskless portfolio Π must grow at the risk-free rate: $d\Pi_t = r\Pi_t\,dt$, which gives, after substituting for Σ_t and A_t, canceling N_t and separating terms in $P^{(1)}$ and $P^{(2)}$:

$$\left(\frac{\partial P^{(1)}}{\partial y}\right)^{-1}\widehat{\mathscr{L}}P^{(1)} = \left(\frac{\partial P^{(2)}}{\partial y}\right)^{-1}\widehat{\mathscr{L}}P^{(2)}, \tag{2.15}$$

with the notation

$$\widehat{\mathscr{L}} = \frac{\partial}{\partial t}+\mathscr{L}_{(X,Y)}-(\mu(y)-r)x\frac{\partial}{\partial x}-r\cdot.$$

Now, the left-hand side of (2.15) contains terms depending on T_1 but not T_2 and *vice versa* for the right-hand side. Thus both sides must be equal to a function that does not depend on expiration date. For reasons explained below, we denote this function by $-\alpha(y) + \beta(y)\Lambda(t,x,y)$, where

$$\Lambda(t,x,y) = \rho\frac{(\mu(y) - r)}{f(y)} + \gamma(t,x,y)\sqrt{1 - \rho^2}, \qquad (2.16)$$

and $\gamma(t,x,y)$ is an arbitrary function. The pricing function $P(t,x,y)$, with the dependence on expiry date supressed, must satisfy the partial differential equation

$$\frac{\partial P}{\partial t} + \frac{1}{2}f^2(y)x^2\frac{\partial^2 P}{\partial x^2} + \rho\beta(y)xf(y)\frac{\partial^2 P}{\partial x\partial y} + \frac{1}{2}\beta^2(y)\frac{\partial^2 P}{\partial y^2}$$
$$+ r\left(x\frac{\partial P}{\partial x} - P\right) + (\alpha(y) - \beta(y)\Lambda(t,x,y))\frac{\partial P}{\partial y} = 0, \quad (2.17)$$

with the terminal condition $P(T,x,y) = h(x)$. In general, there is no closed-form solution for this partial differential equation problem. That said, the Feynman–Kac formula gives a probabilistic representation which will appear in the next section.

We can group the differential operator on the left-hand side of (2.17) in the following way:

$$\underbrace{\frac{\partial}{\partial t} + \frac{1}{2}f^2(y)x^2\frac{\partial^2}{\partial x^2} + r\left(x\frac{\partial}{\partial x} - \cdot\right)}_{\mathscr{L}_{BS}(f(y))} + \underbrace{\rho\beta(y)xf(y)\frac{\partial^2}{\partial x\partial y}}_{\text{Correlation}} +$$

$$\underbrace{\frac{1}{2}\beta^2(y)\frac{\partial^2}{\partial y^2} + \alpha(y)\frac{\partial}{\partial y}}_{\mathscr{L}_Y} \quad - \quad \underbrace{\beta(y)\Lambda\frac{\partial}{\partial y}}_{\text{Premium}}.$$

The first grouping is the Black–Scholes operator (1.40) with volatility level $f(y)$, the second is the term due to the correlation, the third is the infinitesimal generator of the process (Y_t) following the definition (1.74), and finally there is the term due to the market price of volatility risk.

The function γ in (2.16) is the risk premium factor from the *second* source of randomness $(W_t^\perp)$ that drives the volatility: in the perfectly correlated case $|\rho| = 1$ it does not appear, as expected. We can think of Λ in (2.16) as the total risk premium: it is a linear combination of the *stochastic Sharpe ratio* $\frac{(\mu(Y_t) - r)}{f(Y_t)}$ and the *volatility risk premium* γ, weighted by the correlation ρ and its complement $\sqrt{1 - \rho^2}$, respectively.

The reason for this terminology is the calculation

$$dP(t, X_t, Y_t) = \left[\frac{(\mu(Y_t) - r)}{f(Y_t)} \left(X_t f(Y_t) \frac{\partial P}{\partial x} + \rho \beta(Y_t) \frac{\partial P}{\partial y} \right) - rP \right.$$
$$\left. + \gamma(t, X_t, Y_t) \beta(Y_t) \sqrt{1 - \rho^2} \frac{\partial P}{\partial y} \right] dt$$
$$+ \left(X_t f(Y_t) \frac{\partial P}{\partial x} + \rho \beta(Y_t) \frac{\partial P}{\partial y} \right) dW_t^{(0)} + \beta(Y_t) \sqrt{1 - \rho^2} dW_t^\perp,$$

which is obtained using Itô's formula (1.115) and the partial differential equation (2.17) satisfied by P. From this expression, we see that an infinitesimal fractional increase in the volatility risk β increases the infinitesimal rate of return on the option by γ times that fraction, in addition to the increase from the *excess return-to-risk ratio* or *Sharpe ratio* $(\mu(y) - r)/f(y)$.

2.4.2 Pricing with Equivalent Martingale Measures

We give an alternative derivation of the no-arbitrage derivative price using the risk-neutral theory, again for the one-factor model (2.9), but the procedure is valid for general models, including non-Markovian ones.

Suppose first that there *is* an equivalent martingale measure $\mathbb{P}^\star$ under which the discounted stock price $\widetilde{X}_t = e^{-rt} X_t$ is a martingale. Then we know from Chapter 1 that if we price a derivative with maturity time T and square integrable payoff H using the formula

$$V_t = \mathbb{E}^\star \{ e^{-r(T-t)} H \mid \mathscr{F}_t \},$$

for all $t \leq T$, then there is no arbitrage opportunity. Thus V_t is a possible price for the claim. Let us now try to construct equivalent martingale measures.

As in Section 1.4.1, we absorb the drift term of $\widetilde{X}_t$ in its martingale term by setting

$$W_t^{(0)\star} = W_t^{(0)} + \int_0^t \frac{(\mu(Y_s) - r)}{f(Y_s)} ds.$$

Any shift of the second independent Brownian motion of the form

$$W_t^{\perp\star} = W_t^\perp + \int_0^t \gamma_s ds$$

will not change the drift of $\widetilde{X}_t$. By Girsanov's theorem (Section 1.9.2), $W^{(0)\star}$ and $W^{\perp\star}$ are independent standard Brownian motions under a measure $\mathbb{P}^{\star(\gamma)}$ defined by

$$\frac{d\mathbb{P}^{\star(\gamma)}}{d\mathbb{P}} = \exp\left(-\frac{1}{2}\int_0^T ((\theta_s^{(0)})^2 + (\theta_s^{\perp})^2)ds - \int_0^T \theta_s^{(0)}dW_s^{(0)}\right.$$

$$\left. - \int_0^T \theta_s^{\perp}dW_s^{\perp}\right),$$

$$\theta_t^{(0)} = \frac{\mu(Y_t) - r}{f(Y_t)}, \qquad \theta_t^{\perp} = \gamma_t.$$

Here (γ_t) is *any* adapted (and suitably regular) process. Technically, we shall make an assumption on the pair $(\frac{\mu(Y_t) - r}{f(Y_t)}, \gamma_t)$ so that $\mathbb{P}^{\star(\gamma)}$ is well-defined as a probability measure. For example, this will be the case if f is bounded away from zero and $(\mu(Y_t))$ and (γ_t) are bounded. We note that for some of the models listed in the table at the end of Section 2.3.1, $f(Y_t)$ is *not* bounded away from zero, and so Girsanov's theorem may not apply when μ is constant. It is common practice in such cases to assume that $\mu(y)$ is such that the change of equivalent measure is legitimate. In other words, a condition on the stochastic Sharpe ratio $\frac{\mu(Y_t) - r}{f(Y_t)}$ is imposed (boundedness for instance).

Then under $\mathbb{P}^{\star(\gamma)}$, the stochastic differential equations (2.9) become

$$dX_t = rX_t dt + f(Y_t)X_t dW_t^{(0)\star}, \tag{2.18}$$

$$dY_t = (\alpha(Y_t) - \beta(Y_t)\Lambda_t)dt + \beta(Y_t)dW_t^{(1)\star}, \tag{2.19}$$

$$W_t^{(1)\star} = \rho W_t^{(0)\star} + \sqrt{1-\rho^2}W_t^{\perp\star}, $$

where we define the total risk premium process

$$\Lambda_t = \rho\frac{(\mu(Y_t) - r)}{f(Y_t)} + \gamma_t\sqrt{1-\rho^2}. \tag{2.20}$$

Any allowable choice of (γ_t) leads to an equivalent martingale measure $\mathbb{P}^{\star(\gamma)}$ and to the possible *no-arbitrage* prices of the contract with payoff H:

$$V_t = \mathbb{E}^{\star(\gamma)}\{e^{-r(T-t)}H \mid \mathcal{F}_t\}. \tag{2.21}$$

The process (γ_t) is called the volatility risk premium or **the market price of volatility risk** from the *second* source of randomness $W^{\perp}$ that drives the volatility. It parametrizes the space of equivalent martingale measures $\{\mathbb{P}^{\star(\gamma)}\}$. When γ is of the form $\gamma(t, X_t, Y_t)$, we are back to the Markovian setting, and (2.17) is just the Feynman–Kac partial differential equation for (2.21), when $H = h(X_T)$ is a European claim.

Note also that we cannot replicate the claim by trading in stock and bond only. Recall that because we are modeling with two Brownian motions, we are dealing with a larger probability space $\Omega = \mathscr{C}([0,\infty) : \mathbb{R}^2)$ and the filtration $(\mathcal{F}_t)$ is generated by $(W_t^{(0)}, W_t^{\perp})$. As a result, the martingale representation theorem used in Section 1.4.3 says only that the $\mathbb{P}^{\star(\gamma)}$-martingale $M_t = e^{-rt}V_t$ is a stochastic integral with respect to $(W^{(0)\star}, W^{\perp\star})$:

$$M_t = M_0 + \int_0^t \eta_s^{(0)} dW_s^{(0)\star} + \int_0^t \eta_s^{\perp} dW_s^{\perp\star}, \tag{2.22}$$

for some adapted (and suitably bounded) processes $\eta^{(0)}$ and $\eta^{\perp}$. We can translate this into a self-financing trading strategy in stock, bond, *and volatility* (by undoing the discounting and substituting for $dW^{(0)\star}$ and $dW^{\perp\star}$ from (2.18) and (2.19)),

$$dV_t = a_t dX_t + b_t re^{rt} dt + c_t d\sigma_t,$$

for some (a_t, b_t, c_t). Due to the last integral in (2.22), we cannot find a replicating strategy in just the stock and bond ($c_t \neq 0$ in general) as we would like to do, since volatility is not a tradeable asset.

We can however hedge one derivative contract $P^{(1)}$ with the stock and another derivative security $P^{(2)}$, as in equation (2.11). The calculation of Section 2.4.1 (in the Markovian case) yields the hedging ratios (2.13) and (2.14). Again these are nonunique because they depend on γ. This procedure, which is known as Δ–Σ hedging, is usually unsatisfactory because of the higher transaction costs and lesser liquidity associated with trading the second derivative. Therefore, it is an important problem to determine how to hedge as "best as possible" (according to some criteria) with just the stock. We discuss the hedging problem further in Chapter 8.

2.4.3 Market Price of Volatility Risk and Data

From the previous section, we know that (2.21) is a possible *no arbitrage* derivative pricing formula, for any equivalent martingale measure $\mathbb{P}^{\star(\gamma)}$. Much research has investigated the range of possible prices in general settings: for example, the possible derivative prices often fill an interval, and in many cases this interval is quite broad. If the volatility process is unbounded, the range of European call option prices given by (2.21) with $H = (X_T - K)^+$ is between the price of the stock and the intrinsic value of the contract

$$(X_t - K)^+ \leq V_t \leq X_t,$$

and the extremal values are attained for some equivalent martingale measures. When volatility is assumed bounded $\sigma_t \in [\sigma_{\min}, \sigma_{\max}]$, the bounds are

$$C_{BS}(\sigma_{\min}) \leq V_t \leq C_{BS}(\sigma_{\max}),$$

which information is not useful unless one is certain of tight bounds on volatility – that is, volatility is almost constant. Usually worst-case analysis of stochastic volatility models results in such bands that are too wide for practical purposes.

The alternative approach, which we shall follow here, is that the market selects a unique equivalent martingale measure under which derivative contracts are priced. The value of the market's price of volatility risk γ can only be seen in derivative prices since γ does not feature in the real-world model for the stock price (2.9). Note that this approach does not imply we return to a complete markets setting: even though discounted derivative prices are martingales with respect to the particular $\mathbb{P}^{\star(\gamma)}$, they cannot be replicated by trading stock and bond alone. They can however be *super-replicated*: for example, buying the stock at time $t < T$ and holding it till expiration superhedges a short call position because $X_T > (X_T - K)^+$; it may yield a profit but never a loss. However, this (trivial) strategy is very expensive (it costs $\$X_t$).

When we estimate the parameters for our model, we could use econometric methods such as maximum likelihood or method of moments on historical stock price time-series data to find *parametrically specified* functions $(f(y), \alpha(y), \beta(y))$ and the constant ρ, plus the present volatility level in the model (2.9). Then we would need some derivative data to estimate γ, assuming for instance that it is constant.

The common practice, called cross-sectional fitting, is to estimate *all* the parameters from derivative data, usually near-the-money European option prices (or a section of the observed implied volatility surface). This ignores the statistical basis for the modeling, but is easier to implement (especially when there is a formula for option prices under the chosen model) than the time-series methods, which suffer because the (σ_t) process is not directly observable. If today is time $t = 0$ and we denote by ϑ the vector of unknown parameters, then a typical least-squares fit is to observe call option prices $C^{\text{obs}}(K,T)$ for strike prices and expiration dates (K,T) in some set $\mathcal{K}$, and solve

$$\min_{\vartheta} \sum_{(K,T)\in\mathcal{K}} \left(C(K,T;\vartheta) - C^{\text{obs}}(K,T) \right)^2,$$

where $C(K,T;\vartheta)$ is the model-predicted call option price (either from solving the partial differential equation (2.17) with $h(x) = (x - K)^+$, or by

Monte Carlo simulation). This process can be very slow and computationally intensive.

2.4.4 Short-Time Tight Fit vs. Long-Time Approximate Fit: The (K,T,t)-Problem

There is a tradeoff between a tight fit over a short time vs. a rougher fit
over a longer time, and this stability of parameters problem is an important criterion with which to assess any model. A fundamental question in
pricing and hedging derivatives is: how traded call options, quoted in terms
of implied volatilities, can be used to price and hedge more complicated
contracts? One can approach this difficult problem in two different ways:
modeling the evolution of the underlying or modeling the evolution of the
implied volatility surface. In both cases one requires that the model is free
of arbitrage.

Modeling the underlying usually involves the specification of a multifactor Markovian model under the risk-neutral pricing measure. Calibration to
the observed implied volatilities of the parameters of that model, including
the market prices of risk, is a challenging task because of the complex relation between call option prices and model parameters (through a pricing
partial differential equation for instance). The *holy grail* would be a model
that produces stable parameter estimates. The sequencing of difficulties is
often referred to as the "(K,T,t)-problem": for a given present time t and
a fixed maturity T, it is usually easy with low-dimensional models to fit
the skew with respect to strikes K. Getting a good fit of the term structure of implied volatility, that is when a range of observed maturities are
taken into account, is a much harder problem which can be handled with
a sufficient number of random factors and parameters. The main problem remains: the stability with respect to t of these calibrated parameters.
However, this *is* a very important quality if one wants to use the model to
compute no-arbitrage prices of more complex path-dependent derivatives,
since in this case the distribution over time of the underlying is crucial.

Modeling directly the evolution of the implied volatility surface (or similarly the local volatility surface as discussed in Section 2.2.2) is a seemingly
attractive approach, but much more complicated to implement in order to
make sure that the model is free of arbitrage. The choice of a model and
its calibration is also an important issue in this approach. But in order
to use this modeling to price other path-dependent contracts, one has to
identify a corresponding underlying which typically does not lead to a

low-dimensional Markovian evolution. In other words, a nice model with a great fit, but what are you going to do with it?

This way of thinking, which goes by the name *market models*, seems to have sprung from an industry viewpoint of a model's usefulness simply as a mapping from prices to parameters. In other words, just as Black–Scholes maps option prices to implied volatilities which are observed to change daily, a parametric model such as the Heston stochastic volatility model is used to map option prices into model parameters such as the correlation coefficient, rate of mean-reversion and so on, which are then accepted as changing daily as option prices change. The emphasis on assessing the model is purely on its quality of fit to today's implied volatilities or option prices, with no regard to the fact that the supposedly constant parameters may change wildly: each day's values are seen as just the current value of a stochastic process for each parameter, which can be updated daily just as the stock price or the volatility level are updated. This so-called *pernicious practice of recalibration* has given up entirely on the t-problem with sole emphasis on (K, T). Of course it is driven in answer to many traders' mistrusts about any model that does not fit today's data perfectly (the curve has to go through all the data points!) and completely sacrifices stability for goodness-of-(over)fitting. As one speaker at a recent Bachelier Finance Congress pointed out, "To calibrate does not mean to understand. A perfect fit is not a theory."

2.5 General Results on Stochastic Volatility Models

We present a collection of useful general conclusions that can be drawn from the class of stochastic volatility models.

2.5.1 Implied Volatility as a Function of Moneyness

Another reason that implied volatility, introduced in Section 2.1, is a particularly useful measure of the performance of a stochastic volatility model is that implied volatility is often a function of a European option contract's *moneyness*, the ratio K/x of its strike price to the current stock price, as we now show. We consider here stochastic volatility models where the stock price satisfies

$$dX_s = rX_s ds + \sigma_s X_s dW_s^{(0)\star}, \tag{2.23}$$

and when σ_s is a Markov process in its own right under some risk-neutral pricing measure $\mathbb{P}^\star$.

Suppose the present is time t and define $\tilde{X} = X/x$, where $X_t = x$, the current stock price. Then $(\tilde{X}_s)_{s \geq t}$ satisfies the same stochastic differential equation (2.23), with initial value $\tilde{X}_t = 1$, neither of which depend on the number x. The call option price is given by

$$
\begin{aligned}
C(t,x,\sigma_t;K,T) &= \mathbb{E}^{\star(\gamma)}\{e^{-r(T-t)}(X_T - K)^+ \mid X_t = x, \sigma_t\} \\
&= \mathbb{E}^{\star(\gamma)}\{e^{-r(T-t)}(x\tilde{X}_T - K)^+ \mid \tilde{X}_t = 1, \sigma_t\} \\
&= K\mathbb{E}^{\star(\gamma)}\left\{e^{-r(T-t)}\left(\frac{x}{K}\tilde{X}_T - 1\right)^+ \mid \tilde{X}_t = 1, \sigma_t\right\} \\
&= KQ_1(t,x/K;T),
\end{aligned}
$$

for some function Q_1 depending on x/K, but not x and K separately. The implied volatility I is defined from

$$
C(t,x,\sigma_t;K,T) = C_{BS}(t,x;K,T;I),
$$

and from the Black–Scholes formula (1.41) we have

$$
C_{BS}(t,x;K,T;I) = KQ_2(t,x/K,T;I),
$$

for some function Q_2 also depending on x/K, but not x and K separately. Consequently,

$$
KQ_1(t,x/K;T) = KQ_2(t,x/K,T;I),
$$

and therefore, I must be a function of moneyness K/x, but not K and x separately.

This is useful because it tells us that we can obtain the implied volatility curve predicted by a stochastic volatility model in this class by solving, numerically for example, the partial differential equation (2.17) with terminal condition $h(x) = (x - K)^+$ (or $h(x) = (K - x)^+$ for a put) for a fixed strike price K. Then plotting the resulting implied volatilities as a function of moneyness for different starting values x gives the same curve as if we were varying K. A finite-difference algorithm usually gives the numerical solution of (2.17) for a range of current stock prices x, and the "extra" information can be used to generate the implied volatility curve.

2.5.2 Uncorrelated Volatility, Hull–White Formula, and Renault–Touzi Theorem

It turns out that the case when the volatility process (σ_t) is independent of the Brownian motion $W^{(0)\star}$ under the pricing measure $\mathbb{P}^{\star}$ is much easier to handle, and we present some results under this assumption. We refer to such models as **uncorrelated stochastic volatility models**. For diffusion stochastic volatility models of the form (2.18)–(2.19), this corresponds to the case $\rho = 0$ and γ_t independent of $W^{(0)\star}$. In equity markets it is widely believed $\rho < 0$, but some studies suggest ρ is close to zero in foreign-exchange data.

Hull–White Formula

The pricing formula (2.21) for a European option $H = h(X_T)$ can be simplified under our assumptions. We can condition on the path of the volatility process and by iterated expectations. The price of a call option, for example, is given by

$$C_t = \mathbb{E}^{\star(\gamma)} \left\{ \mathbb{E}^{\star(\gamma)} \{ e^{-r(T-t)}(X_T - K)^+ \mid \mathscr{F}_t, \sigma_s, t \le s \le T \} \mid \mathscr{F}_t \right\}.$$

The inner expectation is just the Black–Scholes computation with a time-dependent volatility as in Section 2.2.1. The answer is given by the Black–Scholes formula (1.41) with appropriately averaged volatility, and so

$$C_t = \mathbb{E}^{\star(\gamma)} \{ C_{BS}(t, X_t; K, T; \sqrt{\overline{\sigma^2}}) \mid \mathscr{F}_t \}, \tag{2.24}$$

where

$$\overline{\sigma^2} = \frac{1}{T-t} \int_t^T \sigma_s^2 \, ds. \tag{2.25}$$

We observe that $\sqrt{\overline{\sigma^2}}$ is the root-mean-square time average of (σ_t) over the remaining trajectory of each realization, and the call option price is the average over all possible volatility paths.

Stochastic Volatility Implies Smile

When the volatility price σ_t is independent of $W^{(0)}$, it is known that stochastic volatility models predict European option prices whose implied volatilities smile. This result is an important success of these models, although the fact that one observes smile curves does not necessarily imply that volatility is stochastic. The derivation is instructive because it relates the smile shape directly to the Black–Scholes lognormal model through which implied volatilities are calculated, and we present it in this section. In particular, it identifies the log-moneyness $\log(x/K)$ as an important quantity.

The full statement, due to Renault and Touzi, is as follows. *In an uncorrelated stochastic volatility model, provided $\overline{\sigma^2}$ defined by (2.25) is an L^2 random variable, the implied volatility curve $I(K)$ for fixed t, x, T is a smile,* *that is, it is locally convex around the minimum $K_{\min} = xe^{r(T-t)}$, which is the forward price of the stock.*

To see this, let us fix t and T and start with $\overline{\sigma^2}$ being a Bernoulli random variable

$$\overline{\sigma^2} = \begin{cases} \sigma_1^2 \text{ with probability } p, \\ \sigma_2^2 \text{ with probability } 1-p, \end{cases}$$

under the measure $\mathbb{P}^{\star(\gamma)}$. Then, provided the assumptions listed hold, the Hull–White formula (2.24) applies:

$$C_{BS}(K;I(p,K)) = pC_{BS}(K;\sigma_1) + (1-p)C_{BS}(K;\sigma_2), \tag{2.26}$$

where $C_{BS}(K;\sigma)$ denotes the standard Black–Scholes formula, with only the K and σ arguments displayed. To simplify notation, we give only its volatility argument below. We have written $I = I(p,K)$ to stress the dependence on p and K with x, t and T fixed. Differentiating (2.26) with respect to K gives

$$\frac{\partial C_{BS}}{\partial \sigma}(I(p,K))\frac{\partial I}{\partial K} + \frac{\partial C_{BS}}{\partial K}(I(p,K)) = p\frac{\partial C_{BS}}{\partial K}(\sigma_1) + (1-p)\frac{\partial C_{BS}}{\partial K}(\sigma_2),$$

so that

$$\mathrm{sign}\left(\frac{\partial I}{\partial K}\right) = \mathrm{sign}(g(p)),$$

where, for fixed K,

$$g(p) = p\frac{\partial C_{BS}}{\partial K}(\sigma_1) + (1-p)\frac{\partial C_{BS}}{\partial K}(\sigma_2) - \frac{\partial C_{BS}}{\partial K}(I(p,K)),$$

because $\partial C_{BS}/\partial \sigma > 0$ from the formula (2.2). Clearly, if $p = 0,1$ then $I \equiv \sigma_2, \sigma_1$, respectively, so $g(0) = g(1) = 0$.

The key (and somewhat surprising) step is now to differentiate g with respect to p:

$$g'(p) = \frac{\partial C_{BS}}{\partial K}(\sigma_1) - \frac{\partial C_{BS}}{\partial K}(\sigma_2) - \frac{\partial^2 C_{BS}}{\partial K \partial \sigma}(I(p,K))\frac{\partial I}{\partial p}, \tag{2.27}$$

and we also differentiate (2.26) with respect to p:

$$\frac{\partial C_{BS}}{\partial \sigma}(I(p,K))\frac{\partial I}{\partial p} = C_{BS}(\sigma_1) - C_{BS}(\sigma_2), \tag{2.28}$$

from which we can obtain an expression for $\partial I/\partial p$. Differentiating (2.27) and (2.28) again and substituting for $\partial I/\partial p$ and $\partial^2 I/\partial p^2$, gives

$$g''(p) = \left(\frac{C_{BS}(\sigma_1) - C_{BS}(\sigma_2)}{\frac{\partial C_{BS}}{\partial \sigma}} \right)^2 \left(\frac{\frac{\partial^2 C_{BS}}{\partial K \partial \sigma} \frac{\partial^2 C_{BS}}{\partial \sigma^2}}{\frac{\partial C_{BS}}{\partial \sigma}} - \frac{\partial^3 C_{BS}}{\partial K \partial \sigma^2} \right),$$

where all the partial derivatives of C_{BS} are evaluated at I. Inserting directly from the Black–Scholes formula (1.41) then gives

$$g''(p) = \frac{(C_{BS}(\sigma_1) - C_{BS}(\sigma_2))^2}{\frac{\partial C_{BS}}{\partial \sigma}(I)} \frac{2\tilde{L}}{(T-t)KI^3},$$

where $\tilde{L} = \log(xe^{r(T-t)}/K)$. We know from the calculation (2.2) that $\frac{\partial C_{BS}}{\partial \sigma} > 0$ and so C_{BS} is increasing in the volatility parameter. Since $I > 0$, we conclude that $\text{sign}(g''(p)) = \text{sign}(\tilde{L})$. Thus for $K < K_{\min} = xe^{r(T-t)}$, $\tilde{L} > 0$ and g can only achieve a minimum in $[0,1]$, implying that $g < 0$ for $0 < p < 1$. Hence $\partial I/\partial K < 0$. Similarly for $K > K_{\min}$, $\partial I/\partial K > 0$, and $\partial I/\partial K = 0$ at $K = K_{\min}$. The implied volatility curve is locally convex around $K_{\min}$.

This can be extended by induction on the number of possible values that σ^2 can take. The result then holds for all σ^2 that can be approximated in L^2 by discrete random variables. The full details can be found in the references given in the Notes.

This result says that stochastic volatility European option prices produce the smile curve for *any* volatility process uncorrelated with the Brownian motion driving the price process, and this robustness to specific modeling of the volatility is a very important asset of these models.

2.5.3 Correlated Stochastic Volatility

In general, when volatility is correlated with the Brownian motion $(W_t^{(0)})$ driving the stock price, the situation is more complicated. To illustrate this, we present a generalization of the Hull–White formula in the case of one-factor stochastic volatility driven by a Brownian motion correlated with $(W_t^{(0)})$. Under an equivalent martingale measure $\mathbb{P}^{\star(\gamma)}$, the model is given by (2.18)–(2.19) and can be rewritten

$$\frac{dX_t}{X_t} = r\,dt + \sigma_t \left(\sqrt{1-\rho^2}\,d\hat{W}_t^\star + \rho\,dW_t^{(1)\star} \right), \tag{2.29}$$

$$dY_t = (\alpha(Y_t) - \beta(Y_t)\Lambda_t)\,dt + \beta(Y_t)dW_t^{(1)\star}, \tag{2.30}$$

where here we have decomposed the Brownian motion $W^{(0)\star}$ in terms of $W^{(1)\star}$ and an independent standard Brownian motion $\hat{W}^{\star}$, and we recall that $\sigma_t = f(Y_t)$.

Then by Itô's formula,

$$d\log X_t = (r - \frac{1}{2}\sigma_t^2)\,dt + \sigma_t\left(\sqrt{1-\rho^2}\,d\hat{W}_t^{\star} + \rho\,dW_t^{(1)\star}\right).$$

Now, conditioned on the path of the second Brownian $W^{(1)\star}$, $\log X_t$ is normally distributed with

$$\mathbb{E}^{\star(\gamma)}\left\{\log X_T \mid \mathscr{F}_t, W^{(1)\star}_{[t,T]}\right\} = \log x + \rho\int_t^T \sigma_s dW_s^{(1)\star} - \frac{1}{2}\rho^2\int_t^T \sigma_s^2 ds$$

$$+ (r - \frac{1}{2}(1-\rho^2)\overline{\sigma^2})(T-t)$$

$$\mathrm{var}^{\star(\gamma)}\left\{\log X_T \mid \mathscr{F}_t, W^{(1)\star}_{[t,T]}\right\} = (1-\rho^2)\overline{\sigma^2}(T-t),$$

where $\overline{\sigma^2}$ is defined in (2.25).

Then, using iterated expectations as in Section 2.5.2, the price of a European call, for example, is

$$C_t = \mathbb{E}^{\star(\gamma)}\left\{C_{BS}(t, X_t\xi_t; K, T; \sqrt{(1-\rho^2)\overline{\sigma^2}}) \mid \mathscr{F}_t\right\}, \qquad (2.31)$$

where

$$\xi_t = \exp\left(\rho\int_t^T \sigma_s dW_s^{(1)\star} - \frac{1}{2}\rho^2\int_t^T \sigma_s^2 ds\right).$$

In this form the price is a mixture of Black–Scholes prices with different volatilities *and* different starting values (current stock price).

Using the identity

$$C_{BS}(t, x\xi_t; K, T; \sigma) = \xi_t C_{BS}(t, x; K\xi_t^{-1}; \sigma),$$

which is easily verified from the Black–Scholes formula (1.41), one can rewrite (2.31) as

$$C_t = \mathbb{E}^{\star(\gamma)}\left\{\xi_t C_{BS}(t, X_t; K\xi_t^{-1}, T; \sqrt{(1-\rho^2)\overline{\sigma^2}}) \mid \mathscr{F}_t\right\}. \qquad (2.32)$$

This is an average of the Black–Scholes formula jointly in the volatility and strike price under an equivalent probability with Radon–Nikodym derivative ξ_t, according to Girsanov's theorem (Section 1.4.1).

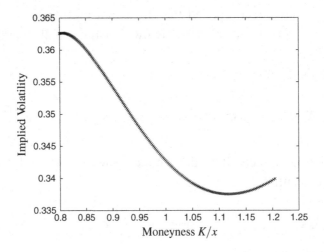

Figure 2.5   Implied volatilities predicted by the correlated stochastic volatility model (2.9), with $\rho = -0.2$, $f(y) = e^y$, and $\alpha = 1$, $\beta = \sqrt{2}$, $m = \log 0.1$, and $\gamma = 0$, $\mu - r = 0.16$ in (2.16). It is computed from a finite-difference solution to the partial differential equation (2.17) with terminal condition $h(x) = (x - K)^+$.

This formula, in either of the forms (2.31) or (2.32), is of practical use for Monte Carlo simulation of prices in a correlated stochastic volatility model because only one Brownian path has to be generated. Unlike the uncorrelated case, it does not reveal any information about the implied volatility curve and a generalization of the Renault–Touzi argument given in Section 2.5.2 is not possible. In fact, numerical computation of implied volatilities from a correlated stochastic volatility model, such as shown in Figure 2.5, tells us that the convexity structure will be much more complicated. This illustrates the increase in complexity of these problems introduced by the skew effect.

2.5.4 Multifactor Stochastic Volatility Models

In the following chapters, we will make the case for models with at least two stochastic volatility factors, which fluctuate on different time scales. The notion of time scales will be treated in detail in Chapter 3. The case with several stocks and multidimensional stochastic volatility will be considered in Chapter 9, Section 9.7.

In this subsection, we discuss briefly models where the volatility is driven by a d-dimensional diffusion process $\mathbf{Y} \in \mathbb{R}^d$. In this case, the Markovian stochastic volatility model (2.9) becomes

$$\left.\begin{array}{rcl} dX_t &=& \mu(\mathbf{Y}_t)X_t\,dt + \sigma_t X_t\,dW_t^{(0)}, \\ \sigma_t &=& f(\mathbf{Y}_t), \\ d\mathbf{Y}_t &=& \alpha(\mathbf{Y}_t)\,dt + \beta(\mathbf{Y}_t)\,d\mathbf{W}_t, \end{array}\right\} \qquad (2.33)$$

where $\alpha : \mathbb{R}^d \to \mathbb{R}^d$, $\beta : \mathbb{R}^d \to \mathbb{R}_+^d \times \mathbb{R}_+^d$ is a *diagonal* matrix with entries $\beta_i(\mathbf{Y}_t)$, and $\mathbf{W} = (W^{(1)},\dots,W^{(d)})$ is a vector of d correlated Brownian motions, which are also correlated with $W^{(0)}$. We denote by ρ_{ij} the correlations between the Brownian motions driving the volatility factors:

$$d\langle W^{(i)}, W^{(j)}\rangle_t = \rho_{ij}\,dt, \quad i,j = 1,\dots,d,$$

and by ρ_i the correlations between shocks to the factor $Y^{(i)}$ and shocks to the stock price:

$$d\langle W^{(0)}, W^{(i)}\rangle_t = \rho_i\,dt, \quad i = 1,\dots,d.$$

It is assumed that the ρ_{ij} and ρ_i satisfy conditions such that the covariance matrix of the Brownian motion $(W^{(0)}, \mathbf{W})$ is positive definite. The function f is smooth, positive, and increasing in each component of $\mathbf{Y}$.

As with one-factor models, it is convenient to decompose each of the Brownian motions $W^{(i)}$ as

$$W^{(i)} = \rho_i W^{(0)} + \sqrt{1 - \rho_i^2}\,W^{(i)\perp},$$

where the d-dimensional Brownian motion $\mathbf{W}^\perp = (W^{(1)\perp},\dots,W^{(d)\perp})$ is independent of $W^{(0)}$. In this case, we have a vector of correlation coefficients $\rho = (\rho_1,\dots,\rho_d)$. The total instantaneous correlation between stock price and volatility shocks is a $\mathbf{Y}_t$-dependent weighted sum of the ρ_i:

$$\text{corr}(dX_t, d\sigma_t) = \sum_i \left(\frac{\frac{\partial f}{\partial y_i}(\mathbf{Y}_t)\beta_i(\mathbf{Y}_t)}{\sqrt{\sum_j \left(\frac{\partial f}{\partial y_i}(\mathbf{Y}_t)\beta_i(\mathbf{Y}_t)\right)^2}} \right) \rho_i.$$

The main concepts and tools discussed in this chapter can be extended to this multidimensional setting. The construction of risk-neutral pricing measures developed in Section 2.4.2 for the case of *one-factor* stochastic volatility is easily generalized when there is a vector of volatility factors. These measures $\mathbb{P}^{\star(\gamma)}$ are parametrized by a d-dimensional market price of volatility risk process $\gamma \in \mathbb{R}^d$ adapted to the filtration generated by the $d+1$ Brownian motions $(W^{(0)}, \mathbf{W})$. The processes $(W^{(0)\star}, W^{(1)\perp\star},\dots,W^{(d)\perp\star})$ defined by

$$W_t^{(0)\star} = W_t^{(0)} + \int_0^t \frac{(\mu(\mathbf{Y}_s) - r)}{f(\mathbf{Y}_s)}\,ds,$$

$$W_t^{(i)\perp\star} = W_t^{(i)\perp} + \int_0^t \gamma_s^{(i)} \, ds, \quad i = 1, \dots, d$$

are Brownian motions under $\mathbb{P}^{\star(\gamma)}$, and the definition of $W^{(0)\star}$ ensures that the discounted stock price is a $\mathbb{P}^{\star(\gamma)}$-martingale. The dynamics under $\mathbb{P}^{\star(\gamma)}$ become

$$dX_t = rX_t \, dt + f(\mathbf{Y}_t)X_t \, dW_t^{(0)\star},$$
$$d\mathbf{Y}_t = (\boldsymbol{\alpha}(\mathbf{Y}_t) - \boldsymbol{\beta}(\mathbf{Y}_t)\underline{\Lambda}_t) \, dt + \boldsymbol{\beta}(\mathbf{Y}_t) \, d\mathbf{W}_t^\star,$$
$$W_t^{(i)\star} = \rho_i W_t^{(0)\star} + \sqrt{1 - \rho_i^2} \, W_t^{(i)\perp\star},$$

where we define the total risk premium processes $\underline{\Lambda}_t = (\Lambda_t^{(1)}, \dots, (\Lambda_t^{(d)})$ by

$$\Lambda_t^{(i)} = \rho_i \frac{(\mu(\mathbf{Y}_t) - r)}{f(\mathbf{Y}_t)} + \gamma_t^{(i)} \sqrt{1 - \rho_i^2}, \quad i = 1, \dots, d.$$

In practice, we may restrict ourselves to the Markovian case when the volatility risk premia are functions of the present values $(t, X_t, \mathbf{Y}_t)$, so that $\underline{\Lambda}_t = \underline{\Lambda}(t, X_t, \mathbf{Y}_t)$. Then, following Section 1.9.3, the pricing function for European options

$$P(t,x,\mathbf{y}) = \mathbb{E}^{\star(\gamma)} \left\{ e^{-r(T-t)} h(X_T) \mid X_t = x, \mathbf{Y}_t = \mathbf{y} \right\}$$

satisfies the Feynman–Kac partial differential equation

$$\frac{\partial P}{\partial t} + \frac{1}{2} f(\mathbf{y})^2 x^2 \frac{\partial^2 P}{\partial x^2} + \sum_{i=1}^d \rho_i \beta_i(\mathbf{y}) x f(\mathbf{y}) \frac{\partial^2 P}{\partial x \partial y_i} +$$
$$\frac{1}{2} \sum_{i,j=1}^d \rho_{ij} \beta_i(\mathbf{y}) \beta_j(\mathbf{y}) \frac{\partial^2 P}{\partial y_i \partial y_j} + r \left(x \frac{\partial P}{\partial x} - P \right)$$
$$+ \sum_{i=1}^d \left(\alpha_i(\mathbf{y}) - \beta_i(\mathbf{y}) \Lambda^{(i)}(t,x,\mathbf{y}) \right) \frac{\partial P}{\partial y_i} = 0,$$

with the terminal condition $P(T,x,y) = h(x)$.

To generalize the correlated Hull–White formula of Section 2.5.3, we decompose the Brownian motion $W^{(0)\star}$ as

$$W^{(0)\star} = \sqrt{1 - c_0^2} \, \hat{W}^\star + \sum_{j=1}^d c_j W^{(j)\star},$$

where $\hat{W}^\star$ is a standard Brownian motion independent of $W^{(j)\star}$, $j = 1, \dots, d$, and the constants $c_0, c_1, \dots, c_d$ are given uniquely by the solution of the equations

$$\sum_{j=1}^{d} \rho_{ij}c_j = \rho_i, \quad i = 1,\ldots,d, \qquad c_0^2 = \sum_{i=1}^{d} \rho_i c_i.$$

That $c_0^2 \in [0,1]$ follows from the positive definiteness of the covariance matrices of $\mathbf{W}$ and $(W^{(0)}, \mathbf{W})$.

Then, it is straightforward to extend the argument of Section 2.5.3 by conditioning on the path of $\mathbf{W}$, to obtain the following expression for the European call option price for instance:

$$C_t = \mathbb{E}^{\star(\gamma)} \left\{ C_{BS}(t, X_t \xi_t; K, T; \sqrt{(1 - c_0^2)\overline{\sigma^2}}) \mid \mathscr{F}_t \right\},$$

where

$$\xi_t = \exp\left(\sum_{j=1}^{d} c_j \int_t^T \sigma_s \, dW_s^{(j)\star} - \frac{1}{2}c_0^2 \int_t^T \sigma_s^2 ds \right),$$

and $\overline{\sigma^2}$ is defined in (2.25).

2.6 Summary and Conclusions

We summarize the main features of the stochastic volatility approach. The positive aspects are:

- It directly models the observed random behavior of market volatility.
- It allows us to reproduce more realistic returns distributions, in particular thicker-than-lognormal tails.
- The asymmetry of the distribution is easily incorporated by correlating the noise sources.
- The smile/skew effect in option prices is exhibited in stochastic volatility models, where the correlation controls the skew.

However, new difficulties are associated with this approach:

- Volatility is not directly observed. As a consequence, estimation of the parameters of a specific model and the current level of volatility is not straightforward.
- Choosing a particular stochastic volatility model is a difficult issue. Economic and empirical evidence provide some guidance on properties such models should have, but model selection is far from unique.
- We have to deal with an incomplete market, which means derivatives cannot be perfectly hedged with just the underlying. In addition, a volatility risk premium has to be estimated from option prices.

These problems can be addressed by exploiting time scales in volatility. This notion is explained in Chapter 3. In Chapter 4, we develop an asymptotic theory based on this observation that yields a method with the following features:

- It applies to a large class of volatility models that are driven by multiscale, multifactor processes and the results are not strongly model-dependent.
- It incorporates nonzero volatility risk premia and nonzero correlation between volatility and asset price shocks that captures the much-observed skew and its term structure.
- The asymptotic analysis yields a simple pricing and hedging theory that corrects the Black–Scholes theory to account for uncertain and changing volatility.
- The parameters needed for the theory are easily "read from the skew." That is, calibration from near-the-money European option implied volatilities is simple and direct. Further, the theory does not need estimation of today's volatility level.

Notes

Empirical evidence for the smile curve of implied volatility appears in many studies, including Rubinstein (1985), using data from before the 1987 crash, and in Jackwerth and Rubinstein (1996), where the post-crash skew is documented.

Work on fitting a local volatility surface, as discussed in Section 2.2, includes Rubinstein (1994), Dupire (1994), Derman and Kani (1994), and Avellaneda *et al.* (1997). For computational methods we also refer to Achdou and Pironneau (2005) and Fengler (2005). An extensive empirical study of the stability of the fitted surfaces can be found in Dumas *et al.* (1998) and Lee (1999) presents an analysis of these results within a stochastic volatility framework. The CEV model is introduced in Cox (1975) (see also Lipton (2001)).

Classical references investigating specific stochastic volatility models are Ball and Roma (1994), Heston (1993), Hull and White (1987), Scott (1987), Stein and Stein (1991), and Wiggins (1987). Survey papers on the subject are Frey (1996) and Hobson (1996). Multidimensional extensions of Heston's model using affine processes are studied in Duffie *et al.* (2000).

Results and extensive references on super-replicating strategies in incomplete markets discussed in Section 2.4.3 are given in detail in Karatzas and Shreve (1998).

The quotation at the end of Section 2.4.4 is taken from the plenary lecture of J.-P. Bouchaud at the 6th World Congress of the Bachelier Finance Society in 2010.

The result of Section 2.5.2 appears in Renault and Touzi (1996) and the proof given here, including the induction step, is in Sircar and Papanicolaou (1999). The correlated Hull–White formula (2.31) of Section 2.5.3 is given by Willard (1996). A generalization of this formula based on Malliavin calculus is derived in Alòs (2006) with an application to the Heston model in Alòs (2009).

For other books on stochastic volatility modeling, we refer to Lewis (2000), Gatheral (2006), and Shephard (2005). For models with jumps we refer to Cont and Tankov (2003) and references therein. A study of moments for stochastic volatility models can be found in Andersen and Piterbarg (2007).

3

Volatility Time Scales

In this chapter, we introduce the characterization of stochastic volatility in terms of its time scales of fluctuation. This is self-contained and uses basic facts of probability theory which can be found in the textbooks referenced in the Notes at the end of the chapter. For most of the rest of the book, we will work with two-factor stochastic volatility models in which one factor is *fast mean-reverting* with respect to the time horizon of a typical options contract, while the other *slow factor* may not appear to be mean-reverting relative to this horizon, though it may be (and often is) mean-reverting over longer horizons. The option pricing problem under these *multiscale* stochastic volatility models is analyzed by asymptotic methods in Chapter 4. Among other applications, speedup of Monte Carlo simulations will be discussed in Section 11.1.

In the first part of the chapter, we will visualize time scales with pictures from numerical simulations. These will be used to illustrate the effect of fast and slow scales in the volatility on stock returns, which, unlike volatility, is the observable quantity we deal with in practice. We also provide illustrations with synthetic data in Section 3.4 and market data in Section 3.5.

3.1 A Simple Picture of Fast and Slow Time Scales

We present some simple examples of mean-reverting processes in the following sections. Mean-reversion refers to the typical time a process takes to return to its long-run mean level, if it exists. This is described by the notion of *ergodicity*, as we shall discuss later. To introduce the heuristic of fast and slow time scales, Figure 3.1 shows four sample paths from simulation of a simple diffusion stochastic volatility model. In the second graph, the characteristic mean-reversion time is 45 days, corresponding to

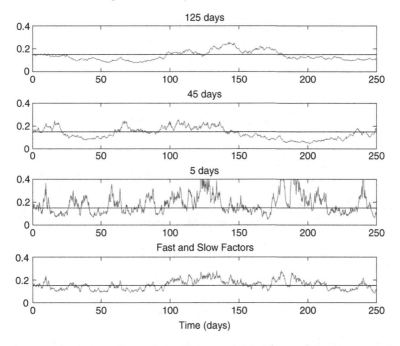

Figure 3.1   Simulated volatility paths over 250 trading days. From top to bottom, the time scales are: slow (characteristic mean-reversion time 125 days), order one (45 days), fast (5 days), and fast + slow (combination of 5 and 125 days). The horizontal lines show an approximate mean level, from where the paths start.

an "order one" time scale with respect to a one- to two-month horizon. The top panel shows a *slowly* fluctuating volatility process, whose mean-reversion time is 125 days. The third sample path is from a *fast* volatility process with a 5-day mean-reversion time. Finally, the fourth graph comes from a *two-factor* volatility model with both the fast and slow factors. We observe *approximately* two cycles in the top slow path, three in the next, and many more in the fast path. In the final graph, there is a superposition of two scales: the path displays globally about three cycles of the slow scale, within which there are many cycles of the fast scale. These interpretations are of course subjective, and only meant to convey a feeling of how time scales may reveal themselves in sample paths.

Figure 3.2 shows some more sample paths from the two-factor fast-and-slow volatility model, but for some different choices of mean-reversion times. Notice that the qualitative behavior is similar. In the asymptotic theory we will present in Chapter 4, it is not required to have a precise estimation of the time scales. The group parameters that arise from the asymptotic theory contain this information, and are calibrated directly from

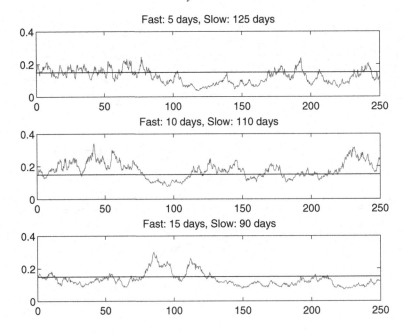

Figure 3.2   Simulated two-factor volatility paths over 250 trading days.

implied volatilities as we describe in Chapter 5. Finally, for reference, we show in Figure 3.3 the daily CBOE volatility index VIX over 2006-07 in the top panel, and 2003-07 in the bottom graph. The VIX is computed from short-term near-the-money S&P 500 implied volatilities and is intended as a measure of market volatility. Although we will not work with the VIX, since it is a form of processed options data, it is useful to obtain a qualitative view of the object we are trying to model. We observe first of all that it certainly appears random in nature, mean-reverting, and seems to exhibit slow and fast time scales.

3.2 Ergodicity and Mean-Reversion

Mean-reversion is closely related to the concept of *ergodicity*. A process (Y_t) is called ergodic if it admits a unique invariant (or stationary) distribution, denoted by Π, and the long-time time average of any measurable bounded function g of the process converges almost surely (a.s.) to the deterministic average with respect to its invariant distribution:

$$\lim_{t \to \infty} \frac{1}{t} \int_0^t g(Y_s)\, ds = \int g(y) \Pi(dy). \tag{3.1}$$

This relation is true on a set of probability one for any initial distribution (or starting point) for Y_0. This property can be viewed as a generalization

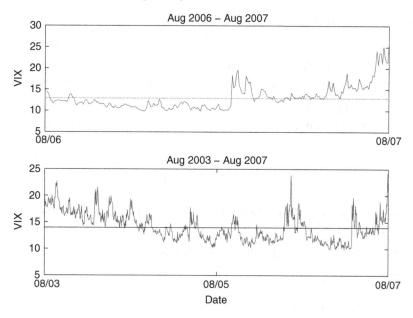

Figure 3.3   Daily VIX volatility index over one year (top) and four years (bottom). The horizontal line is at the level obtained by averaging the log of the VIX over those time periods.

of the classical law of large numbers for the sum of independent identically distributed random variables.

In the context of stochastic volatility models, we are interested in the integrated square volatility as it appeared naturally in Chapter 2, equation (2.25):

$$\overline{\sigma^2} = \frac{1}{T-t} \int_t^T \sigma_s^2 \, ds. \tag{3.2}$$

However, in this application it is not the time horizon T that is large, instead the behavior of $\overline{\sigma^2}$ will be governed by the time scales of fluctuation in the volatility process as we will detail in this chapter.

What is important in order to make precise the notion of fast or slow time scales is an appropriate parsimonious parametrization of the process. We describe a way to take a *Markov* process and construct a one-parameter family of Markov processes where the parameter $\kappa > 0$ characterizes the time scale of fluctuations of the process, slow and fast referring to κ small and κ large, respectively.

Here are the concepts and properties that we will detail for three classes of processes through examples.

Semigroup and Parametrization of Time Scales

We start with a continuous-time Markov process (Y_t) which is time-homogeneous, and lives on a state space $\mathscr{S} \subset \mathbb{R}$, which may be discrete or continuous, compact or not. We always assume that the trajectories of Y_t are right-continuous with left-limits.

We consider its transition semigroup $\mathbf{P}_t$, which acts on bounded measurable functions g by:

$$\mathbf{P}_t g(y) = \mathbb{E}\{g(Y_t) \mid Y_0 = y\}. \tag{3.3}$$

Note that $\mathbf{P}_0 = \mathbf{I}$, the identity operator, and in fact, in the following, we assume that $\mathbf{P}_t$ is strongly continuous, that is $\lim_{t \downarrow 0} \mathbf{P}_t = \mathbf{I}$ uniformly.

The semigroup property follows from the Markov property and time-homogeneity:

$$\begin{aligned}
\mathbf{P}_{t+s} g(y) &= \mathbb{E}\{g(Y_{t+s}) \mid Y_0 = y\} \\
&= \mathbb{E}\{\mathbb{E}\{g(Y_{t+s}) \mid Y_t\} \mid Y_0 = y\} = \mathbb{E}\{\mathbf{P}_s g(Y_t) \mid Y_0 = y\} \\
&= \mathbf{P}_t \mathbf{P}_s g(y).
\end{aligned}$$

For κ a positive parameter, we introduce $\mathbf{P}_t^{(\kappa)} = \mathbf{P}_{\kappa t}$, which is the transition semigroup of a process $(Y_t^{(\kappa)})$ that is equal in distribution to $(Y_{\kappa t})$ as follows from:

$$\mathbb{E}\{g(Y_t^{(\kappa)}) \mid Y_0^{(\kappa)} = y\} = \mathbf{P}_t^{(\kappa)} g(y) = \mathbf{P}_{\kappa t} g(y) = \mathbb{E}\{g(Y_{\kappa t}) \mid Y_0 = y\}.$$

Then the process $Y^{(\kappa)}$ is behaving statistically like the process Y would if it were transitioning κ times as fast.

Infinitesimal Generator

We then define the infinitesimal generator $\mathscr{L}$ of Y by

$$\mathscr{L}g(y) = \lim_{t \downarrow 0} \frac{\mathbf{P}_t g(y) - g(y)}{t}, \tag{3.4}$$

on an appropriate domain of functions g so that the limit exists. From time-homogeneity, it follows that

$$\frac{d}{dt} \mathbf{P}_t g(y) = \mathscr{L}\mathbf{P}_t g(y), \tag{3.5}$$

which is known as the backward Kolmogorov equation for the semigroup $\mathbf{P}_t$, which admits the formal representation $\mathbf{P}_t = e^{t\mathscr{L}}$. Note that $\mathbf{P}_t 1 = 1$ implies $\mathscr{L}1 = 0$.

It can easily be seen that the infinitesimal generator $\mathscr{L}^{(\kappa)}$ of the process $Y^{(\kappa)}$ is given by $\kappa\mathscr{L}$:

$$\mathscr{L}^{(\kappa)}g(y) = \lim_{t\downarrow 0}\frac{\mathbf{P}_t^{(\kappa)}g(y)-g(y)}{t} = \kappa\lim_{t\downarrow 0}\frac{\mathbf{P}_{\kappa t}g(y)-g(y)}{\kappa t} = \kappa\mathscr{L}g(y).$$

Therefore the parameter κ appears as a multiplicative factor at the level of the generator.

Invariant Distribution

An invariant distribution Π of the process (Y_t), if it exists, solves the following problem: find an initial distribution Π for Y_0 such that for any $t > 0$, Y_t has the same distribution. This is also called the stationary or equilibrium distribution. We try to find the distribution of Y_0 such that

$$\frac{d}{dt}\mathbb{E}\{g(Y_t)\} = \frac{d}{dt}\int_{\mathscr{S}}\mathbb{E}\{g(Y_t)\mid Y_0=y\}\Pi(dy) = 0,$$

for any bounded g. Taking the time derivative through the integral and using (3.5), we derive:

$$0 = \int_{\mathscr{S}}\frac{d}{dt}\mathbb{E}\{g(Y_t)\mid Y_0=y\}\Pi(dy) = \int_{\mathscr{S}}\mathscr{L}\mathbf{P}_t g(y)\Pi(dy)$$
$$= \int_{\mathscr{S}}\mathbf{P}_t g(y)\mathscr{L}^*\Pi(dy), \tag{3.6}$$

where the adjoint operator $\mathscr{L}^*$ (with respect to $L^2(\mathbb{R})$) is defined as in Section 1.5.3 by $\int\psi(y)\mathscr{L}\phi(y)\,dy = \int\phi(y)\mathscr{L}^*\psi(y)\,dy$. This adjoint will be given explicitly in the examples which follow. As the last quantity in (3.6) must be zero for any g, an invariant distribution Π, if it exists, must satisfy

$$\mathscr{L}^*\Pi = 0. \tag{3.7}$$

In the case of $Y^{(\kappa)}$, since $(\mathscr{L}^{(\kappa)})^* = \kappa\mathscr{L}^*$, this equation for the invariant distribution $\Pi^{(\kappa)}$ for $Y^{(\kappa)}$ is simply $\kappa\mathscr{L}^*\Pi^{(\kappa)} = 0$, and so Y and $Y^{(\kappa)}$ have the same invariant distributions, if any.

Reversibility and Spectral Gap

We now consider an ergodic Markov process Y whose unique invariant distribution is denoted by Π. When $\int|g(y)|\Pi(dy) < \infty$, we shall use the notation $\langle g\rangle$ for the average of g with respect to Π:

$$\langle g\rangle = \int g(y)\Pi(dy). \tag{3.8}$$

As usual, $g \in L^2(\Pi)$ if $\langle g^2\rangle < \infty$.

By (3.7), zero is an eigenvalue of $\mathscr{L}^*$ (and $\mathscr{L}$), and in fact, the nature of the spectrum of $\mathscr{L}^*$ (or $\mathscr{L}$) near the eigenvalue zero, controls the rate of approach to the invariant distribution. Here, for simplicity, we restrict ourselves to *reversible* processes that have a discrete spectrum with a positive gap, meaning that zero is an isolated eigenvalue. Reversibility means that $\mathscr{L}$ is self-adjoint on the space $L^2(\Pi)$:

$$\langle \phi \mathscr{L} \psi \rangle = \langle \psi \mathscr{L} \phi \rangle, \tag{3.9}$$

for all test functions ϕ and ψ in $L^2(\Pi)$. (Note that this does not imply that the adjoint $\mathscr{L}^*$ with respect to $L^2(\mathbb{R})$ is equal to $\mathscr{L}$.) Our main examples, the OU and CIR processes, finite-state Markov chains, and a continuous-state pure jump process fall in the class of reversible ergodic processes with a discrete spectrum and a positive gap, as we shall see.

Next, we analyze the connection between the spectrum of $\mathscr{L}$ and the rate of convergence to the invariant distribution.

We denote the eigenvalues of $\mathscr{L}$ by $0 = \lambda_0 > \lambda_1 > \lambda_2 > \cdots$. Let ψ_n denote the eigenfunctions with eigenvalues λ_n: $\mathscr{L}\psi_n = \lambda_n \psi_n$. The ψ_n's form an orthonormal basis of $L^2(\Pi)$, and $\psi_0 = 1$.

Given $g \in L^2(\Pi)$, we write the eigenfunction expansion

$$g(y) = \sum_{n \geq 0} c_n \psi_n(y), \tag{3.10}$$

where $c_n = \langle g \psi_n \rangle$. In particular, $c_0 = \langle g \rangle$. Next, we consider the backward Kolmogorov equation (3.5) for the expectation $\mathbf{P}_t g(y) = \mathbb{E}\{g(Y_t) \mid Y_0 = y\}$, namely $\frac{d}{dt}\mathbf{P}_t g(y) = \mathscr{L}\mathbf{P}_t g(y)$. We construct the solution in the form

$$\mathbf{P}_t g(y) = \sum_{n \geq 0} b_n(t) \psi_n(y),$$

for functions $b_n(t)$ to be found. Inserting this form into the Kolmogorov equation gives

$$\sum_{n \geq 0} b_n'(t) \psi_n(y) = \sum_{n \geq 0} b_n(t) \mathscr{L} \psi_n(y) = \sum_{n \geq 0} \lambda_n b_n(t) \psi_n(y),$$

which implies that $b_n(t)$ satisfies the ordinary differential equation $b_n'(t) = \lambda_n b_n(t)$ with initial condition $b_n(0) = c_n$. Therefore, we have

$$\mathbf{P}_t g(y) = \sum_{n \geq 0} c_n e^{\lambda_n t} \psi_n(y). \tag{3.11}$$

We next consider the behavior of $\mathbf{P}_t g$ as $t \to \infty$. From (3.11), we have

$$\mathbf{P}_t g(y) - \langle g \rangle = e^{\lambda_1 t} \sum_{n \geq 1} c_n e^{(\lambda_n - \lambda_1)t} \psi_n(y),$$

where we have used $c_0 = \langle g \rangle$, $\lambda_0 = 0$, and $\psi_0 = 1$. Since $0 > \lambda_1 > \lambda_2 > \cdots$, this implies

$$|\mathbf{P}_t g(y) - \langle g \rangle| \leq C e^{\lambda_1 t},$$

for some constant depending on g and y, which shows exponential convergence of the distribution of Y_t to the invariant distribution at a rate governed by the spectral gap $\lambda = |\lambda_1|$.

A Class of Mean-Reverting Markov Processes

A Markov process Y is in the class of mean-reverting Markov processes we will work with if it is ergodic, reversible, has a discrete spectrum with a spectral gap, and its unique invariant distribution Π admits a mean. We define the *rate of mean-reversion* of Y to be its spectral gap.

Observe that if the spectral gap of Y is λ, then the spectrum of $Y^{(\kappa)}$ has a gap $\kappa\lambda$, as follows by writing the eigenvalue problem for $\kappa\mathscr{L}$. Therefore, the process $Y^{(\kappa)}$ approaches the invariant distribution Π exponentially with its rate of mean-reversion $\kappa\lambda$. When Y is mean-reverting and κ is large, we refer to $Y^{(\kappa)}$ as *fast mean-reverting*.

Poisson Equation

The existence of a unique invariant distribution and a positive spectral gap has important consequences for *Poisson equations* of the form

$$\mathscr{L}\phi = g, \tag{3.12}$$

where $g \in L^2(\Pi)$, and ϕ is the unknown. Such equations come up repeatedly in the asymptotic analysis of Chapter 4.

First, assuming that there is a solution ϕ to (3.12), integrating with respect to Π on both sides gives

$$\langle g \rangle = \langle 1 \mathscr{L}\phi \rangle = \langle \phi(\mathscr{L}1) \rangle = 0,$$

by (3.9). Therefore, g must satisfy the *solvability* or *centering* condition

$$\langle g \rangle = 0. \tag{3.13}$$

We next write the solutions of the Poisson equation using the eigenfunction representation (3.10) of g, where $c_0 = 0$ since g is centered with respect to Π. It is easy to verify that the solutions $\phi \in L^2(\Pi)$ are given by

$$\phi(y) = \sum_{n \geq 1} \frac{c_n}{\lambda_n} \psi_n(y) + c, \tag{3.14}$$

where c is an arbitrary constant.

It is interesting to note that the solution of the Poisson equation can be represented in terms of the semigroup defined in (3.3). Observe that

$$\frac{1}{\lambda_n} = -\int_0^\infty e^{\lambda_n t}\, dt, \quad n \geq 1,$$

where the integral converges because $\lambda_n < 0$. Therefore, we can write ϕ in (3.14) as

$$
\begin{aligned}
\phi(y) &= -\sum_{n \geq 1} c_n \left(\int_0^\infty e^{\lambda_n t}\, dt \right) \psi_n(y) + c \\
&= -\int_0^\infty \sum_{n \geq 1} c_n e^{\lambda_n t} \psi_n(y)\, dt + c \\
&= -\int_0^\infty \mathbf{P}_t g(y)\, dt + c,
\end{aligned}
$$

where the interchange of sum and integral is clearly justified, and we have used the representation (3.11) for $\mathbf{P}_t$.

Application to Stochastic Volatility

Coming back to the integrated square volatility in (3.2), if the volatility process is given by $\sigma_t = f(Y_t^{(\kappa)})$, where Y is mean-reverting and f^2 is Π-integrable, then

$$
\begin{aligned}
\overline{\sigma^2} &= \frac{1}{(T-t)} \int_t^T f^2(Y_s^{(\kappa)})\, ds \\
&\stackrel{\mathscr{D}}{=} \frac{1}{(T-t)} \int_t^T f^2(Y_{\kappa s})\, ds \\
&= \frac{1}{\kappa(T-t)} \int_{\kappa t}^{\kappa T} f^2(Y_s)\, ds \\
&= \frac{1}{(T-t)} \left(T \frac{1}{(\kappa T)} \int_0^{\kappa T} f^2(Y_s)\, ds - t \frac{1}{(\kappa t)} \int_0^{\kappa t} f^2(Y_s)\, ds \right).
\end{aligned}
$$

It follows from (3.1),

$$\lim_{\kappa \uparrow \infty} \overline{\sigma^2} = \int_{\mathscr{S}} f^2(y)\Pi(dy),$$

so in the limit of fast mean-reversion, $\overline{\sigma^2}$ converges to a constant in distribution, and therefore in probability, and in fact almost surely.

In the regime of slow mean-reversion, that is κ small, if f is continuous (and bounded for instance), then

$$\lim_{\kappa \downarrow 0} \frac{1}{(T-t)} \int_t^T f^2(Y_{\kappa s}) \, ds = f^2(Y_t),$$

and therefore $\overline{\sigma^2}$ converges to the spot square volatility in the slow-scale asymptotics κ small.

In the next section, we present some simple examples, which are rich enough to exhibit the main features of volatility we want to describe.

3.3 Examples of Mean-Reverting Processes

We shall work for now with one-factor stochastic volatility models for simplicity. Later in this book, we will study two-factor stochastic volatility models, where one factor is fast and the other is slow. We are primarily interested in the mean-reverting behavior of the fast factor, since the *mean-reversion* property of the slow factor will not have an impact on the problems we will study, just their local characteristics.

We write the volatility process (σ_t) as a positive function $\sigma_t = f(Y_t)$ of the *driving process* (Y_t). The following examples demonstrate the relationship between mean-reversion and time scales in some simple models for (Y_t). They are chosen from three types of continuous-time *ergodic* Markov processes:

- Continuous-time Markov chains on a finite space.
- Pure jump Markov processes on a continuous space.
- Diffusion processes.

It is classical, but beyond the scope of this book to verify, that the four examples we look at are indeed ergodic processes in the sense of (3.1). In each case, we highlight the infinitesimal generator, invariant distribution, reversibility, and rate of mean-reversion, and demonstrate by hand some important properties we will need for the asymptotic theory developed in Chapter 4.

3.3.1 Example: Markov Chain

Markov chains provide convenient tools to model regime-switching. For simplicity, we restrict ourselves to the two-state case.

Suppose (Y_t) is a two-state time-homogeneous Markov chain on the state space $\mathscr{S} = \{y_1, y_2\}$. We can describe this Markov process as follows: if at time t, $Y_t = y_i$ $(i = 1, 2)$, in an infinitesimal time dt, the process switches to the other state with probability $\ell_i dt$, and remains in the same position with probability $1 - \ell_i dt$. Here, ℓ_1, ℓ_2 are positive parameters. The semigroup is the 2×2 transition probability matrix

$$\mathbf{P}_t = \begin{pmatrix} \mathbb{P}\{Y_{s+t} = y_1 \mid Y_s = y_1\} & \mathbb{P}\{Y_{s+t} = y_2 \mid Y_s = y_1\} \\ \mathbb{P}\{Y_{s+t} = y_1 \mid Y_s = y_2\} & \mathbb{P}\{Y_{s+t} = y_2 \mid Y_s = y_2\} \end{pmatrix},$$

and this description implies that for small t,

$$\mathbf{P}_t = \begin{pmatrix} 1 - \ell_1 t & \ell_1 t \\ \ell_2 t & 1 - \ell_2 t \end{pmatrix} + o(t). \tag{3.15}$$

It follows from (3.4) that the infinitesimal generator is

$$\mathcal{L} = \begin{pmatrix} -\ell_1 & \ell_1 \\ \ell_2 & -\ell_2 \end{pmatrix}. \tag{3.16}$$

The invariant distribution of the process (Y_t), if it exists, solves the following problem:

$$\mathcal{L}^* \Pi = 0.$$

Here $\mathcal{L}^*$ is the adjoint operator of $\mathcal{L}$, which is just the matrix transpose of (3.16), and Π is a probability two-vector (π_1, π_2). The unique solution is

$$\pi_1 = \frac{\ell_2}{\ell_1 + \ell_2}, \qquad \pi_2 = \frac{\ell_1}{\ell_1 + \ell_2}.$$

In this case, the "integral" with respect to the invariant distribution $\langle \cdot \rangle$ is

$$\langle g \rangle = \pi_1 g(y_1) + \pi_2 g(y_2).$$

It is also straightforward to check by direct computation that the Markov chain is reversible in that it satisfies (3.9):

$$\langle \psi \mathcal{L} \phi \rangle = \sum_{i=1}^{2} \psi(y_i)(\mathcal{L}\phi)(y_i)\pi_i = \sum_{i=1}^{2} \phi(y_i)(\mathcal{L}\psi)(y_i)\pi_i = \langle \phi \mathcal{L} \psi \rangle,$$

using

$$(\mathcal{L}\phi)(y_1) = -\ell_1(\phi(y_1) - \phi(y_2)), \qquad (\mathcal{L}\phi)(y_2) = \ell_2(\phi(y_1) - \phi(y_2)).$$

An eigenvalue calculation shows that the spectrum of Y is given by $\lambda_0 = 0$ and $\lambda_1 = -(\ell_1 + \ell_2) < 0$, and therefore the spectral gap is $(\ell_1 + \ell_2)$. The corresponding eigenfunctions are

$$\psi_0(y_1) = 1, \quad \psi_0(y_2) = 1; \qquad \psi_1(y_1) = \sqrt{\frac{\ell_1}{\ell_2}}, \quad \psi_1(y_2) = -\sqrt{\frac{\ell_2}{\ell_1}}.$$

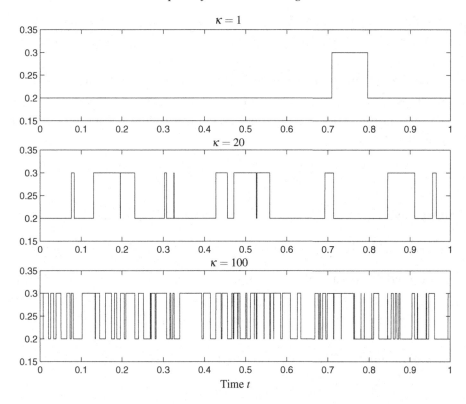

Figure 3.4   Simulated paths of $\sigma_t = f(Y_t^{(\kappa)})$, with $(Y_t^{(\kappa)})$ a two-state symmetric ($\ell_1 = \ell_2$) Markov chain, showing the relation between time scale of fluctuation and κ.

Now we consider the Markov chain $Y^{(\kappa)}$ defined in the previous section by its infinitesimal generator given by $\kappa \mathcal{L}$. From the general theory of the previous section, we know that the rate of mean-reversion of $Y^{(\kappa)}$ is $\kappa(\ell_1 + \ell_2)$. It is a classical fact of the theory of Markov chains that the holding time in state i is exponentially distributed with parameter $\kappa \ell_i$, and the successive holding times are independent. Therefore, the mean return time is proportional to $1/\kappa$, or in other words κ is a measure of the frequency of switching across the mean level, or the rate of mean-reversion.

We show in Figure 3.4 three typical realizations of $\sigma_t = f(Y_t^{(\kappa)})$ with $\kappa = 1$, $\kappa = 20$, and $\kappa = 100$ to illustrate this.

3.3.2 Example: Another Jump Process

Our second example is a simple generalization of the first one, by allowing the process (Y_t) to jump after exponentially distributed holding times with mean one to independent random values which are distributed according to a probability density p on $\mathbb{R}$. We assume that p is integrable so that its mean exists. We refer to (Y_t) as an example of pure jump Markov processes with values in $\mathbb{R}$.

To find the infinitesimal generator, it is convenient to introduce the number of jumps N_t before time t, which is a Poisson process with intensity one, namely a counting process with stationary independent increments that are Poisson-distributed:

$$P\{N_t = k\} = \frac{t^k}{k!} e^{-t}, \qquad k = 0, 1, 2, \ldots$$

Then we have

$$\mathbb{E}\{g(Y_t)\} = \mathbb{E}\{g(Y_t) \mid N_t = 0\} P\{N_t = 0\} + \mathbb{E}\{g(Y_t) \mid N_t \geq 1\} P\{N_t \geq 1\},$$

for any bounded measurable function g. For t small and $Y_0 = y$, we obtain

$$\mathbb{E}\{g(Y_t)\} = g(y) e^{-t} + \left(\int g(z) p(z) dz \right) t e^{-t} + \mathcal{O}(t^2),$$

where we have used $P\{N_t = 0\} = e^{-t}$, $P\{N_t = 1\} = t e^{-t}$, and $P\{N_t \geq 2\} = \mathcal{O}(t^2)$. We deduce that

$$\frac{\mathbb{E}\{g(Y_t)\} - g(y)}{t} = \left(\int g(z) p(z) dz \right) e^{-t} - \left(\frac{1 - e^{-t}}{t} \right) g(y) + \mathcal{O}(t),$$

which gives by taking the limit $t \downarrow 0$

$$\mathcal{L}g(y) = \int (g(z) - g(y)) p(z) \, dz, \tag{3.17}$$

which we observe to be an integral operator in this case.

Let us first compute $\mathcal{L}^*$: for test functions ϕ and ψ, by simple rearrangement we have

$$\int \psi(y) \mathcal{L} \phi(y) dy = \int \psi(y) \left(\int (\phi(z) - \phi(y)) p(z) dz \right) dy$$

$$= \int \phi(y) \left(p(y) \int \psi(z) dz - \psi(y) \right) dy,$$

which shows that

$$\mathcal{L}^* \psi(y) = p(y) \int \psi(z) dz - \psi(y). \tag{3.18}$$

To find the invariant distribution, we look for a density $\Phi(y)$ such that

$$\mathcal{L}^* \Phi(y) = p(y) - \Phi(y) = 0,$$

where we have used $\int \Phi = 1$. We conclude that $\Phi = p$. As in the previous section, we denote by $\langle \cdot \rangle$ the expectation with respect to this distribution:

$$\langle g \rangle = \int g(z) \Phi(z) dz.$$

Furthermore, reversibility can easily be checked by rearrangement of the expression

$$\langle \psi(\mathscr{L}\phi) \rangle = \int \psi(y) \left(\int (\phi(z) - \phi(y)) p(z) dz \right) p(y) dy$$

to obtain $\langle \phi(\mathscr{L}\psi) \rangle$.

We now consider the process $Y^{(\kappa)}$ with generator $\kappa \mathscr{L}$. Figure 3.5 illustrates the behavior of this process for different values of κ. In this case, the spectral gap can be computed as follows: the eigenvalue problem for $\kappa \mathscr{L}$ is $\kappa \mathscr{L} \psi = \lambda \psi$, which by (3.17) becomes

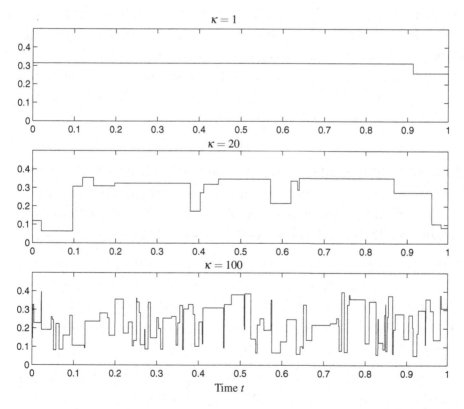

Figure 3.5 Simulated paths of the pure jump process $\sigma_t = Y_t^{(\kappa)}$, where p is the uniform distribution on $[0.05, 0.4]$. The paths top to bottom have increasing rates of mean-reversion κ.

$$\kappa \int \psi(z)p(z)\,dz = (\lambda + \kappa)\psi(y).$$

The left-hand side is constant, therefore either $\lambda = -\kappa$ in which case ψ is centered with respect to p, or else ψ is a constant and $\lambda = 0$. Therefore the spectrum is $\{0, -\kappa\}$ and the spectral gap is κ. In fact, a simple explicit computation, by conditioning on the event the first jump has occurred by time t, shows that

$$\mathbb{E}\{g(Y_t^{(\kappa)}) \mid Y_0^{(\kappa)} = y\} = \langle g \rangle + e^{-\kappa t}(g(y) - \langle g \rangle),$$

for any bounded function g, and the exponential rate is κ. Since we assumed p has a mean, the invariant distribution Φ has a mean, and κ can be viewed as a rate of switching across the mean, or rate of mean-reversion.

3.3.3 Example: The Ornstein–Uhlenbeck Process

We now move to examples of continuous diffusion processes. The simplest among these is the Brownian motion, which has a self-adjoint infinitesimal generator $\mathcal{L} = \mathcal{L}^* = \frac{1}{2}\frac{d^2}{dy^2}$ acting on appropriately regular functions. In this case, there is no invariant distribution as can be seen from trying to find a solution to $\mathcal{L}^*\Pi = 0$, which is a probability measure.

As a first example of a mean-reverting continuous diffusion process, let us now consider (Y_t), an Ornstein–Uhlenbeck process introduced in Section 1.1.6 as the solution of the stochastic differential equation

$$dY_t = (m - Y_t)dt + v\sqrt{2}\,dW_t, \tag{3.19}$$

where (W_t) is a standard Brownian motion. Later, when we scale time by κ to obtain $Y^{(\kappa)}$, we will obtain the general three-parameter OU processes. As we saw in Section 1.1.6, the solution of (3.19) is a Gaussian process which is explicitly given in terms of its starting value $Y_0 = y$ by

$$Y_t = m + (y - m)e^{-t} + v\sqrt{2}\int_0^t e^{-(t-s)}dW_s, \tag{3.20}$$

so that Y_t is $\mathcal{N}(m + (y - m)e^{-t}, v^2(1 - e^{-2t}))$-distributed.

From Section 1.5.1 and formula (1.74), we know that the infinitesimal generator of the Markov process (Y_t) is the differential operator

$$\mathcal{L} = v^2\frac{\partial^2}{\partial y^2} + (m - y)\frac{\partial}{\partial y}. \tag{3.21}$$

We look for an invariant distribution with a density function denoted by $\Phi(y)$, solution of $\mathcal{L}^*\Phi = 0$, where in this case we obtain the adjoint of $\mathcal{L}$ from integration by parts:

$$\mathscr{L}^* = v^2 \frac{\partial^2}{\partial y^2} - \frac{\partial}{\partial y} \left((m-y)\cdot \right). \qquad (3.22)$$

Then, $\Phi(y)$ should satisfy

$$\mathscr{L}^*\Phi = v^2\Phi'' - ((m-y)\Phi)' = 0. \qquad (3.23)$$

Solving this second-order ordinary differential equation with the constraints $\Phi \geq 0$, $\int \Phi = 1$, gives the unique solution which is the density of $\mathscr{N}(m, v^2)$:

$$\Phi(y) = \frac{1}{\sqrt{2\pi v^2}} \exp\left(-\frac{(y-m)^2}{2v^2} \right).$$

As previously, we will denote by $\langle g \rangle$ the average with respect to the invariant density:

$$\langle g \rangle = \int g(z)\Phi(z)\, dz. \qquad (3.24)$$

To establish reversibility of the OU process, we perform the following computation:

$$\langle \psi \mathscr{L}\phi \rangle = v^2 \int \psi(y) \left(\phi''(y) - \frac{(y-m)}{v^2}\phi'(y) \right) \Phi(y)\, dy$$

$$= v^2 \int \psi(y) \frac{d}{dy} \left(\Phi(y)\phi'(y) \right) dy$$

$$= v^2 \int \phi(y) \frac{d}{dy} \left(\Phi(y)\psi'(y) \right) dy$$

$$= \langle \phi \mathscr{L}\psi \rangle,$$

where, in the penultimate step, we have integrated by parts twice and used the fact that the boundary terms are killed by the rapidly decaying Gaussian density Φ. Therefore (3.9) is satisfied.

Before we introduce the time-scale parameter κ as in the previous examples, we compute the spectrum of $\mathscr{L}$. The corresponding eigenvalue problem reads $\mathscr{L}\psi(y) = \lambda \psi(y)$. In order to simplify the computation, it is convenient to make the change of variable $\eta = \frac{y-m}{v}$, which preserves the eigenvalues. The eigenvalue problem for $\Psi(\eta) = \psi(y)$ becomes

$$\Psi'' - \eta\Psi' - \lambda\Psi = 0. \qquad (3.25)$$

For $\lambda = -n$ with $n = 0, 1, 2, \ldots$, it admits the Hermite polynomials H_n as solutions:

$$H_n(\eta) = (-1)^n e^{\eta^2/2} \frac{d^n}{d\eta^n} \left(e^{-\eta^2/2} \right), \qquad (3.26)$$

so that

$$H_0(\eta) = 1, \quad H_1(\eta) = \eta, \quad H_2(\eta) = \eta^2 - 1, \ldots$$

This can easily be checked by first noticing that the functions H_n defined by (3.26) satisfy the relations

$$H_n(\eta) = \eta H_{n-1}(\eta) - H'_{n-1}(\eta), \tag{3.27}$$

and, second, showing by induction that these functions satisfy the differential equations (3.25). We remark that the polynomials (3.26) correspond to the so-called "probabilists' Hermite polynomials" as opposed to the "physicists' Hermite polynomials": $(-1)^n e^{\eta^2} \frac{d^n}{d\eta^n} \left(e^{-\eta^2} \right)$.

The family $(H_n)_{n=0,1,\ldots}$ forms an orthogonal basis of $L^2(\frac{1}{\sqrt{2\pi}} e^{-\eta^2/2})$:

$$\int_{-\infty}^{\infty} H_m(\eta) H_n(\eta) \frac{1}{\sqrt{2\pi}} e^{-\eta^2/2} d\eta = n! \delta_{nm}. \tag{3.28}$$

The orthogonality follows easily from (3.27):

$$(n - m) \int_{-\infty}^{\infty} H_n(\eta) H_m(\eta) e^{-\eta^2/2} d\eta$$

$$= \int (H''_m(\eta) - \eta H'_m(\eta)) H_n(\eta) e^{-\eta^2/2} d\eta$$

$$- \int (H''_n(\eta) - \eta H'_n(\eta)) H_m(\eta) e^{-\eta^2/2} d\eta = 0,$$

by integrations by parts. The $L^2(\frac{1}{\sqrt{2\pi}} e^{-\eta^2/2})$-norm of H_n can be computed as $n!$, by induction and repeated integration by parts. Therefore the orthonormal eigenfunctions of $\mathcal{L}$ in (3.21) are

$$\psi_n(y) = \frac{1}{\sqrt{n!}} H_n \left(\frac{y - m}{v} \right).$$

We conclude that the complete spectrum of $\mathcal{L}$ is given by $\{0, -1, -2, \ldots\}$ with the positive spectral gap of 1, ensuring exponential decorrelation of the Ornstein–Uhlenbeck process.

The process $Y^{(\kappa)}$, with time-scale parameter $\kappa > 0$, is defined by its infinitesimal generator

$$\mathcal{L}^{(\kappa)} = \kappa \mathcal{L} = \kappa v^2 \frac{\partial^2}{\partial y^2} + \kappa (m - y) \frac{\partial}{\partial y}.$$

In distribution, this process is also the solution of the stochastic differential equation

$$dY_t^{(\kappa)} = \kappa (m - Y_t^{(\kappa)}) dt + v \sqrt{2\kappa} \, dW_t, \tag{3.29}$$

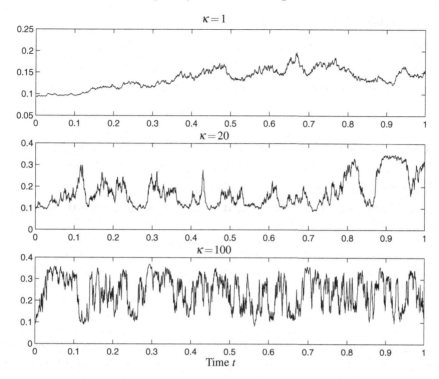

Figure 3.6 Simulated paths of $\sigma_t = f(Y_t^{(\kappa)})$, with $(Y_t^{(\kappa)})$ a mean-reverting OU process defined by (3.29) and $f(y) = 0.35(\arctan y + \pi/2)/\pi + 0.05$, chosen so that $\sigma_t \in (0.05, 0.4)$.

where we note that the drift scales by κ, while the diffusion coefficient scales by $\sqrt{\kappa}$. The invariant distribution of $Y^{(\kappa)}$ is $\mathcal{N}(m, v^2)$, which is independent of κ. Since the spectral gap is κ, its rate of mean-reversion to the mean m is κ.

We illustrate with simulations in Figure 3.6 the effect of κ. Remarkably, the dissimilar paths shown in Figure 3.6 are each from the *same* process with different mean-reversion rates.

Finally, we consider the Poisson equation $\mathcal{L}\phi = g$. For g centered with respect to the invariant distribution Φ, meaning $\langle g \rangle = 0$, we have from the eigenfunction representation (3.14) of the solutions of the Poisson equation,

$$\phi(y) = -\sum_{n \geq 1} \frac{c_n}{n\sqrt{n!}} H_n\left(\frac{y-m}{v}\right) + c,$$

for any constant c, and c_n are the coefficients in the eigenfunction expansion (3.10) of g.

3.3.4 Example: The Cox–Ingersoll–Ross Process

We present another diffusion process example, a CIR process, which is often used in financial models of volatility, interest rates, and default intensities. It is described by the stochastic differential equation

$$dY_t = (m - Y_t)\,dt + v\sqrt{2Y_t}\,dW_t, \tag{3.30}$$

where W is a standard Brownian motion. Later, when we scale time by κ to obtain $Y^{(\kappa)}$, we will obtain the general three-parameter CIR processes.

The main difference with the OU process is the $\sqrt{Y_t}$ term in the diffusion coefficient. As is well known, if the process starts positive, then the condition

$$v^2 \leq m \tag{3.31}$$

ensures that it remains positive at all time. To see this, we remark that

$$\mathscr{L}S \equiv (m - y)S' + v^2 y S'' = 0,$$

for the choice $S(y) = \int_1^y e^{z/v^2} z^{-m/v^2}\,dz$. Let y denote the starting point of Y. Then, for $0 < \varepsilon < y < M$, we stop the process at the first time it hits M or ε by defining the stopping time $\tau_{\varepsilon,M}^y$:

$$\tau_{\varepsilon,M}^y = \tau_\varepsilon^y \wedge \tau_M^y, \quad \tau_M^y = \inf\{t > 0, Y_t = M\}, \quad \tau_\varepsilon^y = \inf\{t > 0, Y_t = \varepsilon\},$$

and by applying Itô's formula, we deduce that for any $t > 0$ we have

$$S\left(Y_{t \wedge \tau_{\varepsilon,M}^y}\right) = S(y) + \int_0^{t \wedge \tau_{\varepsilon,M}^y} S'(Y_u)\,v\sqrt{2Y_u}\,dW_u. \tag{3.32}$$

Taking the variance on both sides, a limit as $t \to \infty$, and using the fact that S' is bounded from below on $[\varepsilon, M]$, we deduce that $\mathbb{E}\left(\tau_{\varepsilon,M}^y\right) < \infty$, and therefore $\tau_{\varepsilon,M}^y < \infty$ almost surely. From the martingale relation (3.32), we deduce that

$$S(y) = \mathbb{E}\left\{S\left(Y_{\tau_{\varepsilon,M}^y}\right)\right\} = S(\varepsilon)\mathbb{P}\left(\tau_\varepsilon^y < \tau_M^y\right) + S(M)\mathbb{P}\left(\tau_\varepsilon^y > \tau_M^y\right).$$

Under the condition (3.31), we have $\lim_{\varepsilon \to 0} S(\varepsilon) = -\infty$, and therefore that $\mathbb{P}\left(\tau_0^y < \tau_M^y\right) = 0$ for all $M > y$, which implies $\mathbb{P}\left(\tau_0^y < \infty\right) = 0$, the desired property: the process Y_t stays strictly positive for all time. (Note that in the case $m < v^2$, one concludes that $\mathbb{P}\left(\tau_0^y < \infty\right) = 1$ so that, almost surely, the process hits zero in a finite time.)

Therefore, by assuming its diffusion coefficient v is not too large compared to the mean level m, Y is positive, which volatilities, interest rates, and default intensities should be. The distribution of the CIR process, as well as its time integral, can be characterized by Laplace transforms. This will be crucial in deriving explicit formulas for bond pricing when using a CIR model for short rates (see Chapter 12), and default probabilities when using a CIR model for default intensities (see Chapter 14).

The infinitesimal generator of the CIR process (3.30) is given by

$$\mathscr{L} = v^2 y \frac{\partial^2}{\partial y^2} + (m - y) \frac{\partial}{\partial y}. \tag{3.33}$$

We look for an invariant distribution Φ, which will be a non-negative solution of the adjoint equation

$$\mathscr{L}^* \Phi = 0,$$

that is

$$v^2 (y\Phi)'' - ((m - y)\Phi)' = 0,$$

on $y > 0$, with the constraint $\int_0^\infty \Phi(y)\, dy = 1$. The unique solution is the Gamma distribution

$$\Phi(y) = \frac{1}{v^{2\theta}\Gamma(\theta)} y^{\theta-1} e^{-(y/v^2)} \mathbf{1}_{(0,\infty)}(y), \tag{3.34}$$

with $\theta = \frac{m}{v^2} \geq 1$, and where Γ denotes the usual Gamma function

$$\Gamma(\theta) = \int_0^\infty y^{\theta-1} e^{-y}\, dy.$$

To establish reversibility of the CIR process, we perform the following computation:

$$\langle \psi \mathscr{L} \phi \rangle = v^2 \int \psi(y) \left(y\phi''(y) - \frac{(y-m)}{v^2}\phi'(y) \right) \Phi(y)\, dy$$

$$= v^2 \int \psi(y) \frac{d}{dy} \left(\Phi(y) y\phi'(y) \right) dy$$

$$= v^2 \int \phi(y) \frac{d}{dy} \left(\Phi(y) y\psi'(y) \right) dy$$

$$= \langle \phi \mathscr{L} \psi \rangle,$$

where, in the penultimate step, we have integrated by parts twice and used that the boundary terms are zero. Therefore (3.9) is satisfied.

We now turn to the study of the spectrum of the CIR process. The eigenvalue problem is $\mathscr{L}\psi(y) = \lambda \psi(y)$. We first make the change of variable $z = y/v^2$, $\psi(y) = \Psi(z)$, which leads to the eigenvalue problem

$$z\Psi'' + (\theta - z)\Psi' - \lambda \Psi = 0.$$

For $\lambda = -n$ with $n = 0, 1, 2, \ldots$, these are the Laguerre differential equations:

$$zL_n'' + (\theta - z)L_n' + nL_n = 0. \tag{3.35}$$

It can be shown that they admit the Laguerre polynomials $L_n^{(\theta)}$ as solutions:

$$L_n^{(\theta)}(z) = \frac{z^{1-\theta}e^z}{n!}\frac{d^n}{dz^n}\left(z^{n+\theta-1}e^{-z}\right), \tag{3.36}$$

so that

$$L_0^{(\theta)}(z) = 1, \quad L_1^{(\theta)}(z) = \theta - z, \ldots$$

The family $\left(L_n^{(\theta)}\right)_{n=0,1,\ldots}$ forms an orthogonal basis of $L^2(\Phi_0^{(\theta)})$, where $\Phi_0^{(\theta)}(z)$ is the invariant distribution (3.34) in the case $v^2 = 1$:

$$\Phi_0^{(\theta)}(z) = \frac{1}{\Gamma(\theta)}z^{\theta-1}e^{-z}, \quad z > 0,$$

and we have

$$\int_0^\infty L_m^{(\theta)}(z)L_n^{(\theta)}(z)\Phi_0^{(\theta)}(z)\,dz = \frac{\Gamma(n+\theta)}{n!\Gamma(\theta)}\delta_{nm}. \tag{3.37}$$

Therefore, the eigenfunctions are

$$\psi_n(y) = \sqrt{\frac{n!\Gamma(\theta)}{\Gamma(n+\theta)}}L_n^{(\theta)}\left(\frac{y}{v^2}\right),$$

which form an orthonormal basis of $L^2(\Phi)$. So, as in the case of the Ornstein–Uhlenbeck process, we have a spectral gap $\lambda = 1$, and the solutions of the Poisson equation $\mathscr{L}\phi = g$ are given by (3.14).

Figure 3.7 illustrates with simulations the effect of the time-scaling parameter κ in the CIR process.

3.3.5 Other One-Dimensional Diffusion Processes

In general, it is not easy to give explicit conditions on the coefficients of a general one-dimensional diffusion

$$dY_t = \alpha(Y_t)\,dt + \beta(Y_t)\,dW_t$$

which ensure that the process has a unique invariant distribution, is ergodic, and has a positive spectral gap. We refer to Karlin and Taylor (1981) for conditions in terms of the so-called scale and speed measures. It is possible to

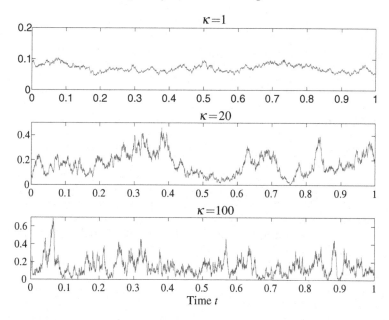

Figure 3.7 Simulated paths of $\sigma_t = Y_t$, with (Y_t) a mean-reverting CIR process defined by (3.30).

give conditions on the coefficients to establish reversibility, and, therefore, a spectral representation, which may have a continuous part, unlike in the OU and CIR examples where the spectrum is discrete. In Chapter 4, we will assume that the process Y driving the fast volatility factor has a spectral gap. Additional properties related to the solutions of Poisson equations will be needed, and will be detailed in Theorem 4.10.

In particular, it will be convenient to work with processes such that the solution $\phi(y)$ of the Poisson equation $\mathcal{L}\phi = g$ is a polynomial for polynomial functions g. This is indeed the case for the OU and CIR processes presented in the previous sections, since in these cases the spectral decomposition of g is finite, and therefore ϕ is also a finite linear combination of polynomials (Hermite for OU and Laguerre for CIR).

In order to be able to use other functions g, one needs to look at specific processes Y. For instance, the following results will be useful.

Lemma 3.1 *Let Y be an Ornstein–Uhlenbeck process with infinitesimal generator $\mathcal{L}$ given by (3.21). Assume that the function g is centered with respect to the invariant distribution of Y, that is $\langle g \rangle = 0$, and g is at most polynomially growing at $\pm\infty$, that is $|g(y)| \leq C_1(1+|y|^n)$ for some constant C_1 and integer n. Then, the Poisson equation $\mathcal{L}\phi = g$ has a solution that is at most polynomially growing if $n \geq 1$:*

$$|\phi(y)| \le C_2(1 + |y|^n),$$

and at most logarithmically growing if $n = 0$:

$$|\phi(y)| \le C_2(1 + \log(1 + |y|)),$$

for some constant C_2.

Proof The Poisson equation can be written

$$\mathscr{L}\phi = v^2\phi'' + (m - y)\phi' = \frac{v^2}{\Phi}(\Phi\phi')' = g,$$

where Φ denotes the density of the invariant distribution $\mathscr{N}(m, v^2)$. Therefore, we deduce

$$\phi'(y) = \frac{1}{v^2\Phi(y)}\int_{-\infty}^{y}g(z)\Phi(z)dz = \frac{-1}{v^2\Phi(y)}\int_{y}^{\infty}g(z)\Phi(z)dz,$$

where we have used $\langle g \rangle = 0$ for the second equality. Without loss of generality, by a simple change of variable, we now assume $m = 0$ and $v = 1$.
   For $y \to +\infty$ and $n \ge 2$, we obtain by integration by parts:

$$\left|\frac{1}{\Phi(y)}\int_{-\infty}^{y}g(z)\Phi(z)dz\right| \le \frac{C}{\Phi(y)}\int_{y}^{\infty}z^n\Phi(z)dz$$

$$= Cy^{n-1} + \frac{C(n-1)}{\Phi(y)}\int_{y}^{\infty}z^{n-2}\Phi(z)dz,$$

for some constant C, using the bound on g. Using

$$\int_{y}^{\infty}z\Phi(z)dz = \Phi(y),$$

and the classical bounds for $\int_{y}^{\infty}\Phi(z)dz$ at $y > 0$,

$$(1 + \frac{1}{y^2})^{-1}\frac{\Phi(y)}{y} = (1 + \frac{1}{y^2})^{-1}\int_{y}^{\infty}(1 + \frac{1}{z^2})\phi(z)dz$$

$$< \int_{y}^{\infty}\Phi(z)dz < \frac{1}{y}\int_{y}^{\infty}z\Phi(z)dz = \frac{\Phi(y)}{y},$$

one deduces that

$$|\phi'(y)| \sim Cy^{n-1}, \quad \text{at} \quad y \to +\infty,$$

for some constant C. A similar estimate holds for $y \to -\infty$. Integrating with respect to y gives the desired bound on ϕ. $\square$

Lemma 3.2 *Let Y be a Cox–Ingersoll–Ross process with infinitesimal generator $\mathcal{L}$ given by (3.33). Assume that the function g is centered with respect to the invariant distribution of Y, that is $\langle g \rangle = 0$, and g is at most polynomially growing at $+\infty$, that is $|g(y)| \le C_1(1+y^n)$ for some constant C_1 and integer n. Then, the solutions of the Poisson equation $\mathcal{L}\phi = g$ can be chosen at most polynomially growing if $n \ge 1$:*

$$|\phi(y)| \le C_2(1+y^n),$$

and at most logarithmically growing if $n = 0$:

$$|\phi(y)| \le C_2(1+\log(1+y)),$$

for some constant C_2.

Proof The Poisson equation can be written

$$\mathcal{L}\phi = v^2 y \phi'' + (m-y)\phi' = \frac{v^2}{\Phi}\left(y\Phi\phi'\right)' = g,$$

where Φ denotes density of the invariant distribution given by (3.34), that is the Gamma distribution on $(0,\infty)$

$$\Phi(y) = \frac{1}{v^{2\theta}\Gamma(\theta)} y^{\theta-1} e^{-(y/v^2)}, \quad \theta = \frac{m}{v^2} \ge 1.$$

Therefore, we deduce

$$\phi'(y) = \frac{1}{v^2 y \Phi(y)} \int_0^y g(z)\Phi(z)dz = \frac{-1}{v^2 y \Phi(y)} \int_y^\infty g(z)\Phi(z)dz,$$

where we have used $\langle g \rangle = 0$ for the second equality. Without loss of generality, by a simple change of variable, we now assume $v^2 = 1$. Using the bound on g as $y \to \infty$ and successive integration by parts, one obtains for constants varying from line to line:

$$\left| \frac{1}{y\Phi(y)} \int_0^y g(z)\Phi(z)\,dz \right|$$
$$\le \frac{C}{y\Phi(y)} \int_y^\infty z^n \Phi(z)\,dz$$
$$= \frac{C}{y\Phi(y)} \left(y^n\Phi(y) + C_1 \int_y^\infty z^{n+\theta-2}e^{-z}\,dz \right)$$
$$\frac{C}{y\Phi(y)} \left(y^n\Phi(y) + C_1 y^{n-1}\Phi(y) + \cdots + C_{k-1}\int_y^\infty z^{n+\theta-k}e^{-z}\,dz \right).$$

For k such that $-1 < n + \theta - k \le 0$ and $y \ge 1$, one then uses

$$\int_y^\infty z^{n+\theta-k} e^{-z} dz \le y^{n+\theta-k} e^{-y},$$

to deduce that

$$|\phi'(y)| \sim C y^{n-1}, \quad \text{as} \quad y \to \infty,$$

for some constant C. Integrating with respect to y gives the desired bound for $\phi(y)$ at $+\infty$.   □

For an OU process, the exponential functions e^{ky} lead to explicit formulas in the asymptotic theory presented in Chapter 4. This example will be important for the "expOU" stochastic volatility model. Note that in this case

$$\langle e^{ky} \rangle = \int e^{ky} \frac{1}{\sqrt{2\pi v^2}} e^{-\frac{(y-m)^2}{2v^2}} = e^{km + \frac{k^2 v^2}{2}}.$$

The situation is different for a CIR process since in that case exponential moments exist only for $k < 1/v^2$.

3.4 Time Scales in Synthetic Returns Data

In reality, we do not observe the volatility process. The actual observable is the stock price and from it we can compute the returns. We demonstrate from simulations that fast mean-reversion can be identified qualitatively in the returns time series, but because κ is the rate of mean-reversion of the *hidden* volatility process, it is very difficult to estimate it precisely from the simulated prices.

3.4.1 The Returns Time Series

If we rewrite the stochastic volatility model (2.8) as

$$\frac{dX_t}{X_t} - \mu \, dt = \sigma_t \, dW_t,$$

we interpret the left-hand side as the de-meaned **returns** which, from the right-hand side, contains the pure fluctuation part of the price.

For simulation, we use a discrete-time version of the stochastic volatility model (2.8): the times are $\{t_n = n\Delta t, n = 0, \dots, N\}$, where $\Delta t = T/N$ is the time spacing and we write the values of the stock and volatility at time t_n

as just (X_n, σ_n). Assuming μ is known, we define the discrete normalized returns process $\{D_n\}_{n=1}^{N}$ by

$$D_n = \frac{1}{\sqrt{\Delta t}} \left(\frac{X_n - X_{n-1}}{X_{n-1}} - \mu \Delta t \right).$$

The Euler simulation of (2.8) is

$$D_n = \sigma_{n-1} \varepsilon_n, \tag{3.38}$$

where $\{\sigma_n\}_{n=0}^{N}$ is obtained from the discrete version of our stochastic volatility models, and $\{\varepsilon_n\}_{n=0}^{N}$ is a sequence of i.i.d. (independent and identically distributed) $\mathcal{N}(0,1)$ random variables.

We present typical returns paths for a pure jump volatility process and for a continuous volatility process. For the first case we use our second model of Section 3.3.2 and for the second case the OU model of Section 3.3.3. The first example in Section 3.3.1, where volatility is a two-state Markov chain, is essentially equivalent to the jump process example, and the fourth example, the CIR process, is qualitatively the same as the OU-based model.

3.4.2 Returns Process with Jump Volatility

We denote by $\{\sigma_n\}_{n=0}^{N}$ the values $\{\sigma_{t_n}\}$ of the simulated jump process described in Section 3.3.2, sampled at the discrete times $\{t_n\}$. We do this for a mean-reversion rate $\kappa = 1$ and a large rate $\kappa = 100$. The initial value for the volatility is chosen in both cases to be the mean volatility $\langle f \rangle$ under the invariant distribution. Since in the case $\kappa = 1$, the volatility process reverts slowly to its mean, this choice has the effect of "freezing" the typical size of the fluctuations of the returns for our time of observation (a few months). In the other large κ case, we know from the ergodic property described in Section 3.3.2 that the volatility process forgets about its starting value. Our choice of starting value is convenient to compare visually the two cases, but a different choice of initial value would not change the qualitative difference in the structure of the fluctuations between the two cases. We generate the returns process from (3.38) using, in this case, a sequence $\{\varepsilon_n\}$ independent of the volatility process. In other words, we deal with the uncorrelated case here.

The first column of Figure 3.8 shows a volatility path with $\kappa = 1$ in the top picture, and the corresponding normalized returns path $\{D_n\}$ below it. The second column is the same for $\kappa = 100$. Looking at the bottom pictures *only*, we observe that the change in the *size of the fluctuations*

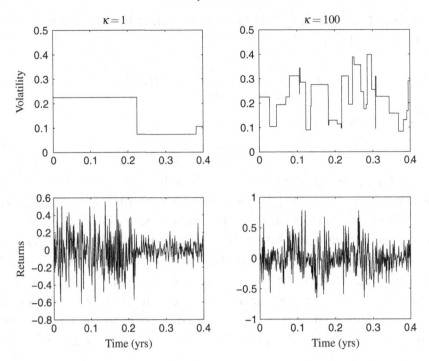

Figure 3.8   Simulated volatility and corresponding returns paths for small and large rates of mean-reversion for the jump volatility model of Section 3.3.2.

is slow on the left. It is highly fluctuating up to about $t = 0.2$ and then exhibits smaller fluctuations for the rest of the time shown. On the right, the *size of the fluctuations* is changing a lot over the same period of time. Since the size of the fluctuations is what we call volatility, fast mean-reversion in volatility can be detected visually in the returns process, but extracting the corresponding mean-reversion rate quantitatively from the returns data is very difficult. However, such an estimate is not needed for the derivatives valuation problems we study in this book with asymptotic approximations.

3.4.3 Returns Process with OU Volatility

For a volatility process driven by (Y_t), a mean-reverting OU process, as described in Section 3.3.3, we incorporate a correlation between volatility and stock price shocks. This is simply done by discretizing the stochastic differential equation (3.19), writing the correlated Brownian motion $(\hat{Z}_t)$ as the sum of two independent Brownians as in equation (2.10). We generate $\{Y_n\}_{n=0}^{N}$ by

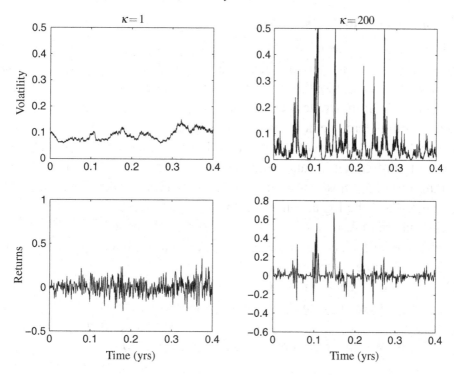

Figure 3.9 Simulated volatility and corresponding returns paths for small and large rates of mean-reversion for the OU model of Section 3.3.3, with the choice $f(y) = e^y$.

$$Y_n = Y_{n-1} + \kappa(m - Y_{n-1})\Delta t + v\sqrt{2\kappa\Delta t}\left(\rho\varepsilon_n + \sqrt{1 - \rho^2}\,\eta_n\right),$$

where $\{\varepsilon_n\}$ and $\{\eta_n\}$ are independent sequences of i.i.d. $\mathcal{N}(0,1)$ random variables. We choose $f(y) = e^y$ and obtain $\sigma_n = f(Y_n)$. The returns are generated from (3.38) using the same sequence $\{\varepsilon_n\}$. To model a negative leverage or skew effect, we choose $\rho = -0.2$. Fixing the mean level m and variance v^2 of the invariant distribution of the OU process driving the volatility leaves us to choose the rate of mean-reversion κ. We again consider a low value $\kappa = 1$ and a high value $\kappa = 200$. Typical paths are shown in Figure 3.9, with $\sigma_0 = \langle f \rangle$ for visual convenience as explained in the previous section.

Again, we observe that the *size of the fluctuations* in the returns process of the bottom left picture is relatively constant over time, while for the large κ picture on the bottom right, this size is changing a lot. This is what we expect qualitatively when we have fast mean-reverting stochastic volatility.

3.5 Time Scales in Market Data

Estimating rates of mean-reversion of the S&P 500 volatility means that we want to identify characteristic time scales in the data. Long time scales (on the order of several months) have been studied extensively (see Engle and Patton (2001) in the context of the Dow Jones index, and other references cited therein). Here, we demonstrate the presence of a well-identified *short* time scale (on the order of a few days). In order to identify this scale, we analyze high-frequency data using the empirical structure function, or variogram, of the log absolute returns (Section 3.5.2). We also illustrate with the spectrum of the log absolute returns in Section 3.5.5.

The material presented in this section is from Fouque *et al.* (2003b), to which we refer for details about the data and statistical analysis, as well as references to other empirical studies.

We are interested in the short-run behavior of the S&P 500. Thus, we look at high-frequency, intraday data so that we can resolve scales on the order of days. At this frequency, a sample corresponding to one year of data is sufficient. We illustrate here with data from 1999–2000.

3.5.1 Short-Term Volatility Statistics of the S&P 500

We introduce the *normalized fluctuations* of the data

$$\bar{D}_n = \frac{2(\bar{X}_n - \bar{X}_{n-1})}{\sqrt{\Delta t}(\bar{X}_n + \bar{X}_{n-1})}. \tag{3.39}$$

Here, $\{\bar{X}_n\}$ are the 5-minute averaged S&P 500 data and Δt is the 5-minute time increment in annualized units.

In the framework of the general class of stochastic volatility models

$$dX_t = \mu X_t dt + \sigma_t X_t dW_t^{(0)}, \tag{3.40}$$

as in Section 3.4.1, the continuous analog of the normalized fluctuations can be written formally as

$$\frac{1}{\sqrt{\Delta t}}\left(\frac{\Delta X_t}{X_t} - \mu \Delta t\right) = \sigma_t \frac{\Delta W_t}{\sqrt{\Delta t}}, \tag{3.41}$$

where σ_t, $t \geq 0$ is the (positive) volatility process, μ is a constant, and ΔW_t is the increment of a Brownian motion. The subtraction of the incremental mean return $\mu \Delta t$ is omitted in the discrete normalized fluctuations $\{\bar{D}_n\}$ because it is negligibly small. Based on (3.41) we model the normalized fluctuation process by

$$\bar{D}_n = \sigma_n \varepsilon_n, \tag{3.42}$$

where $\{\varepsilon_n\}$ is a sequence of i.i.d. Gaussian random variables with mean zero and variance one and $\{\sigma_n\}$ is the volatility at time t_n.

We analyze the log absolute value of the normalized fluctuations

$$F_n = \log|\bar{D}_n| = \log\sigma_n + \log|\varepsilon_n|, \qquad (3.43)$$

which is the sum of the discrete log-volatility process and a discrete white noise process (we do not show here the small regularization used in this data-processing step).

3.5.2 Variogram Analysis

We begin with a study of the *empirical structure function* or *variogram* of F_n:

$$V_j^N = \frac{1}{N}\sum_n (F_{n+j} - F_n)^2, \qquad (3.44)$$

where j is the lag and N is the total number of points.

We show in Figure 3.10 the variogram (dashed line) of the log fluctuations $F_n = \log(|\bar{D}_n|)$. The first observation is the clear periodic component. This is due to the systematic intraday effect, which we discuss and model in Section 3.5.4. If, after removal of this component, the variogram were flat, we would conclude there is no short time scale on the order of days. To gauge the extent of curvature, we fit the variogram to

$$V_j^N \approx 2\gamma^2 + 2v^2(1 - e^{-\kappa j\Delta t}), \qquad (3.45)$$

a convenient three-parameter family, where the vertical intercept $2\gamma^2$ gives twice the variance γ^2 of the noise $\log|\varepsilon_n|$ and $1/\kappa$ quantifies the characteristic time scale by analogy with exponential decorrelation in Markov processes presented in the previous sections.

The expression (3.45) would be exact (in the limit $N \to \infty$), if σ_t was the exponential of an Ornstein–Uhlenbeck process (Y_t), with κ being the rate of mean-reversion and v^2 the variance of its invariant distribution:

$$\sigma_t = e^{Y_t}; \qquad dY_t = \kappa(m - Y_t)\,dt + v\sqrt{2\kappa}\,dW_t^{(1)}. \qquad (3.46)$$

This follows from

$$\text{covariance}(Y_0, Y_t) = v^2 e^{-\kappa t}$$

under the invariant distribution $\mathcal{N}(m, v^2)$.

With this parameterization, v represents the typical size of the fluctuation of Y, and consequently of volatility. Empirically, this is neither extremely

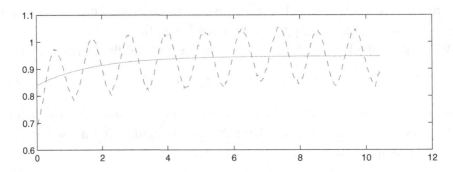

Figure 3.10   The empirical variogram for the S&P 500 (dashed line) for
the 1999–2000 data. The solid line is the exponential fit from which the
rate of mean-reversion can be obtained.

small nor extremely large. The quantity $1/\kappa$ represents the intrinsic time
scale of the factor Y.

For simplicity, we have assumed the Brownian motions $W^{(0)}$ and $W^{(1)}$
are independent (hence (ε_n) and (σ_n) are also independent). We point out
the following.

Intraday variations.   Note in Figure 3.10 the periodic component corresponding to day effects in the data that is not in the model (3.42) but will be
incorporated in Section 3.5.4. The solid line is a fitted exponential obtained
by nonlinear least squares regression.

Result of the fit.   The estimated mean-reversion time, $1/\kappa$ in equation
(3.45), is 1.7 days, which identifies the presence of a short time scale in
S&P 500 volatility. We refer to Fouque *et al.* (2003b) for further details
about the statistics of this estimator.

3.5.3 *Effect of Longer Scales*

A longer sample of data with the variogram extended to larger lags would
reveal the presence of a longer characteristic time scale in the S&P 500, as
documented in numerous studies. This is on the scale of months or longer,
well-separated from the short scale we have identified here. It is natural
to ask how the presence of this scale would affect our empirical analysis. In fact it does not, as we now explain. Let us consider a model with
two components driving the volatility, one fast and one slow. This is a special case of the three-component model analyzed by LeBaron (2001), who
shows that a three-factor multiscale stochastic volatility model can lead to
an apparent power-law variogram at medium lag lengths. Chernov *et al.*

(2003) also suggest a two-factor stochastic volatility model with one factor mean-reverting on a short scale and the other on a longer scale.

We consider a stochastic volatility model of the form

$$\sigma_t = e^{Y_t} e^{Z_t},$$

where (Y_t) and (Z_t) are independent OU processes. The first (Y_t) has characteristic mean-reversion time $1/\kappa$ on the order of days, while the second (Z_t) mean-reverts on the order of months. We denote the rate of mean-reversion and the variance of the equilibrium distribution of (Z_t) by κ_Z and v_Z^2, respectively. The variogram formula (3.45) becomes

$$V_j^N \approx 2\gamma^2 + 2v^2(1 - e^{-\kappa j\Delta t}) + 2v_Z^2(1 - e^{-\kappa_Z j\Delta t}).$$

For the range of lags we are looking at, that is $j\Delta t$ up to a week, $\kappa_Z j\Delta t$ is small and the last term is negligible. Hence the long scale plays no role in our previous variogram estimation. At the opposite extreme, observe that if we had looked at a longer data sample and included larger lags, then the first term would contribute only a constant $(2v^2)$ except at the very short lags.

3.5.4 The Day Effect

The day effect is put into a one-factor stochastic volatility model by replacing (3.42) with

$$\bar{D}_n = f(Y_n)g_n\varepsilon_n, \tag{3.47}$$

where we allow for more general functions f than the exponential. Here Y_n is the discrete OU process, ε_n a unit white noise process, and g_n a deterministic periodic function that models the systematic intraday variations in the volatility. The estimate of g_n from the S&P 500 data is shown in Figure 3.11. In fact, by looking at S&P 500 data for other years, we find that the intraday envelope is stable. This intraday effect will be modeled and analyzed by an averaging argument in Section 9.5.

3.5.5 Spectrum

Instead of using the variogram or structure function, we can go into the frequency domain and analyze the empirical spectrum of the log fluctuations of $\bar{X}_n$ given by (3.43). The analog of (3.45) for the spectrum is the Lorentzian model

$$\Gamma_\omega^N \approx 2\gamma^2 + \frac{2v^2}{\pi} \frac{\kappa}{\kappa^2 + \omega^2}, \tag{3.48}$$

Volatility Time Scales

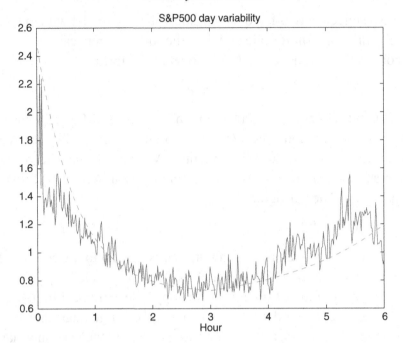

Figure 3.11 Intraday variability in the 1999–2000 data. The dashed line is a four-parameter model envelope.

ignoring again the correlation between the noise $\log(|\varepsilon_n|)$ and the OU process Y_n as well as the day effect that is just an additive impulse function on the right at the frequency of one day.

In Figure 3.12, we show the spectrum for the S&P 500 data. The impulse at the frequency of one day corresponds to the periodic day effect. The fitted Lorentzian model is represented by the smooth solid line and identifies the rate of mean-reversion κ.

3.6 Multiscale Models

Based on data analysis, we conclude that at least two volatility time scales should be present in our stochastic volatility models. For the majority of the book, we will work with two-factor stochastic volatility models driven by a fast mean-reverting diffusion and a slowly varying diffusion. Doing so, we model the dynamics of the underlying asset under the physical measure and we then introduce the risk-neutral pricing measure in order to study option prices and the corresponding implied volatility surface. The spot volatility is given as a function of two factors

$$\sigma_t = f(Y_t, Z_t),$$

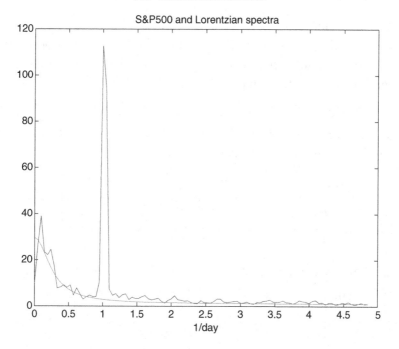

Figure 3.12 Spectrum of $\log|\bar{D}_n|$ for the 1999–2000 S&P 500 data containing the one-day spectral peak. The Lorentzian model (omitting the day effect) is shown by the smooth curve.

with Y a fast factor and Z a slow factor. We will write Y and Z as generic diffusion processes and introduce two small parameters to describe their time scales of fluctuation.

• The process Y_t is a fast mean-reverting process as discussed in Section 3.2, obtained by speeding up an ergodic diffusion:

$$dY_t = \frac{1}{\varepsilon}\alpha(Y_t)\,dt + \frac{1}{\sqrt{\varepsilon}}\beta(Y_t)\,dW_t^{(1)},$$

where κ is replaced from now on by ε^{-1}, with $\varepsilon > 0$ a small parameter representing the short time scale of the process Y_t. Here, $W_t^{(1)}$ is a standard Brownian motion possibly correlated with the Brownian motion driving the underlying asset. Therefore, the fast fluctuation of Y is described by the large drift coefficient $1/\varepsilon$ and the large diffusion coefficient $1/\sqrt{\varepsilon}$.

• The process Z_t is slowly varying, obtained by slowing down a diffusion process with drift and diffusion coefficients denoted, respectively, by $c(z)$ and $g(z)$:

$$dZ_t = \delta c(Z_t)\,dt + \sqrt{\delta}\,g(Z_t)\,dW_t^{(2)},$$

where δ is a small time-scale parameter and $W^{(2)}$ is a standard Brownian motion possibly correlated with $W^{(1)}$ and the Brownian motion driving the underlying asset. Therefore, the slow variation of Z is described by the small drift and diffusion coefficients δ and $\sqrt{\delta}$.

The models we will analyze for derivative pricing by asymptotic approximations in Chapter 4 are the following two-factor stochastic volatility models, written here under the real-world measure $\mathbb{P}$:

$$
\left.
\begin{aligned}
dX_t &= \mu(Y_t,Z_t)X_t\,dt + f(Y_t,Z_t)X_t\,dW_t^{(0)}, \\
dY_t &= \tfrac{1}{\varepsilon}\alpha(Y_t)dt + \tfrac{1}{\sqrt{\varepsilon}}\beta(Y_t)dW_t^{(1)}, \\
dZ_t &= \delta c(Z_t)dt + \sqrt{\delta}\,g(Z_t)dW_t^{(2)}.
\end{aligned}
\right\}
\tag{3.49}
$$

The stock price X is driven by the standard Brownian motion $(W_t^{(0)})$. The standard Brownian motions $\left(W_t^{(0)},W_t^{(1)},W_t^{(2)}\right)$ are correlated as follows:

$$
\begin{aligned}
d\langle W^{(0)},W^{(1)}\rangle_t &= \rho_1\,dt, & (3.50) \\
d\langle W^{(0)},W^{(2)}\rangle_t &= \rho_2\,dt, & (3.51) \\
d\langle W^{(1)},W^{(2)}\rangle_t &= \rho_{12}\,dt, & (3.52)
\end{aligned}
$$

where $|\rho_1| < 1$, $|\rho_2| < 1$, $|\rho_{12}| < 1$, and $1 + 2\rho_1\rho_2\rho_{12} - \rho_1^2 - \rho_2^2 - \rho_{12}^2 > 0$, in order to ensure positive definiteness of the covariance matrix of the three Brownian motions. As discussed in Chapter 2, Section 2.5.4, the correlations are associated with the leverage or skew effect, and we will see in the next chapter that they play a crucial role in our expansion for the option prices.

Notes

For more details on stochastic processes such as the Markov chain, jump process, and the OU process used in this chapter, and their ergodic properties, we refer the reader to Karlin and Taylor (1981), Breiman (1992), Yin and Zhang (1998), and Durrett (2004), for instance. Time scaling of stochastic processes has been studied for years in many applications. It often goes with asymptotic results, and we refer to the Notes at the end of Chapter 4 for further references.

The material presented in Sections 3.4 and 3.5 is from Fouque *et al.* (2003b), to which we refer for details about the data and statistical analysis, as well as references to other empirical studies. For multiscale time-series models we refer to Christoffersen *et al.* (2008b) and references therein.

4

First-Order Perturbation Theory

In this chapter, we present the multiscale perturbation analysis in the case of European options. We take as a starting point the class of multiscale stochastic volatility models introduced in Chapter 3. Recall that these models have two stochastic volatility factors, one fast and one slow. In the Markovian framework that we consider here, option prices are obtained as solutions of partial differential equations. The two time scales translate into two small parameters in these equations and we use a combination of singular and regular perturbation techniques to derive approximations for the option prices. Here we consider only the first-order corrections, while the second-order corrections will be derived in Chapter 9. The approximations we derive for European call options are crucial for the calibration procedure that we discuss in the next chapter. There, we use the implied volatility skew to calibrate the few universal parameters which arise in our first-order perturbation theory. These are the only parameters needed for computing approximations of prices of more complicated path-dependent derivatives discussed in detail in the later chapters. The perturbation theory presented here in the context of *equity* markets will also be the basis for our analysis of the *fixed income* (Chapter 12) and *credit* markets (Chapters 13–14).

4.1 Option Pricing under Multiscale Stochastic Volatility

In this section, we recall the multiscale stochastic volatility models introduced in the previous chapters. Our objective in this chapter is to price European derivatives and therefore we consider these models under a *risk-neutral* pricing probability measure $\mathbb{P}^\star$ as discussed in Section 2.5.4. We then write the pricing partial differential equations to highlight the roles of the different time scales.

4.1.1 Models under Risk-Neutral Measures

We start with the class of two-factor multiscale stochastic volatility models
(3.49) introduced under the real-world measure $\mathbb{P}$ in Section 3.6. Following
the change of measure procedure described in Section 2.5.4, the evolution
of the price X of the underlying asset under the risk-neutral pricing measure
$\mathbb{P}^\star$ is given by the system of stochastic differential equations:

$$
\left.
\begin{aligned}
dX_t &= rX_t\,dt + f(Y_t,Z_t)X_t\,dW_t^{(0)\star}, \\
dY_t &= \left(\tfrac{1}{\varepsilon}\alpha(Y_t) - \tfrac{1}{\sqrt{\varepsilon}}\beta(Y_t)\Lambda_1(Y_t,Z_t)\right)dt + \tfrac{1}{\sqrt{\varepsilon}}\beta(Y_t)\,dW_t^{(1)\star}, \\
dZ_t &= \left(\delta c(Z_t) - \sqrt{\delta}\,g(Z_t)\Lambda_2(Y_t,Z_t)\right)dt + \sqrt{\delta}\,g(Z_t)\,dW_t^{(2)\star}.
\end{aligned}
\right\}
\quad (4.1)
$$

The main aspects of these models are:

- The $\mathbb{P}^\star$-standard Brownian motions $\left(W_t^{(0)\star},W_t^{(1)\star},W_t^{(2)\star}\right)$ are correlated
  as follows:

$$
\begin{aligned}
d\langle W^{(0)\star},W^{(1)\star}\rangle_t &= \rho_1\,dt, \\
d\langle W^{(0)\star},W^{(2)\star}\rangle_t &= \rho_2\,dt, \\
d\langle W^{(1)\star},W^{(2)\star}\rangle_t &= \rho_{12}\,dt,
\end{aligned}
$$

  where $|\rho_1|<1$, $|\rho_2|<1$, $|\rho_{12}|<1$, and $1+2\rho_1\rho_2\rho_{12}-\rho_1^2-\rho_2^2-\rho_{12}^2>0$,
  in order to ensure positive definiteness of the covariance matrix of the
  three Brownian motions.
- The volatility $f(Y_t,Z_t)$ of the underlying asset X is driven by the volatility
  factors Y and Z, where f is a positive function, smooth in z and such that
  $f^2(\cdot,z)$ is integrable with respect to the invariant distribution of Y.
- The instantaneous interest rate r is constant. In Section 9.1 we consider
  generalizations to the cases where *dividends* are paid or where the rate r
  is varying as a function of the stochastic factors Y and Z.
- The value X_t of the underlying asset remains positive, as can be seen by
  applying Itô's formula to deduce

$$
X_t = X_0\exp\left\{\int_0^t\left(r-\tfrac{1}{2}f^2(Y_s,Z_s)\right)ds + \int_0^t f(Y_s,Z_s)\,dW_s^{(0)\star}\right\}.
$$

- The small positive parameter ε corresponds to the short mean-reversion
  time scale of the fast volatility factor Y, as explained in Section 3.6. The
  coefficients $\alpha(y)$ and $\beta(y)$ describe the dynamics of the diffusion process
  Y under the physical measure $\mathbb{P}$. Their particular form does not play a
  role in the perturbation analysis provided that they are defined so that
  the process Y has a unique invariant distribution denoted by Φ and is

mean-reverting as explained in Chapter 3. The typical examples would be an OU process in which case $\alpha(y) = m - y$ and $\beta(y) = v\sqrt{2}$ so that Φ is the density of the normal distribution $\mathcal{N}(m, v^2)$, or a CIR process in which case $\alpha(y) = m - y$, $\beta(y) = v\sqrt{2y}$, and Φ is a Gamma distribution given in Section 3.3.4.

- The small positive parameter δ corresponds to the long time scale $1/\delta$ of the slow volatility factor Z as explained in Section 3.6. The coefficients $c(z)$ and $g(z)$ describe the dynamics of the process Z under the physical measure $\mathbb{P}$. Their particular form will not play a role in the perturbation analysis presented in this chapter.

- The functions $\Lambda_1(y, z)$ and $\Lambda_2(y, z)$ are the combined market prices of volatility risk which determine the risk-neutral pricing measure $\mathbb{P}^\star$ as explained in Chapter 2. We choose them to be functions of the volatility driving factors so that (X, Y, Z) remains a Markov process under $\mathbb{P}^\star$, and the volatility factors (Y, Z) are Markovian by themselves.

In this chapter, we consider a European option with smooth and bounded payoff function $h(x)$ and maturity T. The important case of call and put options will be discussed in the next chapter when we address the issue of calibration. The process (X_t, Y_t, Z_t) is Markovian under the risk-neutral probability measure $\mathbb{P}^\star$ and therefore the price of this option is a function of the present time $t < T$, the present value X_t of the underlying asset, and the present values Y_t and Z_t of the two volatility factors. Denoting by $P^{\varepsilon,\delta}(t, x, y, z)$ the price of the option where we explicitly show the dependence on the two small parameters ε and δ, and denoting by $\mathbb{E}^\star\{\cdot\}$ the expectation with respect to $\mathbb{P}^\star$, we have:

$$P^{\varepsilon,\delta}(t, X_t, Y_t, Z_t) = \mathbb{E}^\star \left\{ e^{-r(T-t)} h(X_T) \mid X_t, Y_t, Z_t \right\}. \tag{4.2}$$

Indeed, this expected value depends on the parameter r and the functions

$$(f, \alpha, \beta, c, g, \Lambda_1, \Lambda_2),$$

which need to be fully specified and estimated in order to use (4.2). These are highly complicated issues and the perturbation approach, presented below, will greatly simplify this problem by approximating the price $P^{\varepsilon,\delta}$ by a quantity which depends only on a few *group market parameters*. This approximation is of the form

$$P^{\varepsilon,\delta} \approx \tilde{P}^{\varepsilon,\delta} := P_{BS} + P^{\varepsilon}_{1,0} + P^{\delta}_{0,1}, \tag{4.3}$$

where P_{BS} is a Black–Scholes price, $P_{1,0}^\varepsilon$ is the first-order *fast* scale correction, and $P_{0,1}^\delta$ is the first-order *slow* scale correction. The formulas are summarized in (4.67).

4.1.2 Pricing Partial Differential Equations

By an application of the multidimensional Feynman–Kac formula presented in Section 1.9.3, the function $P^{\varepsilon,\delta}(t,x,y,z)$ defined in (4.2) is also characterized as the solution of the parabolic partial differential equation

$$\frac{\partial P^{\varepsilon,\delta}}{\partial t} + \mathcal{L}_{(X,Y,Z)} P^{\varepsilon,\delta} - r P^{\varepsilon,\delta} = 0, \tag{4.4}$$

with the terminal condition $P^{\varepsilon,\delta}(T,x,y,z) = h(x)$, and where $\mathcal{L}_{(X,Y,Z)}$ denotes the infinitesimal generator of the Markov process (X_t, Y_t, Z_t) given by (4.1). We define the operator $\mathcal{L}^{\varepsilon,\delta}$ by

$$\mathcal{L}^{\varepsilon,\delta} = \frac{\partial}{\partial t} + \mathcal{L}_{(X,Y,Z)} - r\cdot,$$

so that equation (4.4) and its terminal condition can simply be written as

$$\mathcal{L}^{\varepsilon,\delta} P^{\varepsilon,\delta} = 0, \tag{4.5}$$
$$P^{\varepsilon,\delta}(T,x,y,z) = h(x). \tag{4.6}$$

In order to derive approximations for $P^{\varepsilon,\delta}$ when the parameters (ε,δ) are small, it is convenient to write the operator $\mathcal{L}^{\varepsilon,\delta}$ as a sum of components that are scaled by the different powers of the small parameters (ε,δ) that appear in the infinitesimal generator of (X,Y,Z). This decomposition is

$$\mathcal{L}^{\varepsilon,\delta} = \frac{1}{\varepsilon}\mathcal{L}_0 + \frac{1}{\sqrt{\varepsilon}}\mathcal{L}_1 + \mathcal{L}_2 + \sqrt{\delta}\mathcal{M}_1 + \delta\mathcal{M}_2 + \sqrt{\frac{\delta}{\varepsilon}}\mathcal{M}_3, \tag{4.7}$$

where the operators $\mathcal{L}_i$ and $\mathcal{M}_i$ are defined by:

$$\mathcal{L}_0 = \frac{1}{2}\beta^2(y)\frac{\partial^2}{\partial y^2} + \alpha(y)\frac{\partial}{\partial y}, \tag{4.8}$$

$$\mathcal{L}_1 = \beta(y)\left(\rho_1 f(y,z)x\frac{\partial^2}{\partial x\partial y} - \Lambda_1(y,z)\frac{\partial}{\partial y}\right), \tag{4.9}$$

$$\mathcal{L}_2 = \frac{\partial}{\partial t} + \frac{1}{2}f^2(y,z)x^2\frac{\partial^2}{\partial x^2} + r\left(x\frac{\partial}{\partial x} - \cdot\right), \tag{4.10}$$

$$\mathcal{M}_1 = g(z)\left(\rho_2 f(y,z)x\frac{\partial^2}{\partial x\partial z} - \Lambda_2(y,z)\frac{\partial}{\partial z}\right), \tag{4.11}$$

$$\mathcal{M}_2 = \frac{1}{2}g^2(z)\frac{\partial^2}{\partial z^2} + c(z)\frac{\partial}{\partial z},$$

(4.12)

$$\mathcal{M}_3 = \beta(y)\rho_{12}g(z)\frac{\partial^2}{\partial y\partial z}.$$

(4.13)

Note that:

- $\varepsilon^{-1}\mathcal{L}_0$ is the infinitesimal generator of the process Y under the physical measure $\mathbb{P}$.
- $\mathcal{L}_1$ contains the mixed derivative due to the covariation between X and Y, and the first derivative with respect to y due to the market price of volatility risk Λ_1.
- $\mathcal{L}_2$ contains the time derivative and is the Black–Scholes operator at the volatility level $f(y,z)$, also denoted by $\mathcal{L}_{BS}(f(y,z))$.
- $\mathcal{M}_1$ contains the mixed derivative due to the covariation between X and Z, and the first derivative with respect to z due to the market price of volatility risk Λ_2.
- $\delta\mathcal{M}_2$ is the infinitesimal generator of process Z under the physical measure $\mathbb{P}$.
- $\mathcal{M}_3$ contains the mixed derivative due to the covariation between Y and Z.

Observe that in (4.7), the operator terms that are associated with the parameter ε are diverging in the small ε limit and give rise to a *singular* perturbation problem, while the terms associated with only the parameter δ are small in the small δ limit and give rise to a *regular* perturbation problem about the Black–Scholes operator $\mathcal{L}_2$.

4.2 Formal Regular and Singular Perturbation Analysis

In this section, we give a formal derivation of the price approximation when ε and δ are small. The main theorem stating the accuracy of the approximation is given in Section 4.5, along with its proof in the case where the payoff is smooth and bounded. Observe that the nonsmooth case is important since in particular the payoff of call options is not differentiable at the strike. The proof for the singular perturbation in the nonsmooth case is considerably more involved and can be found in Fouque *et al.* (2003c). In the formal derivation presented here, we choose to expand first with respect to δ and subsequently with respect to ε. This choice is more convenient for the proof than the reverse ordering, which indeed gives the same result.

We first expand $P^{\varepsilon,\delta}$ in powers of $\sqrt{\delta}$:

$$P^{\varepsilon,\delta} = P_0^\varepsilon + \sqrt{\delta}P_1^\varepsilon + \delta P_2^\varepsilon + \cdots.$$

(4.14)

We insert this expansion into the partial differential equation (4.5) and also the terminal condition (4.6), which does not depend on δ (or ε). Using the decomposition (4.7), we collect the terms in the increasing powers of δ, which gives

$$\left(\frac{1}{\varepsilon}\mathcal{L}_0 + \frac{1}{\sqrt{\varepsilon}}\mathcal{L}_1 + \mathcal{L}_2\right) P_0^\varepsilon + \sqrt{\delta}\left\{\left(\frac{1}{\varepsilon}\mathcal{L}_0 + \frac{1}{\sqrt{\varepsilon}}\mathcal{L}_1 + \mathcal{L}_2\right) P_1^\varepsilon\right.$$
$$\left. + \left(\mathcal{M}_1 + \frac{1}{\sqrt{\varepsilon}}\mathcal{M}_3\right) P_0^\varepsilon\right\} + \cdots = 0. \qquad (4.15)$$

Equating to zero first the terms independent of δ and then the terms in $\sqrt{\delta}$ in (4.15), and similarly in the terminal condition leads us to define P_0^ε and P_1^ε as follows.

Definition 4.1   We define P_0^ε as the unique solution to the problem

$$\left(\frac{1}{\varepsilon}\mathcal{L}_0 + \frac{1}{\sqrt{\varepsilon}}\mathcal{L}_1 + \mathcal{L}_2\right) P_0^\varepsilon = 0, \qquad (4.16)$$

$$P_0^\varepsilon(T,x,y,z) = h(x). \qquad (4.17)$$

Definition 4.2   The next term P_1^ε is defined as the unique solution to the problem

$$\left(\frac{1}{\varepsilon}\mathcal{L}_0 + \frac{1}{\sqrt{\varepsilon}}\mathcal{L}_1 + \mathcal{L}_2\right) P_1^\varepsilon = -\left(\mathcal{M}_1 + \frac{1}{\sqrt{\varepsilon}}\mathcal{M}_3\right) P_0^\varepsilon, \qquad (4.18)$$

$$P_1^\varepsilon(T,x,y,z) = 0. \qquad (4.19)$$

Thus P_0^ε is the solution of the homogeneous linear parabolic partial differential equation (4.16) with a terminal condition $h(x)$, while P_1^ε, the first-order term in $\sqrt{\delta}$, solves a similar problem but with a source term and zero terminal condition.

The first-order perturbation theory considered in this chapter involves only P_0^ε and P_1^ε. In the next two subsections, we expand P_0^ε and P_1^ε in powers of $\sqrt{\varepsilon}$.

4.2.1 Zeroth-Order Approximation P_0

We construct an expansion of P_0^ε in powers of $\sqrt{\varepsilon}$:

$$P_0^\varepsilon = P_0 + \sqrt{\varepsilon}P_{1,0} + \varepsilon P_{2,0} + \varepsilon^{3/2}P_{3,0} + \cdots. \qquad (4.20)$$

In the notation $P_{i,j}$, the subindex i corresponds to the power of $\sqrt{\varepsilon}$, while the subindex j corresponds to the power of $\sqrt{\delta}$:

$$P^{\varepsilon,\delta} = \sum_{i \geq 0} \sum_{j \geq 0} (\sqrt{\varepsilon})^i (\sqrt{\delta})^j P_{i,j},$$

and the leading term will simply be denoted by $P_0 = P_{0,0}$. In this chapter, we are interested in the three terms $P_0, P_{1,0}$, and $P_{0,1}$, which will give our first-order approximation (4.3) above, where $P_{1,0}^{\varepsilon}$ and $P_{0,1}^{\delta}$ are defined by absorbing the small parameters into the correction terms:

$$P_{1,0}^{\varepsilon} = \sqrt{\varepsilon}\, P_{1,0}, \qquad P_{0,1}^{\delta} = \sqrt{\delta}\, P_{0,1}.$$

In this and the next subsection, we derive explicit expressions for P_0 and $P_{1,0}^{\varepsilon}$. Inserting the expansion (4.20) in the equation (4.16), one obtains

$$\frac{1}{\varepsilon}\mathcal{L}_0 P_0 + \frac{1}{\sqrt{\varepsilon}}(\mathcal{L}_0 P_{1,0} + \mathcal{L}_1 P_0) + (\mathcal{L}_0 P_{2,0} + \mathcal{L}_1 P_{1,0} + \mathcal{L}_2 P_0)$$
$$+ \sqrt{\varepsilon}(\mathcal{L}_0 P_{3,0} + \mathcal{L}_1 P_{2,0} + \mathcal{L}_2 P_{1,0}) + \cdots = 0. \tag{4.21}$$

The terms of order ε^{-1} give

$$\mathcal{L}_0 P_0 = 0. \tag{4.22}$$

Using the definition of $\mathcal{L}_0$ given in (4.8), we see that this is an ordinary differential equation in y, which has constants as solutions (there are also exponentially growing solutions). We choose P_0 to be constant in y so that $P_0 = P_0(t,x,z)$ and (4.22) is satisfied. This choice is motivated by the fact that the leading-order price should not depend on the current value of the fast factor, as explained at the end of Section 3.2.

The order $\varepsilon^{-1/2}$ terms in (4.21) give the following equation for $P_{1,0}$:

$$\mathcal{L}_0 P_{1,0} + \mathcal{L}_1 P_0 = 0. \tag{4.23}$$

The operator $\mathcal{L}_1$ defined in (4.9) takes derivatives in y, and therefore $\mathcal{L}_1 P_0 = 0$, so that equation (4.23) reduces to

$$\mathcal{L}_0 P_{1,0} = 0. \tag{4.24}$$

As in the case of P_0, we choose $P_{1,0}$ not to depend on y so that $P_{1,0} = P_{1,0}(t,x,z)$ and (4.24) is satisfied.

Next, the order one terms in (4.21) give

$$\mathcal{L}_0 P_{2,0} + \mathcal{L}_2 P_0 = 0, \tag{4.25}$$

where we have used the fact that $\mathcal{L}_1 P_{1,0} = 0$. Equation (4.25) is a Poisson equation for $P_{2,0}$ with respect to the y-variable, and therefore the inhomogeneous (or source) term $\mathcal{L}_2 P_0$ must satisfy the solvability condition (3.13): for there to be a solution, $\mathcal{L}_2 P_0$ has to be in the null complement of the operator $\mathcal{L}_0$. This is equivalent to saying that it must be centered with respect to the invariant distribution of the process with infinitesimal generator $\mathcal{L}_0$: $\langle \mathcal{L}_2 P_0 \rangle = 0$, where the bracket notation means integration with respect to the invariant distribution Φ of the process Y.

Since P_0 does not depend on y, we have $\langle \mathcal{L}_2 P_0 \rangle = \langle \mathcal{L}_2 \rangle P_0$, where

$$\langle \mathcal{L}_2 \rangle = \frac{\partial}{\partial t} + \frac{1}{2}\bar{\sigma}^2(z)x^2\frac{\partial^2}{\partial x^2} + r\left(x\frac{\partial}{\partial x} - \cdot\right),$$

which is the Black–Scholes operator $\mathcal{L}_{BS}(\bar{\sigma}(z))$ where the volatility $\bar{\sigma}(z)$ is the *averaged effective volatility* defined by

$$\bar{\sigma}^2(z) = \langle f^2(\cdot,z) \rangle = \int f^2(y,z)\Phi(dy). \tag{4.26}$$

Therefore the leading-order term P_0 is the Black–Scholes price P_{BS} characterized as follows.

Definition 4.3   The leading-order term $P_0(t,x,z) = P_{BS}(t,x;\bar{\sigma}(z))$ is the Black–Scholes price at volatility $\bar{\sigma}(z)$, which solves

$$\mathcal{L}_{BS}(\bar{\sigma}(z))P_{BS} = 0, \tag{4.27}$$
$$P_{BS}(T,x;\bar{\sigma}(z)) = h(x). \tag{4.28}$$

The above is a partial differential equations problem in the variables (t,x), with z appearing only as a parameter. As explained at the end of Section 3.2, in the limit $\delta \to 0$, the slow factor Z is "frozen" at its starting value z so that the effective volatility is also frozen at its starting level $\bar{\sigma}(z)$.

4.2.2 Fast Time Scale Correction $P^{\varepsilon}_{1,0}$

Next, we derive an expression for $P^{\varepsilon}_{1,0}$. The order $\sqrt{\varepsilon}$ terms in the expansion (4.20) give the following Poisson equation for $P_{3,0}$:

$$\mathcal{L}_0 P_{3,0} + \mathcal{L}_1 P_{2,0} + \mathcal{L}_2 P_{1,0} = 0. \tag{4.29}$$

The solvability condition for this equation is

$$\langle \mathcal{L}_2 P_{1,0} + \mathcal{L}_1 P_{2,0} \rangle = 0.$$

Using the fact that $P_{1,0}$ does not depend on y, we can rewrite this centering condition as

$$\langle \mathcal{L}_2 \rangle P_{1,0} + \langle \mathcal{L}_1 P_{2,0} \rangle = 0. \tag{4.30}$$

From the Poisson equation (4.25) and using the fact that $\langle \mathcal{L}_2 \rangle P_0 = 0$, we can write

$$P_{2,0} = -\mathcal{L}_0^{-1} (\mathcal{L}_2 - \langle \mathcal{L}_2 \rangle) P_0 + c(t,x,z), \tag{4.31}$$

up to an additive function $c(t,x,z)$ which does not depend on y, and which will not play a role in the problem that defines $P_{1,0}$ since the operator $\mathcal{L}_1$ in (4.30) takes derivatives with respect to y. We then have

$$\langle \mathcal{L}_1 P_{2,0} \rangle = -\langle \mathcal{L}_1 \mathcal{L}_0^{-1} (\mathcal{L}_2 - \langle \mathcal{L}_2 \rangle) \rangle P_0.$$

To write the problem defining $P_{1,0}^{\varepsilon} = \sqrt{\varepsilon} P_{1,0}$, it is convenient to introduce the operator

$$\mathcal{A}^{\varepsilon} = \sqrt{\varepsilon} \langle \mathcal{L}_1 \mathcal{L}_0^{-1} (\mathcal{L}_2 - \langle \mathcal{L}_2 \rangle) \rangle, \tag{4.32}$$

which will be computed explicitly below. With this notation, and multiplying (4.30) by $\sqrt{\varepsilon}$, we have $\langle \mathcal{L}_2 \rangle P_{1,0}^{\varepsilon} = \mathcal{A}^{\varepsilon} P_0$. Using again that $\langle \mathcal{L}_2 \rangle = \mathcal{L}_{BS}(\bar{\sigma}(z))$ and $P_0(t,x,z) = P_{BS}(t,x;\bar{\sigma}(z))$, this leads us to define $P_{1,0}^{\varepsilon}$ as follows.

Definition 4.4 The function $P_{1,0}^{\varepsilon}(t,x,z)$ is the classical solution of the inhomogeneous problem

$$\mathcal{L}_{BS}(\bar{\sigma}(z)) P_{1,0}^{\varepsilon} = \mathcal{A}^{\varepsilon} P_{BS}, \tag{4.33}$$

$$P_{1,0}^{\varepsilon}(T,x,z) = 0. \tag{4.34}$$

Note that $P_{1,0}^{\varepsilon}$ is the solution of a Black–Scholes equation with a source term $\mathcal{A}^{\varepsilon} P_{BS}$, but with a zero terminal condition.

Computation of $\mathcal{A}^{\varepsilon}$

We now compute the operator $\mathcal{A}^{\varepsilon}$. In preparation, we introduce the notation

$$D_k = x^k \frac{\partial^k}{\partial x^k}, \tag{4.35}$$

for positive integers k. Then, let $\phi(y,z)$ be a solution of the following Poisson equation with respect to the variable y:

$$\mathcal{L}_0 \phi(y,z) = f^2(y,z) - \bar{\sigma}^2(z). \tag{4.36}$$

With this notation, we have

$$\mathcal{L}_0^{-1} (\mathcal{L}_2 - \langle \mathcal{L}_2 \rangle) = \frac{1}{2} \phi(y,z) x^2 \frac{\partial^2}{\partial x^2} = \frac{1}{2} \phi(y,z) D_2,$$

and therefore

$$
\begin{aligned}
&\mathcal{L}_1\mathcal{L}_0^{-1}(\mathcal{L}_2 - \langle\mathcal{L}_2\rangle)\\
&= \beta(y)\left(\rho_1 f(y,z)x\frac{\partial^2}{\partial x\partial y} - \Lambda_1(y,z)\frac{\partial}{\partial y}\right)\frac{1}{2}\phi(y,z)D_2\\
&= \left(\frac{1}{2}\rho_1\beta(y)f(y,z)\frac{\partial\phi}{\partial y}\right)D_1D_2 - \left(\frac{1}{2}\beta(y)\Lambda_1(y,z)\frac{\partial\phi}{\partial y}\right)D_2. \quad (4.37)
\end{aligned}
$$

Note that ϕ is defined up to an additive function that does not depend on the variable y and which will not affect $\mathscr{A}^\varepsilon$ since the operator $\mathcal{L}_1$ takes derivatives with respect to y.

Finally, by averaging (4.37) in y with respect to the invariant distribution Φ and using the definition (4.32) of the operator $\mathscr{A}^\varepsilon$, we obtain

$$
\mathscr{A}^\varepsilon = -V_3^\varepsilon(z)D_1D_2 - V_2^\varepsilon(z)D_2, \quad (4.38)
$$

where the two group parameters $V_3^\varepsilon(z)$ and $V_2^\varepsilon(z)$ are defined by

$$
V_3^\varepsilon(z) = -\frac{\rho_1\sqrt{\varepsilon}}{2}\left\langle\beta f(\cdot,z)\frac{\partial\phi}{\partial y}(\cdot,z)\right\rangle, \quad (4.39)
$$

$$
V_2^\varepsilon(z) = \frac{\sqrt{\varepsilon}}{2}\left\langle\beta\Lambda_1(\cdot,z)\frac{\partial\phi}{\partial y}(\cdot,z)\right\rangle. \quad (4.40)
$$

It will be useful to relate the sign of $V_3^\varepsilon(z)$ to the sign of the correlation parameter ρ_1. We assume here that the diffusion coefficient $\beta(y)$ is non-negative and that the function $f(y,z)$ is non-negative and increasing in y. Then we have the following result.

Proposition 4.5 $V_3^\varepsilon(z)$ *has the sign of* ρ_1.

Proof Using the relation $(\frac{1}{2}\beta^2\Phi)' - \alpha\Phi = 0$ satisfied by the invariant distribution Φ where derivation is with respect to y, one easily deduces that

$$
\left(\frac{1}{2}\beta^2\phi'\Phi\right)' = (f^2 - \langle f^2\rangle)\,\Phi,
$$

and consequently

$$
\phi' = \frac{2}{\beta^2\Phi}\int_{-\infty}^{\cdot}(f^2 - \langle f^2\rangle)\,\Phi.
$$

Therefore, one has

$$
\begin{aligned}
\langle \beta f \phi' \rangle &= \left\langle \frac{2f}{\beta \Phi} \int_{-\infty}^{\cdot} \left(f^2 - \langle f^2 \rangle \right) \Phi \right\rangle \\
&= \int_{-\infty}^{+\infty} \frac{2f}{\beta} \int_{-\infty}^{\cdot} \left(f^2 - \langle f^2 \rangle \right) \Phi \\
&= - \int_{-\infty}^{+\infty} F \left(f^2 - \langle f^2 \rangle \right) \Phi \\
&= - \left\langle F \left(f^2 - \langle f^2 \rangle \right) \right\rangle,
\end{aligned}
$$

where we have introduced an antiderivative F of $2f/\beta$. Since F and f^2 are both increasing in y, the covariance inequality $\langle F f^2 \rangle \geq \langle F \rangle \langle f^2 \rangle$ implies that $\langle F \left(f^2 - \langle f^2 \rangle \right) \rangle \geq 0$, which combined with the definition (4.39) of $V_3^\varepsilon(z)$ gives the result. $\qquad \square$

4.2.3 Explicit Formula for $P_{1,0}^\varepsilon$ for European Options

In the case of the European options considered here we have in fact an explicit formula for the solution of (4.33), which is a partial differential equation with a source and zero terminal condition.

We recall the definition of D_k in (4.35). By changing variable to $\xi = \log x$, it is easily seen that

$$
D_1 = x \frac{\partial}{\partial x} = \frac{\partial}{\partial \xi},
$$

so we say that D_1 is simply the first *logarithmic derivative* operator. Similarly,

$$
D_2 = x^2 \frac{\partial^2}{\partial x^2} = \frac{\partial^2}{\partial \xi^2} - \frac{\partial}{\partial \xi} = D_1^2 - D_1
$$

is just a quadratic in D_1. Recursively, it follows that D_k given by (4.35) is a polynomial of degree k in D_1. Since polynomials and powers of polynomials commute, we have

$$
D_k^\ell D_{k'}^{\ell'} = D_{k'}^{\ell'} D_k^\ell, \qquad k, \ell, k', \ell' \in \mathbb{Z}_+.
$$

With this notation, we have

$$
\mathscr{L}_{BS}(\bar{\sigma}(z)) = \frac{\partial}{\partial t} + \frac{1}{2} \bar{\sigma}(z)^2 D_2 + r(D_1 - \cdot), \tag{4.41}
$$

and therefore $\mathscr{L}_{BS}(\bar{\sigma}(z))$ and $\mathscr{A}^\varepsilon$ in (4.38) clearly commute. We then have the following explicit formula for $P_{1,0}^\varepsilon$.

Proposition 4.6 *The first-order fast scale correction term $P_{1,0}^\varepsilon(t,x,z)$ is given in terms of P_{BS} by*

$$P_{1,0}^\varepsilon(t,x,z) = -(T-t)\mathscr{A}^\varepsilon P_{BS}(t,x;\bar{\sigma}(z)), \tag{4.42}$$

where $\mathscr{A}^\varepsilon$ is given explicitly by (4.38).

Proof We first show that $P_{1,0}^\varepsilon$ given by (4.42) solves the equation (4.33):

$$\mathscr{L}_{BS}(\bar{\sigma}(z))\Big(-(T-t)\mathscr{A}^\varepsilon P_{BS}\Big) = \mathscr{A}^\varepsilon P_{BS} - (T-t)\mathscr{L}_{BS}(\bar{\sigma}(z))\mathscr{A}^\varepsilon P_{BS}$$
$$= \mathscr{A}^\varepsilon P_{BS} - (T-t)\mathscr{A}^\varepsilon \mathscr{L}_{BS}(\bar{\sigma}(z))P_{BS}$$
$$= \mathscr{A}^\varepsilon P_{BS},$$

since $\mathscr{L}_{BS}$ and $\mathscr{A}^\varepsilon$ commute, and $\mathscr{L}_{BS}(\bar{\sigma}(z))P_{BS} = 0$. This establishes (4.33). In order to complete the proof we need to show that $P_{1,0}^\varepsilon$ given by (4.42) satisfies the zero terminal condition (4.34). In the case of smooth and bounded payoffs, the Black–Scholes price P_{BS} is a smooth function of x and $\mathscr{A}^\varepsilon P_{BS} = \Big(V_3^\varepsilon(z)D_2D_1 + V_2^\varepsilon(z)D_2\Big)P_{BS}$ remains bounded as $t \to T$ for fixed values of x and z. Therefore $(T-t)\mathscr{A}^\varepsilon P_{BS} \to 0$ as $t \to T$, which is the desired zero terminal condition. $\qquad\square$

Returning to the expansion (4.14), we have

$$P^{\varepsilon,\delta} = (P_0 + \sqrt{\varepsilon}P_{1,0} + \varepsilon P_{2,0} + \cdots) + \sqrt{\delta}P_1^\varepsilon + \delta P_2^\varepsilon + \cdots;$$

so far we have identified $P_0 = P_{BS}$ and $P_{1,0}^\varepsilon = \sqrt{\varepsilon}P_{1,0}$, with the formula for the latter given in (4.42). In order to complete the first-order approximation in $\sqrt{\varepsilon}$ and $\sqrt{\delta}$ it remains to identify the first term in the expansion of P_1^ε.

4.2.4 Slow Time Scale Correction $P_{0,1}^\delta$

We next expand P_1^ε in powers of $\sqrt{\varepsilon}$ as follows:

$$P_1^\varepsilon = P_{0,1} + \sqrt{\varepsilon}P_{1,1} + \varepsilon P_{2,1} + \varepsilon^{3/2}P_{3,1} + \cdots, \tag{4.43}$$

and we derive an explicit expression for $P_{0,1}^\delta = \sqrt{\delta}P_{0,1}$. Substituting the expansion (4.43) for P_1^ε and the expansion (4.20) for P_0^ε into equation (4.18) and grouping the powers of $\sqrt{\varepsilon}$, we have

$$\frac{1}{\varepsilon}\mathcal{L}_0 P_{0,1} + \frac{1}{\sqrt{\varepsilon}}(\mathcal{L}_0 P_{1,1} + \mathcal{L}_1 P_{0,1} + \mathcal{M}_3 P_0)$$
$$+ (\mathcal{L}_0 P_{2,1} + \mathcal{L}_1 P_{1,1} + \mathcal{L}_2 P_{0,1}) + \mathcal{M}_1 P_0 + \mathcal{M}_3 P_{1,0}$$
$$+ \sqrt{\varepsilon}(\mathcal{L}_0 P_{3,1} + \mathcal{L}_1 P_{2,1} + \mathcal{L}_2 P_{1,1} + \mathcal{M}_1 P_{1,0} + \mathcal{M}_3 P_{2,0})$$
$$+ \cdots = 0. \tag{4.44}$$

From the highest-order term in ε^{-1},

$$\mathcal{L}_0 P_{0,1} = 0, \tag{4.45}$$

and, as we did for P_0 and $P_{1,0}$, we choose $P_{0,1}$ to be constant in y.

We next look at the terms of order $\varepsilon^{-1/2}$ in the expansion (4.44). Observe that $\mathcal{L}_1 P_{0,1} = 0$ and $\mathcal{M}_3 P_0 = 0$ since the operators $\mathcal{L}_1$ and $\mathcal{M}_3$ given in (4.13) take derivatives in y, and neither $P_{0,1}$ nor P_0 depend on y. Therefore the terms of order $\varepsilon^{-1/2}$ in the expansion (4.44) reduce to

$$\mathcal{L}_0 P_{1,1} = 0, \tag{4.46}$$

which allows us to choose $P_{1,1}$ independent of y.

Evaluating the terms of order one and using the fact that $\mathcal{M}_3 P_{1,0} = \mathcal{L}_1 P_{1,1} = 0$, we find

$$\mathcal{L}_0 P_{2,1} + \mathcal{L}_2 P_{0,1} + \mathcal{M}_1 P_0 = 0. \tag{4.47}$$

This is a Poisson equation in y for $P_{2,1}$ and the associated solvability condition is

$$\langle \mathcal{L}_2 P_{0,1} + \mathcal{M}_1 P_0 \rangle = 0. \tag{4.48}$$

Again, since P_0 and $P_{0,1}$ do not depend on y, condition (4.48), after multiplying by $\sqrt{\delta}$, is

$$\langle \mathcal{L}_2 \rangle P_{0,1}^\delta = -\sqrt{\delta} \langle \mathcal{M}_1 \rangle P_0.$$

This is an inhomogeneous Black–Scholes partial differential equation for $P_{0,1}^\delta$, and it remains to compute the source term on the right-hand side. Averaging $\mathcal{M}_1$ defined in (4.11) with respect to the invariant distribution Φ gives

$$\sqrt{\delta}\langle \mathcal{M}_1 \rangle P_0 = \sqrt{\delta}\left(-g(z)\langle \Lambda_2(\cdot, z)\rangle \frac{\partial}{\partial z} + \rho_2 g(z)\langle f(\cdot, z)\rangle D_1 \frac{\partial}{\partial z} \right) P_0$$
$$= 2\mathcal{A}^\delta P_{BS}, \tag{4.49}$$

where we define the operator

$$\mathcal{A}^\delta = V_0^\delta(z)\frac{\partial}{\partial \sigma} + V_1^\delta(z)D_1\frac{\partial}{\partial \sigma}, \tag{4.50}$$

and the *group parameters* (V_0^δ, V_1^δ) by

$$V_0^\delta(z) = -\frac{g(z)\sqrt{\delta}}{2}\langle\Lambda_2(\cdot,z)\rangle\bar{\sigma}'(z), \tag{4.51}$$

$$V_1^\delta(z) = \frac{\rho_2 g(z)\sqrt{\delta}}{2}\langle f(\cdot,z)\rangle\bar{\sigma}'(z). \tag{4.52}$$

Here, we converted $\frac{\partial}{\partial z}$ derivatives acting on $P_0(t,x,z)$ into $\frac{\partial}{\partial\sigma}$ derivatives acting on $P_{BS}(t,x;\sigma)$ using $P_0(t,x,z) = P_{BS}(t,x;\bar{\sigma}(z))$, which brings $\bar{\sigma}'(z)$ into the definitions (4.51)–(4.52).

Note that the terminal condition (4.6) for $P^{\varepsilon,\delta}$ has already been satisfied by the leading term P_0 in (4.28), and thus we define $P_{0,1}^\delta$ as follows.

Definition 4.7   The first-order slow scale correction $P_{0,1}^\delta(t,x,z)$ is the unique classical solution of the problem

$$\mathcal{L}_{BS}(\bar{\sigma}(z))P_{0,1}^\delta = -2\mathcal{A}^\delta P_{BS}, \tag{4.53}$$

$$P_{0,1}^\delta(T,x,z) = 0. \tag{4.54}$$

Observe from the definition (4.52) of $V_1^\delta(z)$ that, if the diffusion coefficient g and the function f are non-negative, and if $\bar{\sigma}(z)$ is nondecreasing in z, then $V_1^\delta(z)$ has the sign of ρ_2.

4.2.5 Explicit Formula for $P_{0,1}^\delta$ for European Options

In fact, in the European derivatives case, $P_{0,1}^\delta$ is given in terms of derivatives (or Greeks) of P_{BS} with respect to x and σ.

To motivate the derivation, we recall from (1.51) that, in the general case of a European derivative whose price $P(t,x;\sigma)$ satisfies the Black–Scholes partial differential equation $\mathcal{L}_{BS}(\sigma)P = 0$, with a terminal condition $P(T,x;\sigma) = h(x)$, we have the following relation between the *Vega* and the *Gamma*:

$$\frac{\partial P}{\partial\sigma} = (T-t)\sigma x^2\frac{\partial^2 P}{\partial x^2}. \tag{4.55}$$

We will use (4.55) to convert from Vegas to Gammas and back again to demonstrate the explicit solution of (4.53). In particular, we will use for $P_{BS} = P_{BS}(t,x;\bar{\sigma}(z))$, that

$$\mathcal{A}^\delta P_{BS} = \left(V_0^\delta(z)\frac{\partial}{\partial\sigma} + V_1^\delta(z)D_1\frac{\partial}{\partial\sigma}\right)P_{BS}$$

$$= \bar{\sigma}(z)(T-t)\left(V_0^\delta(z)D_2 P_{BS} + V_1^\delta(z)D_1 D_2 P_{BS}\right). \tag{4.56}$$

Proposition 4.8 *The first-order slow scale correction $P_{0,1}^{\delta}(t,x,z)$ is given in terms of $P_{BS}(t,x;\bar{\sigma}(z))$ by*

$$P_{0,1}^{\delta} = (T-t)\mathscr{A}^{\delta}P_{BS}. \tag{4.57}$$

Proof We first check that $P_{0,1}^{\delta}$ given in (4.57) solves the equation (4.53):

$$
\begin{aligned}
\mathscr{L}_{BS}(\bar{\sigma}(z))P_{0,1}^{\delta} \\
&= \mathscr{L}_{BS}(\bar{\sigma}(z))\left[(T-t)\mathscr{A}^{\delta}P_{BS}\right] \\
&= \mathscr{L}_{BS}(\bar{\sigma}(z))\left[\bar{\sigma}(z)(T-t)^2\left(V_0^{\delta}(z)D_2 P_{BS} + V_1^{\delta}(z)D_1 D_2 P_{BS}\right)\right] \\
&= -2\bar{\sigma}(z)(T-t)\left(V_0^{\delta}(z)D_2 P_{BS} + V_1^{\delta}(z)D_1 D_2 P_{BS}\right) \\
&\quad + \bar{\sigma}(z)(T-t)^2\mathscr{L}_{BS}(\bar{\sigma}(z))\left(V_0^{\delta}(z)D_2 + V_1^{\delta}(z)D_1 D_2\right)P_{BS} \\
&= -2\left(V_0^{\delta}(z)\frac{\partial}{\partial\sigma} + V_1^{\delta}(z)D_1\frac{\partial}{\partial\sigma}\right)P_{BS} \\
&= -2\mathscr{A}^{\delta}P_{BS},
\end{aligned}
$$

where we used (4.56) in the first step to convert Vegas into Gammas, and in the last step to turn Gammas back into Vegas. We also used the fact that $\mathscr{L}_{BS}$ commutes with $\left(V_0^{\delta}(z)D_2 + V_1^{\delta}(z)D_1 D_2\right)$, and that $\mathscr{L}_{BS}(\bar{\sigma}(z))P_{BS} = 0$. Finally, as in the proof of Proposition 4.6, the zero terminal condition (4.54) is ensured by the boundedness of $\frac{\partial}{\partial\sigma}P_{BS}$ and $D_1\frac{\partial}{\partial\sigma}P_{BS}$. $\quad\square$

In summary, our first-order approximation so far is

$$
\begin{aligned}
\tilde{P}^{\varepsilon,\delta} = P_{BS} &+ (T-t)\left[V_0^{\delta}(z)\frac{\partial}{\partial\sigma} + V_1^{\delta}(z)D_1\left(\frac{\partial}{\partial\sigma}\right) + V_2^{\varepsilon}(z)D_2\right. \\
&\left. + V_3^{\varepsilon}(z)D_1 D_2\right]P_{BS}. \tag{4.58}
\end{aligned}
$$

4.3 Parameter Reduction

Our first-order approximation (4.58) requires five market group parameters

$$\left\{\bar{\sigma}(z), V_0^{\delta}(z), V_1^{\delta}(z), V_2^{\varepsilon}(z), V_3^{\varepsilon}(z)\right\},$$

which depend on the current level z of the slow volatility factor. The volatility level $\bar{\sigma}(z)$, defined in (4.26), is needed to compute the leading-order term P_0 given in Definition 4.3 as the Black–Scholes price with constant volatility $\bar{\sigma}(z)$. In principle, this parameter can be estimated statistically from historical returns data, and the other four parameters will be *calibrated* to the observed implied volatility surface. The estimation of $\bar{\sigma}(z)$ is a delicate issue and therefore, for options pricing, it is highly desirable to avoid the use of historical returns data, and rely only on options data. In fact, this is possible and it is due to the fact that the parameter $V_2^\varepsilon(z)$ plays a particular role and can be *absorbed* in the volatility level, in other words, it corresponds to a *volatility level correction*. This is because $V_2^\varepsilon(z)$ is associated with a second derivative which can be incorporated in the diffusion part of the infinitesimal generator driving the Black–Scholes equation.

At this point, it is worth recalling that the explicit formula (4.58) has been derived in the case of European options where no other boundary conditions are imposed other than the terminal condition at maturity T. However, this approximation has been constructed by solving partial differential equations (4.27), (4.33), and (4.53). Denoting the combined first-order terms by $P_1^{\varepsilon,\delta}$, that is

$$P_1^{\varepsilon,\delta} = P_{1,0}^\varepsilon + P_{0,1}^\delta, \tag{4.59}$$

and using the linearity of the two problems (4.33) and (4.53), $P_1^{\varepsilon,\delta}$ solves the problem

$$\mathscr{L}_{BS}(\bar{\sigma}(z))P_1^{\varepsilon,\delta} = -\mathscr{H}^{\varepsilon,\delta}P_{BS}, \tag{4.60}$$
$$P_1^{\varepsilon,\delta}(T,x,z) = 0,$$

where the combined source operator is given by

$$\mathscr{H}^{\varepsilon,\delta} = 2V_0^\delta(z)\frac{\partial}{\partial\sigma} + 2V_1^\delta(z)D_1\frac{\partial}{\partial\sigma} + V_2^\varepsilon(z)D_2 + V_3^\varepsilon(z)D_1D_2. \tag{4.61}$$

Observe that the original pricing partial differential equation is three-dimensional in the state variables (x,y,z), while the computation of our first-order approximation requires us to solve two one-dimensional partial differential equations in x, (4.27) and (4.60), where z is a fixed parameter and the variable y plays no role.

Now we can derive the parameter reduction that absorbs $\bar{\sigma}$ and V_2^ε into a corrected volatility level $\sigma^\star$ defined by

$$\sigma^\star(z) = \sqrt{\bar{\sigma}^2(z) + 2V_2^\varepsilon(z)}, \tag{4.62}$$

where we do not show the dependence of $\sigma^\star$ on ε to simplify the notation, and where we assume that $V_2^\varepsilon(z)$ is small enough so that $\bar{\sigma}^2(z) + 2V_2^\varepsilon(z) \geq 0$. We define a new approximation $P_{BS}^\star + P_1^\star$ which has the same order of accuracy as $P_0 + P_1^{\varepsilon,\delta}$ (as proved below in Theorem 4.11), but does not need separate estimation of $\bar{\sigma}$ from historical data. Instead, it depends on the combination $\sigma^\star$ of $\bar{\sigma}$ and V_2^ε in (4.62).

The zeroth-order term, denoted by $P_{BS}^\star$, is the Black–Scholes price at volatility $\sigma^\star(z)$ which solves the problem

$$\mathcal{L}_{BS}(\sigma^\star(z))P_{BS}^\star = 0, \tag{4.63}$$
$$P_{BS}^\star(T,x,z) = h(x).$$

We next define the first-order correction $P_1^\star$ as the solution to the problem

$$\mathcal{L}_{BS}(\sigma^\star(z))P_1^\star = -\mathcal{H}_\star P_{BS}^\star, \tag{4.64}$$
$$P_1^\star(T,x,z) = 0,$$

where, in the source term, we define the operator $\mathcal{H}_\star$ by

$$\mathcal{H}_\star = 2V_0^\delta(z)\frac{\partial}{\partial\sigma} + 2V_1^\delta(z)D_1\frac{\partial}{\partial\sigma} + V_3^\varepsilon(z)D_1D_2, \tag{4.65}$$

which is simply $\mathcal{H}^{\varepsilon,\delta}$ in (4.61) without the second derivative term in x associated with $V_2^\varepsilon(z)$, which has been moved into $\mathcal{L}_{BS}(\sigma^\star(z))$.

In the present case with European options the first-order correction $P_1^\star$ is given in terms of $P_{BS}^\star$ by

$$P_1^\star = (T-t)\left(V_0^\delta(z)\frac{\partial P_{BS}^\star}{\partial\sigma} + V_1^\delta(z)D_1\frac{\partial P_{BS}^\star}{\partial\sigma} + V_3^\varepsilon(z)D_1D_2P_{BS}^\star\right). \tag{4.66}$$

The derivation of this formula follows the lines of the derivation of the expression for P_1 in (4.58), without the V_2^ε term.

4.4 First-Order Approximation: Summary and Discussion

In terms of the group market parameters $(\sigma^\star(z), V_0^\delta(z), V_1^\delta(z), V_3^\varepsilon(z))$ defined by, respectively, formulas (4.62), (4.51), (4.52), and (4.39), the price approximation $\tilde{P}^{\varepsilon,\delta}$ for European options is given to the same order by

$$\boxed{\tilde{P}^\star := P_{BS}^\star + (T-t)\left(V_0^\delta(z)\frac{\partial P_{BS}^\star}{\partial\sigma} + V_1^\delta(z)D_1\frac{\partial P_{BS}^\star}{\partial\sigma} + V_3^\varepsilon(z)D_1D_2P_{BS}^\star\right).} \tag{4.67}$$

This formula involves Greeks of the Black–Scholes price $P_{BS}^\star$ evaluated at $\sigma^\star(z)$. Using the notation introduced in Chapter 1 for the *Delta*, *Gamma*, and *Vega*, and with a slight abuse of this notation, we can express our price approximation as

$$\tilde{P}^{\star} = P_{BS}^{\star} + (T-t)\left\{V_0^{\delta}(z)\mathcal{V} + V_1^{\delta}(z)x\Delta\mathcal{V} + V_3^{\varepsilon}(z)x\Delta\left(x^2\Gamma\right)\right\}. \quad (4.68)$$

The fast time scale contribution to $\tilde{P}^{\star}$ is contained in $\sigma^{\star}$ and V_3^{ε}. These two parameters depend on z and are complicated functions of the original model parameters (they will be computed explicitly in the case of the Heston model in Chapter 10). The parameter V_3^{ε} is small of order $\sqrt{\varepsilon}$: the superscript refers to the ε-dependence and the subscript refers to the order of the corresponding differential operator with respect to x. The parameter V_2^{ε} now absorbed into $\sigma^{\star}$ contains the influence of the market price of risk Λ_1, while V_3^{ε} incorporates the effect of the correlation (or skew) ρ_1.

In order to compute the contribution due to the fast time scale, only the *group market parameters* $\sigma^{\star}(z)$ and $V_3^{\varepsilon}(z)$ are needed rather than the full specification of a particular stochastic volatility model. In the next chapter, we will explain how these parameters can be estimated (calibrated) from, for instance, call option data that gives implied volatilities.

The slow time scale contribution to $\tilde{P}^{\star}$ is given by (4.57). We have again separated the influence of the correlation ρ_2 and the influence of the market price of risk Λ_2 in the two parameters $V_0^{\delta}(z)$ and $V_1^{\delta}(z)$. These two parameters, small of order $\sqrt{\delta}$, depend on z and are again complicated functions of the original model parameters. The superscript refers to the δ-dependence and the subscript refers as above to the order of the corresponding differential operator with respect to x.

In order to compute the contribution due to the slow time scale only the *group market parameters* $V_1^{\delta}(z)$ and $V_0^{\delta}(z)$ are needed. In the next chapter we will show how these parameters, along with $\sigma^{\star}(z)$ and $V_3^{\varepsilon}(z)$, are calibrated to the observed implied volatilities.

In the next section we present the proof of the order of accuracy of our approximation, which is not crucial to understand our approach to calibration and therefore can be skipped by readers primarily interested in calibration.

4.5 Accuracy of First-Order Approximation

So far in this chapter we have given a formal derivation of the first-order price approximation based on formal regular and singular perturbation arguments. This analysis led us to make choices for the functions P_0, $P_{1,0}$, and $P_{0,1}$, which give the price approximation (4.58). In this section, we state the accuracy of this first-order approximation and give a complete proof in the case with smooth and bounded payoffs. This is then extended to the parameter-reduced approximation (4.67). The singular case with continuous and piecewise smooth payoffs (calls and puts for instance) is treated in Fouque *et al.* (2003c).

Our strategy is to write an equation for the residual, that is the difference between the true model price $P^{\varepsilon,\delta}$ and our approximation $\tilde{P}^{\varepsilon,\delta}$, and use its probabilistic representation to obtain a bound. The dependency on the process X will appear through the successive derivatives of the Black–Scholes leading-order price and will be easily controlled by assuming that the payoff is smooth with bounded derivatives. The dependency on the volatility factors Y and Z will be controlled by assuming that expectations of the form $\mathbb{E}^\star|J(Y_t,Z_t)|$ for various functions J are uniformly bounded with respect to the small parameters ε and δ for fixed starting points (x,y,z). In particular, the set of functions J must include the solutions $\phi(y,z)$ of Poisson equations such as (4.36), and the coefficients $\alpha(y),\beta(y),c(z),g(z)$. This will be the case if these functions are at most polynomially growing, thanks to the following result.

Lemma 4.9 *We assume that the market prices of risk $\Lambda_1(y,z)$ and $\Lambda_2(y,z)$ are bounded, and that:*

(i) *The process $Y^{(1)}$ with infinitesimal generator $\mathscr{L}_0$ given by (4.8) admits moments of any order uniformly in t:*

$$\sup_t \mathbb{E}\left\{|Y_t^{(1)}|^k\right\} \leq C(k).$$

(ii) *The process $Z^{(1)}$ with infinitesimal generator $\mathscr{M}_2$ given by (4.12) admits moments of any order uniformly in $t \leq T$:*

$$\sup_{t\leq T} \mathbb{E}\left\{|Z_t^{(1)}|^k\right\} \leq C(T,k).$$

Note that these assumptions hold for the OU and CIR processes, in particular the first assumption which is ensured by the existence of moments of the invariant distribution.

Then, if $J(y,z)$ is at most polynomially growing, for every (y,z) and $t \leq T$, there exists a positive constant C such that for any $\varepsilon \leq 1,\ \delta \leq 1$

$$\mathbb{E}^\star_{x,y,z}|J(Y_t,Z_t)| \leq C.$$

Proof It is enough to show the result for $J(y,z) = y^k$ and $J(y,z) = z^k$ for any integer k. We start with Z, whose dynamics is given in (4.1). First, we consider Z under $\mathbb{P}$: that is, without the market price of risk Λ_2. It is simply given by $Z^{(1)}_{\delta t}$, where $Z^{(1)}$ is the process with infinitesimal generator $\mathcal{M}_1$. Therefore, by our assumption on $Z^{(1)}$, we have

$$\mathbb{E}\left\{|Z_t|^k\right\} \leq \sup_{\delta \leq 1}\mathbb{E}\left\{|Z^{(1)}_{\delta t}|^k\right\} \leq C(T,k).$$

Now we consider Z under $\mathbb{P}^\star$. Using Girsanov's theorem, we introduce the $\mathbb{P}$-martingale

$$M_t^{(\Lambda_2)} = e^{-\int_0^t \Lambda_2(Y_s,Z_s)dW_s^{(2)} - \frac{1}{2}\int_0^t \Lambda_2(Y_s,Z_s)^2 ds},$$

and we write

$$\mathbb{E}^\star\left\{|Z_t|^k\right\} = \mathbb{E}\left\{|Z_t|^k M_t^{(\Lambda_2)}\right\}$$

$$= \mathbb{E}\left\{|Z_t|^k e^{\frac{1}{2}\int_0^t \Lambda_2(Y_s,Z_s)^2 ds}\sqrt{M_t^{(2\Lambda_2)}}\right\}$$

$$\leq \sqrt{\mathbb{E}\left\{|Z_t|^{2k} e^{\int_0^t \Lambda_2(Y_s,Z_s)^2 ds}\right\}}\sqrt{\mathbb{E}\left\{M_t^{(2\Lambda_2)}\right\}}$$

$$= \sqrt{\mathbb{E}\left\{|Z_t|^{2k} e^{\int_0^t \Lambda_2(Y_s,Z_s)^2 ds}\right\}}$$

$$\leq \sqrt{\mathbb{E}\left\{|Z^{(1)}_{\delta t}|^{2k}\right\} e^{t||\Lambda_2||_\infty^2}} \leq C,$$

where we have used the fact that $M_t^{(2\Lambda_2)}$ is a $\mathbb{P}$-martingale.

We turn now to the fast volatility factor Y. First, we consider Y under $\mathbb{P}$: that is, without the market price of risk Λ_1. It is simply given by $Y^{(1)}_{t/\varepsilon}$, where $Y^{(1)}$ is the process with infinitesimal generator $\mathcal{L}_0$. Therefore, by our assumption on $Y^{(1)}$, we have

$$\mathbb{E}\left\{|Y_t|^k\right\} \leq \sup_{\varepsilon \leq 1}\mathbb{E}\left\{|Y^{(1)}_{t/\varepsilon}|^k\right\} \leq C(k).$$

Now we consider Y under $\mathbb{P}^\star$. Using Girsanov's theorem, we introduce the $\mathbb{P}$-martingale

$$M_t^{(\Lambda_1)} = e^{-\int_0^t \Lambda_1(Y_s,Z_s)dW_s^{(1)} - \frac{1}{2}\int_0^t \Lambda_1(Y_s,Z_s)^2 ds},$$

and by repeating the argument given above for Z, we obtain

$$\mathbb{E}^\star\left\{|Y_t|^k\right\} = \mathbb{E}\left\{|Y_t|^k M_t^{(\Lambda_1)}\right\}$$

$$\leq \sqrt{\mathbb{E}\left\{|Y^{(1)}_{t/\varepsilon}|^{2k}\right\} e^{t||\Lambda_2||_\infty^2}} \leq C. \qquad \square$$

Our accuracy result is as follows. We give it in a general form so that it can be applied to OU and CIR volatility factors. The assumptions on the volatility function $f(y,z)$ can be relaxed with particular choices of volatility factors (such as f being an exponential function in the fast factor chosen to be an OU process, or f being the square-root function in the fast factor chosen to be a CIR process as in the Heston model presented in detail in Chapter 10).

Theorem 4.10 *We assume:*

(i) *Existence and uniqueness of* (X,Y,Z) *given for fixed* (ε,δ) *as the solution of the system of stochastic differential equations (3.49) under* $\mathbb{P}$.

(ii) *Existence and uniqueness of* (X,Y,Z) *given for fixed* (ε,δ) *as the solution of stochastic differential equations (4.1) under* $\mathbb{P}^\star$.

(iii) *The market prices of risk* $\Lambda_1(y,z)$ *and* $\Lambda_2(y,z)$ *are bounded and assumptions* (i) *and* (ii) *in Lemma 4.9 hold.*

(iv) *The process* $Y^{(1)}$ *with infinitesimal generator* $\mathcal{L}_0$ *has a unique invariant distribution and is mean-reverting as defined in Section 3.2.*

(v) *The function* $f(y,z)$ *is smooth in* z *and such that the solution* ϕ *to the Poisson equation (4.36) is at most polynomially growing.*

(vi) *The payoff function* $h(x)$ *and its derivatives are smooth and bounded.*

Then, for fixed (t,x,y,z), *there exists a positive constant* C *such that for any* $\varepsilon \le 1, \delta \le 1$

$$|P^{\varepsilon,\delta} - \tilde{P}^{\varepsilon,\delta}| \le C(\varepsilon + \delta),$$

where the price $P^{\varepsilon,\delta}$ *is given by an expectation in (4.2), and the price approximation* $\tilde{P}^{\varepsilon,\delta}$ *is given explicitly by (4.58).*

Proof In order to establish the accuracy of the approximation we introduce the following higher-order approximation for $P^{\varepsilon,\delta}$:

$$
\begin{aligned}
\widehat{P}^{\varepsilon,\delta} \\
&= \tilde{P}^{\varepsilon,\delta} + \varepsilon \left(P_{2,0} + \sqrt{\varepsilon} P_{3,0} \right) + \sqrt{\delta} \left(\sqrt{\varepsilon} P_{1,1} + \varepsilon P_{2,1} \right) \\
&= \left(P_0 + \sqrt{\varepsilon} P_{1,0} + \varepsilon P_{2,0} + \varepsilon^{3/2} P_{3,0} \right) + \sqrt{\delta} \left(P_{0,1} + \sqrt{\varepsilon} P_{1,1} + \varepsilon P_{2,1} \right).
\end{aligned}
$$

$$(4.69)$$

The functions P_0, $P_{1,0}^\varepsilon = \sqrt{\varepsilon} P_{1,0}$, and $P_{0,1}^\delta = \sqrt{\delta} P_{0,1}$ are given respectively in Definitions 4.3, 4.4, and 4.7. We choose a particular $P_{2,0}$ according to (4.31), and $P_{3,0}$ is a solution of the Poisson equation (4.29). In order to characterize $P_{1,1}$ and $P_{2,1}$ we need to go further in the expansion (4.43) of P_1^ε. From (4.47) and (4.48) we define

$$P_{2,1} = -\mathcal{L}_0^{-1} \left[(\mathcal{L}_2 - \langle \mathcal{L}_2 \rangle) P_{0,1} + (\mathcal{M}_1 - \langle \mathcal{M}_1 \rangle) P_0 \right].$$

Canceling the terms of order $\sqrt{\varepsilon}$ in equation (4.44), we get

$$\mathcal{L}_0 P_{3,1} + \mathcal{L}_1 P_{2,1} + \mathcal{L}_2 P_{1,1} + \mathcal{M}_1 P_{1,0} + \mathcal{M}_3 P_{2,0} = 0.$$

This is a Poisson equation for $P_{3,1}$ and its solvability condition reads

$$\langle \mathcal{L}_2 \rangle P_{1,1} + \langle \mathcal{L}_1 P_{2,1} + \mathcal{M}_1 P_{1,0} + \mathcal{M}_3 P_{2,0} \rangle = 0.$$

This is a Black–Scholes equation for $P_{1,1}$ with a source which is completely characterized by our previous choices of $P_{1,0}, P_{2,0}$, and $P_{2,1}$. We choose $P_{1,1}$ to be the solution of this equation with a zero terminal condition.

We next introduce the *residual*

$$R^{\varepsilon,\delta} = P^{\varepsilon,\delta} - \widehat{P}^{\varepsilon,\delta}, \tag{4.70}$$

and apply the operator $\mathcal{L}^{\varepsilon,\delta}$ defined in (4.7). Using the pricing equation (4.5) we have $\mathcal{L}^{\varepsilon,\delta} P^{\varepsilon,\delta} = 0$, and then we compute $\mathcal{L}^{\varepsilon,\delta}$ acting on $\widehat{P}^{\varepsilon,\delta}$ given by (4.69) to obtain:

$$
\begin{aligned}
\mathcal{L}^{\varepsilon,\delta} R^{\varepsilon,\delta} &+ \frac{1}{\varepsilon} (\mathcal{L}_0 P_0) + \frac{1}{\sqrt{\varepsilon}} (\mathcal{L}_0 P_{1,0} + \mathcal{L}_1 P_0) \\
&+ (\mathcal{L}_0 P_{2,0} + \mathcal{L}_1 P_{1,0} + \mathcal{L}_2 P_0) + \sqrt{\varepsilon} (\mathcal{L}_0 P_{3,0} + \mathcal{L}_1 P_{2,0} + \mathcal{L}_2 P_{1,0}) \\
&+ \sqrt{\delta} \left(\frac{1}{\varepsilon} (\mathcal{L}_0 P_{0,1}) + \frac{1}{\sqrt{\varepsilon}} (\mathcal{L}_0 P_{1,1} + \mathcal{L}_1 P_{0,1} + \mathcal{M}_3 P_0) \right) \\
&+ \sqrt{\delta} (\mathcal{L}_0 P_{2,1} + \mathcal{L}_1 P_{1,1} + \mathcal{L}_2 P_{0,1} + \mathcal{M}_1 P_0 + \mathcal{M}_3 P_{1,0}) \\
&+ \varepsilon R_1^\varepsilon + \sqrt{\varepsilon \delta} R_2^\varepsilon + \delta R_3^\varepsilon = 0, \tag{4.71}
\end{aligned}
$$

where R_1^ε, R_2^ε, and R_3^ε are given by

$$R_1^\varepsilon = \mathcal{L}_2 P_{2,0} + \mathcal{L}_1 P_{3,0} + \sqrt{\varepsilon} \mathcal{L}_2 P_{3,0}, \tag{4.72}$$

$$
\begin{aligned}
R_2^\varepsilon &= \mathcal{L}_2 P_{1,1} + \mathcal{L}_1 P_{2,1} + \mathcal{M}_1 P_{1,0} + \mathcal{M}_3 P_{2,0} \\
&\quad + \sqrt{\varepsilon} (\mathcal{L}_2 P_{2,1} + \mathcal{M}_1 P_{2,0} + \mathcal{M}_3 P_{3,0}) + \varepsilon \mathcal{M}_1 P_{3,0}, \tag{4.73}
\end{aligned}
$$

$$
\begin{aligned}
R_3^\varepsilon &= \mathcal{M}_1 P_{0,1} + \mathcal{M}_2 P_0 + \mathcal{M}_3 P_{1,1} \\
&\quad + \sqrt{\varepsilon} (\mathcal{M}_1 P_{1,1} + \mathcal{M}_2 P_{1,0} + \mathcal{M}_3 P_{2,1}) + \varepsilon (\mathcal{M}_1 P_{2,1} + \mathcal{M}_2 P_{2,0}). \tag{4.74}
\end{aligned}
$$

From the properties of the functions $P_{i,j}$ established in the previous sections, the functions $R_1^\varepsilon, R_2^\varepsilon$, and R_3^ε are smooth functions of t, x, y, and z that are, for $\varepsilon \le 1$ and $\delta \le 1$, bounded by smooth functions of t, x, y, z independent of ε and δ, uniformly bounded in t, x, z and at most linearly growing in y through the solution of the Poisson equation (4.36).

With our choice of $(P_0, P_{1,0}, P_{2,0}, P_{3,0}, P_{0,1}, P_{1,1}, P_{2,1})$, the term of order $1/\varepsilon$ in (4.71) cancels by (4.22), the term of order $1/\sqrt{\varepsilon}$ cancels by (4.23), the term of order one cancels by (4.25), the term of order $\sqrt{\varepsilon}$ cancels by (4.29). Moreover, the term of order $\sqrt{\delta}/\varepsilon$ cancels by (4.45), the term of order $\sqrt{\delta}/\sqrt{\varepsilon}$ cancels by (4.46), and finally, the term of order $\sqrt{\delta}$ cancels by (4.47). Therefore we find

$$\mathscr{L}^{\varepsilon,\delta} R^{\varepsilon,\delta} + \varepsilon R_1^\varepsilon + \sqrt{\varepsilon\delta} R_2^\varepsilon + \delta R_3^\varepsilon = 0. \tag{4.75}$$

Using the terminal conditions for $P^{\varepsilon,\delta}, P_0, P_{1,0}$, and $P_{0,1}$ at time T, we can write

$$
\begin{aligned}
R^{\varepsilon,\delta}&(T,x,y,z) \\
&= P^{\varepsilon,\delta}(T,x,y,z) - \widehat{P}^{\varepsilon,\delta}(T,x,y,z) \\
&= -\varepsilon(P_{2,0} + \sqrt{\varepsilon}P_{3,0})(T,x,y,z) - \sqrt{\varepsilon}\sqrt{\delta}(P_{1,1} + \sqrt{\varepsilon}P_{2,1})(T,x,y,z) \\
&= \varepsilon G_1^\varepsilon(x,y,z) + \sqrt{\varepsilon\delta} G_2^\varepsilon(x,y,z),
\end{aligned}
\tag{4.76}
$$

where the functions G_1^ε and G_2^ε are defined by

$$
\begin{aligned}
G_1^\varepsilon(x,y,z) &= -(P_{2,0} + \sqrt{\varepsilon}P_{3,0})(T,x,y,z), \\
G_2^\varepsilon(x,y,z) &= -(P_{1,1} + \sqrt{\varepsilon}P_{2,1})(T,x,y,z),
\end{aligned}
$$

which are independent of t, and have in the other variables the same properties as the functions $R_1^\varepsilon, R_2^\varepsilon$, and R_3^ε discussed above. Using the Feynman–Kac probabilistic representation formula for the solution $R^{\varepsilon,\delta}$ of equation (4.75) with source and terminal condition (4.76), one can write

$$R^{\varepsilon,\delta}$$

$$
\begin{aligned}
&= \varepsilon \mathbb{E}^\star \left\{ e^{-r(T-t)} G_1^\varepsilon(X_T, Y_T, Z_T) + \int_t^T e^{-r(s-t)} R_1^\varepsilon(s, X_s, Y_s, Z_s)\,ds \mid X_t, Y_t, Z_t \right\} \\
&\quad + \sqrt{\varepsilon\delta}\, \mathbb{E}^\star \Big\{ e^{-r(T-t)} G_2^\varepsilon(X_T, Y_T, Z_T) \\
&\qquad\qquad + \int_t^T e^{-r(s-t)} R_2^\varepsilon(s, X_s, Y_s, Z_s)\,ds \mid X_t, Y_t, Z_t \Big\} \\
&\quad + \delta \mathbb{E}^\star \left\{ \int_t^T e^{-r(s-t)} R_3^\varepsilon(s, X_s, Y_s, Z_s)\,ds \mid X_t, Y_t, Z_t \right\},
\end{aligned}
\tag{4.77}
$$

where (X, Y, Z) is the original process described in (4.1) under the risk-neutral probability measure $\mathbb{P}^{\star}$. Using the assumptions on the model, one can deduce that the functions $(R_1^{\varepsilon}, R_2^{\varepsilon}, R_3^{\varepsilon}, G_1^{\varepsilon}, G_2^{\varepsilon})$ are bounded in x and at most polynomially growing in (y, z). By Lemma 4.9 we get

$$|R^{\varepsilon,\delta}| \leq C_1 \varepsilon + C_2 \sqrt{\varepsilon \delta} + C_3 \delta \leq C_4 (\varepsilon + \delta).$$

Finally, from the definition (4.70) of the residual $R^{\varepsilon,\delta}$ and the definition (4.69) of $\widehat{P}^{\varepsilon,\delta}$, we have

$$
\begin{aligned}
|P^{\varepsilon,\delta} - \tilde{P}^{\varepsilon,\delta}| &\leq |R^{\varepsilon,\delta}| + |\widehat{P}^{\varepsilon,\delta} - \tilde{P}^{\varepsilon,\delta}| \\
&\leq C_4(\varepsilon + \delta) + \varepsilon|P_{2,0} + \sqrt{\varepsilon}P_{3,0}| + \sqrt{\varepsilon\delta}|P_{1,1} + \sqrt{\varepsilon}P_{2,1}| \\
&\leq C(\varepsilon + \delta),
\end{aligned}
$$

which is the desired result. $\square$

Finally, for the parameter-reduced approximation $P_{BS}^{\star} + P_1^{\star}$ introduced in Section 4.3, we obtain the same accuracy of approximation as in Theorem 4.10:

Theorem 4.11   *When the payoff function $h(x)$ and its derivatives are smooth and bounded, for fixed (t, x, y, z), there exists a constant $C > 0$ such that for any $\varepsilon \leq 1, \delta \leq 1$*

$$|P^{\varepsilon,\delta} - (P_{BS}^{\star} + P_1^{\star})| \leq C(\varepsilon + \delta).$$

Proof   We first observe that with the definition of $\sigma^{\star}$ and $\mathscr{L}_{BS}(\sigma)$ we have

$$\mathscr{L}_{BS}(\sigma^{\star}) = \mathscr{L}_{BS}(\bar{\sigma}) + \frac{1}{2} 2V_2^{\varepsilon} D_2.$$

Denoting $P_0 = P_{BS}(t, x; \bar{\sigma}(z))$ simply by P_{BS}, we have that $P_0 - P_{BS}^{\star} = P_{BS} - P_{BS}^{\star}$ satisfies

$$
\begin{aligned}
\mathscr{L}_{BS}(\bar{\sigma})(P_{BS} - P_{BS}^{\star}) &= V_2^{\varepsilon} D_2 P_{BS}^{\star}, & (4.78) \\
(P_{BS} - P_{BS}^{\star})(T, x, z) &= 0. & (4.79)
\end{aligned}
$$

Since the source term is $\mathscr{O}(\sqrt{\varepsilon})$ because of the V_2^{ε} factor, the difference $P_{BS} - P_{BS}^{\star}$ is also $\mathscr{O}(\sqrt{\varepsilon})$. Note that by taking derivatives with respect to x in (4.78)–(4.79) we obtain similarly that the derivatives of the difference $P_{BS} - P_{BS}^{\star}$ are also $\mathscr{O}(\sqrt{\varepsilon})$. Next we write

$$|P^{\varepsilon,\delta} - (P_{BS}^{\star} + P_1^{\star})| \leq |P^{\varepsilon,\delta} - (P_{BS} + P_1)| + |P_{BS} + P_1 - (P_{BS}^{\star} + P_1^{\star})|.$$

$$(4.80)$$

From (4.63), (4.60), and (4.64), it follows that

$$\mathscr{L}_{BS}(\bar{\sigma})(P_{BS}+P_1) = -\mathscr{H}^{\varepsilon,\delta}P_{BS},$$
$$\mathscr{L}_{BS}(\sigma^\star)(P_{BS}^\star+P_1^\star) = -\mathscr{H}_\star P_{BS}^\star.$$

The residual

$$E = (P_{BS}+P_1) - (P_{BS}^\star+P_1^\star)$$

satisfies the equation

$$\mathscr{L}_{BS}(\bar{\sigma})(E) = -\mathscr{H}^{\varepsilon,\delta}P_{BS} - (\mathscr{L}_{BS}(\sigma^\star) - V_2^\varepsilon D_2)(P_{BS}^\star+P_1^\star) \quad (4.81)$$
$$= -\mathscr{H}^{\varepsilon,\delta}P_{BS} + \mathscr{H}_\star P_{BS}^\star + V_2^\varepsilon D_2(P_{BS}^\star+P_1^\star) \quad (4.82)$$
$$= \mathscr{H}_\star(P_{BS}^\star - P_{BS}) + V_2^\varepsilon D_2(P_{BS}^\star - P_{BS} + P_1^\star) \quad (4.83)$$
$$= \mathscr{O}(\varepsilon+\delta), \quad (4.84)$$

where we used the fact that $\mathscr{H}_\star = \mathscr{O}(\sqrt{\varepsilon}+\sqrt{\delta})$, $V_2^\varepsilon = \mathscr{O}(\sqrt{\varepsilon})$, $P_{BS}^\star - P_{BS} = \mathscr{O}(\sqrt{\varepsilon})$, $\frac{\partial^k}{\partial x^k}(P_{BS}^\star - P_{BS}) = \mathscr{O}(\sqrt{\varepsilon})$, and $\frac{\partial^k}{\partial x^k}P_1^\star = \mathscr{O}(\sqrt{\varepsilon}+\sqrt{\delta})$. Since E vanishes at the terminal time T we can conclude that $E = \mathscr{O}(\varepsilon+\delta)$ and the theorem follows from (4.80) and Theorem 4.10. $\qquad\square$

We conclude that with this parameter reduction only the parameters

$$\{\sigma^\star(z), V_0^\delta(z), V_1^\delta(z), V_3^\varepsilon(z)\}$$

are needed in order to get the price approximation $P_{BS}^\star + P_1^\star$. In the next chapter we show how to calibrate these important market group parameters from the implied volatility surface.

Remark on the Case of Call Payoffs In Fouque *et al.* (2003c), we use a regularization argument to treat the case of call options with a fast factor. The regular perturbation due to the slow factor does not introduce additional difficulty. In fact, it follows that

$$P^{\varepsilon,\delta} = P^\star + \mathscr{O}(\varepsilon\log|\varepsilon|+\delta).$$

This error bound is not optimal, and by taking into account the second-order correction due to the fast factor, one can show that the error of the first-order approximation is in fact of order $(\varepsilon+\delta)$. A full detailed proof would involve combining the second-order correction derived in Section 9.3 with the regularization argument given in Fouque *et al.* (2003c). We do not show these lengthy details here.

Notes

The derivation of the first-order approximation with only the short time scale ε associated with the *fast* volatility factor Y was presented in Fouque *et al.* (2000) in the case with a smooth payoff function. The proof of convergence for the case of nonsmooth European call options appears in Fouque *et al.* (2003c). The argument given in Lemma 4.9 is in Cotton *et al.* (2004).

The long time scale associated with the small parameter δ corresponds to the *slow* factor Z. In the case of a single slow volatility factor such an expansion has been considered in Fournie *et al.* (1997), Lee (1999), and Sircar and Papanicolaou (1999), for instance. See also Lewis (2000) and Hull and White (1988) for related regular perturbation expansions, and Zhu and Avellaneda (1997) for approximations based on large strike price limits.

The derivation of the first-order perturbation approximation with both the fast and the slow volatility factors present appears in Fouque *et al.* (2003a).

The second-order perturbation approximation is presented in Section 9.3 for the fast factor (and implemented in Section 10.2.2 in the case of a fast mean-reverting Heston model), and in Section 9.4 for the slow factor.

A mathematical theory for the type of singular perturbation analysis that is used in this chapter can be found in Blankenship and Papanicolaou (1978). The first-order perturbation due to the correlation between X and Y and between X and Z appears to be closely related to financial applications and to our knowledge does not appear in other fields of application. This is discussed in Fouque *et al.* (2004). Some important aspects of the price approximation derived in this chapter are that we combine regular and singular perturbations, that we go beyond the zero-order approximation and obtain a first-order approximation which does not depend on the current value for the fast volatility factor Y.

The reduction of parameters in Section 4.3 appeared first in Fouque *et al.* (2004b), and has been developed in Fouque and Kollman (2011).

Similar multiscale perturbation methods have also been studied in Conlon and Sullivan (2005) by using a Fourier transform, in Howison (2005) by using matched asymptotic expansions, in Souza and Zubelli (2007) by using inner–outer perturbation expansions, in Fukasawa (2011) by martingale expansions, in Alòs (2009) by a decomposition formula, and in Fouque *et al.* (2010) by using spectral methods. Asymptotic results for implied volatility at extreme strikes are surveyed in Benaim *et al.* (2009).

Applications have been proposed in Bayraktar and Yang (2011) for unifying credit and equity derivatives, in Hikspoors and Jaimungal (2008) for commodity derivatives, in Chi *et al.* (2010) for insurance products, in Nayak and Papanicolaou (2007) for fitting the implied volatility surface, and in Choi *et al.* (2010) for a generalization of the CEV model.

Other related expansion techniques for various financial models have been introduced in Hagan *et al.* (2002) using WKB expansion for the SABR model, in Gobet and Miri (2010) using Malliavin calculus, in Hu and Knessl (2010) using the ray method of geometrical optics, and in Bergomi (2009) using times scales in forward variances.

5

Implied Volatility Formulas and Calibration

By *calibration* we mean adjusting the model parameters so that the model option prices reproduce in the best possible way the market prices for a given set of observed derivatives. Our approach consists of using, instead of the model prices, their approximations derived in the previous chapter. We then calibrate the group market parameters produced by our first-order approximation instead of the parameters of a fully specified stochastic volatility model. This *reduction of parameters* is a main feature of our method. Its full strength resides in the fact that these parameters are exactly those needed to price other derivatives to the same level of approximation, or to approximate hedging strategies, as will be shown in the following chapters.

To summarize the task ahead, we recall that our starting point is the following class of stochastic volatility models for the underlying under the real-world measure:

$$\left. \begin{aligned} dX_t &= \mu X_t dt + f(Y_t, Z_t) X_t \, dW_t^{(0)}, \\ dY_t &= \frac{1}{\varepsilon}\alpha(Y_t)dt + \frac{1}{\sqrt{\varepsilon}}\beta(Y_t)\,dW_t^{(1)}, \\ dZ_t &= \delta c(Z_t)dt + \sqrt{\delta}\,g(Z_t)\,dW_t^{(2)}, \end{aligned} \right\} \tag{5.1}$$

as introduced in Section 3.6. For the derivative pricing problem associated with the underlying X, we consider these models under a risk-neutral measure $\mathbb{P}^\star$:

$$\left. \begin{aligned} dX_t &= rX_t dt + f(Y_t, Z_t) X_t \, dW_t^{(0)\star}, \\ dY_t &= \left(\frac{1}{\varepsilon}\alpha(Y_t) - \frac{1}{\sqrt{\varepsilon}}\beta(Y_t)\Lambda_1(Y_t, Z_t)\right) dt + \frac{1}{\sqrt{\varepsilon}}\beta(Y_t)\,dW_t^{(1)\star}, \\ dZ_t &= \left(\delta\,c(Z_t) - \sqrt{\delta}\,g(Z_t)\Lambda_2(Y_t, Z_t)\right) dt + \sqrt{\delta}\,g(Z_t)\,dW_t^{(2)\star}, \end{aligned} \right\} \tag{5.2}$$

where Λ_1 and Λ_2 are market prices of volatility risk. We seek the price $P^{\varepsilon,\delta}$ of a European derivative with payoff $h(X_T)$ on expiration date T, which can be expressed as the risk-neutral expected discounted payoff:

$$P^{\varepsilon,\delta}(t,X_t,Y_t,Z_t) = \mathbb{E}^\star \left\{ e^{-r(T-t)}h(X_T) \mid X_t,Y_t,Z_t \right\}, \qquad (5.3)$$

where the function $P^{\varepsilon,\delta}(t,x,y,z)$ solves the partial differential equation (4.5) involving as coefficients the functions appearing in the model (5.2). Notice from (5.3) that pricing requires the current values of the volatility factors Y and Z, which are not directly observed, in addition to calibration of the model (5.2). Calibration to historical data and options data involves choosing and fitting the functions $f,\alpha,\beta,c,g,\Lambda_1,\Lambda_2$. This is computationally intensive and requires good initial guesses in order that the fitted parameters are realistic representations of volatility behavior. The perturbation theory developed in the previous chapter vastly simplifies this calibration problem. As stated in Theorem 4.11, for nice payoff functions h,

$$P^\star = P^\star_{BS} + (T-t)\left(V_0^\delta \frac{\partial P^\star_{BS}}{\partial \sigma} + V_1^\delta D_1\left(\frac{\partial P^\star_{BS}}{\partial \sigma}\right) + V_3^\varepsilon D_1 D_2 P^\star_{BS}\right) \quad (5.4)$$

is a price approximation with accuracy $P^{\varepsilon,\delta} = P^\star + O(\varepsilon+\delta)$. Observe that only the *group market parameters*

$$(\sigma^\star, V_0^\delta, V_1^\delta, V_3^\varepsilon) \qquad (5.5)$$

are needed to compute the approximate price $P^\star$ (here, for simplicity, we omit the dependence on z). In this chapter, we show how to calibrate these parameters from the term structure of implied volatilities obtained from European call options.

In later chapters, we show that these same parameters are exactly those needed to price more complicated instruments such as American options or path-dependent options to the same level of accuracy. Thus, in the regime where ε and δ are small, corresponding to stochastic volatility models with fast and slow volatility factors, we can handle a wide class of pricing problems which would be very complicated without the perturbation theory.

5.1 Approximate Call Prices and Implied Volatilities

We consider a European call option with strike K and maturity T. The corresponding payoff function is $h(x) = (x-K)^+$ and the Black–Scholes price (1.3.4) at volatility $\sigma^\star$ is

$$P_{BS}^\star = xN(d_1^\star) - Ke^{-r\tau}N(d_2^\star), \tag{5.6}$$

where we use the notation $\tau = T - t$, and

$$d_{1,2}^\star = \frac{\log(x/K) + \left(r \pm \frac{1}{2}\sigma^{\star 2}\right)\tau}{\sigma^\star \sqrt{\tau}}. \tag{5.7}$$

We will now use formula (5.4) in the case where $P_{BS}^\star$ is given by (5.6). Notice that in this case the payoff function h is not smooth. However, it is continuous and piecewise smooth, and we refer to the remark at the end of Section 4.5 regarding the accuracy of approximation in this case.

It is convenient to rewrite (5.4) as

$$P^\star = P_{BS}^\star + \left\{ \tau V_0^\delta + \tau V_1^\delta D_1 + \frac{V_3^\varepsilon}{\sigma^\star} D_1 \right\} \frac{\partial P_{BS}^\star}{\partial \sigma}, \tag{5.8}$$

where we have used the relation

$$\frac{\partial P_{BS}^\star}{\partial \sigma} = \tau \sigma^\star D_2 P_{BS}^\star$$

between the *Vega* and the *Gamma*, and where $P_{BS}^\star$ and $\partial P_{BS}^\star / \partial \sigma$ are evaluated at $\sigma = \sigma^\star$. Recall also from Chapter 1 that in the case of a call, *Vega* is explicitly given by

$$\frac{\partial C_{BS}}{\partial \sigma} = \frac{x\sqrt{\tau}e^{-d_1^2/2}}{\sqrt{2\pi}}. \tag{5.9}$$

As explained in Section 2.1, we convert the price $P^\star$ to an implied volatility I defined by

$$C_{BS}(I) = P^\star, \tag{5.10}$$

where the left-hand side is the Black–Scholes call price at the volatility level I, and the right-hand side is the call price (5.3) that follows from the fully specified stochastic volatility model. It follows from (5.9) that we can invert for this (unique) implied volatility.

The next step consists of expanding the difference $I - \sigma^\star$ between the implied volatility I and the volatility used to compute $P_{BS}^\star$, in powers of $\sqrt{\varepsilon}$ and $\sqrt{\delta}$:

$$I - \sigma^\star = \sqrt{\varepsilon} I_{1,0} + \sqrt{\delta} I_{0,1} + \cdots, \tag{5.11}$$

where the argument below shows that indeed there is no order one term on the right-hand side. We then expand both sides of equation (5.10) by using (5.8) and (5.11):

$$C_{BS}(\sigma^\star) + (\sqrt{\varepsilon}I_{1,0} + \sqrt{\delta}I_{0,1})\frac{\partial C_{BS}(\sigma^\star)}{\partial \sigma} + \cdots =$$

$$P_{BS}^\star + \left\{ \tau V_0^\delta + \tau V_1^\delta D_1 + \frac{V_3^\varepsilon}{\sigma^\star}D_1 \right\}\frac{\partial P_{BS}^\star}{\partial \sigma} + \cdots .$$

From its definition, we know that $P_{BS}^\star = C_{BS}(\sigma^\star)$, and we can now match separately the $\sqrt{\varepsilon}$ terms and the $\sqrt{\delta}$ terms:

$$\sqrt{\varepsilon}I_{1,0}\frac{\partial C_{BS}(\sigma^\star)}{\partial \sigma} = \frac{V_3^\varepsilon}{\sigma^\star}D_1\frac{\partial P_{BS}^\star}{\partial \sigma}, \tag{5.12}$$

$$\sqrt{\delta}I_{0,1}\frac{\partial C_{BS}(\sigma^\star)}{\partial \sigma} = \left\{ \tau V_0^\delta + \tau V_1^\delta D_1 \right\}\frac{\partial P_{BS}^\star}{\partial \sigma}, \tag{5.13}$$

where we recall that V_3^ε is proportional to $\sqrt{\varepsilon}$, and V_0^δ and V_1^δ are proportional to $\sqrt{\delta}$.

By differentiating the expression for *Vega* in (5.9) with respect to x, we deduce

$$D_1\frac{\partial C_{BS}}{\partial \sigma} = \left(1 - \frac{d_1}{\sigma\sqrt{\tau}}\right)\frac{\partial C_{BS}}{\partial \sigma}. \tag{5.14}$$

From the relations (5.12), (5.13), and (5.14) evaluated at $\sigma = \sigma^\star$, we find

$$\sqrt{\varepsilon}I_{1,0} = \frac{V_3^\varepsilon}{\sigma^\star}\left(1 - \frac{d_1^\star}{\sigma^\star\sqrt{\tau}}\right)$$

$$= \frac{V_3^\varepsilon}{2\sigma^\star}\left(1 - \frac{2r}{\sigma^{\star 2}}\right) + \left(\frac{V_3^\varepsilon}{\sigma^{\star 3}}\right)\frac{\log(K/x)}{\tau}, \tag{5.15}$$

$$\sqrt{\delta}I_{0,1} = \tau V_0^\delta + \tau V_1^\delta\left(1 - \frac{d_1^\star}{\sigma^\star\sqrt{\tau}}\right)$$

$$= \tau\left\{ V_0^\delta + \frac{V_1^\delta}{2}\left(1 - \frac{2r}{\sigma^{\star 2}}\right)\right\} + \left(\frac{V_1^\delta}{\sigma^{\star 2}}\right)\log(K/x). \tag{5.16}$$

In terms of the reduced variable *log-moneyness to maturity ratio* defined by

$$\text{LMMR} = \frac{\log(K/x)}{T - t}, \tag{5.17}$$

the first-order approximation for the implied volatility,

$$I \approx \sigma^\star + \sqrt{\varepsilon}I_{1,0} + \sqrt{\delta}I_{0,1},$$

takes the simple form

$$I \approx b^{\star} + \tau b^{\delta} + \left(a^{\varepsilon} + \tau a^{\delta} \right) \text{LMMR}, \qquad (5.18)$$

where the parameters $(b^{\star}, b^{\delta}, a^{\varepsilon}, a^{\delta})$ are defined in terms of $(\sigma^{\star}, V_0^{\delta}, V_1^{\delta}, V_3^{\varepsilon})$ by

$$b^{\star} = \sigma^{\star} + \frac{V_3^{\varepsilon}}{2\sigma^{\star}} \left(1 - \frac{2r}{\sigma^{\star 2}} \right), \ a^{\varepsilon} = \frac{V_3^{\varepsilon}}{\sigma^{\star 3}}, \qquad (5.19)$$

$$b^{\delta} = V_0^{\delta} + \frac{V_1^{\delta}}{2} \left(1 - \frac{2r}{\sigma^{\star 2}} \right), \ a^{\delta} = \frac{V_1^{\delta}}{\sigma^{\star 2}}. \qquad (5.20)$$

In Figure 5.1, we show an example of an implied volatility surface predicted by this model. Notice the negative slope in the strike variable which becomes progressively more pronounced for short maturities. For fixed maturity the implied volatility is actually affine in the log-moneyness, $\log(K/x)$, for the approximation (5.18).

Since V_3^{ε} is of order $\sqrt{\varepsilon}$ whereas V_0^{δ} and V_1^{δ} are both of order $\sqrt{\delta}$, the relations (5.19) and (5.20) show that the intercept $b^{\star}$ is of order one, the slope coefficients a^{ε} and a^{δ} are of order $\sqrt{\varepsilon}$ and $\sqrt{\delta}$, respectively, and b^{δ} is of order $\sqrt{\delta}$.

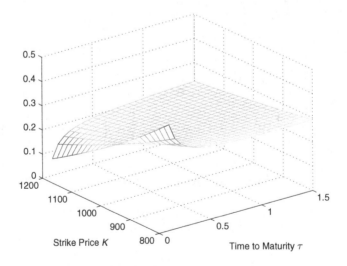

Figure 5.1   The implied volatility surface approximation described in (5.18) for $(a^{\varepsilon}, a^{\delta}, b^{\delta}, b^{\star}) = (-0.0791, -0.1183, 0.0141, 0.2328)$. For a fixed time to maturity the surface is affine in $\log(K/x)$.

In practice, formula (5.18) will be fitted to the observed term structure of implied volatility, as we discuss in detail in the following sections. From the fitted parameters $(a^{\varepsilon}, b^{\star}, a^{\delta}, b^{\delta})$, we are able to calibrate the market group parameters $(\sigma^{\star}, V_0^{\delta}, V_1^{\delta}, V_3^{\varepsilon})$ by inverting the relations above, retaining only the terms up to order $\sqrt{\varepsilon}$ and $\sqrt{\delta}$. From the first equation in (5.19), we deduce the following quadratic equation for $\sigma^{\star}$:

$$\frac{a^{\varepsilon} \sigma^{\star 2}}{2} + \sigma^{\star} - (b^{\star} + ra^{\varepsilon}) = 0.$$

Note that we have $\sigma^{\star} = b^{\star} + \mathcal{O}(\sqrt{\varepsilon})$ and $a^{\varepsilon} = \mathcal{O}(\sqrt{\varepsilon})$, therefore, to the same order of approximation:

$$\frac{a^{\varepsilon} b^{\star 2}}{2} + \sigma^{\star} - (b^{\star} + ra^{\varepsilon}) = 0.$$

We then find

$$\sigma^{\star} = b^{\star} + a^{\varepsilon}\left(r - \frac{b^{\star 2}}{2}\right).$$

From (5.20) it follows that the parameter V_3^{ε} can be written

$$V_3^{\varepsilon} = a^{\varepsilon} \sigma^{\star 3} = a^{\varepsilon} b^{\star 3} + \cdots,$$

where again only terms up to order $\sqrt{\varepsilon}$ are shown. The derivation of expressions for V_0^{δ} and V_1^{δ} from (5.20) is similar.

To the accuracy of our first-order approximation, the *calibration formulas* are:

$$\sigma^{\star} = b^{\star} + a^{\varepsilon}\left(r - \frac{b^{\star 2}}{2}\right), \quad V_3^{\varepsilon} = a^{\varepsilon} b^{\star 3}, \qquad (5.21)$$

$$V_0^{\delta} = b^{\delta} + a^{\delta}\left(r - \frac{b^{\star 2}}{2}\right), \quad V_1^{\delta} = a^{\delta} b^{\star 2}, \qquad (5.22)$$

and hence all the parameters that are needed for pricing can be obtained from the fitting of the term structure of implied volatility.

Note that by the result of Proposition 4.5 at the end of Section 4.2.2, if f is nondecreasing in y then the slope a^{ε} has the sign of ρ_1 as expected. Similarly, using the remark at the end of Section 4.2.4, one sees that in general a^{δ} has the sign of ρ_2.

5.1.1 Relation to Historical Data

We showed above how to get the volatility level $\sigma^\star$ from the implied volatility term structure. This volatility level was introduced in (4.62) by

$$\sigma^\star = \sqrt{\bar{\sigma}^2(z) + 2V_2^\varepsilon(z)}, \qquad (5.23)$$

where $\bar{\sigma}$ is obtained by averaging the two-factor volatility with respect to the invariant distribution of the fast factor:

$$\bar{\sigma}^2 = \bar{\sigma}^2(z) = \langle f^2(\cdot, z) \rangle.$$

Note that this parameter is therefore evaluated at the current level z of the slow volatility factor. The volatility $\bar{\sigma}$ can be obtained by constructing an estimator that computes averages of squared returns data over a time period that is short compared to the slow factor and long compared to the fast factor. The main advantage of constructing the price approximation relative to the volatility level $\sigma^\star$ is that we avoid the complications associated with constructing such an estimator and can rely on the term structure only. From (5.19) and (5.23), we find that $b^\star$ can be expressed by

$$b^\star = \bar{\sigma} + \frac{V_2^\varepsilon}{\bar{\sigma}} + \frac{V_3^\varepsilon}{2\bar{\sigma}}\left(1 - \frac{2r}{\bar{\sigma}^2}\right) + \cdots,$$

where only terms up to order $\sqrt{\varepsilon}$ are shown. It follows that $b^\star$ is a correction to $\bar{\sigma}$; the corrections due to V_2^ε and V_3^ε come from the market price of risk and from the leverage effect associated with the correlation ρ_1, respectively.

5.2 Calibration Procedure

In this section, we discuss the practical details of calibration of parameters to market implied volatility data. On a given day t, we have a finite set of implied volatilities $\{I(T_i, K_{ij})\}$ which are parameterized by expiration dates T_i and strike prices K_{ij}. Notice that options of different maturities may, and typically do, have different strikes. Clearly there is no unique way to fit the formula (5.18) to the usually large dataset to recover estimates of the four parameters $(a^\varepsilon, b^\star, a^\delta, b^\delta)$. However, the data is sparse in the T direction as compared with the K direction and our experience is that it is best to first fit the skew maturity by maturity and then across the term structure, as we now explain.

For fixed T_i, we linearly regress the implied volatilities for strikes $(K_{ij})_j$ on the corresponding LMMRs

$$(\text{LMMR})_{ij} = \frac{\log(K_{ij}/x)}{\tau_i},$$

where x is the present value of the underlying and $\tau_i = T_i - t$, the time to maturity for this subset of options. We use the least-squares criterion to obtain the estimates $\hat{a}_i$ and $\hat{b}_i$ which solve

$$\min_{a_i,b_i} \sum_j \left(I(T_i, K_{ij}) - (a_i(\text{LMMR})_{ij} + b_i)\right)^2. \tag{5.24}$$

Then we regress these estimates on affine functions of τ_i to obtain estimates of the intercept $\hat{a}^{\varepsilon}$ and slope $\hat{a}^{\delta}$ which solve

$$\min_{a_0,a_1} \sum_i \left\{\hat{a}_i - (a_0 + a_1 \tau_i)\right\}^2, \tag{5.25}$$

and the intercept $\hat{b}^{\star}$ and slope $\hat{b}^{\delta}$ which solve

$$\min_{b_0,b_1} \sum_i \left\{\hat{b}_i - (b_0 + b_1 \tau_i)\right\}^2.$$

This two-step fitting procedure gives us estimates of the four group market parameters we need. Some key issues that we comment on in the context of S&P 500 data below are goodness-of-fit and stability over time of the estimated parameters.

5.3 Illustration with S&P 500 Data

We illustrate the fitting procedure using S&P 500 options data obtained from the Wharton Research Data Service (WRDS) database. Our dataset contains implied volatilities from closing S&P 500 European option prices. However, as discussed in Figlewski (2010, section 4), the data needs some treatment to remove inconsistencies and unreliable entries. These include very low priced out-of-the-money contracts where bid–ask spreads are more than 100% of the midpoint price; quotes existing for deep out-of-the-money or in-the-money options which have not been trading recently; and a pronounced "jump" in the implied volatility curve at the point where implied volatilities switch from being computed from put option prices to call option prices. We roughly follow the data-cleaning procedure described in Figlewski (2010), with some modifications, and describe its impact on the test case of April 19, 2005 data.

5.3.1 Data Cleaning

From the WRDS database, we extract daily closing bid and ask quotes for European options on the S&P 500 index, along with each option's strike and expiration date and the WRDS-computed implied volatility. On the illustration date April 19, 2005, there are originally 566 option entries.

- We first filter out datapoints which have
  - bid quotes less than $0.50 (this removes 70 entries on the illustration date);
  - no implied volatility value (this removes a further 90);
  - incorrect expiration dates (an occasional problem, but none on the test date).
- In the WRDS data there is a jump between the implied volatility curves coming from calls or puts with the same maturity. To smooth this out, following the general procedure described in Figlewski (2010), we blend the implied volatility values for options with strikes between certain cutoffs L and H, with $L < H$.
  - *Choosing cutoffs*: For a given maturity, let $\mathcal{K}$ be the set of strikes for which both call and put implied volatilities are available, and let x denote the day's closing level of the S&P 500 index. We set $L = \max(0.85x, \min(\mathcal{K}))$ and $H = \min(1.15x, \max(\mathcal{K}))$. Essentially we set the cutoffs to $+/-15\%$ of at-the-money, and then modify them if necessary since we can only blend when we have a put–call pair. Occasionally there is still a handful of strikes between L and H for which only one type of option is traded, in which case these unpaired values are discarded since they cannot be blended.
  - We discard *puts* with strikes above H and *calls* with strikes below L, that is those that are too far *in*-the-money to be sufficiently liquid. On the test date, this removes 63 datapoints.
  - *Blending*: For each maturity and strike $K \in (L, H)$, let $I_p(K)$ and $I_c(K)$ denote the put and call implied volatilities in the database where they both exist. With $w = w(K) = (H - K)/(H - L)$, we set the implied volatility value $I(K)$ for that strike to be

$$I(K) = wI_p(K) + (1 - w)I_c(K).$$

  On the test date, there are 260 options in the blending region, one put and one call for each strike. Blending halves this number and therefore reduces the data by a further 130 points. So finally, there are 213 datapoints left on April 19, 2005, which is 37.6% of the original

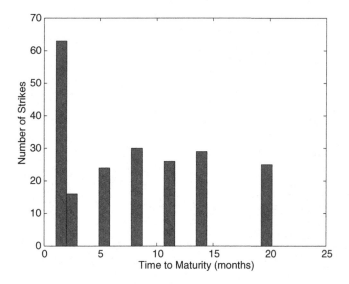

Figure 5.2   Number of strikes for each maturity: S&P 500 options on April 19, 2005.

data size for that day. Figure 5.2 shows the number of strikes for each maturity.

– In summary: for strikes $\leq L$, we use only puts, for strikes $\geq H$, we use only calls, and for strikes between L and H, we blend the two.

• For the month of April 2005, there are $12,173$ original datapoints and we remove about 14% that have low bid prices, 19% where no implied volatility is given, and 10% that are too far in-the-money. The final number of datapoints after blending is 4339, which is 35.6% of the original.

We next proceed to examine the fits of the asymptotic formulas on the data.

5.3.2 Fast Volatility Factor Only

We first look at the performance using the fast scale theory only, ignoring the slow scale. Figure 5.3 shows the fit using only the fast-factor approximation

$$I \approx b^\star + a^\varepsilon (\text{LMMR}),$$

which corresponds to assuming $\delta = 0$. Each strand in Figure 5.3 comes from options of different maturities (with the shortest maturities on the leftmost strand, and the maturity increasing going clockwise). Clearly the

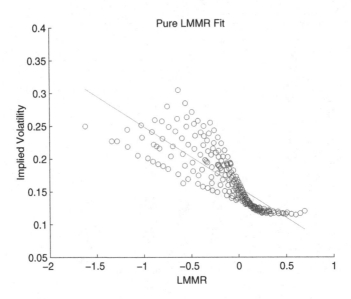

Figure 5.3   S&P 500 implied volatilities as a function of LMMR on April 19, 2005. The circles are from S&P 500 data, and the line $b^\star + a^\varepsilon (LMMR)$ shows the result using the estimated parameters from only an LMMR (fast factor) fit.

single-factor theory struggles to capture the range of maturities and so runs through the middle.

5.3.3 Slow Volatility Factor Only

In Figure 5.4, we show the result of the calibration using only the slow-factor approximation, which corresponds to assuming $\varepsilon = 0$:

$$I \approx \bar{\sigma} + b^\delta \tau + a^\delta (\mathrm{LM}),$$

where we denote $\mathrm{LM} = \tau \mathrm{LMMR}$, the log-moneyness. We plot the maturity adjusted implied volatility defined by $I - b^\delta \tau$. The fit as a function of the regressor LM is shown in Figure 5.4, with the maturities increase going counterclockwise from the top-leftmost strand. Again, the single-factor theory struggles to capture the range of maturities.

5.3.4 Fast and Slow Volatility Factors

Finally, in fitting the two-factor volatility approximation (5.18), we follow the two-step procedure of fitting the skew to obtain $\hat{a}_i$ and $\hat{b}_i$ for different maturities τ_i. These are then fitted to an affine functions of τ to give estimates of a^ε, $b^\star$, a^δ, and b^δ. A plot of this second term-structure fit is

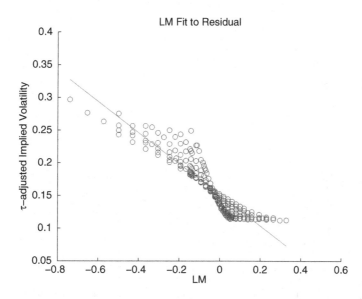

Figure 5.4 τ-adjusted implied volatility $I - b^\delta \tau$ as a function of LM. The circles are from S&P 500 data on April 19, 2005, and the line $\sigma^\star + a^\delta (LM)$ shows the fit using the estimated parameters from only a slow factor fit.

shown in Figure 5.5. Observe that in the regime where our approximation is valid the parameters a^ε, a^δ, and b^δ are expected to be small, while $b^\star$ is the leading-order magnitude of volatility: this is also what we see in Figure 5.5.

In Figure 5.6, we show the calibrated multiscale approximation (5.18) to all the data on April 19, 2005. We see the ability to capture the range of maturities is much improved. The largest misfitting is at the level of the shortest maturities (the leftmost strand). One way to handle this discrepancy uses a periodic modulation of the fast time scale with period corresponding to the monthly expiration cycles of traded options, and we discuss this in Section 5.4. The rightmost *turn of the skew* also comes from the shortest maturity and can be captured using the second-order asymptotic theory derived in Section 9.3 and implemented in this chapter in Section 5.5.2.

The picture remains largely similar in the aftermath of the financial crisis. Figure 5.7 shows the analogs of Figures 5.5 and 5.6 using data from September 19, 2009.

5.3.5 Parameter Stability

A crucial feature of a model is *time stability* of its fitted parameters, since this time stability means that the model captures well the aspects of the dynamics of the underlying, which is essential to ensure consistency when

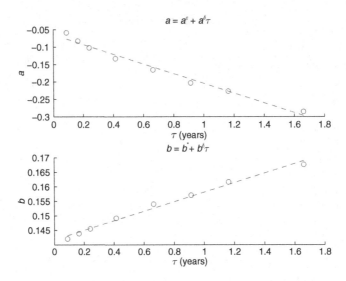

Figure 5.5   Term-structure fits on April 19, 2005. The circles in the top plot are the slope coefficients $\hat{a}_i$ of LMMR fitted in the first step of the regression. The solid line is the straight line $(a^\varepsilon + a^\delta \tau)$ fitted in the second step of the regression. The bottom plot shows the corresponding picture for the skew intercepts $\hat{b}_i$ fitted to the straight line $b^* + b^\delta \tau$. The estimates are $(a^\varepsilon, a^\delta, b^\delta, b^*) = (-0.0646, -0.1397, 0.0164, 0.1417)$.

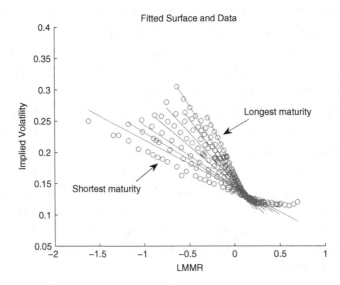

Figure 5.6   Data and calibrated fit on April 19, 2005. The circles are from S&P 500 data, and the lines are the formula (5.18) using $(a^\varepsilon, b^*, a^\delta, b^\delta)$ estimated from the full fast and slow factor fit. The average *relative* fitting error is 3.75%.

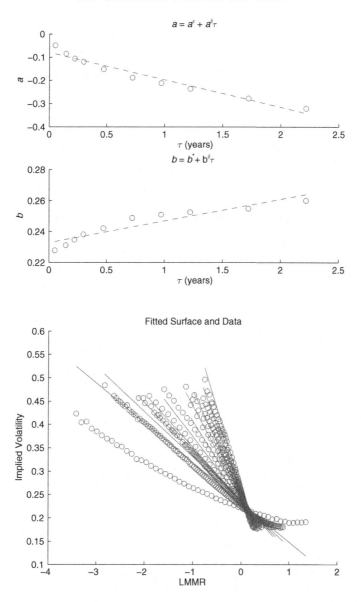

Figure 5.7   Term-structure fits (top) and calibrated fit (bottom) on September 28, 2009. The estimates are $(a^\varepsilon, a^\delta, b^\delta, b^*) = (-0.0791, -0.1183, 0.0141, 0.2328)$ and the average *relative* fitting error is 4.65%.

pricing and hedging path-dependent options, for example. Practically, this means that we are looking at the time evolution (with respect to t) of the estimated parameters $(a^\varepsilon, b^\star, a^\delta, b^\delta)$. This is shown in Figure 5.8 over the period January 2000 through October 2009, using S&P 500 options data, but excluding the shortest maturity. We show also the corresponding

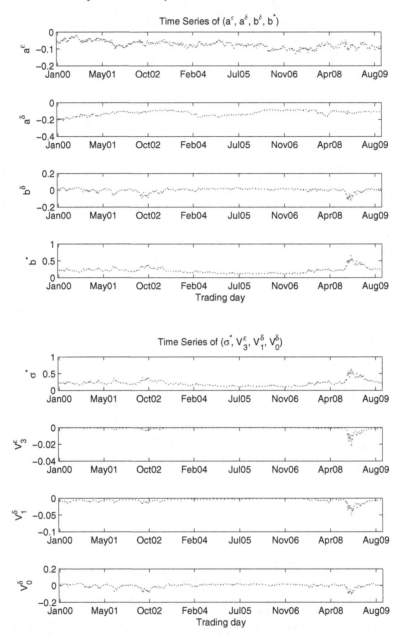

Figure 5.8 Stability of fitted parameters. The top picture shows daily cal-ibrated values of $(a^\varepsilon, b^\star, a^\delta, b^\delta)$, with the corresponding $(\sigma^\star, V_0^\delta, V_1^\delta, V_3^\varepsilon)$ in the bottom picture.

plots for the market group parameters $(\sigma^\star, V_0^\delta, V_1^\delta, V_3^\varepsilon)$ computed from the calibration formulas (5.21)–(5.22).

In these, one sees clearly the financial crisis that erupted in September 2008: in particular, its impact is felt more in $(b^\star, b^\delta)$ than in $(a^\varepsilon, a^\delta)$, indicating a more dramatic jump in implied volatility levels rather than skew slope.

The mean values of the group parameters over this period are:

$$\sigma^\star = 0.2054, \qquad V_0^\delta = 0.0008, \qquad V_1^\delta = -0.0059, \qquad V_3^\varepsilon = -0.0010.$$

The first gives an indication of the "average" S&P 500 volatility level of 20%. The other parameters, the V's, are small as the asymptotic theory demands.

5.4 Maturity Cycles

In the previous calibration we did not use the shortest maturity options, which have the steepest skew. When this maturity is included, an interesting periodicity reveals itself in the fitted slopes of the skew. To see this clearly, we consider for the moment just a fast-factor fit to the first three maturities. We keep the shortest maturity option until three days to expiration, in general the data comprises one-month, two-month, and three-month options, which we fit to the formula

$$I \approx b^\star + a^\varepsilon \text{LMMR}$$

to obtain daily slope estimates a^ε. Figure 5.9 shows these through the year 2007. The ticks on the graph correspond to the expiration dates of S&P 500 options, which is the third Friday of each month. There is clearly a periodic phenomenon coinciding with the expiration dates. These dates (which are known in advance) therefore seem to have some bearing on the evolution of the implied volatility surface. This pattern contains information about

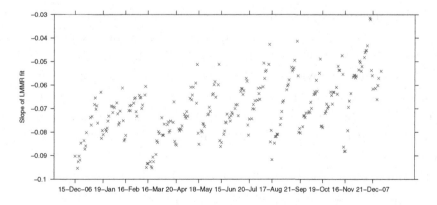

Figure 5.9 Fitted skew slopes a^ε through 2007 using only the first three maturities. The ticks on the graph correspond to the expiration dates of S&P 500 options.

the systematic behavior of market option prices which is not easy to see directly from implied volatility data, but is revealed clearly once the regression with respect to LMMR has been carried out to filter out stochastic volatility effects.

The "jumps" in the fitted skew slope, when the cycle returns to the beginning (it is not continuous), coincide with the closest-to-maturity options disappearing from the data. The fact that some options are expiring has a mild "feedback effect" on the entire options market (with the common underlying), probably through the volatility of the underlying itself. When those options disappear, the shortest length of options in the data jumps from near zero to about 30 days. As a consequence of there not being a very steep skew visible on that day, the estimated skew slope becomes much less negative as the next expiration date is suddenly much further away.

Rather than proposing a mechanism that might explain this, we model the empirical observation directly through a periodic variation in the *local speed* of the fast process driving volatility: it is still fast, but beomes faster periodically leading up to the expiration dates. In the following sections we present the theoretical aspects of this extension and in Section 5.4.5 we illustrate that this greatly improves the stability of the parameter fitting. The "calendar effect" we observe in S&P 500 options is important primarily for options expiring on the next few maturity dates. This cycle effect may not be present in different markets with different expiration date structures, e.g. OTC markets where there is an almost continuous range of maturity dates.

5.4.1 Perturbation with Time-Varying Coefficients

We outline the extension of the asymptotic theory of Chapter 4 to a specific instance of time-dependent parameters in the fast volatility process. Here we choose to introduce variation in the time scale coefficient ε. For our data fitting below, we will make this time scale dependent on the time until the next maturity date, and we will find that the periodic pattern seen in Figure 5.9 is consistent with a local mean reversion speed that increases as we approach each expiration date, and then drops back down until the next one is neared.

Our model under the risk-neutral measure becomes

$$dX_t = rX_t \, dt + f(Y_t, Z_t)X_t \, dW_t^{(0)\star}, \tag{5.26}$$

$$dY_t = \left(\frac{1}{\varepsilon v(t)} \alpha(Y_t) - \frac{1}{\sqrt{\varepsilon v(t)}} \beta(Y_t) \Lambda_1(Y_t) \right) dt + \frac{1}{\sqrt{\varepsilon v(t)}} \beta(Y_t) \, dW_t^{(1)\star},$$

$$dZ_t = \Big(\delta c(Z_t) - \sqrt{\delta}\, g(Z_t)\Lambda_2(Y_t,Z_t)\Big)\,dt + \sqrt{\delta}\, g(Z_t)\,dW_t^{(2)\star}, \qquad (5.27)$$

where we have introduced the time-dependent scaling factor $v(t)$ into the fast factor Y. Typically this factor will become small as t approaches a maturity date. As usual, the standard Brownian motions $\left(W_t^{(0)\star}, W_t^{(1)\star}\right)$ are correlated according to the following cross-variation:

$$d\langle W^{(0)\star}, W^{(1)\star}\rangle_t = \rho_1 dt, \qquad (5.28)$$

where $|\rho_1| < 1$. The volatility of the underlying asset X is $f(Y_t,Z_t)$, where f is a positive function satisfying the conditions in Theorem 4.10. Other aspects of the model are described in more detail in Section 4.1. The special case with $v(t) = 1$ corresponds to the time-independent theory discussed in Chapter 4.

As before, we seek an approximation for option prices in the regime where ε and δ are small. The price of a European derivative is given by

$$P^\varepsilon(t,X_t,Y_t,Z_t) = \mathbb{E}^\star\left\{e^{-r(T-t)}h(X_T) \mid X_t, Y_t, Z_t\right\}, \qquad (5.29)$$

and is also characterized as the solution of the parabolic partial differential equation

$$\mathcal{L}^{\varepsilon,\delta}(t)P^\varepsilon = 0, \qquad (5.30)$$
$$P^\varepsilon(T,x,y) = h(x),$$

where the operator $\mathcal{L}^{\varepsilon,\delta}(t)$ can be expressed as a sum of components:

$$\mathcal{L}^\varepsilon(t) = \frac{1}{\varepsilon v(t)}\mathcal{L}_0 + \frac{1}{\sqrt{\varepsilon v(t)}}\mathcal{L}_1 + \mathcal{L}_2 + \sqrt{\delta}\mathcal{M}_1 + \delta\mathcal{M}_2 + \sqrt{\frac{\delta}{\varepsilon v(t)}}\mathcal{M}_3,$$

with the operators $\mathcal{L}_i$ being defined in (4.8), (4.9), and (4.10) and the $\mathcal{M}_i$ in (4.11), (4.12), and (4.13). Thus, we find that the problem described by (5.30) corresponds exactly to the one described by (4.5) with the replacements

$$\mathcal{L}_0 \longmapsto \frac{1}{v(t)}\mathcal{L}_0, \quad \mathcal{L}_1 \longmapsto \frac{1}{\sqrt{v(t)}}\mathcal{L}_1, \quad \mathcal{M}_3 \longmapsto \frac{1}{\sqrt{v(t)}}\mathcal{M}_3.$$

The argument presented in Section 4.2 can now be repeated with some slight modifications.

We first expand the price as in Section 4.2.1:

$$P^\varepsilon = P_0 + \varepsilon^{1/2}P_{1,0} + \delta^{1/2}P_{0,1} + \cdots.$$

With our usual definition of the constant effective volatility

$$\bar{\sigma}(z) = \sqrt{\langle f^2(\cdot,z)\rangle},$$

one can follow the lines of the derivation of the first-order approximation in Chapter 4, and obtain that the leading-order term P_0 is the Black–Scholes price P_{BS} which solves

$$\mathscr{L}_{BS}(\bar{\sigma}(z))P_{BS} = 0,$$
$$P_{BS}(T,x) = h(x).$$

Next, one finds that the first correction $P_{1,0}^{\varepsilon} = \sqrt{\varepsilon}P_{1,0}$ is modified because the operator $\mathscr{A}^{\varepsilon}$ in (4.32) is replaced by

$$\sqrt{\varepsilon}\left\langle \frac{1}{\sqrt{v(t)}}\mathscr{L}_1\left(\frac{1}{v(t)}\mathscr{L}_0\right)^{-1}(\mathscr{L}_2 - \langle\mathscr{L}_2\rangle)\right\rangle$$
$$= \sqrt{\varepsilon v(t)}\langle\mathscr{L}_1\mathscr{L}_0^{-1}(\mathscr{L}_2 - \langle\mathscr{L}_2\rangle)\rangle = \sqrt{v(t)}\mathscr{A}^{\varepsilon}.$$

It follows that $P_{1,0}^{\varepsilon}$ solves the inhomogeneous problem

$$\mathscr{L}_{BS}(\bar{\sigma})P_{1,0}^{\varepsilon} = -\sqrt{v(t)}\left(V_2^{\varepsilon}D_2P_{BS} + V_3^{\varepsilon}D_1D_2P_{BS}\right),$$
$$P_{1,0}^{\varepsilon}(T,x) = 0.$$

Using the argument in Section 4.2.3, one can easily check that the solution is given explicitly in terms of P_{BS} by

$$P_{1,0}^{\varepsilon} = \left(\int_t^T \sqrt{v(s)}\,ds\right)\left(V_2^{\varepsilon}D_2P_{BS} + V_3^{\varepsilon}D_1D_2P_{BS}\right).$$

We define now the time-averaged quantity

$$\overline{v_{t,T}^{1/2}} = \frac{1}{T-t}\int_t^T \sqrt{v(s)}\,ds.$$

It is straightforward to see that the slow scale correction is unaffected by the time-varying coefficient in the fast factor, and therefore the first-order approximation in (4.58) becomes

$$\tilde{P}^{\varepsilon,\delta} = P_{BS} + (T-t)\left[V_0^{\delta}(z)\frac{\partial}{\partial\sigma} + V_1^{\delta}(z)D_1\left(\frac{\partial}{\partial\sigma}\right) + \overline{v_{t,T}^{1/2}}\left(V_2^{\varepsilon}(z)D_2\right.\right.$$

$$\left.\left. + V_3^{\varepsilon}(z)D_1D_2\right)\right]P_{BS}. \tag{5.31}$$

Clearly, the effect of the time-dependent parameter at this level is the replacement

$$V_2^{\varepsilon}(z) \longmapsto \overline{v_{t,T}^{1/2}}V_2^{\varepsilon}(z), \qquad V_3^{\varepsilon}(z) \longmapsto \overline{v_{t,T}^{1/2}}V_3^{\varepsilon}(z).$$

Following the parameter-reduction argument given in Section 4.3, one can deduce that the first-order price approximation takes the form

$$\tilde{P}^{\star} := P_{BS}(\sigma_{t,T}^{\star}) + (T-t)\left[V_0^{\delta}(z)\frac{\partial}{\partial\sigma} + V_1^{\delta}(z)D_1\left(\frac{\partial}{\partial\sigma}\right)\right.$$

$$\left. + \overline{v_{t,T}^{1/2}}V_3^{\varepsilon}(z)D_1D_2\right]P_{BS}(\sigma_{t,T}^{\star}), \qquad (5.32)$$

where the (time-dependent) corrected volatility $\sigma_{t,T}^{\star}$ is given by

$$\sigma_{t,T}^{\star} = \sqrt{\bar{\sigma}^2 + 2\overline{v_{t,T}^{1/2}}V_2^{\varepsilon}}. \qquad (5.33)$$

This generalization involves the time-dependent Black–Scholes operator with deterministic time-varying square volatility $\bar{\sigma}^2 + 2\sqrt{v(t)}V_2^{\varepsilon}$, and the fact that the corresponding Black–Scholes price is obtained as the Black–Scholes price with time-averaged square volatility as explained in Section 2.2.1. The accuracy results given in Section 4.5 can be generalized to this time-dependent case and the price correction due to the slow factor is not affected by the time variation in the parameters to the accuracy of our approximation.

5.4.2 Calendar Time

We will choose $v(t)$ to be a power of the time to next maturity date. Let T_n denote the maturity dates (for example, the third Fridays of the month in the case of S&P 500 options). We define $n(t)$, the time to the next maturity, by

$$n(t) = \inf\{n : T_n - t \ge \Delta t\}, \qquad (5.34)$$

with $\Delta t > 0$ a small cutoff which will prevent blowup of v^{-1} at maturity dates. Then our maturity cycles model is

$$v(t) = c_p\left(T_{n(t)} - t\right)^p, \qquad p > 0, \qquad (5.35)$$

where p is a parameter and c_p is a normalization constant to be chosen. We suppose that the option expiration dates are approximately ΔT apart (for example, one month). Then c_p is chosen so that

$$\frac{\int_0^{\Delta T}(v(s))^{1/2}\,ds}{\Delta T} = 1. \qquad (5.36)$$

This gives

$$c_p^{1/2} = \frac{1}{(\Delta T)^{p/2}}\left(1 + \frac{p}{2}\right).$$

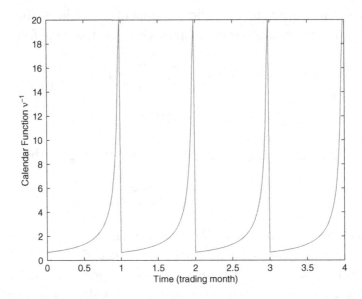

Figure 5.10   The calendar function $1/v(t)$ defined in (5.35), with $p = 1$, as a function of trading month.

The choice $p = 0$ corresponds to a constant rate of volatility mean-reversion, as we had in Chapter 4. The data presented in Section 5.4.5 is consistent with the choice $p = 1$. In Figure 5.10 we plot the "calendar function" v^{-1}; in this case, it is small at the beginning of the trading month and becomes very large close to the expiration dates. A very large v^{-1} corresponds to a very rapid mean-reversion rate of the Y process.

For large times to maturity the time-dependent case corresponds approximately to the constant parameter case since the monthly variation in the parameter averages out. To show this, assume that the period between the expiration dates ΔT is constant and that the time to maturity is decomposed as $T - t = m_0 \Delta T + \eta$ with m_0 an integer and $0 \leq \eta < \Delta T$. Also assume for simplicity the cutoff $\Delta t = 0$ in (5.34). We then find, in view of (5.35) and (5.36),

$$
\overline{v_{t,T}^{1/2}} = \frac{1}{T-t} \int_t^T (v(s))^{1/2}\, ds = \frac{1}{T-t} \int_0^{m_0 \Delta T + \eta} (v(T-s))^{1/2}\, ds
$$

$$
= \frac{\sqrt{c_p}}{m_0 \Delta T + \eta} \left[\int_t^{t+\eta} (t+\eta - s)^{p/2}\, ds + m_0 \int_0^{\Delta T} (\Delta T - s)^{p/2}\, ds \right]
$$

$$
= \frac{1}{(\Delta T)^{p/2}} \frac{\left(\eta^{1+\frac{p}{2}} + m_0 (\Delta T)^{1+\frac{p}{2}} \right)}{m_0 \Delta T + \eta}.
$$

Therefore, when $T - t$ is within one ΔT of maturity and so $m_0 = 0$, we have

$$\overline{v_{t,T}^{1/2}} = \left(\frac{T-t}{\Delta T}\right)^{p/2},$$

and for longer maturities when $m_0 \Delta T >> \eta$,

$$\overline{v_{t,T}^{1/2}} \approx 1,$$

so we return to the time-independent theory.

5.4.3 Calibration with Time-Dependent Parameters

In this section we generalize the analysis presented in Section 5.1 to the case when the price approximation is given by (5.32). Thus, we consider a European call option with strike K and maturity T so that the payoff function is $h(x) = (x - K)^+$. The call price at the *constant* volatility level σ is denoted $C_{BS}(\sigma)$ and given by the Black–Scholes formula in (5.6) for the case $\sigma = \sigma^\star$. The implied volatility I associated with our stochastic volatility model solves

$$C_{BS}(I) = P^\varepsilon,$$

the volatility level implied by the Black–Scholes formula for the price given by the fully specified stochastic volatility model.

Consider first the case of *only* the fast volatility factor in the model. In the case of a call, our price approximation takes the form $P^\varepsilon \approx P_{BS}(\sigma_{t,T}^\star) + P_{1,0}^\varepsilon(\sigma_{t,T}^\star)$. We next identify the approximation for the implied volatility skew that follows from this first-order price approximation. As before, we expand as

$$I = \sigma_{t,T}^\star + \sqrt{\varepsilon} I_{1,0} + \cdots,$$

then the expression (5.15) generalizes to

$$\sqrt{\varepsilon} I_{1,0} = \left(\frac{V_3^\varepsilon}{2\sigma_{t,T}^\star}\left(1 - \frac{2r}{(\sigma_{t,T}^\star)^2}\right) + \left(\frac{V_3^\varepsilon}{(\sigma_{t,T}^\star)^3}\right)\frac{\log(K/x)}{\tau}\right)\overline{v_{t,T}^{1/2}},$$

and (5.18) to

$$I \approx b_{t,T}^\star + a_{t,T}^\star \, \text{LMMR} \, \overline{v_{t,T}^{1/2}},$$

where

$$a_{t,T}^\varepsilon = \frac{V_3^\varepsilon}{(\sigma_{t,T}^\star)^3},$$

$$b_{t,T}^\star = \sigma_{t,T}^\star + \frac{V_3^\varepsilon}{2\sigma_{t,T}^\star}\left(1 - \frac{2r}{(\sigma_{t,T}^\star)^2}\right)\overline{v_{t,T}^{1/2}}.$$

Recall that V_2^ε and V_3^ε are of order $\sqrt{\varepsilon}$. Moreover, that

$$\sigma^\star = \sqrt{\bar\sigma^2 + 2V_2^\varepsilon},$$

$$\sigma_{t,T}^\star = \sqrt{\bar\sigma^2 + 2\overline{v_{t,T}^{1/2}}V_2^\varepsilon},$$

so that we can write

$$a_{t,T}^\varepsilon = \frac{V_3^\varepsilon}{(\sigma^\star)^3} + \mathcal{O}(\varepsilon) = a^\varepsilon + \mathcal{O}(\varepsilon), \tag{5.37}$$

$$b_{t,T}^\star = \bar\sigma + \frac{V_2^\varepsilon}{\bar\sigma}\overline{v_{t,T}^{1/2}} + \frac{V_3^\varepsilon}{2\sigma^\star}\left(1 - \frac{2r}{(\sigma^\star)^2}\right)\overline{v_{t,T}^{1/2}} + \mathcal{O}(\varepsilon) \tag{5.38}$$

$$= \bar\sigma + (b^\star - \bar\sigma)\overline{v_{t,T}^{1/2}} + \mathcal{O}(\varepsilon), \tag{5.39}$$

with a^ε and $b^\star$ defined in (5.19).

Recall that in the case where the slow factor is also present, the price approximation can be written in the form (5.32). Since the price correction that is due to the slow factor is not affected by the time dependence in the parameters, we find that the calibration formulas (5.21) and (5.22) for the group market parameters associated with the slow factor, V_0^δ and V_1^δ, remain unchanged. Now, incorporating back the slow factor leads to the following first-order approximation for the implied volatility:

$$I \approx \bar\sigma + (b^\star - \bar\sigma)\overline{v_{t,T}^{1/2}} + a^\varepsilon\,\text{LMMR}\,\overline{v_{t,T}^{1/2}} + b^\delta\tau + a^\delta\text{LM}$$

$$= \bar\sigma + \Delta b\,\overline{v_{t,T}^{1/2}} + a^\varepsilon\,\text{LMMR}_v + b^\delta\tau + a^\delta\text{LM}, \tag{5.40}$$

with

$$\text{LMMR}_v = \text{LMMR}\,\overline{v_{t,T}^{1/2}},$$

$$\Delta b = b^\star - \bar\sigma.$$

Notice that by observing the implied volatilities for various maturities and values for the moneyness we can estimate the parameters a^ε, $b^\star$, and $\bar\sigma$ (as well as b^δ and a^δ). To the accuracy of our first-order approximation we then obtain *calibration formulas* for V_2^ε and V_3^ε from (5.37), (5.38), and (5.39) by

$$\frac{V_2^\varepsilon}{\bar{\sigma}} = \Delta b + a^\varepsilon \left(r - \frac{b^{\star 2}}{2} \right), \quad V_3^\varepsilon = a^\varepsilon (b^\star)^3.$$

We assume that $v(t)$ is a known function, therefore these parameters are sufficient to obtain a first-order approximation for $\sigma_{t,T}^\star$. Thus, from the calibration of the implied volatility term structure we can identify the parameters that are needed to compute the price approximation in (5.32).

In the next sections we continue our example using S&P 500 data to illustrate some practical aspects of the calibration. As mentioned earlier, these data are consistent with the model (5.35) for $p = 1$. We therefore examine the form for the price and the implied volatility in this special case. In our discussion below we assume that the intervals between the maturity dates are constant and equal to ΔT. We have

$$\overline{v_{t,T}^{1/2}} = \begin{cases} \sqrt{\frac{T-t}{\Delta T}} & \text{for } T - t < \Delta T, \\ 1 & \text{for } T - t \gg \Delta T, \end{cases}$$

from which it follows that

$$\text{LMMR}_v = \begin{cases} \dfrac{\log(K/x)}{\sqrt{(T-t)\Delta T}} & \text{for } T - t < \Delta T, \\ \dfrac{\log(K/x)}{T-t} & \text{for } T - t \gg \Delta T. \end{cases}$$

In fact, a common observation is that for fixed $K \neq x$ the observed term structure (variation with time to maturity $T - t$) behaves like $(T - t)^{-1/2}$ rather than $(T - t)^{-1}$. The time-dependent theory presented above provides a bridge to this observation and below we examine this transition in more detail by returning to the S&P 500 data, first looking at some practical aspects of the data fitting.

5.4.4 Practical Fitting with Time-Dependent Parameters

We will now fit the implied volatility surface to (5.40) to obtain estimates of $(a^\varepsilon, b^\star, \bar{\sigma}, b^\delta, a^\delta)$. As usual, on a given day t, we have a finite set of implied volatility $\{I(T_i, K_{ij})\}$ data which are parameterized by expiration dates T_i and strike prices K_{ij}. Again there are far more observation points in the K dimension than the T dimension, and we find it convenient to implement a two-step regression procedure where we first carry out a linear regression with respect to moneyness for fixed t and T.

First, for each maturity date T_i, we linearly regress the implied volatilities for the various strikes $(K_{ij})_j$ on the variable

$$(\text{LM})_{ij} = \log(K_{ij}/x),$$

with $X_t = x$ being the present value for the underlying. We use the least-squares criterion to obtain the coefficients $\hat{a}_i$ and $\hat{b}_i$ which solve

$$\min_{a_i,b_i} \sum_j \left(I(T_i, K_{ij}) - (a_i(\text{LM})_{ij} + b_i) \right)^2.$$

This step gives estimates

$$\hat{a}_i \approx a^\varepsilon \frac{\overline{v_{t,T_i}^{1/2}}}{\tau_i} + a^\delta, \qquad \hat{b}_i \approx b_{t,T_i}^\star = \bar{\sigma} + \Delta b \overline{v_{t,T_i}^{1/2}} + b^\delta \tau_i,$$

for each time to maturity $\tau_i = T_i - t$.

Next, we obtain estimates $\widehat{\Delta b}$, $\widehat{\bar{\sigma}}$ and $\widehat{b^\delta}$ by regressing on the factors $\overline{v_{t,T_i}^{1/2}}$ and τ_i as

$$\min_{\sigma,\Delta b,b^\delta} \sum_i \left\{ \hat{b}_i - (\sigma + \Delta b \overline{v_{t,T_i}^{1/2}} + b^\delta \tau_i) \right\}^2. \tag{5.41}$$

Finally, to get estimates $\widehat{a^\delta}$ and $\widehat{a^\varepsilon}$, we regress the estimated skew slopes $\hat{a}_i$ on $\overline{v_{t,T_i}^{1/2}}/\tau_i$:

$$\min_{a^\delta,a^\varepsilon} \left\{ \hat{a}_i - \left(a^\varepsilon \frac{\overline{v_{t,T_i}^{1/2}}}{\tau_i} + a^\delta \right) \right\}^2. \tag{5.42}$$

This then gives the group market parameters we need. Notice that we obtain estimates for each day t and can examine the stability properties by plotting these as a function of time and the goodness-of-fit properties by plotting the fit with respect to the calibrated model for a given day.

5.4.5 Maturity Cycles for S&P 500 Data

We illustrate the performance of the fitting, with the maturity cycles, on one day's data in Figures 5.11 and 5.12.

We argue that the above data analysis illustrates how just as implied volatility is a convenient transformation for capturing the deviations of options prices from the Black–Scholes theory, the LMMR representation is useful for picking out secondary phenomena relative to the basic skew captured by a stochastic volatility theory. We next illustrate this further by looking at aspects of the skew that cannot be captured by the first-order theory described above, but that can be handled by higher-order terms in the asymptotic price approximation.

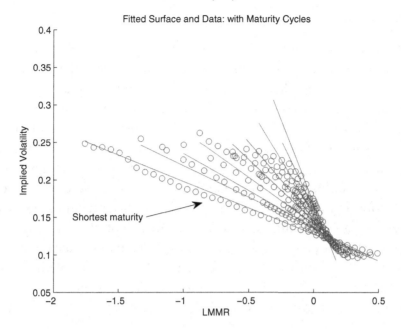

Figure 5.11   Fit of maturity cycles approximation formula (5.40) on S&P 500 options data from July 26, 2006. Notice the better fit to the shortest maturity.

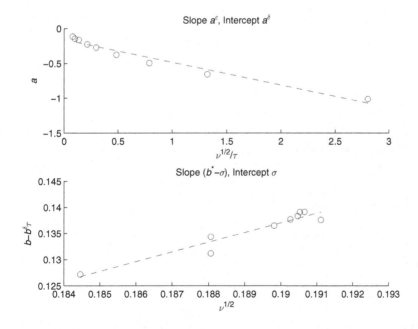

Figure 5.12   Term-structure fit of maturity cycles approximation formula (5.40) on S&P 500 options data from July 26, 2006. The top plot shows the fit of $\hat{a}$ against $\frac{\overline{v_{t,T}^{1/2}}}{\tau}$ in the formula (5.42). The bottom plot illustrates the fit (5.41) of $\hat{b}$ by plotting $\hat{b} - b^\delta\tau$ against $\overline{v_{t,T}^{1/2}}$.

5.5 Higher-Order Corrections

We have discussed fitting of the implied volatility surface for a wide range
of maturities. Fitting of the surface for very long or short-dated options led
us to generalize the model by introducing, respectively, time-varying coefficients and a slow volatility factor. In this section we examine the implied
volatility surface for extreme strikes giving large values for the moneyness. The surface often turns or shows "wings" for extreme strikes, see
for instance Figure 5.13. The first-order perturbation theory gives skews
that are affine in log-moneyness and does not capture these wings well. A
natural extension is now to introduce higher-order terms in the expansion.
The next set of terms turns out to give models that allow for skews that
are quartic (fourth-order) polynomials in log-moneyness. Recall that in our
perturbation approach we first define a class of stochastic volatility models
containing fast and slow volatility factors. We then expand the corresponding pricing equation with respect to the two small parameters ε and δ, with

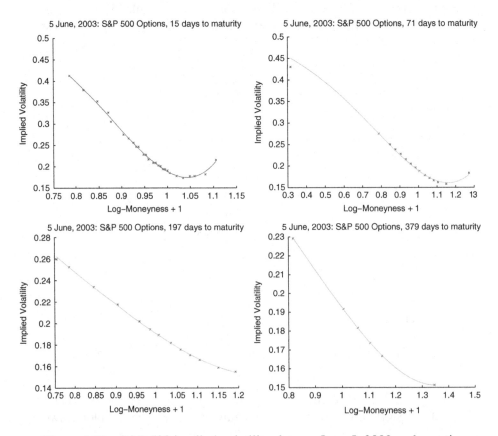

Figure 5.13   S&P 500 implied volatility data on June 5, 2003 and quartic
fits to the asymptotic theory for four maturities.

ε being the time scale of the fast factor and δ being the reciprocal of the time scale of the slow factor. In the notation introduced in Chapter 4, we write the price as

$$P^{\varepsilon,\delta} = P_0 + \sqrt{\varepsilon}P_{1,0} + \sqrt{\delta}P_{0,1} + \varepsilon P_{2,0} + \sqrt{\varepsilon\delta}P_{1,1} + \delta P_{0,2} + \cdots,$$

with $P_0 + \sqrt{\varepsilon}P_{1,0} + \sqrt{\delta}P_{0,1}$ corresponding to the first-order theory. In the second-order theory we include the next three terms as well. Indeed, this increases the number of parameters or degrees of freedom in the model for the implied volatility. The first-order theory involves the four group market parameters $(a^\varepsilon, b^\star, a^\delta, b^\delta)$, whereas the second-order theory gives eleven parameters. By including the second-order terms we thus improve the quality of the fit to the skew and the accuracy of the pricing formulas. Here, we describe the approximation for the implied volatility that follows from including the second-order terms and how the extended model captures the observed wings in the skew. The theoretical derivation is presented in Sections 9.3 and 9.4. Notice also that in this section we consider the case with time-independent parameters.

The derivation of the second-order approximation to the implied volatility presented in Section 9.3.4 shows that outside of a small terminal layer (very close to expiration), the formula (5.18) for the implied volatilities now generalizes to

$$I \approx \sum_{j=0}^{4} a_j(\tau)(\text{LMMR})^j + \frac{1}{\tau}\Phi_t, \tag{5.43}$$

where as before $\tau = T - t$, $\text{LMMR} = \log(K/x)/(T-t)$, and Φ_t varies with the fast volatility factor Y_t. The LMMR coefficients are all third-order polynomials in the time to maturity and have the form:

$$a_1(\tau) = \sum_{k=0}^{3} a_{1,k}\tau^k, \quad a_2(\tau) = \sum_{k=0}^{3} a_{2,k}\tau^k,$$

$$a_3(\tau) = \sum_{k=2}^{3} a_{3,k}\tau^k, \quad a_4(\tau) = \sum_{k=2}^{3} a_{4,k}\tau^k.$$

The coefficients $a_{j,k}$ are now the relevant parameters and we discuss next practical aspects of fitting these to the observed implied volatility surface.

5.5.1 Second-Order Fitting

We describe how to calibrate the model (5.43) to the observed implied volatility records. Again, we employ a two-stage fitting procedure that recognizes the thinness of data in the maturity dimension, relative to the many available strikes. The procedure now also makes use of the decomposition of the implied volatility into first and second-order parts. Since the second-order terms are correction terms that are associated with additional degrees of freedom, we first fit the data to the leading or first-order model exactly as in Section 5.2, and then we fit the residual to the additional degrees of freedom provided by the second-order theory. In other words, by viewing the wings as small corrections to the linear skew, we avoid the "tail wagging the dog" phenomenon.

The first stage in the fitting procedure is thus as described in Section 5.2. That is, for each day t we carry out the least squares regression in (5.24) to obtain $\hat{a}_i^I$ and $\hat{b}_i^I$ by

$$\min_{a_i^I, b_i^I} \sum_j \left(I(T_i, K_{ij}) - (a_i^I (\text{LMMR})_{ij} + b_i^I) \right)^2,$$

for a given maturity τ_i. Note that the coefficient $\hat{b}_i^I$ contains a small, rapidly fluctuating component due to the term Φ_t. Here, however, our primary focus is the behavior of the surface in the moneyness dimension rather than the maturity dimension, as was our focus above.

Then we identify second-order effects in the implied volatility by computing the residuals

$$I^{II}(T_i, K_{ij}) = \frac{I(T_i, K_{ij}) - (a_i^I (\text{LMMR})_{ij} + b_i^I)}{((\text{LMMR})_{ij} + 1)^2}.$$

Observe that we introduced a shift in the denominator to avoid divide by zero issues. The residual is then fitted to a quadratic polynomial in LMMR as above,

$$\min_{c_i^{II}, a_i^{II}, b_i^{II}} \sum_j \left(I^{II}(T_i, K_{ij}) - (c_i^{II} (\text{LMMR})_{ij}^2 + a_i^{II} (\text{LMMR})_{ij} + b_i^{II}) \right)^2.$$

The final stage of the fitting procedure consists of fitting these coefficients that are parameterized by the time to maturity τ_i to a cubic polynomial in τ, again using the least squares criterion. The coefficients of the polynomial $a_1(\tau)$ solve

$$\min_{a_{1,k}} \sum_i \left(\hat{a}_i^I - \sum_{k=0}^3 a_{1,k} \tau_i^k \right)^2,$$

and similarly for the other coefficients.

5.5.2 Capturing the S&P 500 Wings

We continue our example with the S&P 500 data and now carry out the second-order fitting procedure described in the previous section.

First, we show some typical quartic fits of S&P 500 implied volatilities for a few maturities in Figure 5.13. These plots show that the quartic produced by second-order approximation becomes important in capturing the turn of the skew, in particular for short maturities.

Figure 5.14 shows the fits of the observed $a_j(\tau_i)$'s as introduced in (5.43) to their calibrated term-structure formulas for S&P 500 data on June 5, 2003.

The final step is to recover the parameters needed for pricing from the estimates of $\{a_{j,k}\}$, the analog of the group market parameters (5.5) in the first-order theory which are obtained via the calibration formulas (5.21) and (5.22). In the second-order theory these relations are no longer linear, and a nonlinear inversion algorithm is required based on formulas such as (9.53) for the fast factor presented in Chapter 9. This aspect has to be treated case by case in order to take advantage of the particular features of the market under study. For instance, in FX markets, the correlation between the

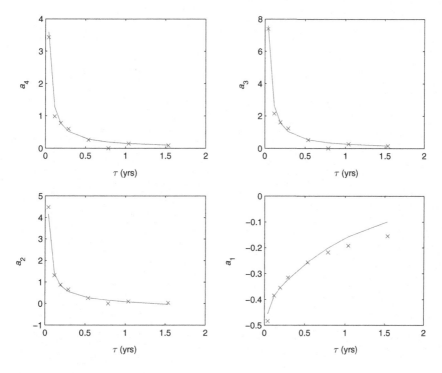

Figure 5.14   S&P 500 term-structure fit using second-order approximation. Data from June 5, 2003.

underlying and its volatility tends to be zero, which reduces the complexity of the implementation of the second-order theory.

Notes

The first-order calibration with fast and slow scales appeared in Fouque *et al.* (2003a), the time-dependent parameter derivation and calibration appeared in Fouque *et al.* (2004a), and the second-order calibration appeared in Fouque *et al.* (2004b). We thank Dan Lacker for research assistance with the data calibration in this chapter.

Phenomena with market dynamics that change as the time to the next "witching hour" gets small are well known. In fact, there are also special end-of-the-year effects, but we do not deal with those here.

6

Application to Exotic Derivatives

In this chapter, we show how to compute the first-order approximations for other claims (digital, barrier, and Asian options). This is where we see that the only parameters needed in these approximations are the ones calibrated from the observed implied volatilities as discussed in Chapter 5.

6.1 European Binary Options

We start with an example of a binary or digital option, with no early exercise feature. We consider a cash-or-nothing call that pays a fixed amount Q on date T if $X_T > K$, and zero if $X_T \leq K$. Its payoff function is

$$h(x) = Q\mathbf{1}_{\{x>K\}}, \tag{6.1}$$

where $\mathbf{1}_A$ denotes the indicator function of a set A.

6.1.1 Approximation Formula

Applying the perturbation theory of Chapter 4, we compute the stochastic volatility-corrected price

$$\widetilde{P}(t,x,z) = P_0(t,x,z) + P_1(t,x,z).$$

The leading term $P_0(t,x,z) = P_{BS}^\star(t,x)$ is simply the Black–Scholes price of the contract with volatility $\sigma^\star$ defined in (4.62) and calibrated in Section 5.1. From (1.69) in Chapter 1, it is given by

$$P_{BS}^\star(t,x) = Qe^{-r(T-t)}N(d_2), \tag{6.2}$$

where

$$d_2 = \frac{\log(x/K) + \left(r - \frac{1}{2}\sigma^{\star 2}\right)(T-t)}{\sigma^\star\sqrt{T-t}}.$$

The correction $P_1(t,x,z)$ satisfies

$$\mathscr{L}_{BS}(\sigma^\star)P_1 = -\mathscr{H}_\star P_{BS}^\star,$$
$$P_1(T,x,z) = 0,$$

as in Section 4.3, using the notation $\mathscr{L}_{BS}$ for the Black–Scholes differential operator, defined in (1.40). Recall from (4.65) that $\mathscr{H}_\star$ is given by

$$\mathscr{H}_\star = 2V_0^\delta \frac{\partial}{\partial\sigma} + 2V_1^\delta x\frac{\partial}{\partial x}\left(\frac{\partial}{\partial\sigma}\right) + V_3^\varepsilon x\frac{\partial}{\partial x}\left(x^2\frac{\partial^2}{\partial x^2}\right)$$

$$\equiv 2V_0^\delta \frac{\partial}{\partial\sigma} + 2V_1^\delta D_1\left(\frac{\partial}{\partial\sigma}\right) + V_3^\varepsilon D_1 D_2,$$

where the small parameters V_0^δ, V_1^δ, and V_3^ε are calibrated to the implied volatility surface of European call options as explained in Chapter 5.

We know from the analysis of European contracts in Chapter 4 that the solution $P_1(t,x,z)$ is given by

$$P_1(t,x,z) = (T-t)\left(V_0^\delta \frac{\partial}{\partial\sigma} + V_1^\delta D_1\left(\frac{\partial}{\partial\sigma}\right) + V_3^\varepsilon D_1 D_2\right)P_{BS}^\star,$$

where $P_{BS}^\star$ is the Black–Scholes binary price given by (6.2). Direct computation yields the explicit formula

$$P_1(t,x,z) = \frac{Q\tau e^{-r\tau - d_2^2/2}}{\sqrt{2\pi}}\left[-V_0^\delta\left(\frac{d_2}{\sigma^\star} + \sqrt{\tau}\right)\right.$$

$$\left. + \left(\frac{V_1^\delta}{\sigma^\star} + \frac{V_3^\varepsilon}{\sigma^{\star 2}\tau}\right)\left(\frac{d_2^2 - 1}{\sigma^\star\sqrt{\tau}} + d_2\right)\right], \qquad (6.3)$$

where $\tau = T - t$ as usual.

6.1.2 Accuracy of the Approximation

The payoff of a binary option is discontinuous, and therefore less regular than that of a call option. In order to establish the rate of convergence of the singular perturbation approximation, one can use the regularization method introduced in Fouque *et al.* (2003c), which consists of controlling the successive derivatives of the Black–Scholes price with the regularized payoff, parameterized by a small parameter η. In the case of the call, the first derivative can be taken on the original payoff, while the remaining derivatives are taken on the smoothing kernel. Adapting the proof given in Fouque *et al.* (2003c) to the case of the digital payoff, all the derivatives have to hit the kernel, and therefore it produces an extra $\eta^{-1/2}$ factor. Consequently, the problem of optimal bounding of the error terms $(\eta, \varepsilon \log |\eta|, \frac{\varepsilon^{3/2}}{\sqrt{\eta}})$ is now to optimally bound $(\eta, \frac{\varepsilon \log |\eta|}{\sqrt{\eta}}, \frac{\varepsilon^{3/2}}{\eta})$. Substituting $\eta = \varepsilon^q$, we reduce to the max–min problem

$$\max \min \left\{ q, 1 - \frac{q}{2}, \frac{3}{2} - q \right\},$$

which admits the solution $q = 2/3$.

The regular perturbation with respect to δ gives an order of accuracy $\mathcal{O}(\delta)$, and therefore pointwise in (t, x, y, z),

$$P^{\varepsilon, \delta} - (P_0 + P_1) = \mathcal{O}(\varepsilon^{2/3} \log |\varepsilon| + \delta).$$

This argument was introduced in Fouque *et al.* (2006) in the context of defaultable bonds. In fact, as discussed in the remark at the end of Section 4.5 in the case of call payoffs, the above error estimate may not be optimal.

6.2 Barrier Options

We demonstrate now how the asymptotic theory handles path-dependent securities by computing the stochastic volatility correction for a down-and-out barrier call option, introduced in Section 1.2.3. Recall that this contract gives the holder the right to buy the underlying asset on expiration date T for strike price K unless the asset price has hit the barrier B at any time before T, in which case the contract expires worthless. In what follows, we shall assume $B < K$. Again, there is an explicit formula in the constant volatility case.

6.2.1 First-Order Correction

In the stochastic volatility environment, the price $P(t, x, y, z)$ of the barrier option satisfies (4.5) with $P(T, x, y, z) = (x - K)^+$, and the boundary condition $P(t, B, y, z) = 0$ at $x = B$. The asymptotic calculations of Chapter 4 are not affected by the smaller fixed domain $\{x > B\}$ of this problem, as long as we keep track of this boundary condition. Our first-order approximation is

$$\widetilde{P}(t, x, z) = P_0(t, x, z) + P_1(t, x, z),$$

where $P_0(t, x, z) = P^\star_{BS}(t, x)$ is the Black–Scholes barrier price with volatility parameter $\sigma^\star$ which depends on z. As discussed in Section 1.7, the Black–Scholes price $P^\star_{BS}$ is obtained by the method of images and given by

$$P_{BS}^\star(t,x) = C_{BS}(t,x) - \left(\frac{x}{B}\right)^{1-k} C_{BS}(t,B^2/x), \qquad (6.4)$$

where $C_{BS}(t,x)$ is the Black–Scholes formula *for a call option*, with the z-dependent volatility parameter $\sigma^\star$, and where k is defined by $k = 2r/\sigma^{\star 2}$.

From Section 4.3, the stochastic volatility correction $P_1(t,x,z)$ satisfies the partial differential equation

$$\mathscr{L}_{BS}(\sigma^\star)P_1 = -\left(2V_0^\delta \frac{\partial}{\partial\sigma} + 2V_1^\delta D_1 \left(\frac{\partial}{\partial\sigma}\right) + V_3^\varepsilon D_1 D_2\right) P_{BS}^\star$$

$$\text{in } \{x > B, t < T\}, \qquad (6.5)$$

$$P_1(t,B,z) = 0, \qquad (6.6)$$

$$P_1(T,x,z) = 0, \qquad (6.7)$$

where we used again the notation $D_n = x^n \partial^n/\partial x^n$. Because of the boundary condition, this does not seem to have a simple explicit solution in terms of the Greeks of the Black–Scholes price of the contract, as was the case for European contracts. However, we can transform the current source with zero boundary condition problem into a *boundary value* problem with no source. From here, it can be reduced to a one-dimensional integral, as we show below. This is much simpler to compute numerically than solving the full source problem (6.5).

To do this, we first set $P_1^{(B)} = P_1 - P_1^{(A)}$, where

$$P_1^{(A)} = (T-t)\left[2V_0^\delta \frac{\partial}{\partial\sigma} + 2V_1^\delta D_1 \left(\frac{\partial}{\partial\sigma}\right) + V_3^\varepsilon D_1 D_2\right] P_{BS}^\star. \qquad (6.8)$$

This is motivated by the form of the source term in (6.5) and the approaches we applied in the European case. It is simply the source from (6.5) multiplied by $-(T-t)$, so that on applying the Black–Scholes operator, the time derivative hitting this factor returns this source. Using the relationships

$$\mathscr{L}_{BS}(\sigma^\star)\left(\frac{\partial}{\partial\sigma}P_{BS}^\star\right) = -\sigma^\star D_2 P_{BS}^\star, \qquad (6.9)$$

$$\mathscr{L}_{BS}(\sigma^\star)\left(D_1 \frac{\partial}{\partial\sigma}P_{BS}^\star\right) = -\sigma^\star D_1 D_2 P_{BS}^\star, \qquad (6.10)$$

which can be obtained from differentiating the Black–Scholes partial differential equation with respect to the volatility parameter, we find that $P_1^{(B)}$ solves the following problem in $x > B$:

$$\mathscr{L}_{BS}(\sigma^\star)P_1^{(B)} = (T-t)\left[2V_0^\delta\sigma^\star D_2 + 2V_1^\delta\sigma^\star D_1 D_2\right]P_{BS}^\star,$$

$$P_1^{(B)}(t,B,z) = g^{(B)}(t,z),$$

$$P_1^{(B)}(T,x,z) = 0,$$

where we have also used $\mathscr{L}_{BS}(\sigma^\star)(D_1 D_2 P_{BS}^\star) = 0$, as usual. The boundary condition function $g^{(B)}(t,z)$ is given by

$$g^{(B)}(t,z) = -\lim_{x\downarrow B}P_1^{(A)}(t,x,z),$$

since P_1 is zero at the boundary $x = B$.

So far, we have replaced one source problem with another, but the new source involves only the operators D_1 and D_2 which can be removed by the following transformation. We define $P_1^{(D)} = P_1^{(B)} - P_1^{(C)}$, where

$$P_1^{(C)} = -\frac{1}{2}(T-t)^2\left[2V_0^\delta\sigma^\star D_2 + 2V_1^\delta\sigma^\star D_1 D_2\right]P_{BS}^\star. \tag{6.11}$$

Then $P_1^{(D)}$ solves the following problem in $x > B$:

$$\mathscr{L}_{BS}(\sigma^\star)P_1^{(D)} = 0,$$

$$P_1^{(D)}(t,B,z) = g(t,z),$$

$$P_1^{(D)}(T,x,z) = 0,$$

where the boundary condition function $g(t,z)$ is given by

$$g(t,z) = g^{(B)}(t,z) - \lim_{x\downarrow B}P_1^{(C)}(t,x,z).$$

This problem for $P_1^{(D)}$ can be transformed to a constant coefficient problem associated with the backward heat equation by setting

$$\xi = \log x,$$

$$P_1^{(D)}(t,x,z) = e^{-\frac{1}{8}\sigma^{\star 2}(1+k)^2(T-t)+\frac{1}{2}(1-k)\xi}u(t,\xi,z).$$

If we define $L = \log B$, then we find that u solves

$$\frac{\partial u}{\partial t} + \frac{1}{2}\sigma^{\star 2}\frac{\partial^2 u}{\partial\xi^2} = 0, \quad \text{for } \xi > L, t < T,$$

$$u(T,\xi) = 0,$$

$$u(t,L) = \tilde{g}(t,z),$$

where the boundary term is given by

$$\tilde{g}(t,z) = e^{\frac{1}{8}\sigma^{\star 2}(1+k)^2(T-t)}B^{(k-1)/2}g(t,z). \tag{6.12}$$

The probabilistic representation of u is simply

$$u(t,\xi,z) = \mathbb{E}\left\{\tilde{g}(\tau_t,z)\mathbf{1}_{\{\tau_t \le T\}} \mid B_t = \xi > L\right\},$$

where (B_t) is a Brownian motion with $\langle B \rangle_t = \sigma^{*2}t$, and τ_t is the first time after t that it hits L. Using the distribution for the hitting time τ_t, this expectation is given by the integral

$$u(t,\xi,z) = \frac{1}{\sigma^{\star}\sqrt{2\pi}} \int_t^T \frac{(\xi-L)}{(s-t)^{3/2}} \exp\left(-\frac{(\xi-L)^2}{2\sigma^{\star 2}(s-t)}\right)\tilde{g}(s,z)\,ds. \quad (6.13)$$

Finally, using the expression for $\tilde{g}$ in (6.12) we then obtain

$$P_1^{(D)} = x^{\frac{1}{2}(1-k)}e^{-\frac{1}{8}\sigma^{*2}(1+k)^2(T-t)}u(t,\log x, z). \quad (6.14)$$

To summarize, the correction is given by

$$P_1 = P_1^{(A)} + P_1^{(C)} + P_1^{(D)}, \quad (6.15)$$

where the first two components are given explicitly in (6.8), (6.11) and the third component by (6.14) which contains the integral term in (6.13).

6.2.2 Explicit Expressions

To evaluate the expression for the correction in (6.15), we express the components in terms of Black–Scholes call option prices and Greeks. From (6.8), $P_1^{(A)}$ involves two Greeks of the Black–Scholes barrier price. These are computed from the definition of $P_{BS}^{\star}$ in (6.4):

$$\frac{\partial}{\partial\sigma}P_{BS}^{\star}(t,x)$$

$$= \mathcal{V}_{BS}(t,x) - \left(\frac{x}{B}\right)^{1-k}\mathcal{V}_{BS}\left(t,\frac{B^2}{x}\right) - \frac{2k}{\sigma^{\star}}\left(\frac{x}{B}\right)^{1-k}\log\left(\frac{x}{B}\right)C_{BS}\left(t,\frac{B^2}{x}\right),$$

$$D_1\frac{\partial}{\partial\sigma}P_{BS}^{\star}(t,x)$$

$$x\mathcal{V}_{BS}'(t,x) + \left(\frac{x}{B}\right)^{1-k}\left[(k-1)\mathcal{V}_{BS}\left(t,\frac{B^2}{x}\right) + \left(\frac{B}{x}\right)^2 x\mathcal{V}_{BS}'\left(t,\frac{B^2}{x}\right)\right]$$

$$-\frac{2k}{\sigma^{\star}}\left(\frac{x}{B}\right)^{1-k}\left(1-k\log\left(\frac{x}{B}\right)\right)C_{BS}\left(t,\frac{B^2}{x}\right)$$

$$+\frac{2k}{\sigma^{\star}}\left(\frac{x}{B}\right)^{-(k+1)}\log\left(\frac{x}{B}\right)x\Delta_{BS}\left(t,\frac{B^2}{x}\right),$$

where Δ_{BS}, $\mathcal{V}_{BS}$, and $\mathcal{V}'_{BS} = \frac{\partial}{\partial x}\mathcal{V}_{BS}$ are the Greeks of a call option, given in Section 1.3.5, and evaluated with volatility $\sigma^\star$.

The component $P_1^{(C)}$ in (6.11) further involves the two Greeks

$$D_2 P_{BS}^\star(t,x) = x^2 \Gamma_{BS}(t,x) + k(1-k)\left(\frac{x}{B}\right)^{1-k} C_{BS}\left(t, \frac{B^2}{x}\right)$$

$$- 2k\left(\frac{x}{B}\right)^{-1-k} x\Delta_{BS}\left(t, \frac{B^2}{x}\right) + \left(\frac{x}{B}\right)^{-3-k} x^2 \Gamma_{BS}\left(t, \frac{B^2}{x}\right),$$

$$D_1 D_2 P_{BS}^\star(t,x)$$

$$= 2x^2 \Gamma_{BS}(t,x) + x^3 \Gamma'_{BS}(t,x) + k(1-k)^2 \left(\frac{x}{B}\right)^{1-k} C_{BS}\left(t, \frac{B^2}{x}\right)$$

$$+ k(3k-1)\left(\frac{x}{B}\right)^{-1-k} x\Delta_{BS}\left(t, \frac{B^2}{x}\right) + (k-1)\left(\frac{x}{B}\right)^{-3-k} x^2 \Gamma_{BS}\left(t, \frac{B^2}{x}\right)$$

$$- \left(\frac{x}{B}\right)^{-5-k} x^3 \Gamma'_{BS}\left(t, \frac{B^2}{x}\right),$$

where we also use the Greeks Γ_{BS} and $\Gamma'_{BS} = \frac{\partial}{\partial x}\Gamma_{BS}$.

The boundary function $\tilde{g}(t,z)$ which appears in the integral term (6.13) is given by

$$\tilde{g}(t,z) = e^{\frac{1}{8}\sigma^{\star 2}(1+k)^2 \tau} B^{(k-1)/2} \times$$

$$\left[\left\{\left(\frac{4kV_1^\delta}{\sigma^\star} - V_3^\varepsilon k(1-k)^2\right)\tau + \sigma^\star k(1-k)((1-k)V_1^\delta + V_0^\delta)\tau^2\right\} C_{BS}(t,B)\right.$$

$$+ \left\{V_3^\varepsilon k(1-3k)\tau + (V_1^\delta - 2V_0^\delta k)\sigma^\star \tau^2\right\} B\Delta_{BS}(t,B)$$

$$+ 2V_1^\delta(1-k)\tau \mathcal{V}_{BS}(t,B) - 4V_1^\delta \tau B \mathcal{V}'_{BS}(t,B)$$

$$+ \left\{-V_3^\varepsilon(1+k)\tau + (2V_0^\delta + V_1^\delta)\sigma^\star \tau^2\right\} B^2 \Gamma_{BS}(t,B)\right].$$

In conclusion, using the stable parameters $(\sigma^\star, V_0^\delta, V_1^\delta, V_3^\varepsilon)$ to price an option ensures that no arbitrage (up to the order of accuracy of the approximation) is introduced.

6.3 Asian Options

To illustrate how to apply our perturbation method to Asian options, for simplicity, we show only the correction due to the fast mean-reverting stochastic volatility factor, and we present the case of an Asian average-strike call option with no early exercise. Recall from Section 1.7 that the payoff of this contract on date T is like that of a call option whose strike price is the average of the stock price between time $t = 0$ and T.

This contract involves the new process

$$I_t = \int_0^t X_s \, ds,$$

and as shown in Section 1.7, it introduces a new spatial variable I in the partial differential equation for the pricing function. We review the essential steps of the extension of this to stochastic volatility models and then we show how to apply the fast mean-reversion asymptotics to this new partial differential equation. In fact, we shall see that the main equation for the stochastic volatility correction is modified only slightly.

We start with the stochastic volatility model for the price process (X_t) and volatility $(f(Y_t))$ given by (2.9), to which we add

$$dI_t = X_t \, dt, \tag{6.16}$$

with $I_0 = 0$, assuming that the starting time of the average specified in the contract is $t = 0$. Under the risk-neutral probability $\mathbb{P}^{\star(\gamma)}$ the process (X_t, Y_t, I_t) remains a Markov process, and is described by equations (2.18) and (2.19). The equation (6.16) for (I_t) is unchanged under the change of measure. The price $P(t, x, y, I)$ of the contract at time $0 \leq t < T$ is given by

$$P(t,x,y,I) = \mathbb{E}^{\star(\gamma)} \left\{ e^{-r(T-t)} \left(X_T - \frac{I_T}{T} \right)^+ \Big| X_t = x, Y_t = y, I_t = I \right\}. \tag{6.17}$$

It is also obtained as the solution of the partial differential equation

$$\frac{\partial P}{\partial t} + \frac{1}{2} f(y)^2 x^2 \frac{\partial^2 P}{\partial x^2} + r \left(x \frac{\partial P}{\partial x} - P \right) + \rho \frac{\beta(y)}{\sqrt{\varepsilon}} x f(y) \frac{\partial^2 P}{\partial x \partial y} + \frac{1}{2} \frac{\beta^2(y)}{\varepsilon} \frac{\partial^2 P}{\partial y^2}$$
$$+ \left(\frac{\alpha(y)}{\varepsilon} - \frac{\beta(y)}{\sqrt{\varepsilon}} \Lambda(y) \right) \frac{\partial P}{\partial y} + x \frac{\partial P}{\partial I} = 0, \tag{6.18}$$

which is as in equation (2.17) with the extra term $x \partial P / \partial I$ coming from (6.16). The terminal condition is

$$P(T, x, y, I) = \left(x - \frac{I}{T} \right)^+.$$

Under fast mean-reversion, we proceed as in Chapter 4 and write (6.18) as

$$\left(\frac{1}{\varepsilon} \mathcal{L}_0 + \frac{1}{\sqrt{\varepsilon}} \mathcal{L}_1 + \mathcal{L}_2 \right) P = 0,$$

where $\mathscr{L}_0, \mathscr{L}_1$, and $\mathscr{L}_2$ are given by (4.8), (4.9), and (4.10) and we define

$$\hat{\mathscr{L}}_2 = \mathscr{L}_2 + x\frac{\partial}{\partial I}.$$

Now the calculations of Section 4.2 go through exactly, with $\hat{\mathscr{L}}_2$ replacing $\mathscr{L}_2$. Again the corrected price is given by

$$\widetilde{P}(t,x,I) = P_0(t,x,I) + \widetilde{P}_1(t,x,I),$$

where P_0 solves

$$\langle\hat{\mathscr{L}}_2\rangle P_0 = 0,$$

with terminal condition $P_0(T,x,I) = \left(x - \frac{I}{T}\right)^+$. Notice that

$$\langle\hat{\mathscr{L}}_2\rangle = \langle\mathscr{L}_2\rangle + x\frac{\partial}{\partial I} = \mathscr{L}_{BS}(\bar{\sigma}) + x\frac{\partial}{\partial I}.$$

In other words, P_0 is the Black–Scholes Asian price with constant volatility $\bar{\sigma}$. This is usually solved numerically.

The correction $\widetilde{P}_1(t,x,I)$ satisfies

$$\langle\hat{\mathscr{L}}_2\rangle\widetilde{P}_1 = \hat{\mathscr{A}}^{\varepsilon} P_0,$$

where

$$\hat{\mathscr{A}}^{\varepsilon} = \sqrt{\varepsilon}\left\langle\mathscr{L}_1\mathscr{L}_0^{-1}\left(\hat{\mathscr{L}}_2 - \langle\hat{\mathscr{L}}_2\rangle\right)\right\rangle,$$

analogous to (4.32). Since the additive extra term in $\hat{\mathscr{L}}_2$ does not depend on y, we have

$$\hat{\mathscr{L}}_2 - \langle\hat{\mathscr{L}}_2\rangle = \mathscr{L}_2 - \langle\mathscr{L}_2\rangle,$$

which implies that $\hat{\mathscr{A}}^{\varepsilon} = \mathscr{A}^{\varepsilon}$ involving the V's introduced before. Therefore,

$$\left(\mathscr{L}_{BS}(\bar{\sigma}) + x\frac{\partial}{\partial I}\right)\widetilde{P}_1(t,x,I) = \mathscr{A}^{\varepsilon} P_0(t,x,I),$$

with zero terminal condition. Notice from (4.38) that the computation of the right-hand side $\mathscr{A}^{\varepsilon} P_0(t,x,I)$ involves only x-derivatives. This equation is solved numerically too.

Incorporating also a slow varying factor and implementing the parameter-reduction technique of Section 4.3, we then obtain that the correction $P_1^{\star}$ satisfies

$$\left(\mathscr{L}_{BS}(\sigma^{\star}(z))+x\tfrac{\partial}{\partial I}\right)P_1^{\star} \;=\; \left.\begin{array}{r}-\left(2V_0^{\delta}(z)\tfrac{\partial}{\partial\sigma}+2V_1^{\delta}(z)D_1\tfrac{\partial}{\partial\sigma}\right.\\[2mm]\left.+V_3^{\varepsilon}(z)D_1D_2\right)P_{BS}^{\star},\\[2mm]P_1^{\star}(T,x,z,I) \;=\; 0,\end{array}\right\}\quad(6.19)$$

where $P_{BS}^{\star}$ is the price of the Asian option evaluated at the volatility level $\sigma^{\star}(z)$.

Notes

Pricing of the exotic contracts discussed here under the constant volatility Black–Scholes model is solved, explicitly or numerically, in Wilmott *et al.* (1996) by differential equations methods, or in Musiela and Rutkowski (2002) by probabilistic representations, for instance.

For the case of Asian options, we refer to Fouque and Han (2004a) for more details and also for the use of a dimension reduction technique. For other bounds and approximation techniques for Asian options we refer to Forde and Jacquier (2010).

The correction for barrier options with fast and slow factors appeared in Fouque *et al.* (2006). Fast factor stochastic volatility asymptotic approximations for barrier, lookback, and passport options were analyzed in Ilhan *et al.* (2004).

7

Application to American Derivatives

We now look at American-style derivative contracts, introduced in Section 1.2.2. Recall that they give the holder the right of early exercise, and so the date that the contract is terminated is not known beforehand, unlike in the European case. As we reviewed in Section 1.6, for an American put option under the Black–Scholes model, the *no-arbitrage* pricing function satisfies a free boundary value problem characterized by the system of equations (1.92) with boundary conditions (1.93)–(1.96). This is much more difficult than the European pricing problem, and there are no explicit solutions in general. It has to be solved numerically (a notable exception is the case of infinite-horizon, or perpetual, American put options).

In this chapter, we show that the asymptotic method for correcting the Black–Scholes price for multiscale stochastic volatility can be extended to contracts with the early exercise feature, simplifying considerably the three-dimensional free boundary problems that arise in these models. Furthermore, the correction depends only on the universal group parameters $(\sigma^\star, V_0^\delta, V_1^\delta, V_3^\varepsilon)$ we have estimated from the implied volatility surface, as discussed in Chapter 5. We concentrate on the American put option, which is the best known American-style derivative. The procedure may be adapted to the American version of the other derivatives we have considered in Chapter 6.

7.1 American Options Valuation under Stochastic Volatility

We assume the stochastic volatility model (4.1) and that the market selects a unique pricing measure $\mathbb{P}^\star$ which is reflected in liquidly traded around-the-money European option prices. Prices of other derivative securities must be priced with respect to this measure, if there are to be no arbitrage opportunities. Therefore, the American put price is given by

$$P^{\varepsilon,\delta}(t,x,y,z) = \sup_{t \leq \tau \leq T} \mathbb{E}^{\star}\left\{ e^{-r(\tau-t)}(K-X_{\tau})^{+} | X_t = x, Y_t = y, Z_t = z \right\},$$

$$(7.1)$$

where the supremum is taken over all stopping times with respect to the filtration generated by the Brownian motions $\left(W_t^{(0)\star}, W_t^{(1)\star}, W_t^{(2)\star} \right)$, taking values in $[t,T]$.

As usual with optimal stopping problems, we look for a solution to a free boundary problem analogous to (1.92), with the additional spatial variables y and z. The free boundary is now a surface $F^{\varepsilon,\delta}(t,x,y,z) = 0$, which we write as $x = x_b^{\varepsilon,\delta}(t,y,z)$, and which has to be determined as part of the problem:

$$P^{\varepsilon,\delta}(t,x,y,x) = K - x, \qquad \text{for } x < x_b(t,y,z), \qquad (7.2)$$
$$\mathcal{L}^{\varepsilon,\delta} P^{\varepsilon,\delta} = 0, \qquad \text{for } x > x_b(t,y,z), \qquad (7.3)$$

with

$$P^{\varepsilon,\delta}(T,x,y,z) = (K-x)^{+}, \qquad (7.4)$$

and where the operator $\mathcal{L}^{\varepsilon,\delta}$ is given by (4.7)–(4.13).

We look for a solution $P^{\varepsilon,\delta}$ with $P^{\varepsilon,\delta}$ and its first derivatives $\frac{\partial P^{\varepsilon,\delta}}{\partial x}$, $\frac{\partial P^{\varepsilon,\delta}}{\partial y}$, and $\frac{\partial P^{\varepsilon,\delta}}{\partial z}$ continuous across the boundary $x_b(t,y,z)$, so that

$$P^{\varepsilon,\delta}(t,x_b(t,y,z),y,z) = (K - x_b(t,y,z)), \qquad (7.5)$$

$$\frac{\partial P^{\varepsilon,\delta}}{\partial x}\Big|_{x=x_b(t,y,z)} = -1, \qquad (7.6)$$

$$\frac{\partial P^{\varepsilon,\delta}}{\partial y}\Big|_{x=x_b(t,y,z)} = 0, \qquad (7.7)$$

$$\frac{\partial P^{\varepsilon,\delta}}{\partial z}\Big|_{x=x_b(t,y,z)} = 0, \qquad (7.8)$$

where we have removed the positive part on the right-hand side of (7.5) because it is clear that the optimal exercise strategy will involve exercising at or in-the-money, so $x_b(t,y,z) \leq K$. These are shown in Figure 7.1.

7.2 Stochastic Volatility Correction for American Put

As in Chapter 4, we look for an asymptotic solution of the form

$$P^{\varepsilon,\delta}(t,x,y,z) = P_0(t,x,y,z) + \sqrt{\varepsilon}P_{1,0}(t,x,y,z) + \sqrt{\delta}P_{0,1}(t,x,y,z) + \cdots,$$

$$(7.9)$$

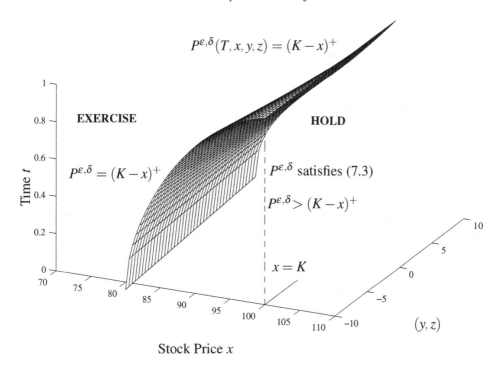

Figure 7.1   The full problem for the American put under stochastic volatility. The free boundary conditions on the surface are given by (7.5)–(7.6).

$$x_b^{\varepsilon,\delta}(t,y,z) = x_0(t,y,z) + \sqrt{\varepsilon}\,x_{1,0}(t,y,z) + \sqrt{\delta}\,x_{0,1}(t,y,z) + \cdots, \qquad (7.10)$$

where we have expanded the formula for the free boundary surface as well. We will denote by $P_{BS}^A(t,x;\sigma)$ the Black–Scholes American put option price, as a function of the time, stock price, and volatility parameter. It is well known that, inside its continuation region, P_{BS}^A is smooth in (t,x), and we shall assume that P_{BS}^A and $\frac{\partial P_{BS}^A}{\partial x}$ are continuously differentiable in σ.

Our strategy for constructing a solution will be to expand the equations and boundary conditions in powers of ε and δ, substituting the expansions (7.9) and (7.10). The formal construction of the expansion is similar to that of the European case in Section 4.2, but we need to pay special attention to the free boundary conditions. To derive the first few terms in the expansion most directly, we expand and compare powers simultaneously in ε and δ, which leads to the same as expanding first in δ and then in ε as we presented in the derivation for European options.

The expansion of the partial differential equation $\mathscr{L}^{\varepsilon,\delta}P^{\varepsilon,\delta} = 0$ **in the hold region** is formally, as in the European case,

$\mathscr{L}^{\varepsilon,\delta}P^{\varepsilon,\delta}$

$$= \left(\frac{1}{\varepsilon}\mathscr{L}_0 + \frac{1}{\sqrt{\varepsilon}}\mathscr{L}_1 + \mathscr{L}_2 + \sqrt{\delta}\mathscr{M}_1 + \delta\mathscr{M}_2 + \sqrt{\frac{\delta}{\varepsilon}}\mathscr{M}_3\right)\sum_{j,k\geq 0}\varepsilon^{j/2}\delta^{k/2}P_{j,k}$$

$$= \sum_{j\geq -2}\sum_{k\geq 0}\varepsilon^{j/2}\delta^{k/2}\Bigg(\mathscr{L}_0 P_{j+2,k} + \mathscr{L}_1 P_{j+1,k} + \mathscr{L}_2 P_{j,k} + \mathscr{M}_1 P_{j,k-1}$$

$$+ \mathscr{M}_2 P_{j,k-2} + \mathscr{M}_3 P_{j+1,k-1}\Bigg)$$

$$= 0, \tag{7.11}$$

where the terms with negative indices on the P are understood to be zero. We also expand the known value in the exercise region (7.2) to give

$$P_0(t,x,y,z) + \sqrt{\varepsilon}P_{1,0}(t,x,y,z) + \sqrt{\delta}P_{0,1}(t,x,y,z) + \cdots$$

$$= K - x, \text{ in } \{x < x_b^{\varepsilon,\delta}(t,y,z)\},$$

and the boundary conditions (7.5) and (7.6):

$$P_0(t,x_0,y,z) + \sqrt{\varepsilon}\left(P_{1,0}(t,x_0,y,z) + x_{1,0}\frac{\partial P_0}{\partial x}(t,x_0,y,z)\right)$$

$$+ \sqrt{\delta}\left(P_{0,1}(t,x_0,y,z) + x_{0,1}\frac{\partial P_0}{\partial x}(t,x_0,y,z)\right) + \cdots$$

$$= K - x_0 - \sqrt{\varepsilon}x_{1,0} - \sqrt{\delta}x_{0,1} - \cdots,$$

$$\frac{\partial P_0}{\partial x}(t,x_0,y,z) + \sqrt{\varepsilon}\left(\frac{\partial P_{1,0}}{\partial x}(t,x_0,y,z) + x_{1,0}\frac{\partial^2 P_0}{\partial x^2}(t,x_0,y,z)\right)$$

$$+ \sqrt{\delta}\left(\frac{\partial P_{0,1}}{\partial x}(t,x_0,y,z) + x_{0,1}\frac{\partial^2 P_0}{\partial x^2}(t,x_0,y,z)\right) + \cdots$$

$$= -1. \tag{7.12}$$

In the latter expansion above, the second-order partial derivatives are taken to mean the one-sided derivatives into the region $x > x_0(t,y,z)$; however, only the principal term in this expansion will be used in the construction of the first-order approximation to the American put price below.

We are interested in the first three terms in (7.9), which we obtain by setting successively the individual terms to zero as follows.

- Terms of order ε^{-1} in (7.11) and order one in the boundary conditions:

$$\begin{aligned}
\mathscr{L}_0 P_0 &= 0, & \text{in } x > x_0(t,y,z), \\
P_0(t,x,y,z) &= K - x, & \text{in } x < x_0(t,y,z), \\
P_0(t,x_0(t,y,z),y,z) &= K - x_0(t,y,z), \\
\frac{\partial P_0}{\partial x}(t,x_0(t,y,z),y,z) &= -1.
\end{aligned} \tag{7.13}$$

Note that, in the first two expressions, we choose the dividing boundary to be the leading-order term of $x_b^{\varepsilon,\delta}$ in order to construct a P_0 that does not depend on ε and δ. From the first equation in (7.13), we seek $P_0 = P_0(t,x,z)$ independent of y in the region $x > x_0(t,y,z)$. From the second equation in (7.13), P_0 is also independent of y on $x < x_0(t,y,z)$, and so, by continuity, it is independent of y at the boundary x_0, and therefore everywhere. Since, from the last equation in (7.13), P_0 is changing with x at the boundary x_0, but is independent of y, we must have that x_0 is independent of y as well: $x_0 = x_0(t,z)$.

- Terms of order $\varepsilon^{-1/2}$ in (7.11) and order $\varepsilon^{1/2}$ in the boundary conditions:

$$\mathcal{L}_0 P_{1,0} + \mathcal{L}_1 P_0^{\;0} = 0, \quad \text{in } x > x_0(t,z),$$
$$P_{1,0}(t,x,y,z) = 0, \quad \text{in } x < x_0(t,z),$$

$$P_{1,0}(t,x_0,y,z) = -x_{1,0}\frac{\partial P_0}{\partial x}(t,x_0,z)^{\;-1} - x_{1,0} = 0, \quad (7.14)$$

where again the dividing boundary in the first two expressions is chosen to be x_0. Following the same argument as for P_0, we seek $P_{1,0} = P_{1,0}(t,x,z)$, independent of y.

- Terms of order $\varepsilon^{-1}\delta^{1/2}$ in (7.11) and order $\delta^{1/2}$ in the boundary conditions:

$$\mathcal{L}_0 P_{0,1} = 0, \quad \text{in } x > x_0(t,z),$$
$$P_{0,1}(t,x,y,z) = 0, \quad \text{in } x < x_0(t,z),$$

$$P_{0,1}(t,x_0,y,z) = -x_{0,1}\frac{\partial P_0}{\partial x}(t,x_0,z)^{\;-1} - x_{0,1} = 0, \quad (7.15)$$

and therefore we seek $P_{0,1} = P_{0,1}(t,x,z)$, independent of y.

- Terms of order one in (7.11):

$$\mathcal{L}_0 P_{2,0} + \mathcal{L}_1 P_{1,0}^{\;0} + \mathcal{L}_2 P_0 = 0,$$

which is a Poisson equation for $P_{2,0}$ in its full domain in y for fixed (t,x,z) such that $x > x_0(t,z)$. The centering condition implies

$$\langle \mathcal{L}_2 \rangle P_0 = 0 \quad \text{in } x > x_0(t,z). \quad (7.16)$$

Recall that $\langle \mathcal{L}_2 \rangle$ is the Black–Scholes operator with effective volatility $\bar{\sigma}(z)$ defined in (4.26). Together with the terminal condition $P_0(T,x,z) = (K-x)^+$, the free boundary conditions in (7.13), and the value in the exercise region, we see that $P_0(t,x,z)$ is exactly the Black–Scholes American put price with volatility $\bar{\sigma}(z)$, and associated optimal exercise boundary

$x_0(t,z)$. There is no known explicit solution for $P_0(t,x,z)$ or $x_0(t,z)$, and they have to be computed numerically.

- Terms of order $\varepsilon^{1/2}$ in (7.11):

$$\mathcal{L}_0 P_{3,0} + \mathcal{L}_1 P_{2,0} + \mathcal{L}_2 P_{1,0} = 0,$$

which is a Poisson equation for $P_{3,0}$ in its full domain in y for fixed (t,x,z) such that $x > x_0(t,z)$. The centering condition implies

$$\langle \mathcal{L}_2 \rangle P_{1,0} = \mathcal{A} P_0, \quad \text{in } x > x_0(t,z), \tag{7.17}$$

where the operator $\mathcal{A} = \mathcal{A}^\varepsilon / \sqrt{\varepsilon}$ and $\mathcal{A}^\varepsilon$ is defined in (4.32), and given explicitly in (4.38). We see that $P_{1,0}$ is determined by the boundary value problem (7.17) with fixed boundary $x_0(t,z)$, boundary condition (7.14), and terminal condition $P_{1,0}(T,x,z) = 0$.

Again, $P_{1,0}$ is found numerically after obtaining the numerical solution P_0. Observe that, in contrast to the European case, the function $-(T - t)\mathcal{A} P_0(t,x,z)$ satisfies the equation (7.17) in the hold region $x > x_0(t,z)$, as in Proposition 4.6, but it does not satisfy the zero boundary condition required for $P_{1,0}$.

- Terms of order $\delta^{1/2}$ in (7.11):

$$\mathcal{L}_0 P_{1,1} + \mathcal{L}_2 P_{0,1} + \mathcal{M}_1 P_0 = 0,$$

which is a Poisson equation for $P_{1,1}$ in its full domain in y for fixed (t,x,z) such that $x > x_0(t,z)$. The centering condition implies

$$\langle \mathcal{L}_2 \rangle P_{0,1} = -\langle \mathcal{M}_1 \rangle P_0, \quad \text{in } x > x_0(t,z), \tag{7.18}$$

where $\langle \mathcal{M}_1 \rangle = 2\mathcal{A}^\delta / \sqrt{\delta}$ and $\mathcal{A}^\delta$ is given explicitly in (4.50). We see that $P_{0,1}$ is determined by the boundary value problem (7.18) with fixed boundary $x_0(t,z)$, boundary condition (7.15), and terminal condition $P_{0,1}(T,x,z) = 0$.

As before, $P_{1,0}$ is found numerically after obtaining the numerical solution P_0. Observe that, in contrast to the European case, the function $\frac{1}{2}(T - t)\langle \mathcal{M}_1 \rangle P_0(t,x,z)$ satisfies the equation (7.18) in the hold region $x > x_0(t,z)$, as in Proposition 4.8, but it does not satisfy the zero boundary condition required for $P_{1,0}$.

We remark that the original three-dimensional free boundary problem for $P^{\varepsilon,\delta}(t,x,y,z)$ has been approximated by a one-dimensional Black–Scholes American option free boundary problem for $P_0(t,x,z)$, with z appearing only as a parameter, and the correction terms $P_{1,0}$ and $P_{0,1}$ solving fixed boundary value problems in one space dimension with zero Dirichlet boundary condition on the boundary x_0 determined from the free boundary problem for P_0.

In determining the approximation, we have not used the boundary conditions (7.7) and (7.8). But since P_0, $P_{1,0}$, and $P_{0,1}$ do not depend on y, (7.7) is satisfied up to order (ε, δ). Expanding (7.8) gives

$$\frac{\partial P^{\varepsilon, \delta}}{\partial z}(t, x, y, z) \big|_{x=x_b(t, y, z)}$$

$$= \left[\frac{\partial P_0}{\partial z}(t, x, z) + \sqrt{\varepsilon} \left(\frac{\partial P_{1,0}}{\partial z}(t, x, z) + x_{1,0} \frac{\partial^2 P_0}{\partial x \partial z}(t, x, z) \right) \right.$$

$$\left. + \sqrt{\delta} \left(\frac{\partial P_{0,1}}{\partial z}(t, x, z) + x_{0,1} \frac{\partial^2 P_0}{\partial x \partial z}(t, x, z) \right) + \cdots \right]_{x=x_0}.$$

We recall that P_0 is the Black–Scholes American put price evaluated at the volatility $\bar{\sigma}(z)$:

$$P_0(t, x, z) = P_{BS}^A(t, x; \bar{\sigma}(z)).$$

Since $P_{BS}^A(t, x; \sigma) = K - x$ at any point x on its associated optimal exercise boundary, the Vega of the Black–Scholes American put price is zero on that boundary, and hence

$$\frac{\partial P_0}{\partial z} \big|_{x=x_0(t, z)} = 0.$$

Similarly, $\frac{\partial^2 P_0}{\partial x \partial z} \big|_{x=x_0(t, z)} = 0$. The correction terms $P_{1,0}$ and $P_{0,1}$ are zero on the boundary x_0, and therefore $\frac{\partial P_{1,0}}{\partial z} \big|_{x=x_0} = \frac{\partial P_{0,1}}{\partial z} \big|_{x=x_0} = 0$, and we conclude that the approximation $P_0(t, x, z) + \sqrt{\varepsilon} P_{0,1}(t, x, z) + \sqrt{\delta} P_{0,1}(t, x, z)$, which depends on (t, x) and on z as a parameter, satisfies the boundary conditions (7.7) and (7.8).

We can also use (7.12) to find the next terms in the expansion (7.10) for the optimal exercise boundary $x_b^{\varepsilon, \delta}(t, y, z)$:

$$x_{1,0}(t, z) = -\frac{\frac{\partial P_{1,0}}{\partial x} \big|_{x=x_0}}{\frac{\partial^2 P_0}{\partial x^2} \big|_{x=x_0}}, \qquad x_{0,1}(t, z) = -\frac{\frac{\partial P_{0,1}}{\partial x} \big|_{x=x_0}}{\frac{\partial^2 P_0}{\partial x^2} \big|_{x=x_0}},$$

where the second-order partial derivative is taken to mean the one-sided derivative into the region $x > x_0(t, z)$, and it is known that there is a strict discontinuity in the second derivative (the Gamma) of the Black–Scholes American put option price across the exercise boundary, so the denominator is not zero. Observe that the corrections to the boundary are determined after the corrections to the price, and indeed they are not needed if we are just interested in approximations to the American put price.

The mathematical proof of accuracy of the formal asymptotic expansion in the American case remains an open issue that we do not address here. One possible approach would be to combine the analysis for the European case presented in Section 4.5 and the penalization method for free boundary value problems.

7.3 Parameter Reduction

We define a new approximation $P_{BS}^{A,\star} + P_1^\star$ which has the same order of accuracy as $P_0 + \sqrt{\varepsilon}\,P_{0,1} + \sqrt{\delta}\,P_{0,1}$, but which depends on the combination $\sigma^\star$ of $\bar{\sigma}$ and V_2^ε in (4.62). The zeroth-order term, denoted by $P_{BS}^{A,\star} = P_{BS}^A(t,x;\sigma^\star(z))$, is the Black–Scholes American put price at volatility $\sigma^\star(z)$; we denote its associated optimal exercise boundary by $x^\star(t,z)$.

Next, we define the first-order correction $P_1^\star$ as the solution to the fixed boundary value problem

$$
\left.
\begin{aligned}
\mathscr{L}_{BS}(\sigma^\star(z))P_1^\star &= -\mathscr{H}_\star P_{BS}^{A,\star}, \quad x > x^\star(t,z), \\
P_1^\star(t,x^\star(t,x),z) &= 0, \\
P_1^\star(T,x,z) &= 0,
\end{aligned}
\right\}
\tag{7.19}
$$

where, in the source term, we have the operator $\mathscr{H}_\star$, defined in (4.65) for the correction to the European option price, and which we recall is given by:

$$
\mathscr{H}_\star = 2V_0^\delta(z)\frac{\partial}{\partial\sigma} + 2V_1^\delta(z)x\frac{\partial}{\partial x}\left(\frac{\partial}{\partial\sigma}\right) + V_3^\varepsilon(z)x\frac{\partial}{\partial x}\left(x^2\frac{\partial^2}{\partial x^2}\right).
$$

It depends on the parameters $(V_0^\delta, V_1^\delta, V_3^\varepsilon)$, and takes partial derivatives in x and the volatility σ.

It follows from the assumption that $P_{BS}^A(t,x;\sigma)$ is continuously differentiable in σ that $P_{BS}^{A,\star}$ and P_0 differ by order $\sqrt{\varepsilon}$, since $(\sigma^\star)^2 = \bar{\sigma}^2 + 2V_2^\varepsilon$ and $V_2^\varepsilon = \mathcal{O}(\sqrt{\varepsilon})$. Defining the difference

$$
E = (P_0 + \sqrt{\varepsilon}\,P_{0,1} + \sqrt{\delta}\,P_{0,1}) - (P_{BS}^{A,\star} + P_1^\star),
$$

and following the calculation (4.84) in the proof of Theorem 4.10 given in the European case, one easily obtains that $\mathscr{L}_{BS}(\bar{\sigma})E = \mathcal{O}(\varepsilon + \delta)$ in the intersection of the two hold regions $x > x_0(t,z)$ and $x > x^\star(t,z)$.

We now consider the value of E at the boundary of this intersection. Since, for fixed (t,z), we have

$$
\begin{cases}
x^\star(t,z) = \sup\{x > 0 \mid P_{BS}^A(t,x;\sigma^\star(z)) = (K-x)\}, \\
x_0(t,z) = \sup\{x > 0 \mid P_{BS}^A(t,x;\bar{\sigma}(z)) = (K-x)\},
\end{cases}
\tag{7.20}
$$

and using smooth pasting and strict convexity in the hold regions, it follows that $x^\star(t,z)$ and $x_0(t,z)$ differ by $\mathcal{O}(\sqrt{\varepsilon})$.

Since E vanishes at the boundary of the union of the hold regions and its first partial derivative with respect to x is $\mathcal{O}(\sqrt{\varepsilon} + \sqrt{\delta})$ there, the difference at the boundary of the intersection of the hold regions is $\mathcal{O}(\varepsilon + \delta)$. We now show this explicitly using the monotonicity of the price with respect to volatility. Let us suppose that $V_2^\varepsilon > 0$, so that $\sigma^\star > \bar{\sigma}$. The case $V_2^\varepsilon < 0$ can be handled by reversing $\sigma^\star$ and $\bar{\sigma}$ in the following arguments, and the case $V_2^\varepsilon = 0$ is trivial. It is known that P_{BS}^A is increasing with volatility, and so, by (7.20), we have: $x^\star(t,z) < x_0(t,z)$. By a Taylor expansion,

$$P_{BS}^A(t,x_0;\sigma^\star(z)) = P_{BS}^A(t,x^\star;\sigma^\star(z)) + \frac{\partial P_{BS}^A}{\partial x}(t,x^\star;\sigma^\star(z))(x_0 - x^\star)$$

$$+ \mathcal{O}(|x_0 - x^\star|^2)$$

$$= K - x^\star + (-1)(x_0 - x^\star) + \mathcal{O}(\varepsilon)$$

$$= K - x_0 + \mathcal{O}(\varepsilon).$$

We find also that $P_1^*(t,x_0,z) - P_1^*(t,x^\star,z) = \mathcal{O}(\sqrt{\varepsilon})$, so that $E(t,x_0(t,z),z) = \mathcal{O}(\varepsilon + \delta)$.

In conclusion, the difference

$$E = (P_0 + \sqrt{\varepsilon} P_{0,1} + \sqrt{\delta} P_{0,1}) - (P_{BS}^{A,\star} + P_1^\star)$$

satisfies $\mathcal{L}_{BS}(\bar{\sigma})E = \mathcal{O}(\varepsilon + \delta)$ in the intersection of the hold regions, moreover, it differs by $\mathcal{O}(\varepsilon + \delta)$ at the boundary of this intersection so that we can conclude $E = \mathcal{O}(\varepsilon + \delta)$.

7.4 Summary

In conclusion, our approximation to the American put option price (7.1) under multiscale stochastic volatility is given by $P_{BS}^{A,\star} + P_1^\star$, where $P_{BS}^{A,\star} = P_{BS}^A(t,x;\sigma^\star(z))$ is the Black–Scholes American put price at volatility $\sigma^\star(z)$, and $P_1^\star$ is the solution to the fixed boundary value problem (7.19). This approximation depends on the parameters $(\sigma^\star, V_0^\delta, V_1^\delta, V_3^\varepsilon)$ which are calibrated from European option data, as explained in Chapter 5, and the full stochastic volatility model does not need to be calibrated.

In situations where the options on an underlying are primarily American, the approximation above can be used as part of a numerical calibration procedure to obtain the parameters $(\sigma^\star, V_0^\delta, V_1^\delta, V_3^\varepsilon)$ from American option data.

Notes

Details of the derivation of the free boundary formulation of the American pricing problem for the constant volatility case are given, for example, in Shreve (2004, chapter 8). Numerical solution of the Black–Scholes problem is discussed in Lamberton and Lapeyre (1996), in Wilmott *et al.* (1996), and Achdou and Pironneau (2005). An asymptotic analysis of the exercise boundary near expiration can be found in Kuske and Keller (1998), and its convexity is further analyzed in Chen *et al.* (2008). Properties of the exercise boundary in the case of stochastic volatility are studied in Touzi (1999). The penalization method for this type of problem can be found in Bensoussan and Lions (1982).

The fast scale asymptotics for American options first appeared in Fouque *et al.* (2001a). The combination of slow and fast scales with parameter reduction appears here for the first time.

8

Hedging Strategies

In this chapter, we look at how the perturbation analysis helps with the risk management problem of hedging a derivative position. As discussed at the end of Section 2.4.2, financial institutions often want to eliminate or reduce their exposure to a contingent claim written on an asset by trading in the underlying asset. In an incomplete market, a perfect hedge is not possible and the goal is to find an acceptable tradeoff between the risk of a failed hedge and the cost of implementing the hedge. The statistical performance of a strategy is measured by the investor's subjective probability $\mathbb{P}$.

In the first section we briefly recall the Black–Scholes Delta hedging strategy under constant volatility. In Section 8.2 we compute the cost of the same strategy under stochastic volatility driven by a fast and a slow factor. The perturbation method enables us to identify the first-order terms of the cost in (8.13). In Section 8.3 we propose to correct the hedging ratio in such a way that cost become unbiased with a reduced variance. The strategy requires dynamically estimating the effective volatility from historical returns data, and dynamically calibrating the correction parameters from implied volatility data. In Section 8.4 we propose a more practical strategy which consists of using frozen parameters estimated at time zero, and we quantify the additional cost. Even though the parameters are frozen, the strategy still requires a continuous dynamic hedging and the problem of transaction cost is not addressed here. In Section 8.5 we propose two strategies based only on parameters calibrated from implied volatilities as described in Chapter 5. In the last section we show that our price approximation can be derived by using a martingale argument similar to the one used in the construction of hedging strategies in this chapter. This argument does not require that the price of the derivative be given as the solution of a partial differential equation, therefore applicable to

non-Markovian models, while the price approximation and its corrections are characterized as solutions of partial differential equations as in Chapter 4.

8.1 Black–Scholes Delta Hedging

From Section 1.3.2 we know that in a constant volatility environment $\bar{\sigma}$, where the risky asset price $\bar{X}_t$ is a geometric Brownian motion with growth rate μ under $\mathbb{P}$, a short position in a European derivative which pays $h(\bar{X}_T)$ can be perfectly hedged by managing the self-financing porfolio

$$\left[\frac{\partial P_{BS}}{\partial x}(t,\bar{X}_t)\right]\bar{X}_t + \left[e^{-rt}\left(P_{BS}(t,\bar{X}_t) - \bar{X}_t\frac{\partial P_{BS}}{\partial x}(t,\bar{X}_t)\right)\right]e^{rt},$$

made, at time t, of $e^{-rt}\left(P_{BS}(t,\bar{X}_t) - \bar{X}_t\frac{\partial P_{BS}}{\partial x}(t,\bar{X}_t)\right)$ units of the riskless asset and Delta units of the risky asset. Recall that this is simply because the value of such a portfolio at time t is $P_{BS}(t,\bar{X}_t)$, which is precisely $h(\bar{X}_T)$ at maturity, and its variation

$$d\left(P_{BS}(t,\bar{X}_t)\right) = \frac{\partial P_{BS}}{\partial x}(t,\bar{X}_t)d\bar{X}_t$$
$$+ \left(\frac{\partial P_{BS}}{\partial t}(t,\bar{X}_t) + \frac{1}{2}\bar{\sigma}^2\bar{X}_t^2\frac{\partial^2 P_{BS}}{\partial x^2}(t,\bar{X}_t)\right)dt$$

is exactly the variation due to the market

$$\left[\frac{\partial P_{BS}}{\partial x}(t,\bar{X}_t)\right]d\bar{X}_t + \left[e^{-rt}\left(P_{BS}(t,\bar{X}_t) - \bar{X}_t\frac{\partial P_{BS}}{\partial x}(t,\bar{X}_t)\right)\right]d\left(e^{rt}\right),$$

by Itô's formula and the fact that P_{BS} satisfies the Black–Scholes equation.

8.2 The Strategy and its Cost

Let us imagine for a moment that we are following the same strategy in the stochastic volatility environment, $\bar{X}$ being replaced by the price process (X_t) associated with the volatility driving processes (Y_t) and (Z_t) which are modeled under the physical measure $\mathbb{P}$ by

$$\begin{aligned}
dX_t &= \mu X_t dt + f(Y_t, Z_t) X_t dW_t^{(0)}, \\
dY_t &= \frac{1}{\varepsilon}\alpha(Y_t)dt + \frac{1}{\sqrt{\varepsilon}}\beta(Y_t)dW_t^{(1)}, \\
dZ_t &= \delta\, c(Z_t)dt + \sqrt{\delta}\, g(Z_t)dW_t^{(2)},
\end{aligned} \right\} \tag{8.1}$$

as introduced in Section 3.6. The price P_{BS} is computed at the local effective volatility $\bar{\sigma} = \bar{\sigma}(z)$ estimated from historical returns data, so that

$P_{BS}(t,x) = P_0(t,x,z)$, the leading-order term in our price approximation derived in Chapter 4.

Such a strategy replicates the derivative at maturity since $P_0(T, X_T, Z_T) = h(X_T)$, but it is not self-financing. The portfolio has

$$a_t = \frac{\partial P_0}{\partial x}(t, X_t, Z_t) \qquad (8.2)$$

stocks and

$$b_t = e^{-rt}\left(P_0(t, X_t, Z_t) - X_t\frac{\partial P_0}{\partial x}(t, X_t, Z_t)\right)$$

bonds at time t, so that its value is

$$a_t X_t + b_t e^{rt} = P_0(t, X_t, Z_t).$$

Using Itô's formula on $P_0(t, X_t, Z_t)$, its infinitesimal change is given by

$$dP_0(t, X_t, Z_t) = \left(\frac{\partial P_0}{\partial t} + \frac{1}{2}f^2(Y_t, Z_t)X_t^2\frac{\partial^2 P_0}{\partial x^2}\right.$$
$$\left. + \sqrt{\delta}\,\rho_2\, g(Z_t)f(Y_t, Z_t)X_t\frac{\partial^2 P_0}{\partial x\partial z} + \frac{\delta}{2}g^2(Z_t)\frac{\partial^2 P_0}{\partial z^2}\right)dt$$
$$+ a_t dX_t + \frac{\partial P_0}{\partial z}dZ_t,$$

where we have used (8.2), and, to simplify the notation, we have dropped the argument (t, X_t, Z_t) in the derivatives of P_0. The infinitesimal change due to the market (the self-financing part) is given by

$$a_t dX_t + rb_t e^{rt}dt,$$

and consequently the infinitesimal P&L (positive or negative cost) induced by the strategy is given by the difference

$$dP_0(t, X_t, Z_t) - a_t dX_t - rb_t e^{rt}dt$$
$$= \frac{1}{2}\left(f^2(Y_t, Z_t) - \bar{\sigma}^2(Z_t)\right)X_t^2\frac{\partial^2 P_0}{\partial x^2}dt$$
$$+ \sqrt{\delta}\left(\rho_2\, g(Z_t)f(Y_t, Z_t)X_t\frac{\partial^2 P_0}{\partial x\partial z} + \frac{\sqrt{\delta}}{2}g^2(Z_t)\frac{\partial^2 P_0}{\partial z^2}\right)dt + \frac{\partial P_0}{\partial z}dZ_t$$
$$= \frac{1}{2}\left(f^2(Y_t, Z_t) - \bar{\sigma}^2(Z_t)\right)X_t^2\frac{\partial^2 P_0}{\partial x^2}dt$$
$$+ \sqrt{\delta}\,\rho_2\, g(Z_t)f(Y_t, Z_t)X_t\frac{\partial^2 P_0}{\partial x\partial z}dt + \sqrt{\delta}\,g(Z_t)\frac{\partial P_0}{\partial z}dW_t^{(2)} + \mathcal{O}(\delta)$$

where we have used the Black–Scholes equation satisfied by P_0, and then identified the terms of order at most $\sqrt{\delta}$.

The corresponding *cumulative P&L* up to time t, and discounted to time 0, is

$$
E_0(t) = \frac{1}{2} \int_0^t e^{-rs} \left(f^2(Y_s, Z_s) - \bar{\sigma}^2(Z_s) \right) X_s^2 \frac{\partial^2 P_0}{\partial x^2} ds
$$
$$
+ \sqrt{\delta} \rho_2 \int_0^t e^{-rs} g(Z_s) f(Y_s, Z_s) X_s \frac{\partial^2 P_0}{\partial x \partial z} ds +
$$
$$
+ \sqrt{\delta} \int_0^t e^{-rs} g(Z_s) \frac{\partial P_0}{\partial z} dW_s^{(2)} + \mathcal{O}(\delta), \tag{8.3}
$$

and the *total cost* is $E_0(T)$ in addition to the initial cost $P_0(0, X_0, Z_0)$. In other words, the strategy has to be further financed or money taken out (consumed) according to the behavior of these integrals.

The first integral is small due to the *averaging effect* caused by the rapid oscillations of the fast volatility factor Y. This fact is made precise in the following section. The second integral is a bias, and the third integral is a mean-zero martingale, both small of order $\sqrt{\delta}$, due to the *small noise effect* caused by the slow variations of Z_t.

8.2.1 Averaging Effect

In this subsection we study the first integral denoted by $E_0^\varepsilon(t)$, appearing in the cost (8.3):

$$
E_0^\varepsilon(t) = \frac{1}{2} \int_0^t e^{-rs} \left(f^2(Y_s, Z_s) - \bar{\sigma}^2(Z_s) \right) X_s^2 \frac{\partial^2 P_0}{\partial x^2}(s, X_s, Z_s) ds. \tag{8.4}
$$

In the spirit of the central limit theorem, we can look at the size of the fluctuations of this small cost. The following calculation shows that it can be written as

$$
E_0^\varepsilon(t) = \sqrt{\varepsilon} \left(B_t^\varepsilon + M_t^\varepsilon \right) + \mathcal{O}(\varepsilon + \delta),
$$

where B_t^ε is a systematic bias and M_t^ε a mean-zero martingale, both of which will be computed explicitly below.

In order to make this argument rigorous and to evaluate the cost (8.4), one can use the definition (4.36) of the function $\phi(y, z)$ to write

$$
f^2(Y_s, Z_s) - \bar{\sigma}^2(Z_s) = (\mathcal{L}_0 \phi)(Y_s, Z_s),
$$

where $\mathcal{L}_0$ is the infinitesimal generator of $Y^{(1)}$. Using Itô's formula and the operator notation introduced in Section 4.1.2, one gets

$$d\left(\phi(Y_s, Z_s)\right) = \left(\frac{1}{\varepsilon}\mathcal{L}_0 + \delta \mathcal{M}_2 + \sqrt{\frac{\delta}{\varepsilon}}\mathcal{M}_3\right)\phi(Y_s, Z_s)ds$$

$$+\frac{1}{\sqrt{\varepsilon}}\beta(Y_s)\frac{\partial\phi}{\partial y}(Y_s, Z_s)dW_s^{(1)}$$

$$+\sqrt{\delta}g(Z_s)\frac{\partial\phi}{\partial z}(Y_s, Z_s)dW_s^{(2)}.$$

The cumulative cost (8.4) is then given by

$$E_0^\varepsilon(t) = \frac{\varepsilon}{2}\int_0^t e^{-rs}X_s^2\frac{\partial^2 P_0}{\partial x^2}(s, X_s, Z_s)$$

$$\times\left\{d\left(\phi(Y_s, Z_s)\right) - \left(\delta\mathcal{M}_2 + \sqrt{\frac{\delta}{\varepsilon}}\mathcal{M}_3\right)\phi(Y_s, Z_s)\,ds\right.$$

$$\left.-\frac{1}{\sqrt{\varepsilon}}\beta(Y_s)\frac{\partial\phi}{\partial y}(Y_s, Z_s)dW_s^{(1)} - \sqrt{\delta}g(Z_s)\frac{\partial\phi}{\partial z}(Y_s, Z_s)dW_s^{(2)}\right\}.$$

The first integral, with respect to $d\left(\phi(Y_s, Z_s)\right)$, is computed as follows:

$$d\left(e^{-rs}X_s^2\frac{\partial^2 P_0}{\partial x^2}(s, X_s, Z_s)\phi(Y_s, Z_s)\right) = e^{-rs}X_s^2\frac{\partial^2 P_0}{\partial x^2}(s, X_s, Z_s)d\left(\phi(Y_s, Z_s)\right)$$

$$+\phi(Y_s, Z_s)d\left(e^{-rs}X_s^2\frac{\partial^2 P_0}{\partial x^2}(s, X_s, Z_s)\right) + d\left\langle e^{-rs}X^2\frac{\partial^2 P_0}{\partial x^2}(t, X, Z), \phi(Y, Z)\right\rangle_s.$$

The covariation term is given by

$$d\left\langle e^{-rs}X^2\frac{\partial^2 P_0}{\partial x^2}(t, X, Z), \phi(Y, Z)\right\rangle_s$$

$$= e^{-rs}\left(\rho_1\frac{\beta(Y_s)}{\sqrt{\varepsilon}}\frac{\partial\phi}{\partial y} + \rho_2\sqrt{\delta}g(Z_s)\frac{\partial\phi}{\partial z}\right)\frac{\partial}{\partial x}\left(x^2\frac{\partial^2 P_0}{\partial x^2}\right)f(Y_s, Z_s)X_s ds$$

$$+e^{-rs}\left(\rho_{12}\frac{\beta(Y_s)}{\sqrt{\varepsilon}}\frac{\partial\phi}{\partial y}\sqrt{\delta}g(Z_s) + \delta g^2(Z_s)\frac{\partial\phi}{\partial z}\right)\frac{\partial}{\partial z}\left(x^2\frac{\partial^2 P_0}{\partial x^2}\right)ds,$$

where, to simplify the notation, we have dropped the argument (s, X_s, Z_s) in the partial derivatives of P_0 and the argument (Y_s, Z_s) of ϕ. We then deduce the cumulative cost (8.4):

$E_0^\varepsilon(t) =$

$$
\frac{\varepsilon}{2}\Bigg\{ e^{-rt}X_t^2\frac{\partial^2 P_0}{\partial x^2}(t,X_t,Z_t)\phi(Y_t,Z_t) - X_0^2\frac{\partial^2 P_0}{\partial x^2}(0,X_0,Z_0)\phi(Y_0,Z_0)
$$

$$
- \int_0^t \phi d\left(e^{-rs}X_s^2\frac{\partial^2 P_0}{\partial x^2}\right)\Bigg\}
$$

$$
-\frac{1}{2}\int_0^t e^{-rs}\left(\rho_1\sqrt{\varepsilon}\,\beta(Y_s)\frac{\partial\phi}{\partial y} + \rho_2\varepsilon\sqrt{\delta}\,g(Z_s)\frac{\partial\phi}{\partial z}\right)
$$

$$
\times\frac{\partial}{\partial x}\left(x^2\frac{\partial^2 P_0}{\partial x^2}\right)f(Y_s,Z_s)X_s ds
$$

$$
-\frac{1}{2}\int_0^t e^{-rs}\left(\rho_{12}\sqrt{\varepsilon\delta}\,\beta(Y_s)g(Z_s)\frac{\partial\phi}{\partial y} + \varepsilon\delta\,g^2(Z_s)\frac{\partial\phi}{\partial z}\right)\frac{\partial}{\partial z}\left(x^2\frac{\partial^2 P_0}{\partial x^2}\right)ds
$$

$$
-\frac{1}{2}\int_0^t e^{-rs}X_s^2\frac{\partial^2 P_0}{\partial x^2}\left(\delta\varepsilon\mathcal{M}_2 + \sqrt{\delta\varepsilon}\mathcal{M}_3\right)\phi ds
$$

$$
-\frac{\sqrt{\varepsilon}}{2}\int_0^t e^{-rs}X_s^2\frac{\partial^2 P_0}{\partial x^2}\beta(Y_s)\frac{\partial\phi}{\partial y}dW_s^{(1)}
$$

$$
-\frac{\varepsilon\sqrt{\delta}}{2}\int_0^t e^{-rs}X_s^2\frac{\partial^2 P_0}{\partial x^2}g(Z_s)\frac{\partial\phi}{\partial z}dW_s^{(2)}. \tag{8.5}
$$

The terms in the first line are at most of order ε since Y is not involved in the differentials anymore and $x^2\frac{\partial^2 P_0}{\partial x^2}$, and its derivatives, are bounded. The averaging effect will also take place in the first term in the integral on the second line but now the function $\beta f\frac{\partial\phi}{\partial y}$ has no reason to be centered with respect to the invariant distribution of Y. This tells us that the first term on the second line is of order $\sqrt{\varepsilon}$. The same argument applies to the first *martingale term* with respect to $dW_s^{(1)}$ in the last line. The function $\beta\frac{\partial\phi}{\partial y}$ is not centered with respect to the invariant distribution of Y but nevertheless it is a zero-mean martingale. All the other terms are $\mathcal{O}(\varepsilon+\delta)$. To summarize,

$$
E_0^\varepsilon(t) = \sqrt{\varepsilon}\,(B_t^\varepsilon + M_t^\varepsilon) + \mathcal{O}(\varepsilon+\delta) \tag{8.6}
$$

where

$$
B_t^\varepsilon = -\frac{\rho_1}{2}\int_0^t e^{-rs}\beta(Y_s)\frac{\partial\phi}{\partial y}(Y_s,Z_s)f(Y_s,Z_s)
$$

$$
\left[x\frac{\partial}{\partial x}\left(x^2\frac{\partial^2 P_0}{\partial x^2}\right)\right](s,X_s,Z_s)ds, \tag{8.7}
$$

$$M_t^\varepsilon = -\frac{1}{2}\int_0^t e^{-rs}\beta(Y_s)\frac{\partial\phi}{\partial y}(Y_s,Z_s)\left(x^2\frac{\partial^2 P_0}{\partial x^2}\right)(s,X_s,Z_s)dW_s^{(1)}. \quad (8.8)$$

This identifies the drift or bias B_t^ε, and the mean-zero martingale part M_t^ε of the cost. Note that the expected bias $\mathbb{E}\{B_t^\varepsilon\}$ can be of either sign and that the uncorrected Black–Scholes strategy is not *mean self-financing* to the order $\sqrt{\varepsilon}$, which would be the case if the only remaining term of that order was the martingale term. In Section 8.3, we present a correction that achieves exactly this. It is combined with the correction needed to compensate for the small noise effect presented in the following section.

8.2.2 Small Noise Effect and Summary of the Cost

The cost, denoted by $E_0^\delta(t)$, due to the volatility factor Z, is given in the second line of (8.3) by

$$\begin{aligned}
E_0^\delta(t) &= \sqrt{\delta}\rho_2\int_0^t e^{-rs}g(Z_s)f(Y_s,Z_s)X_s\frac{\partial^2 P_0}{\partial x\partial z}ds \\
&\quad + \sqrt{\delta}\int_0^t e^{-rs}g(Z_s)\frac{\partial P_0}{\partial z}dW_s^{(2)} + \mathcal{O}(\delta).
\end{aligned} \quad (8.9)$$

This cost can be written

$$E_0^\delta(t) = \sqrt{\delta}\left(B_t^\delta + M_t^\delta\right) + \mathcal{O}(\delta), \quad (8.10)$$

where we define the bias

$$B_t^\delta = \rho_2\int_0^t e^{-rs}g(Z_s)f(Y_s,Z_s)X_s\frac{\partial^2 P_0}{\partial x\partial z}(s,X_s,Z_s)ds \quad (8.11)$$

and the zero-mean martingale

$$M_t^\delta = \int_0^t e^{-rs}g(Z_s)\frac{\partial P_0}{\partial z}(s,X_s,Z_s)dW_s^{(2)}. \quad (8.12)$$

To summarize, using (8.6) and (8.10), the cost $E_0(t)$ induced by a Black–Scholes hedging strategy, given in (8.3), can be written

$$E_0(t) = \sqrt{\varepsilon}(B_t^\varepsilon + M_t^\varepsilon) + \sqrt{\delta}\left(B_t^\delta + M_t^\delta\right) + \mathcal{O}(\varepsilon + \delta), \quad (8.13)$$

where the biases B^ε and B^δ are given respectively by (8.7) and (8.11), and the zero-mean martingales M^ε and M^δ are given respectively by (8.8) and (8.12).

We are now in a position to correct the Black–Scholes hedging strategy in order to compensate for the biases B^{ε} and B^{δ}.

8.3 Mean Self-Financing Hedging Strategy

We propose to use the perturbation method to remove the biases in the cumulative cost highlighted in the calculation above. More precisely, the biases B^{ε} and B^{δ} will be pushed to the next order by adding small corrections to the Black–Scholes delta, which has the effect of centering the biasing terms. The important feature is that the corrections are not simply the delta of the corrections to the price because that contains features of the market price of volatility risk which are not relevant to the hedging problem. Here $\mathbb{P}$ is the controlling probability measure. However, the hedging correction will depend on the market parameters V_3^{ε} and V_1^{δ} introduced in (4.39) and (4.52) respectively, which can be calibrated from the implied volatility surface as described in Chapter 5.

The analysis separates the effect of the correlation and the effect of the volatility risk premia so that some information from the skew, which is computed in the risk-neutral world, is useful for a real-world risk management problem.

We shall **correct the hedging strategy** in the following way: manage the portfolio made of

$$a_t = \frac{\partial\left(P_{BS}+Q_{1,0}^{\varepsilon}+Q_{0,1}^{\delta}\right)}{\partial x}(t,X_t,Z_t)$$

shares of the risky asset and

$$b_t = e^{-rt}\left((P_{BS}+Q_{1,0}^{\varepsilon}+Q_{0,1}^{\delta})(t,X_t,Z_t)\right.$$
$$\left. -X_t\frac{\partial\left(P_{BS}+Q_{1,0}^{\varepsilon}+Q_{0,1}^{\delta}\right)}{\partial x}(t,X_t,Z_t)\right)$$

shares of the riskless asset, where $P_{BS}(t,x) = P_0(t,x,z)$ is computed at $\bar{\sigma} = \bar{\sigma}(z)$, and $Q_{1,0}^{\varepsilon}(t,x,z)$ and $Q_{0,1}^{\delta}(t,x,z)$ solve respectively the following partial differential equations

$$\mathcal{L}_{BS}(\bar{\sigma}(z))Q_{1,0}^{\varepsilon} = -V_3^{\varepsilon}(z)x\frac{\partial}{\partial x}\left(x^2\frac{\partial^2 P_{BS}}{\partial x^2}\right), \qquad (8.14)$$

$$\mathscr{L}_{BS}(\bar{\sigma}(z))Q^{\delta}_{0,1} = -2V^{\delta}_1(z)x\frac{\partial}{\partial x}\left(\frac{\partial P_{BS}}{\partial \sigma}\right), \tag{8.15}$$

with zero terminal conditions $Q^{\varepsilon}_{1,0}(T,x,z) = Q^{\delta}_{0,1}(T,x,z) = 0$, and where the parameters $V^{\varepsilon}_3(z)$ and $V^{\delta}_1(z)$ are given in (4.39) and (4.52).

As we know from Chapter 4, $Q^{\varepsilon}_{1,0}$ and $Q^{\delta}_{0,1}$ are given explicitly by

$$Q^{\varepsilon}_{1,0} = (T-t)V^{\varepsilon}_3(z)x\frac{\partial}{\partial x}\left(x^2\frac{\partial^2 P_{BS}}{\partial x^2}\right), \tag{8.16}$$

$$Q^{\delta}_{0,1} = (T-t)V^{\delta}_1(z)x\frac{\partial}{\partial x}\left(\frac{\partial P_{BS}}{\partial \sigma}\right). \tag{8.17}$$

Notice that indeed $Q^{\varepsilon}_{1,0}(t,x,z)$ is small of order $\sqrt{\varepsilon}$.

The hedging ratio a_t is now given by

$$
\frac{\partial P_{BS}}{\partial x} + (T-t)V^{\varepsilon}_3(z)\frac{\partial}{\partial x}\left(x\frac{\partial}{\partial x}\left(x^2\frac{\partial^2 P_{BS}}{\partial x^2}\right)\right)
$$

$$
+ (T-t)V^{\delta}_1(z)\frac{\partial}{\partial x}\left(x\frac{\partial}{\partial x}\left(\frac{\partial P_{BS}}{\partial \sigma}\right)\right),
$$

which is the usual Black–Scholes Delta corrected by a combination of Greeks up to fourth-order derivatives. With this new hedging ratio, the infinitesimal cost of the hedging strategy, denoted by $dE^Q_1(t)$, is given by

$$dE^Q_1(t) = d(P_{BS} + Q^{\varepsilon}_{1,0} + Q^{\delta}_{0,1})(t, X_t, Z_t) - a_t dX_t - rb_t e^{rt} dt.$$

Repeating the calculation of the previous section with the new hedging strategy, we find

$$dE^Q_1(t) = \frac{1}{2}\left(f^2(Y_t, Z_t) - \bar{\sigma}^2(Z_t)\right)X^2_t\frac{\partial^2(P_{BS} + Q^{\varepsilon}_{1,0} + Q^{\delta}_{0,1})}{\partial x^2}dt$$

$$+ \sqrt{\delta}\rho_2 g(Z_t)f(Y_t, Z_t)X_t\frac{\partial^2 P_{BS}}{\partial x \partial z}dt + \sqrt{\delta}g(Z_t)\frac{\partial P_{BS}}{\partial z}dW^{(2)}_t$$

$$+ \mathscr{L}_{BS}(\bar{\sigma}(Z_t))(Q^{\varepsilon}_{1,0} + Q^{\delta}_{0,1})dt + \mathscr{O}(\varepsilon + \delta),$$

where we have used the Black–Scholes equation satisfied by P_{BS} and we have grouped terms of higher order in $\mathscr{O}(\varepsilon + \delta)$. The cumulative cost discounted at time zero is given by

$$E_1^Q(t) = \frac{1}{2} \int_0^t e^{-rs} \left(f^2(Y_s, Z_s) - \bar{\sigma}^2(Z_s) \right) X_s^2 \frac{\partial^2 (P_{BS} + Q_{1,0}^\varepsilon + Q_{0,1}^\delta)}{\partial x^2} ds$$

$$+ \sqrt{\delta} \, \rho_2 \int_0^t e^{-rs} g(Z_s) f(Y_s, Z_s) X_s \frac{\partial^2 P_{BS}}{\partial x \partial z} ds$$

$$+ \sqrt{\delta} \int_0^t e^{-rs} g(Z_s) \frac{\partial P_{BS}}{\partial z} dW_s^{(2)}$$

$$- \int_0^t e^{-rs} V_3^\varepsilon(Z_s) X_s \frac{\partial}{\partial x} \left(x^2 \frac{\partial^2 P_{BS}}{\partial x^2} \right) ds$$

$$- 2 \int_0^t e^{-rs} V_1^\delta(Z_s) X_s \frac{\partial}{\partial x} \left(\frac{\partial P_{BS}}{\partial \sigma} \right) ds + \mathcal{O}(\varepsilon + \delta), \qquad (8.18)$$

where we have used equations (8.14) and (8.15) satisfied by $Q_{1,0}^\varepsilon$ and $Q_{0,1}^\delta$. Using the averaging effect in the first integral and the definitions (8.7), (8.8), (8.11), (8.12) introduced in the previous section, one deduces that

$$E_1^Q(t) =$$

$$\sqrt{\varepsilon} \left(B_t^\varepsilon + M_t^\varepsilon \right) + \sqrt{\delta} \left(B_t^\delta + M_t^\delta \right) - \int_0^t e^{-rs} V_3^\varepsilon(Z_s) X_s \frac{\partial}{\partial x} \left(x^2 \frac{\partial^2 P_{BS}}{\partial x^2} \right) ds$$

$$- 2 \int_0^t e^{-rs} V_1^\delta(Z_s) X_s \frac{\partial}{\partial x} \left(\frac{\partial P_{BS}}{\partial \sigma} \right) ds + \mathcal{O}(\varepsilon + \delta).$$

We then observe that, with this choice of corrections in the hedging ratio and the definitions of V_3^ε and V_1^δ, we have

$$\sqrt{\varepsilon} B_t^\varepsilon - \int_0^t e^{-rs} V_3^\varepsilon(Z_s) X_s \frac{\partial}{\partial x} \left(x^2 \frac{\partial^2 P_{BS}}{\partial x^2} \right) ds$$

$$= -\frac{\rho_1 \sqrt{\varepsilon}}{2} \int_0^t e^{-rs} \left\{ \beta(Y_s) f(Y_s, Z_s) \frac{\partial \phi}{\partial y}(Y_s, Z_s) - \left\langle \beta f \frac{\partial \phi}{\partial y} \right\rangle (Z_s) \right\}$$

$$\times \left[x \frac{\partial}{\partial x} \left(x^2 \frac{\partial^2 P_{BS}}{\partial x^2} \right) \right] (s, X_s, Z_s) ds \qquad (8.19)$$

and

$$\sqrt{\delta} B_t^\delta - 2 \int_0^t e^{-rs} V_1^\delta(Z_s) X_s \frac{\partial}{\partial x} \left(\frac{\partial P_{BS}}{\partial \sigma} \right) ds$$

$$= \rho_2 \sqrt{\delta} \int_0^t e^{-rs} g(Z_s) \{ f(Y_s, Z_s) - \langle f \rangle (Z_s) \} \bar{\sigma}'(Z_s)$$

$$\times \left[x \frac{\partial}{\partial x} \left(\frac{\partial P_{BS}}{\partial \sigma} \right) \right] (s, X_s, Z_s) ds. \qquad (8.20)$$

Due to the choice of our corrections, these two integrals are now centered with respect to the invariant distribution of Y, and the averaging effect shows that (8.19) is of order ε, and (8.20) is of order $\sqrt{\varepsilon\delta}$. Indeed, the argument requires us to introduce the functions $\phi_{1,0}(y,z)$ and $\phi_{0,1}(y,z)$, solutions of the Poisson equations

$$\mathcal{L}_0\phi_{1,0}(y,z) = \beta(y)\frac{\partial\phi}{\partial y}(y,z)f(y,z) - \left\langle\beta(\cdot)f(\cdot,z)\frac{\partial\phi}{\partial y}(\cdot,z)\right\rangle,$$
$$\mathcal{L}_0\phi_{0,1}(y,z) = f(y,z) - \langle f(\cdot,z)\rangle.$$

In the end, all the terms in the corrected cumulative cost (8.18) are of order $(\varepsilon+\delta)$ except the remaining mean-zero martingale terms from (8.13), so that

$$E_1^Q(t) = \sqrt{\varepsilon}M_t^\varepsilon + \sqrt{\delta}M_t^\delta + \mathcal{O}(\varepsilon+\delta).$$

Doing so we have removed the systematic biases in (8.13) and therefore we have **reduced the variance** of this small "nonhedgable" part of the risk due to stochastic volatility. The strategy is now mean self-financing to order $(\varepsilon+\delta)$.

In conclusion, it is important to notice that implementing this strategy requires knowing the effective volatility $\bar\sigma(z)$ and the parameters $V_1^\delta(z)$ and $V_3^\varepsilon(z)$. The calibration of the last two parameters comes from the observed implied volatility surface as discussed in Chapter 5, however $\bar\sigma(z)$ needs to be estimated from historical data, which is not a simple matter. In the following, we propose other strategies which are easier to implement, but with additional costs that we compute.

8.4 A Strategy with Frozen Parameters

In the previous section we have assumed that the local effective volatility $\bar\sigma(z)$ was available dynamically in order to compute P_{BS} and its derivatives at the right volatility level $\bar\sigma(Z_t)$. As we mentioned above, this is not easy to implement. Here we propose a different hedging strategy which consists of using the coefficients $\bar\sigma$, V_1^δ, and V_3^ε calibrated at time zero and frozen until maturity time. This is indeed more practical, but as we will see this strategy produces an additional cost of order $\sqrt{\delta}$. We suggest that this is a good compromise for short maturities, while for longer maturities the parameters should be "recalibrated" several times.

8.4.1 The Black–Scholes Strategy with Frozen Volatility

We start by computing the cost of an uncorrected Black–Scholes strategy where the volatility is constant given by $\bar\sigma_0 = \bar\sigma(Z_0)$, estimated at time zero.

The Black–Scholes price, denoted by $P_{BS}^{(0)}$, is computed at the local effective volatility $\bar{\sigma}_0$ estimated at time zero from historical returns data, so that $P_{BS}^{(0)}(t,x) = P_0(t,x,Z_0)$. Such a strategy replicates the derivative at maturity since $P_0(T,X_T,z) = h(X_T)$ independently of z, but it is not self-financing. The portfolio has

$$a_t = \frac{\partial P_{BS}^{(0)}}{\partial x} \tag{8.21}$$

stocks and

$$b_t = e^{-rt}\left(P_{BS}^{(0)} - X_t \frac{\partial P_{BS}^{(0)}}{\partial x}\right)$$

bonds at time t, so that its value is

$$a_t X_t + b_t e^{rt} = P_{BS}^{(0)}(t,X_t).$$

Using Itô's formula on $P_{BS}^{(0)}(t,X_t)$, its infinitesimal change is given by

$$dP_{BS}^{(0)}(t,X_t) = \left(\frac{\partial P_{BS}^{(0)}}{\partial t}(t,X_t) + \frac{1}{2}f^2(Y_t,Z_t)X_t^2 \frac{\partial^2 P_{BS}^{(0)}}{\partial x^2}(t,X_t)\right)dt + a_t dX_t,$$

where we have used (8.21). The infinitesimal change due to the market (the self-financing part) is given by

$$a_t dX_t + rb_t e^{rt} dt.$$

Consequently, the infinitesimal cost (positive or negative) of the strategy is given by the difference

$$dP_{BS}^{(0)}(t,X_t) - a_t dX_t - rb_t e^{rt} dt = \frac{1}{2}\left(f^2(Y_t,Z_t) - \bar{\sigma}_0^2\right)X_t^2 \frac{\partial^2 P_{BS}^{(0)}}{\partial x^2}(t,X_t)dt,$$

where we have used the Black–Scholes equation satisfied by $P_{BS}^{(0)}$.

The corresponding *cumulative cost* up to time t, and discounted to time 0, is

$$E_{\bar{\sigma}_0}(t) = \frac{1}{2}\int_0^t e^{-rs}\left(f^2(Y_s,Z_s) - \bar{\sigma}_0^2\right)X_s^2 \frac{\partial^2 P_{BS}^{(0)}}{\partial x^2}(s,X_s)ds, \tag{8.22}$$

and the *total cost* is $E_{\bar{\sigma}_0}(T)$ in addition to the initial cost $P_{BS}^{(0)}(0,X_0)$.

In order to separate the effects of the fast and slow factors, we rewrite the cost $E_{\bar{\sigma}_0}(t)$ as

$$E_{\bar{\sigma}_0}(t) = \frac{1}{2} \int_0^t e^{-rs} \left(f^2(Y_s, Z_s) - \bar{\sigma}^2(Z_s) \right) X_s^2 \frac{\partial^2 P_{BS}^{(0)}}{\partial x^2}(s, X_s) ds$$

$$+ \frac{1}{2} \int_0^t e^{-rs} \left(\bar{\sigma}^2(Z_s) - \bar{\sigma}_0^2 \right) X_s^2 \frac{\partial^2 P_{BS}^{(0)}}{\partial x^2}(s, X_s) ds. \qquad (8.23)$$

The first integral, denoted by $E_{\bar{\sigma}_0}^{\varepsilon}(t)$, is small due to the *averaging effect* caused by the rapid oscillations of the fast volatility factor Y. It is very similar to the cost $E_0^{\varepsilon}(t)$ studied in Section 8.2.1, the only difference being that the price $P_{BS}^{(0)}(t, X_t)$, and its derivatives, do not vary with Z_t. Following the lines of the computation in Section 8.2.1, one easily deduces that

$$E_{\bar{\sigma}_0}^{\varepsilon}(t) = \sqrt{\varepsilon} \left(B_t^{\varepsilon}(\bar{\sigma}_0) + M_t^{\varepsilon}(\bar{\sigma}_0) \right) + \mathcal{O}(\varepsilon + \delta), \qquad (8.24)$$

where $B_t^{\varepsilon}(\bar{\sigma}_0)$ is a systematic bias and $M_t^{\varepsilon}(\bar{\sigma}_0)$ a mean-zero martingale, given by

$$B_t^{\varepsilon}(\bar{\sigma}_0) = -\frac{\rho_1}{2} \int_0^t e^{-rs} \beta(Y_s) \frac{\partial \phi}{\partial y}(Y_s, Z_s) f(Y_s, Z_s)$$

$$\times \left[x \frac{\partial}{\partial x} \left(x^2 \frac{\partial^2 P_{BS}^{(0)}}{\partial x^2} \right) \right] (s, X_s) ds,$$

$$M_t^{\varepsilon}(\bar{\sigma}_0) = -\frac{1}{2} \int_0^t e^{-rs} \beta(Y_s) \frac{\partial \phi}{\partial y}(Y_s, Z_s) \left(x^2 \frac{\partial^2 P_{BS}^{(0)}}{\partial x^2} \right) (s, X_s) dW_s^{(1)}.$$

The second integral in (8.23) is small due to the *small noise effect* caused by the slow variations of Z_t from its initial value Z_0 and the assumed smoothness of $\bar{\sigma}(z)$. In the following sections we make these two facts precise.

8.4.2 The Small Noise Effect

In this section we study the second integral appearing in (8.23). We denote this part of the cost by $E_{\bar{\sigma}_0}^{\delta}(t)$. It is given by

$$E_{\bar{\sigma}_0}^{\delta}(t) = \frac{1}{2} \int_0^t e^{-rs} \left(\bar{\sigma}^2(Z_s) - \bar{\sigma}_0^2 \right) X_s^2 \frac{\partial^2 P_{BS}^{(0)}}{\partial x^2} ds,$$

where we have dropped the argument (s, X_s) in the derivatives of $P_{BS}^{(0)}$. Applying Itô's formula to the smooth function $\bar{\sigma}^2$ one gets

$$d\bar{\sigma}^2(Z_s) = 2\bar{\sigma}(Z_s)\bar{\sigma}'(Z_s) \left(\delta c(Z_s) ds + \sqrt{\delta} g(Z_s) dW_s^{(2)} \right)$$

$$+ \delta(\bar{\sigma}\bar{\sigma}')'(Z_s) g^2(Z_s) ds,$$

and consequently

$$\bar{\sigma}^2(Z_s) - \bar{\sigma}_0^2 = 2\sqrt{\delta} \int_0^s \bar{\sigma}(Z_u)\bar{\sigma}'(Z_u)g(Z_u)dW_u^{(2)} + \mathcal{O}(\delta),$$

$$= 2\sqrt{\delta}\,\bar{\sigma}_0\bar{\sigma}_0'g_0 W_s^{(2)} + \mathcal{O}(\delta),$$

by using the assumed smoothness of $\bar{\sigma}\bar{\sigma}'g$, and the notations $\bar{\sigma}(Z_0) = \bar{\sigma}_0$, $\bar{\sigma}'(Z_0) = \bar{\sigma}_0'$, and $g(Z_0) = g_0$. Therefore the cost $E_{\bar{\sigma}_0}^\delta(t)$ can be written

$$E_{\bar{\sigma}_0}^\delta(t) = \sqrt{\delta}\,\bar{\sigma}_0\bar{\sigma}_0'g_0 \int_0^t e^{-rs}W_s^{(2)}X_s^2\frac{\partial^2 P_{BS}^{(0)}}{\partial x^2}ds + \mathcal{O}(\delta).$$

Integrating by parts gives

$$d\left(-(T-s)e^{-rs}W_s^{(2)}X_s^2\frac{\partial^2 P_{BS}^{(0)}}{\partial x^2}\right)$$

$$= e^{-rs}W_s^{(2)}X_s^2\frac{\partial^2 P_{BS}^{(0)}}{\partial x^2}ds - (T-s)d\left(e^{-rs}W_s^{(2)}X_s^2\frac{\partial^2 P_{BS}^{(0)}}{\partial x^2}\right),$$

and therefore

$$E_{\bar{\sigma}_0}^\delta(t) = -\sqrt{\delta}\,\bar{\sigma}_0\bar{\sigma}_0'g_0(T-t)e^{-rt}W_t^{(2)}X_t^2\frac{\partial^2 P_{BS}^{(0)}}{\partial x^2}(t,X_t)$$

$$+ \sqrt{\delta}\,\bar{\sigma}_0\bar{\sigma}_0'g_0 \int_0^t (T-s)d\left(e^{-rs}W_s^{(2)}X_s^2\frac{\partial^2 P_{BS}^{(0)}}{\partial x^2}\right) + \mathcal{O}(\delta). \quad (8.25)$$

Using the notation $D_k = x^k\partial^k/\partial x^k$, the stochastic integration by parts formula gives

$$d\left(e^{-rs}W_s^{(2)}D_2P_{BS}^{(0)}\right) = e^{-rs}D_2P_{BS}^{(0)}dW_s^{(2)} + W_s^{(2)}d\left(e^{-rs}D_2P_{BS}^{(0)}\right)$$

$$+ d\left\langle W_s^{(2)}, e^{-rs}D_2P_{BS}^{(0)}\right\rangle_s. \quad (8.26)$$

We then compute

$$d\left(e^{-rs}D_2P_{BS}^{(0)}\right)$$

$$= e^{-rs}\left(-rD_2P_{BS}^{(0)}ds + \frac{\partial}{\partial t}D_2P_{BS}^{(0)}ds + \frac{\partial}{\partial x}D_2P_{BS}^{(0)}dX_s + \frac{1}{2}\frac{\partial^2}{\partial x^2}D_2P_{BS}^{(0)}d\langle X\rangle_s\right)$$

$$= e^{-rs}\left(-rD_2P_{BS}^{(0)} + \frac{\partial}{\partial t}D_2P_{BS}^{(0)} + \mu D_1D_2P_{BS}^{(0)} + \frac{1}{2}f^2D_2D_2P_{BS}^{(0)}\right)ds$$

$$+ e^{-rs}fD_1D_2P_{BS}^{(0)}dW_t^{(0)}$$

$$= e^{-rs}\left((\mu - r)D_1 D_2 P_{BS}^{(0)} + \frac{1}{2}(f^2 - \bar{\sigma}_0^2)D_2^2 P_{BS}^{(0)}\right)ds$$

$$+ e^{-rs}f D_1 D_2 P_{BS}^{(0)} dW_s^{(0)}, \tag{8.27}$$

where we have used the fact that the operators $\partial/\partial t$, D_1, and D_2 commute, and the fact that $P_{BS}^{(0)}(t,x)$ satisfies the Black–Scholes equation with volatility $\bar{\sigma}_0$. Substituting (8.27) into (8.26) one gets

$$d\left(e^{-rs}W_s^{(2)}D_2 P_{BS}^{(0)}\right) = e^{-rs}D_2 P_{BS}^{(0)} dW_s^{(2)} \tag{8.28}$$

$$+ e^{-rs}W_s^{(2)}\left((\mu - r)D_1 D_2 P_{BS}^{(0)} + \frac{1}{2}(f^2 - \bar{\sigma}_0^2)D_2^2 P_{BS}^{(0)}\right)ds$$

$$+ e^{-rs}W_s^{(2)}f D_1 D_2 P_{BS}^{(0)} dW_s^{(0)} + \rho_2 e^{-rs}f D_1 D_2 P_{BS}^{(0)} ds, \tag{8.29}$$

and therefore from (8.25) we get

$$E_{\bar{\sigma}_0}^{\delta}(t) = -\sqrt{\delta}\,\bar{\sigma}_0 \bar{\sigma}_0' g_0 (T-t)e^{-rt}W_t^{(2)}X_t^2 \frac{\partial^2 P_{BS}^{(0)}}{\partial x^2}(t, X_t)$$

$$+ \sqrt{\delta}\,\bar{\sigma}_0 \bar{\sigma}_0' g_0 \int_0^t (T-s)e^{-rs}D_2 P_{BS}^{(0)} dW_s^{(2)}$$

$$+ \sqrt{\delta}\,\bar{\sigma}_0 \bar{\sigma}_0' g_0 \int_0^t (T-s)e^{-rs}W_s^{(2)}$$

$$\times \left((\mu - r)D_1 D_2 P_{BS}^{(0)} + \frac{1}{2}(f^2 - \bar{\sigma}_0^2)D_2^2 P_{BS}^{(0)}\right)ds$$

$$+ \sqrt{\delta}\,\bar{\sigma}_0 \bar{\sigma}_0' g_0 \int_0^t (T-s)e^{-rs}W_s^{(2)}f D_1 D_2 P_{BS}^{(0)} dW_s^{(0)}$$

$$+ \rho_2 \sqrt{\delta}\,\bar{\sigma}_0 \bar{\sigma}_0' g_0 \int_0^t (T-s)e^{-rs}f D_1 D_2 P_{BS}^{(0)} ds + \mathcal{O}(\delta).$$

The first term vanishes at maturity time T and therefore does not contribute to the total cost $E_{\bar{\sigma}_0}^{\delta}(T)$. The second and fourth terms are mean-zero martingales of order $\sqrt{\delta}$.

By repeating the *averaging small noise* argument, used in (8.22) and which consists of decomposing $(f^2(y,z) - \bar{\sigma}_0^2)$ into $(f^2(y,z) - \bar{\sigma}^2(z)) + (\bar{\sigma}^2(z) - \bar{\sigma}_0^2)$, up to terms of at most order $(\sqrt{\varepsilon\delta} + \delta)$, the third term reduces to the bias

$$\sqrt{\delta}\,\bar{\sigma}_0 \bar{\sigma}_0' g_0 (\mu - r)\int_0^t (T-s)e^{-rs}W_s^{(2)}D_1 D_2 P_{BS}^{(0)} ds.$$

Finally, the fifth term is also a bias of order $\sqrt{\delta}$.

To summarize we have

$$E_{\bar{\sigma}_0}^{\delta}(t) = \sqrt{\delta}\left(B_t^{\delta}(\bar{\sigma}_0) + M_t^{\delta}(\bar{\sigma}_0)\right) + \sqrt{\delta}\,\mathcal{O}(T-t) + \mathcal{O}(\varepsilon+\delta), \quad (8.30)$$

with

$$B_t^{\delta}(\bar{\sigma}_0) = \bar{\sigma}_0\bar{\sigma}_0'g_0 \int_0^t (T-s)e^{-rs}\left[(\mu-r)W_s^{(2)} + p_2 f\right]D_1 D_2 P_{BS}^{(0)}\,ds,$$

$$M_t^{\delta}(\bar{\sigma}_0) = \bar{\sigma}_0\bar{\sigma}_0'g_0 \int_0^t (T-s)e^{-rs}\left[D_2 P_{BS}^{(0)}\,dW_s^{(2)} + W_s^{(2)}fD_1 D_2 P_{BS}^{(0)}\,dW_s^{(0)}\right].$$

In the following section we show that the biases $B_t^{\varepsilon}(\bar{\sigma}_0)$ and $B_t^{\delta}(\bar{\sigma}_0)$ in (8.6) and (8.10) can be reduced by correcting the strategy.

8.4.3 Bias Reduction

We show here how to correct the strategy (8.21) in order to reduce the biases $B_t^{\varepsilon}(\bar{\sigma}_0)$ and $B_t^{\delta}(\bar{\sigma}_0)$ appearing in the costs (8.6) and (8.10).

For that we write

$$\sqrt{\varepsilon}\,B_t^{\varepsilon}(\bar{\sigma}_0)$$

$$= -\frac{\rho_1\sqrt{\varepsilon}}{2}\int_0^t e^{-rs}\beta(Y_s)\frac{\partial\phi}{\partial y}(Y_s,Z_s)f(Y_s,Z_s)D_1 D_2 P_{BS}^{(0)}\,ds$$

$$= -\frac{\rho_1\sqrt{\varepsilon}}{2}\int_0^t e^{-rs}\left(\beta(Y_s)\frac{\partial\phi}{\partial y}(Y_s,Z_s)f(Y_s,Z_s) - \left\langle\beta\frac{\partial\phi}{\partial y}f\right\rangle(Z_s)\right)D_1 D_2 P_{BS}^{(0)}\,ds$$

$$-\frac{\rho_1\sqrt{\varepsilon}}{2}\int_0^t e^{-rs}\left(\left\langle\beta\frac{\partial\phi}{\partial y}f\right\rangle(Z_s) - \left\langle\beta\frac{\partial\phi}{\partial y}f\right\rangle(Z_0)\right)D_1 D_2 P_{BS}^{(0)}\,ds$$

$$-\frac{\rho_1\sqrt{\varepsilon}}{2}\left\langle\beta\frac{\partial\phi}{\partial y}f\right\rangle(Z_0)\int_0^t e^{-rs}D_1 D_2 P_{BS}^{(0)}\,ds, \quad (8.31)$$

and we observe that the first term is of order ε by the averaging effect, the second term is of order $\sqrt{\varepsilon\delta}$ by the small noise effect, and consequently, up to order $\sqrt{\varepsilon}+\sqrt{\delta}$, the bias $\sqrt{\varepsilon}\,B_t^{\varepsilon}(\bar{\sigma}_0)$ can be replaced by the last term.

Similarly we have

$$\sqrt{\delta}\,B_t^{\delta}(\bar{\sigma}_0)$$

$$= \bar{\sigma}_0\bar{\sigma}_0'g_0\sqrt{\delta}\int_0^t (T-s)e^{-rs}\left[(\mu-r)W_s^{(2)} + p_2 f\right]D_1 D_2 P_{BS}^{(0)}\,ds$$

$$= (\mu-r)\bar{\sigma}_0\bar{\sigma}_0'g_0\sqrt{\delta}\int_0^t (T-s)e^{-rs}W_s^{(2)}D_1 D_2 P_{BS}^{(0)}\,ds$$

$$+ p_2\bar{\sigma}_0\bar{\sigma}_0'g_0\sqrt{\delta}\int_0^t (T-s)e^{-rs}\left(f(Y_s,Z_s) - \langle f\rangle(Z_s)\right)D_1 D_2 P_{BS}^{(0)}\,ds$$

$$+\rho_2\bar{\sigma}_0\bar{\sigma}_0'g_0\sqrt{\delta}\int_0^t(T-s)e^{-rs}(\langle f\rangle(Z_s)-\langle f\rangle(Z_0))D_1D_2P_{BS}^{(0)}ds$$

$$+\rho_2\bar{\sigma}_0\bar{\sigma}_0'g_0\sqrt{\delta}\langle f\rangle(Z_0)\int_0^t(T-s)e^{-rs}D_1D_2P_{BS}^{(0)}ds, \tag{8.32}$$

where the second term is of order ε and the third term is of order $\sqrt{\varepsilon\delta}$. Consequently, up to order $\sqrt{\varepsilon}+\sqrt{\delta}$, the bias $\sqrt{\delta}B_t^{\delta}(\bar{\sigma}_0)$ reduces to the first and last terms.

We now correct the hedging strategy in the following way: manage the portfolio made of

$$a_t = \frac{\partial\left(P_{BS}^{(0)}+Q_{1,0}^{(0)\varepsilon}+Q_{0,1}^{(0)\delta}\right)}{\partial x}(t,X_t)$$

shares of the risky asset and

$$b_t = e^{-rt}\left((P_{BS}^{(0)}+Q_{1,0}^{(0)\varepsilon}+Q_{0,1}^{(0)\delta})(t,X_t)-X_t\frac{\partial\left(P_{BS}^{(0)}+Q_{1,0}^{(0)\varepsilon}+Q_{0,1}^{(0)\delta}\right)}{\partial x}(t,X_t)\right)$$

shares of the riskless asset, where $P_{BS}^{(0)}(t,x)=P_0(t,x,Z_0)$ is computed at $\bar{\sigma}_0=\bar{\sigma}(Z_0)$, and $Q_{1,0}^{(0)\varepsilon}(t,x)$ and $Q_{0,1}^{(0)\delta}(t,x)$ solve respectively the following partial differential equations

$$\mathcal{L}_{BS}(\bar{\sigma}_0)Q_{1,0}^{(0)\varepsilon}=-V_3^\varepsilon(Z_0)x\frac{\partial}{\partial x}\left(x^2\frac{\partial^2P_{BS}^{(0)}}{\partial x^2}\right), \tag{8.33}$$

$$\mathcal{L}_{BS}(\bar{\sigma}_0)Q_{0,1}^{(0)\delta}=-2V_1^\delta(Z_0)x\frac{\partial}{\partial x}\left(\frac{\partial P_{BS}^{(0)}}{\partial\sigma}\right), \tag{8.34}$$

with zero terminal conditions $Q_{1,0}^{(0)\varepsilon}(T,x)=Q_{0,1}^{(0)\delta}(T,x)=0$, and where the parameters $V_3^\varepsilon(Z_0)$ and $V_1^\delta(Z_0)$ are frozen at $z=Z_0$. The small corrections $Q_{1,0}^{(0)\varepsilon}$ and $Q_{0,1}^{(0)\delta}$ are given explicitly by

$$Q_{1,0}^{(0)\varepsilon}=(T-t)V_3^\varepsilon(Z_0)x\frac{\partial}{\partial x}\left(x^2\frac{\partial^2P_{BS}^{(0)}}{\partial x^2}\right), \tag{8.35}$$

$$Q_{0,1}^{(0)\delta}=(T-t)V_1^\delta(Z_0)x\frac{\partial}{\partial x}\left(\frac{\partial P_{BS}^{(0)}}{\partial\sigma}\right). \tag{8.36}$$

The corrected hedging ratio is given by

$$
\frac{\partial P_{BS}^{(0)}}{\partial x} + (T-t)V_3^{\varepsilon}(Z_0)\frac{\partial}{\partial x}\left(x\frac{\partial}{\partial x}\left(x^2\frac{\partial^2 P_{BS}^{(0)}}{\partial x^2}\right)\right)
$$
$$
+ (T-t)V_1^{\delta}(Z_0)\frac{\partial}{\partial x}\left(x\frac{\partial}{\partial x}\left(\frac{\partial P_{BS}^{(0)}}{\partial \sigma}\right)\right),
$$

which is the Black–Scholes Delta at frozen volatility $\bar{\sigma}_0$, corrected by a combination of Greeks up to fourth-order derivatives.

Indeed, these corrections have been chosen so that, when computing the cost associated with this hedging strategy, they produce the terms needed to *center* the biases. Denoting the corrected cost by $E_{\bar{\sigma}_0}^{Q}(t)$, using (8.6), (8.10), (8.31), (8.32), and the same technique as in Section 8.3, we obtain

$$
E_{\bar{\sigma}_0}^{Q}(t) = \sqrt{\varepsilon}M_t^{\varepsilon}(\bar{\sigma}_0) + \sqrt{\delta}M_t^{\delta}(\bar{\sigma}_0)
$$
$$
+ (\mu - r)\bar{\sigma}_0\bar{\sigma}_0'g_0\sqrt{\delta}\int_0^t (T-s)e^{-rs}W_s^{(2)}D_1D_2P_{BS}^{(0)}ds
$$
$$
+ \sqrt{\delta}\,\mathcal{O}(T-t) + \mathcal{O}(\varepsilon + \delta)
$$
$$
- \frac{\rho_1\sqrt{\varepsilon}}{2}\left\langle\beta\frac{\partial\phi}{\partial y}f\right\rangle(Z_0)\int_0^t e^{-rs}D_1D_2P_{BS}^{(0)}ds
$$
$$
- V_3^{\varepsilon}(Z_0)\int_0^t e^{-rs}D_1D_2P_{BS}^{(0)}ds
$$
$$
+ \rho_2\bar{\sigma}_0\bar{\sigma}_0'g_0\sqrt{\delta}\langle f\rangle(Z_0)\int_0^t (T-s)e^{-rs}D_1D_2P_{BS}^{(0)}ds
$$
$$
- 2V_1^{\delta}(Z_0)\int_0^t e^{-rs}D_1\frac{\partial P_{BS}^{(0)}}{\partial\sigma}ds,
$$

where the last four terms cancel by the definition of $V_3^{\varepsilon}(Z_0)$, the definition of $V_1^{\delta}(Z_0)$, and the relation

$$
\bar{\sigma}_0(T-s)D_2P_{BS}^{(0)}(s,x) = \frac{\partial P_{BS}^{(0)}}{\partial\sigma}(s,x).
$$

To summarize, up to order $(\varepsilon + \delta + \sqrt{\delta}(T-t))$ which becomes $(\varepsilon + \delta)$ at maturity, the cumulative cost has been reduced to the sum of two mean-zero martingales of order $\sqrt{\varepsilon}$ and $\sqrt{\delta}$, and a nonreducible bias of order $\sqrt{\delta}$ and proportional to $\mu - r$. It is remarkable that this small "nonhedgable" bias is the only price to pay for not dynamically estimating the parameters

$\bar{\sigma}(z)$, $V_1^\delta(z)$, and $V_3^\varepsilon(z)$, but, instead, freezing them at their calibrated initial values.

Note that in this strategy the coefficients are not dynamically calibrated but nevertheless it is a dynamic hedging strategy in the sense that the hedging portfolio is continuously rebalanced according to the hedging ratio given above. The issue of trading cost is not addressed here.

8.5 Strategies Based on Implied Volatilities

In Section 8.3 we have proposed a mean self-financing hedging strategy which removes all biases up to order $(\varepsilon + \delta)$, but which requires the dynamic estimation of the effective volatility $\bar{\sigma}(z)$ and the dynamic calibration of the coefficients $V_1^\delta(z)$ and $V_3^\varepsilon(z)$.

In Section 8.4 we proposed a strategy which consists of using frozen coefficients where the volatility $\bar{\sigma}_0$ is estimated at time zero from historical returns data, and $V_1^\delta(Z_0)$ and $V_3^\varepsilon(Z_0)$ are calibrated from the implied volatilities at time zero. We have seen that the price to pay is a bias of order $\sqrt{\delta}$.

In this section we propose two strategies which are based only on implied volatilities and obtained by replacing statistical estimates of the volatility $\bar{\sigma}(z)$ with the corrected implied volatility $\sigma^\star(z)$ introduced in Section 4.3. In the first one we simply replace $\bar{\sigma}(z)$ with $\sigma^\star(z)$ in the strategy described in Section 8.3. This makes this strategy based solely on implied volatilities data but creates a bias which we compute. In the second strategy we adopt the point of view of being as close as possible to the derivative price at all times, and we compute the additional cost.

8.5.1 Strategy Based on Dynamically Calibrated Parameters

Our motivation here is to make the strategy introduced in Section 8.3 more practical by replacing the dynamically estimated volatility $\bar{\sigma}(z)$ with the dynamically calibrated volatility $\sigma^\star(z)$. It is expected that a bias will be created since $\sigma^\star(z)$ contains the parameter V_2^ε of order $\sqrt{\varepsilon}$ which is due to the market price of volatility risk Λ_1 irrelevant to the hedging problem under $\mathbb{P}$.

Our hedging strategy consists of managing the portfolio made of

$$a_t = \frac{\partial \left(P_{BS}^\star + Q_{1,0}^{\star\varepsilon} + Q_{0,1}^{\star\delta} \right)}{\partial x}(t, X_t, Z_t)$$

shares of the risky asset and

$$b_t = e^{-rt}\left((P^\star_{BS}+Q^{\star\varepsilon}_{1,0}+Q^{\star\delta}_{0,1})(t,X_t,Z_t)-X_t\frac{\partial\left(P^\star_{BS}+Q^{\star\varepsilon}_{1,0}+Q^{\star\delta}_{0,1}\right)}{\partial x}(t,X_t,Z_t)\right)$$

shares of the riskless asset, where $P^\star_{BS}(t,x)=P_0(t,x,z)$ is computed at $\sigma^\star=\sigma^\star(z)$, calibrated from implied volatilities as explained in Chapter 5, and $Q^{\star\varepsilon}_{1,0}(t,x,z)$ and $Q^{\star\delta}_{0,1}(t,x,z)$ solve respectively the following partial differential equations

$$\mathscr{L}_{BS}(\sigma^\star)Q^{\star\varepsilon}_{1,0}=-V^\varepsilon_3 x\frac{\partial}{\partial x}\left(x^2\frac{\partial^2 P^\star_{BS}}{\partial x^2}\right), \tag{8.37}$$

$$\mathscr{L}_{BS}(\sigma^\star)Q^{\star\delta}_{0,1}=-2V^\delta_1 x\frac{\partial}{\partial x}\left(\frac{\partial P^\star_{BS}}{\partial\sigma}\right), \tag{8.38}$$

with zero terminal conditions $Q^{\star\varepsilon}_{1,0}(T,x,z)=Q^{\star\delta}_{0,1}(T,x,z)=0$. The corresponding corrected hedging ratio is now given by

$$\boxed{\begin{aligned}&\frac{\partial P^\star_{BS}}{\partial x}+(T-t)V^\varepsilon_3\frac{\partial}{\partial x}\left(x\frac{\partial}{\partial x}\left(x^2\frac{\partial^2 P^\star_{BS}}{\partial x^2}\right)\right)\\&\qquad\qquad+(T-t)V^\delta_1\frac{\partial}{\partial x}\left(x\frac{\partial}{\partial x}\left(\frac{\partial P^\star_{BS}}{\partial\sigma}\right)\right).\end{aligned}}$$

The derivation of the associated hedging cost follows the lines of Section 8.3 with the only difference, up to the terms of order $(\sqrt{\varepsilon}+\sqrt{\delta})$, that $\bar{\sigma}^2(z)$ is replaced by $\sigma^{\star2}(z)$, which creates in (8.18) the additional cost

$$\int_0^t e^{-rs}V^\varepsilon_2(Z_s)X^2_s\frac{\partial^2 P^\star_{BS}}{\partial x^2}(s,X_s)ds.$$

In the end, the new cumulative cost, denoted by $E^{Q\star}_1(t)$, is given by

$$E^{Q\star}_1(t)=\int_0^t e^{-rs}V^\varepsilon_2(Z_s)X^2_s\frac{\partial^2 P^\star_{BS}}{\partial x^2}(s,X_s)ds+\sqrt{\varepsilon}M^\varepsilon_t+\sqrt{\delta}M^\delta_t+\mathcal{O}(\varepsilon+\delta),$$

where M^ε and M^δ are the mean-zero martingales given in (8.8) and (8.12).

We have seen in Chapter 5 that the parameters $(\sigma^\star,V^\delta_0,V^\delta_1,V^\varepsilon_3)$ can be calibrated from the implied volatilities without using any historical returns data. However, the parameter V^ε_2 associated with the market price of fast volatility risk cannot be calibrated and consequently the bias introduced above cannot be compensated as was the case in Section 8.3.

Note that the value of the hedging portfolio $\left(P_{BS}^{\star}+Q_{1,0}^{\star\varepsilon}+Q_{0,1}^{\star\delta}\right)$ is closer to the price of the derivative than its value $\left(P_{BS}+Q_{1,0}^{\varepsilon}+Q_{0,1}^{\delta}\right)$ in the strategy proposed in Section 8.3 in the sense that, from Chapter 4, in the first case we have

$$P^{\varepsilon,\delta}-\left(P_{BS}^{\star}+Q_{1,0}^{\star\varepsilon}+Q_{0,1}^{\star\delta}\right)=(T-t)V_0^{\delta}\frac{\partial P_{BS}^{\star}}{\partial\sigma}+\mathscr{O}(\varepsilon+\delta),$$

while in the second case we have

$$P^{\varepsilon,\delta}-\left(P_{BS}+Q_{1,0}^{\varepsilon}+Q_{0,1}^{\delta}\right)$$
$$=(T-t)V_0^{\delta}\frac{\partial P_{BS}}{\partial\sigma}+(T-t)V_2^{\varepsilon}x^2\frac{\partial^2 P_{BS}}{\partial x^2}+\mathscr{O}(\varepsilon+\delta),$$
$$=(T-t)V_0^{\delta}\frac{\partial P_{BS}^{\star}}{\partial\sigma}+(T-t)V_2^{\varepsilon}x^2\frac{\partial^2 P_{BS}^{\star}}{\partial x^2}+\mathscr{O}(\varepsilon+\delta).$$

This remark actually leads us to the strategy designed from the point of view of staying as close as possible to the price of the derivative at all times before maturity. This is discussed in the following section.

8.5.2 Staying Close to the Price

In some circumstances it might be desirable to manage a portfolio which stays close to the price of the derivative at times before maturity. Indeed, such a strategy creates an additional cost which we quantify here. The remark at the end of the previous section shows that the strategy associated with the hedging ratio

$$a_t=\frac{\partial\left(P_{BS}^{\star}+Q_{1,0}^{\star\varepsilon}+Q_{0,1}^{\star\delta}\right)}{\partial x}(t,X_t,Z_t)$$

leads to a hedging portfolio whose value differs from the value of the derivative by a term of order $\sqrt{\delta}$ and other terms of higher order. Using the parameter V_0^{δ} we can modify the correction $Q_{0,1}^{\star\delta}$ in the following way: define $\widetilde{Q_{0,1}^{\star\delta}}$ as the solution to the problem

$$\mathscr{L}_{BS}(\sigma^{\star})\widetilde{Q_{0,1}^{\star\delta}}=-2V_0^{\delta}\frac{\partial P_{BS}^{\star}}{\partial\sigma}-2V_1^{\delta}x\frac{\partial}{\partial x}\left(\frac{\partial P_{BS}^{\star}}{\partial\sigma}\right),$$
$$\widetilde{Q_{0,1}^{\star\delta}}(T,x,z)=0.$$

We now consider the strategy associated with the new hedging ratio

$$a_t = \frac{\partial \left(P_{BS}^\star + Q_{1,0}^{\star\varepsilon} + \widetilde{Q_{0,1}^{\star\delta}} \right)}{\partial x} (t, X_t, Z_t).$$

The value of the corresponding hedging portfolio is given by

$$\left(P_{BS}^\star + Q_{1,0}^{\star\varepsilon} + \widetilde{Q_{0,1}^{\star\delta}} \right),$$

which is exactly the price approximation obtained in Section 4.3, and therefore equal to the price of the derivative up to order $(\varepsilon + \delta)$.

Following the lines of the derivation of the cost in Section 8.3, one easily deduces that the new cost, denoted by $E_1^{\widetilde{Q^\star}}(t)$, is given by

$$E_1^{\widetilde{Q^\star}}(t) = \int_0^t e^{-rs} V_2^\varepsilon(Z_s) X_s^2 \frac{\partial^2 P_{BS}^\star}{\partial x^2} ds - 2 \int_0^t e^{-rs} V_0^\delta \frac{\partial P_{BS}^\star}{\partial \sigma} ds$$
$$+ \sqrt{\varepsilon} M_t^\varepsilon + \sqrt{\delta} M_t^\delta + \mathcal{O}(\varepsilon + \delta),$$

where the first bias, of order $\sqrt{\varepsilon}$, is associated with the market price of fast volatility risk Λ_1, and the second bias, of order $\sqrt{\delta}$, is associated with the market price of slow volatility risk Λ_2.

8.6 Martingale Approach to Pricing

The goal of this section is to show that the approximation of the price of a European derivative obtained in Chapter 4 can also be derived by using a martingale approach. The argument is very similar to the one we have used in the previous sections to reduce the Black–Scholes hedging biases. It is also important in order to handle non-Markovian models.

8.6.1 Main Argument

In order to explain the martingale approach, we develop it here in the context of the multiscale stochastic volatility models given by (4.1) under the risk-neutral pricing measure $\mathbb{P}^\star$.

The price at time t of a European derivative with terminal payoff function h, denoted by $P_t^{\varepsilon,\delta}$, is given by

$$P_t^{\varepsilon,\delta} = \mathbb{E}^\star \{ e^{-r(T-t)} h(X_T) | \mathscr{F}_t \},$$

where the conditional expectation is taken under the equivalent martingale measure $\mathbb{P}^\star$ chosen by the market and with respect to the past of (X, Y, Z) up to time t denoted by $\mathscr{F}_t$. Note that this is also the natural filtration of the Brownian motions $(W^{(0)}, W^{(1)}, W^{(2)})$. We will not use here the Markov property of the process (X, Y, Z) as we did in the partial differential equation approach presented in Chapter 4.

The price $P_t^{\varepsilon,\delta}$ is also characterized by the fact that the process $M^{\varepsilon,\delta}$ defined by

$$M_t^{\varepsilon,\delta} = e^{-rt} P_t^{\varepsilon,\delta} = \mathbb{E}^\star \{ e^{-rT} h(X_T) | \mathscr{F}_t \} \tag{8.39}$$

is a martingale with a terminal value given by

$$M_T^{\varepsilon,\delta} = e^{-rT} h(X_T).$$

On the other hand, for a given function $Q^{\varepsilon,\delta}(t, x, y, z)$ which may depend on ε and δ, we consider the process $N^{\varepsilon,\delta}$ defined by

$$N_t^{\varepsilon,\delta} = e^{-rt} Q^{\varepsilon,\delta}(t, X_t, Y_t, Z_t). \tag{8.40}$$

If the function $Q^{\varepsilon,\delta}$ satisfies $Q^{\varepsilon,\delta}(T, x, y, z) = h(x)$ at the final time T, we have

$$M_T^{\varepsilon,\delta} = N_T^{\varepsilon,\delta}. \tag{8.41}$$

The method consists of finding $Q^{\varepsilon,\delta}(t, x, y, z)$ such that $N^{\varepsilon,\delta}$ can be **decomposed** as

$$N_t^{\varepsilon,\delta} = \widetilde{M_t^{\varepsilon,\delta}} + R_t^{\varepsilon,\delta}, \tag{8.42}$$

where $\widetilde{M^{\varepsilon,\delta}}$ is a martingale and $R_t^{\varepsilon,\delta}$ is of order $\varepsilon + \delta$. In finding such functions $Q^{\varepsilon,\delta}$, we will require that the terms of order $\sqrt{\varepsilon}$ and $\sqrt{\delta}$ are absorbed in the martingale part, and that $Q^{\varepsilon,\delta}$ depends essentially on t and x, in the sense that it does not depend on y at all, and on z only as a parameter.

Supposing that this has been established, by taking a conditional expectation with respect to $\mathscr{F}_t$ on both sides of the equality

$$N_T^{\varepsilon,\delta} = \widetilde{M_T^{\varepsilon,\delta}} + R_T^{\varepsilon,\delta},$$

one obtains

$$\mathbb{E}^\star \{ N_T^{\varepsilon,\delta} | \mathscr{F}_t \} = \widetilde{M_t^{\varepsilon,\delta}} + \mathbb{E}^\star \{ R_T^{\varepsilon,\delta} | \mathscr{F}_t \},$$

by the martingale property of $\widetilde{M^{\varepsilon,\delta}}$. From (8.41) and the martingale property of $M^{\varepsilon,\delta}$ we deduce that the left-hand side is also equal to $M_t^{\varepsilon,\delta}$. From the decomposition (8.42) we have $\widetilde{M_t^{\varepsilon,\delta}} = N_t^{\varepsilon,\delta} - R_t^{\varepsilon,\delta}$, and therefore

$$M_t^{\varepsilon,\delta} = N_t^{\varepsilon,\delta} + \mathbb{E}^\star \{ R_T^{\varepsilon,\delta} | \mathscr{F}_t \} - R_t^{\varepsilon,\delta}.$$

Multiplying by e^{rt} and using (8.39) and (8.40) one deduces

$$P_t^{\varepsilon,\delta} = Q^{\varepsilon,\delta}(t,X_t,Y_t,Z_t) + \mathscr{O}(\varepsilon+\delta),$$

which is the desired **approximation result**.

Indeed it remains to determine $Q^{\varepsilon,\delta}$ leading to the decomposition (8.42) of $N^{\varepsilon,\delta}$. The technique is basically the same as we have used in this chapter and in particular in Section 8.3. The only difference is that the computation is done in the risk-neutral world under $\mathbb{P}^\star$. We outline it in the next section and we show that $Q^{\varepsilon,\delta}$ can be chosen equal to $P_0 + \sqrt{\varepsilon}P_{1,0} + \sqrt{\delta}P_{0,1} = P_0 + P_1^{\varepsilon,\delta}$ where P_0 and $P_1^{\varepsilon,\delta}$ have been obtained in Section 4.2.

8.6.2 Decomposition Result

Assuming that we are looking for a function $Q^{\varepsilon,\delta}(t,x,z)$ which does not depend on y, from the definition (8.40) of $N^{\varepsilon,\delta}$ and Itô's formula, we deduce

$$dN_t^{\varepsilon,\delta}$$

$$= e^{-rt}\left(\frac{\partial}{\partial t} + \frac{1}{2}f(Y_t,Z_t)^2 X_t^2 \frac{\partial^2}{\partial x^2} + rX_t\frac{\partial}{\partial x} - r\right)Q^{\varepsilon,\delta}(t,X_t,Z_t)dt$$

$$- \sqrt{\delta}e^{-rt}g(Z_t)\Lambda_2(Y_t,Z_t)\frac{\partial Q^{\varepsilon,\delta}}{\partial z}(t,X_t,Z_t)dt$$

$$+ \rho_2\sqrt{\delta}e^{-rt}g(Z_t)f(Y_t,Z_t)X_t\frac{\partial^2 Q^{\varepsilon,\delta}}{\partial x\partial z}(t,X_t,Z_t)dt$$

$$+ e^{-rt}f(Y_t,Z_t)X_t\frac{\partial Q^{\varepsilon,\delta}}{\partial x}(t,X_t,Z_t)dW_t^{(0)\star}$$

$$+ \sqrt{\delta}e^{-rt}g(Z_t)\frac{\partial Q^{\varepsilon,\delta}}{\partial z}(t,X_t,Z_t)dW_t^{(2)\star} + \mathscr{O}(\delta), \qquad (8.43)$$

where we assume *a priori* enough smoothness on the function $Q^{\varepsilon,\delta}$. In this section we also assume that the market price of volatility risk Λ_1 depends only on y, so that the process Y given by

$$dY_t = \left(\frac{1}{\varepsilon}\alpha(Y_t) - \frac{1}{\sqrt{\varepsilon}}\beta(Y_t)\Lambda_1(Y_t)\right)dt + \frac{1}{\sqrt{\varepsilon}}\beta(Y_t)dW_t^{(1)\star}$$

is a Markov process which admits the infinitesimal generator $\varepsilon^{-1}\mathscr{L}_0^\varepsilon$, where

$$\mathscr{L}_0^\varepsilon = \frac{\beta^2(y)}{2}\frac{\partial^2}{\partial y^2} + \left(\alpha(y) - \sqrt{\varepsilon}\beta(y)\Lambda_1(y)\right)\frac{\partial}{\partial y} \qquad (8.44)$$

$$= \mathscr{L}_0 - \sqrt{\varepsilon}\beta(y)\Lambda_1(y)\frac{\partial}{\partial y}$$

is a perturbation of the infinitesimal generator $\mathcal{L}_0$. Assuming that $|\Lambda_1(y)|$ is bounded, the process Y has a unique invariant distribution, denoted by Φ_ε, which is a perturbation of the invariant distribution Φ associated with the unperturbed infinitesimal generator $\mathcal{L}_0$. The density $\Phi_\varepsilon(y)$ is given by

$$\Phi_\varepsilon(y) = J_\varepsilon \Phi(y) e^{-2\sqrt{\varepsilon}\widetilde{\Lambda}_\beta(y)}, \tag{8.45}$$

where $\widetilde{\Lambda}_\beta$ is an antiderivative of Λ_1/β which is at most linear at infinity by assuming that β is bounded away from zero, and J_ε is a normalization constant depending on ε. This is easily seen by checking that Φ_ε, given by (8.45), satisfies the adjoint equation $(\mathcal{L}_0^\varepsilon)^*\Phi_\varepsilon = 0$ which, in our case, reduces to

$$\left(\frac{\beta^2}{2}\Phi_\varepsilon\right)'' - \left((\alpha - \sqrt{\varepsilon}\beta\Lambda_1)\Phi_\varepsilon\right)' = 0,$$

where the derivatives are taken with respect to y, and Φ. We will denote by $\langle\cdot\rangle_\varepsilon$ the expectation with respect to this invariant distribution Φ_ε

$$\langle g\rangle_\varepsilon = J_\varepsilon \int g(y)\Phi(y)e^{-2\sqrt{\varepsilon}\widetilde{\Lambda}_\beta(y)}dy,$$

and define the local effective volatility $\bar{\sigma}_\varepsilon(z)$ by

$$\bar{\sigma}_\varepsilon^2(z) = \langle f^2(\cdot,z)\rangle_\varepsilon. \tag{8.46}$$

We rewrite (8.43) by introducing the Black–Scholes operator $\mathcal{L}_{BS}(\bar{\sigma}_\varepsilon(z))$:

$$dN_t^{\varepsilon,\delta} =$$

$$e^{-rt}\left(\mathcal{L}_{BS}(\bar{\sigma}_\varepsilon(Z_t)) + \frac{1}{2}\left(f(Y_t,Z_t)^2 - \bar{\sigma}_\varepsilon^2(Z_t)\right)X_t^2\frac{\partial^2}{\partial x^2}\right)Q^{\varepsilon,\delta}dt$$

$$+\sqrt{\delta}e^{-rt}g(Z_t)\left(\rho_2 f(Y_t,Z_t)X_t\frac{\partial^2 Q^{\varepsilon,\delta}}{\partial x\partial z} - \Lambda_2(Y_t,Z_t)\frac{\partial Q^{\varepsilon,\delta}}{\partial z}\right)dt$$

$$+e^{-rt}f(Y_t,Z_t)X_t\frac{\partial Q^{\varepsilon,\delta}}{\partial x}dW_t^{(0)\star} + \sqrt{\delta}e^{-rt}g(Z_t)\frac{\partial Q^{\varepsilon,\delta}}{\partial z}dW_t^{(2)\star} + \mathcal{O}(\delta).$$

$$\tag{8.47}$$

At this point the technique is identical to what we have done in Sections 8.2 and 8.3, starting from equation (8.3). The main difference is that $\bar{\sigma}(z)$ is replaced by $\bar{\sigma}_\varepsilon(z)$ and the averages are with respect to Φ_ε instead of Φ.

We successively define

- The solution $Q_0^\varepsilon(t,x,z) = Q_{BS}^\varepsilon(t,x,\bar{\sigma}_\varepsilon(z))$ of the Black–Scholes equation

$$\mathcal{L}_{BS}(\bar{\sigma}_\varepsilon(z))Q_0^\varepsilon = 0, \tag{8.48}$$

with the terminal condition $Q_0^\varepsilon(T,x,z) = h(x)$.

- A solution $\phi_\varepsilon(y,z)$ of the Poisson equation

$$\mathscr{L}_0^\varepsilon \phi_\varepsilon(y,z) = f(y,z)^2 - \bar{\sigma}_\varepsilon^2(z). \tag{8.49}$$

Using (8.44) one can deduce

$$\frac{\partial \phi_\varepsilon}{\partial y}(y,z) = \frac{2}{\beta^2(y)\Phi_\varepsilon(y)} \int_{-\infty}^y \left(f^2(s,z) - \langle f^2(\cdot,z)\rangle_\varepsilon\right)\Phi_\varepsilon(s)ds, \tag{8.50}$$

as well as estimates similar to the estimates presented in Section 3.3.5.
- The z-dependent quantities

$$V_0^{\varepsilon,\delta} = -\frac{g\sqrt{\delta}}{2}\langle \Lambda_2\rangle_\varepsilon \bar{\sigma}_\varepsilon',$$

$$V_1^{\varepsilon,\delta} = \frac{\rho_2 g\sqrt{\delta}}{2}\langle f\rangle_\varepsilon \bar{\sigma}_\varepsilon',$$

$$V_3^\varepsilon = -\frac{\rho_1\sqrt{\varepsilon}}{2}\left\langle \beta f\frac{\partial \phi_\varepsilon}{\partial y}\right\rangle_\varepsilon.$$

- The source term

$$H^{\varepsilon,\delta} = \left[2V_0^{\varepsilon,\delta}\frac{\partial}{\partial\sigma} + 2V_1^{\varepsilon,\delta}x\frac{\partial}{\partial x}\left(\frac{\partial}{\partial\sigma}\right) + V_3^\varepsilon x\frac{\partial}{\partial x}\left(x^2\frac{\partial^2}{\partial x^2}\right)\right]Q_{BS}^\varepsilon. \tag{8.51}$$

- The solution $\widetilde{Q_1^{\varepsilon,\delta}}(t,x,z)$ of the equation

$$\mathscr{L}_{BS}(\bar{\sigma}_\varepsilon(z))\widetilde{Q_1^{\varepsilon,\delta}} = -H^{\varepsilon,\delta}, \tag{8.52}$$

$$\widetilde{Q_1^{\varepsilon,\delta}}(T,x,z) = 0.$$

- Finally, the function $Q^{\varepsilon,\delta}(t,x,z)$ is defined by

$$Q^{\varepsilon,\delta} = Q_0^\varepsilon + \widetilde{Q_1^{\varepsilon,\delta}}, \tag{8.53}$$

and satisfies the terminal condition $Q^{\varepsilon,\delta}(T,x) = h(x)$ necessary to (8.41).

For this choice of $Q^{\varepsilon,\delta}$ and $H^{\varepsilon,\delta}$, the equation (8.47) can be rewritten

$$dN_t^{\varepsilon,\delta}$$

$$= e^{-rt}\left[\frac{1}{2}\left(f(Y_t,Z_t)^2 - \bar{\sigma}_\varepsilon^2(Z_t)\right)X_t^2\frac{\partial^2 Q^{\varepsilon,\delta}}{\partial x^2} - V_3^\varepsilon X_t\frac{\partial}{\partial x}\left(X_t^2\frac{\partial^2 Q_{BS}^\varepsilon}{\partial x^2}\right)\right]dt$$

$$+ \sqrt{\delta}e^{-rt}g(Z_t)\left(\rho_2 f(Y_t,Z_t)X_t\frac{\partial^2 Q^{\varepsilon,\delta}}{\partial x\partial z} - \Lambda_2(Y_t,Z_t)\frac{\partial Q^{\varepsilon,\delta}}{\partial z}\right)dt$$

$$- e^{-rt} \left(2V_1^{\varepsilon,\delta} X_t \frac{\partial}{\partial x} \left(\frac{\partial Q_{BS}^{\varepsilon}}{\partial \sigma} \right) + 2V_0^{\varepsilon,\delta} \frac{\partial Q_{BS}^{\varepsilon}}{\partial \sigma} \right) dt$$

$$+ e^{-rt} f(Y_t, Z_t) X_t \frac{\partial Q^{\varepsilon,\delta}}{\partial x} dW_t^{(0)\star} + \sqrt{\delta} e^{-rt} g(Z_t) \frac{\partial Q^{\varepsilon,\delta}}{\partial z} dW_t^{(2)\star} + \mathcal{O}(\delta).$$

$$(8.54)$$

Following the technique described in Section 8.2.1, after an integration with respect to t, the averaging effect applied to the first term creates a bias of order $\sqrt{\varepsilon}$ which is centered by the V_3^{ε} term. Consequently, the terms in the first line of (8.54) reduce to additional martingale terms and terms of higher orders. As in Section 8.2.2, the biases of order $\sqrt{\delta}$ appearing in the second line of (8.54) are centered by the $V_0^{\varepsilon,\delta}$ and $V_1^{\varepsilon,\delta}$ terms appearing in the third line. Consequently, the second and third lines in (8.54) reduce also to additional martingale terms and terms of higher order.

In sum, we have shown that, with the choice (8.53), $N_t^{\varepsilon,\delta} = e^{-rt} Q^{\varepsilon,\delta}$ admits the desired decomposition (8.42) into a sum of martingales and terms of order $\varepsilon + \delta$.

8.6.3 *Comparison with the PDE Approach*

We show in this section how to derive, from $Q^{\varepsilon,\delta}(t,x,z)$ given by (8.53), the approximate price $P_0(t,x,z) + P_1(t,x,z)$ which we have obtained in Chapter 4 by working directly on the partial differential equation satisfied by the modeled price $P^{\varepsilon,\delta}(t,x,y,z)$. These two approximations are functions of (t,x,z) only, they have the same order of approximation $\mathcal{O}(\varepsilon + \delta)$, and they have the same terminal condition $h(x)$.

By expanding (8.45) with respect to $\sqrt{\varepsilon}$, including the normalization constant J_ε, one can easily see that the invariant distribution of the process Y with infinitesimal generator $\mathcal{L}_0^{\varepsilon}$ is related to the invariant distribution of the process with the unperturbed infinitesimal generator $\mathcal{L}_0$, by

$$\langle \bullet \rangle_\varepsilon = \langle \bullet \rangle - 2\sqrt{\varepsilon} \langle \widetilde{\Lambda_\beta}(\bullet - \langle \bullet \rangle) \rangle + \mathcal{O}(\varepsilon). \tag{8.55}$$

Consequently

$$\mathcal{L}_{BS}(\bar{\sigma}_\varepsilon(z)) = \mathcal{L}_{BS}(\bar{\sigma}(z)) + \frac{1}{2} \left(\langle f^2 \rangle_\varepsilon - \langle f^2 \rangle \right) x^2 \frac{\partial^2}{\partial x^2}$$

$$= \mathcal{L}_{BS}(\bar{\sigma}(z)) - \sqrt{\varepsilon} \langle \widetilde{\Lambda_\beta}(f^2 - \langle f^2 \rangle) \rangle x^2 \frac{\partial^2}{\partial x^2} + \mathcal{O}(\varepsilon)$$

$$= \mathcal{L}_{BS}(\bar{\sigma}(z)) + \frac{\sqrt{\varepsilon}}{2} \langle \beta \Lambda_1 \frac{\partial \phi}{\partial y} \rangle x^2 \frac{\partial^2}{\partial x^2} + \mathcal{O}(\varepsilon), \tag{8.56}$$

by using the definition of $\widetilde{\Lambda_\beta}$ as an antiderivative of Λ_1/β, the relation (8.50), and an integration by parts. We see here that the market price of risk Λ_1 induces a correction to the effective local volatility $\bar{\sigma}(z)$.

Replacing ϕ_ε by ϕ and $\langle \cdots \rangle_\varepsilon$ by $\langle \cdots \rangle$ in the definitions of $V_0^{\varepsilon,\delta}, V_1^{\varepsilon,\delta}, V_3^\varepsilon$ given in the previous section, one can replace these quantities in (8.51) by $V_0^\delta(z), V_1^\delta(z), V_3^\varepsilon(z)$ defined in Sections 4.2.2 and 4.2.4, without affecting the $\sqrt{\varepsilon}$ and $\sqrt{\delta}$-order terms. Still without affecting this order we can then replace $Q_0^\varepsilon(t,x,z)$ by $P_0(t,x,z) = P_{BS}(t,x;\bar{\sigma}(z))$, the solution of the Black–Scholes equation with the local effective volatility $\bar{\sigma}(z)$. Then using (8.52), (8.56), and the definition (4.40) of $V_2^\varepsilon(z)$, we obtain that $Q^{\varepsilon,\delta}$ defined by (8.53) solves

$$\mathcal{L}_{BS}(\bar{\sigma}(z))Q^{\varepsilon,\delta} = -\left[2V_0^\delta(z)\frac{\partial}{\partial\sigma} + 2V_1^\delta(z)x\frac{\partial}{\partial x}\left(\frac{\partial}{\partial\sigma}\right) + V_2^\varepsilon(z)x^2\frac{\partial^2}{\partial x^2} \right.$$
$$\left. + V_3^\varepsilon(z)x\frac{\partial}{\partial x}\left(x^2\frac{\partial^2}{\partial x^2}\right) \right] P_{BS} + \mathcal{O}(\varepsilon+\delta).$$

Except for the terms of order $\varepsilon+\delta$ collected in $\mathcal{O}(\varepsilon+\delta)$ this is the same equation defining P_0+P_1 obtained in Chapter 4, since the sources and the terminal conditions are the same. Therefore, by an argument similar to the one detailed in Section 4.3, $Q^{\varepsilon,\delta}$ and P_0+P_1 differ only at the order $\varepsilon+\delta$.

Observe that in practice we will use $P_{BS}^\star + P_1^\star$ introduced in Section 4.3, since it gives the same order of approximation and it only requires the parameters $(\sigma^\star, V_0, V_1, V_3)$ which are easily calibrated as explained in Chapter 5. Nevertheless, the martingale approach presented here shows the universality of the procedure and can be fully exploited in non-Markovian situations as presented in the following section.

8.7 Non-Markovian Models of Volatility

In this section we indicate how the martingale approach presented in Section 8.6 can be used to handle non-Markovian models of stochastic volatility. Writing the most general model which would fit within our theory would be much too technical. Instead, for simplicity and without loss of generality, we consider only the fast scale and we build a class of models on top of the fast mean-reverting factor Y. We consider volatility driving processes $\widetilde{Y}_t = (Y_t, \xi_t)$ where the additional factor is an independent ergodic process (ξ_t) with a *rate of mixing* of the same order as the rate of mean-reversion of Y.

8.7.1 Setting: An Example

In order to illustrate this setting, we consider a concrete example where ξ is an independent stationary centered Gaussian process with covariance denoted by Γ:

$$\mathbb{E}\{\xi_s \xi_{s+t}\} = \Gamma(t).$$

Unless $\Gamma(t)$ is proportional to $e^{-a|t|}$ for some constant $a > 0$, in which case ξ would also be an OU process, in general ξ is not Markovian. A convenient way to write such a process is by using its spectral representation

$$\xi_t = \int e^{i\omega t} \sqrt{2\pi \hat{\Gamma}(\omega)} \, \widetilde{W}(d\omega), \tag{8.57}$$

where $\widetilde{W}$ is a *Gaussian white noise* in the frequency domain, independent of the Brownian motions driving X and Y, and $\hat{\Gamma}$ is the Fourier transform of the covariance function. One has

$$\mathbb{E}\{\xi_s \xi_{s+t}\} = 2\pi \int \int e^{i\omega s} e^{-i\omega'(s+t)} \sqrt{\hat{\Gamma}(\omega)\hat{\Gamma}(\omega')} \, \mathbb{E}\{\widetilde{W}(d\omega)\overline{\widetilde{W}(d\omega')}\}$$

$$= 2\pi \int e^{-i\omega t} \hat{\Gamma}(\omega) d\omega$$

$$= \Gamma(t),$$

where the decay at infinity of Γ is controlled by the behavior of its Fourier transform $\hat{\Gamma}$ at low frequencies, which we suppose smooth.

An interesting particular case is obtained by considering a function $f(y, \xi)$ which depends on the sum $y + \xi$ only. In this case, the volatility σ_t is driven by the process $\widetilde{Y} = Y + \xi^\varepsilon$ and under the risk-neutral probability $\mathbb{P}^\star$ the model is now as follows:

$$dX_t = rX_t dt + f(\widetilde{Y}_t)X_t dW_t^{(0)\star},$$
$$\widetilde{Y}_t = Y_t + \xi_t^\varepsilon,$$
$$dY_t = \left(\frac{1}{\varepsilon}\alpha(Y_t) - \frac{1}{\sqrt{\varepsilon}}\beta(Y_t)\Lambda_1(Y_t, \xi_t^\varepsilon) \right) dt + \frac{1}{\sqrt{\varepsilon}}\beta(Y_t) dW_t^{(1)\star},$$
$$\xi_t^\varepsilon = \xi_{t/\varepsilon} = \int e^{i\omega t/\varepsilon} \sqrt{2\pi\hat{\Gamma}(\omega)} \widetilde{W}(d\omega), \tag{8.58}$$

where $W^{(0)\star}$ and $W^{(1)\star}$ are independent of the white noise $\widetilde{W}$.

We denote by $\langle g \rangle_\varepsilon$ the average of a function $g(y, \xi)$ with respect to the invariant distribution of the process (Y, ξ^ε)

$$\langle g \rangle_\varepsilon = \int \int g(y, \xi) \Phi_\varepsilon(y, \xi) p_\Gamma(\xi) dy d\xi, \tag{8.59}$$

where Φ_ε is the density of Y given by (8.45), which may now depend on ξ through $\Lambda_1(y,\xi)$, and p_Γ is the $\mathcal{N}(0,\Gamma(0))$ density of the Gaussian invariant distribution of the process (ξ_t).

8.7.2 Asymptotics in the Non-Markovian Case

The major difference with the Markovian models considered up to this section is that no-arbitrage prices of derivatives are not obtained as solutions of partial differential equations since we have "lost" the Markov property and the associated tools such as infinitesimal generators. These prices are obtained as conditional expectations with respect to the past of (X,Y,ξ^ε) defined in (8.58). This filtration will be denoted by $(\mathcal{F}_t^\varepsilon)$. For instance, the price at time t of a European derivative contract with payoff function h at maturity T is given by

$$P^\varepsilon(t) = \mathbb{E}^\star\{e^{-r(T-t)}h(X_T)|\mathcal{F}_t^\varepsilon\},$$

or characterized by the fact that

$$M_t^\varepsilon = e^{-rt}P^\varepsilon(t),$$

is a martingale with respect to $(\mathcal{F}_t^\varepsilon)$ with terminal value $M_T^\varepsilon = h(X_T^\varepsilon)$.

As in Section 8.6.1 we seek a function $Q^\varepsilon(t,x)$ satisfying the terminal condition

$$Q^\varepsilon(T,x) = h(x),$$

and such that the process N^ε defined by

$$N_t^\varepsilon = e^{-rt}Q^\varepsilon(t,X_t^\varepsilon),$$

admits the decomposition (8.42)

$$N_t^\varepsilon = \widetilde{M_t^\varepsilon} + R_t^\varepsilon,$$

where $\widetilde{M^\varepsilon}$ is a martingale absorbing the order $\sqrt{\varepsilon}$ and R_t^ε is of order ε. Following the martingale argument given in Section 8.6.1, one can easily deduce the approximation result

$$P^\varepsilon(t) = Q^\varepsilon(t,X_t^\varepsilon) + \mathcal{O}(\varepsilon),$$

where it remains to construct such a function Q^ε.

In the setting of Section 8.7.1 one can still write equations similar to (8.43) and (8.47) for dN_t^ε by replacing Y by (Y,ξ^ε). This gives

$$dN_t^\varepsilon = e^{-rt}\left(\mathcal{L}_{BS}(\bar\sigma_\varepsilon) + \frac{1}{2}\left(f(Y_t,\xi_t^\varepsilon)^2 - \bar\sigma_\varepsilon^2\right)(X_t)^2 \frac{\partial^2}{\partial x^2}\right) Q^\varepsilon(t,X_t)dt$$

$$+e^{-rt}\left(\frac{\partial Q^\varepsilon}{\partial x}(t,X_t)\right) f(Y_t,\xi_t^\varepsilon) X_t dW_t^{(0)\star}, \tag{8.60}$$

where $\bar\sigma_\varepsilon^2$ is the average (8.59) of f^2.

We then define, as in (8.48), the solution $Q_0^\varepsilon(t,x)$ of the Black–Scholes equation with constant volatility $\bar\sigma_\varepsilon$. Since we are not in a Markovian setting we cannot write anymore the Poisson equation defining the function $\phi_\varepsilon(y,\xi)$ as in (8.49). Instead, $\phi_\varepsilon(Y_t,\xi_t^\varepsilon)$ is replaced by the random quantity $\widetilde\phi_\varepsilon(t)$ defined by the *conditional shift*

$$\widetilde\phi_\varepsilon(t) = -\frac{1}{\varepsilon}\mathbb{E}^\star\left\{\int_t^T \left(f(Y_s,\xi_s^\varepsilon)^2 - \bar\sigma_\varepsilon^2\right) ds \Big| \mathscr{F}_t^\varepsilon\right\}. \tag{8.61}$$

It is very similar to the way the solutions of the Poisson equations are constructed in the Markovian case since, after a change of variable $u = s/\varepsilon$ in (8.61), it becomes

$$\widetilde\phi_\varepsilon(t) = -\mathbb{E}^\star\left\{\int_{t/\varepsilon}^{T/\varepsilon} \left(f(Y_u^{(1)},\xi_u)^2 - \bar\sigma_\varepsilon^2\right) du \Big| \mathscr{F}_{t/\varepsilon}\right\},$$

which is the non-Markovian equivalent of

$$\phi_\varepsilon(y) = -\int_0^{+\infty} \mathbb{E}^\star\left\{\left(f(Y_t^{(1)})^2 - \bar\sigma_\varepsilon^2\right) | Y_0 = y\right\} dt,$$

solution of the Poisson equation (8.49). Exponential decay of the covariance function Γ, for instance, and the properties of the process Y ensure that $\widetilde\phi_\varepsilon(t)$ is well-defined by (8.61).

It is easily checked with the definition (8.61) that

$$\mathcal{M}_t^\varepsilon = \widetilde\phi_\varepsilon(t) - \frac{1}{\varepsilon}\int_0^t \left(f(Y_s,\xi_s^\varepsilon)^2 - \bar\sigma_\varepsilon^2\right) ds \tag{8.62}$$

is a martingale.

The second term in dN_t^ε given by (8.60) can be rewritten

$$\frac{e^{-rt}}{2}\left(f(Y_t,\xi_t^\varepsilon)^2 - \bar\sigma_\varepsilon^2\right) X_t^2 \frac{\partial^2 Q^\varepsilon}{\partial x^2}(t,X_t)dt$$

$$= \frac{\varepsilon e^{-rt}}{2} X_t^2 \frac{\partial^2 Q^\varepsilon}{\partial x^2}(t,X_t)\left\{d\widetilde\phi_\varepsilon(t) - d\mathcal{M}_t^\varepsilon\right\}.$$

Repeating the argument given in Section 8.2.1, we perform the integration by parts

$$X_t^2 \frac{\partial^2 Q^\varepsilon}{\partial x^2}(t, X_t) d\widetilde{\phi}_\varepsilon(t) = d\left(X_t^2 \frac{\partial^2 Q^\varepsilon}{\partial x^2}(t, X_t) \widetilde{\phi}_\varepsilon(t)\right)$$

$$- \widetilde{\phi}_\varepsilon(t) d\left(X_t^2 \frac{\partial^2 Q^\varepsilon}{\partial x^2}(t, X_t)\right) - d\left\langle X^2 \frac{\partial^2 Q^\varepsilon}{\partial x^2}(t, X), \widetilde{\phi}_\varepsilon \right\rangle_t,$$

where, from the martingale decomposition (8.62), the covariation term is given by

$$d\left\langle X^2 \frac{\partial^2 Q^\varepsilon}{\partial x^2}(t, X), \widetilde{\phi}_\varepsilon \right\rangle_t = d\left\langle X^2 \frac{\partial^2 Q^\varepsilon}{\partial x^2}(t, X), \mathcal{M}^\varepsilon \right\rangle_t$$

$$= X_t f(Y_t, \xi_t^\varepsilon) \frac{\partial}{\partial x}\left(x^2 \frac{\partial^2 Q^\varepsilon}{\partial x^2}\right)(t, X_t) d\langle W^\star, \mathcal{M}^\varepsilon \rangle_t.$$

We define $\psi_\varepsilon(t)$ by

$$\frac{\sqrt{\varepsilon}}{2} d\langle W^\star, \mathcal{M}^\varepsilon \rangle_t = \psi_\varepsilon(t) dt,$$

so that it plays the role of $\sqrt{\varepsilon}\rho_1 \beta(Y_t)\phi'(Y_t)$ in the Markovian case. Putting all the nonmartingale terms of (8.60) together we see that, in order to center the $\sqrt{\varepsilon}$-order terms, $Q^\varepsilon = Q_0^\varepsilon + Q_1^\varepsilon$ should be chosen such that

$$\mathcal{L}_{BS}(\bar{\sigma}_\varepsilon)Q^\varepsilon = \langle f\psi_\varepsilon \rangle_\varepsilon \frac{\partial}{\partial x}\left(x^2 \frac{\partial^2 Q_0^\varepsilon}{\partial x^2}\right)$$

with the terminal condition $Q^\varepsilon(T, x) = Q_0^\varepsilon(T, x) = h(x)$.

Defining the constant V_3^ε by

$$V_3^\varepsilon = -\langle f\psi_\varepsilon \rangle_\varepsilon,$$

we conclude that Q^ε is characterized as in the Markovian case treated in Section 8.6.2, the difference being in the way the parameters $\bar{\sigma}_\varepsilon$ and V_3^ε are related to the model we started with. Applying the technique of expansion of the invariant distribution presented in Section 8.6.3, one can deduce that a possible choice of approximated price is again given by $P_0 + P_1$ characterized in Chapter 4.

Notes

Results about hedging in incomplete markets can be found in Karatzas and Shreve (1998) and El Karoui and Quenez (1995) and references cited therein. Variance-minimizing strategies and mean self-financing portfolios are described in the survey article of Schweizer (1999), which has references on the history. Hedging strategies for uncorrelated stochastic volatility computed using asymptotic analysis are described in Sircar and Papanicolaou (1999). For practical aspects of using (8.3), namely Gamma-weighted hedging, we refer to Dupire (2010).

The results presented in this chapter generalize, to the case of two volatility time scales, the results given in chapter 7 of Fouque *et al.* (2000) in the case of only one fast time scale. The martingale approach given in Section 8.6 generalizes the results presented in section 10.4 of Fouque *et al.* (2000) and Fouque *et al.* (2001b). The method of averaging by conditional shifts in non-Markovian situations presented in Section 8.7 is described in detail in Kushner (1984).

9
Extensions

We present in this chapter several extensions to the asymptotic approach developed in the previous chapters. In Section 9.1, we extend the results of Chapter 4 to the case where the short rate is also varying, driven by the same factors driving the volatility, and the stock is paying dividends. In Sections 9.3 and 9.4, we derive the second-order corrections generated by the fast and the slow factors. These second terms produce the smile of implied volatilities as illustrated with the Heston model in Chapter 10. In Section 9.5, we show that a periodic daily component can easily be incorporated in the model without additional difficulty in the asymptotic analysis. In Section 9.6, we introduce jumps in the fast volatility factor. We indicate how to generalize the perturbation method to multidimensional models in Section 9.7.

9.1 Dividends and Varying Interest Rates

In this section, we present generalizations of the first-order approximation to the cases where dividends are paid and/or where the interest rate is varying as a function of the factors (Y, Z). We also present the probabilistic representations of these approximations.

9.1.1 Dividends

In this section we indicate how the perturbation theory explained above can easily be modified to incorporate dividend modeling into the stock price model. For simplicity we consider a continuous dividend yield D, which is the fraction of the stock price received by a stockholder per unit of time. The evolution (4.1) of the stock price then becomes

$$dX_t = (r - D)X_t \, dt + f(Y_t, Z_t)X_t \, dW_t^{(0)\star},$$

where the stream of payment due to dividends is taken into account under the risk-neutral measure $\mathbb{P}^\star$. The corresponding Black–Scholes operator for this model is then

$$\mathcal{L}_{BS}^D(\sigma) = \frac{\partial}{\partial t} + \frac{1}{2}\sigma^2 x^2 \frac{\partial^2}{\partial x^2} + (r - D)x\frac{\partial}{\partial x} - r,$$

which is as our usual definition (1.40) with $r - D$ replacement in the $\partial/\partial x$ term only.

The calculations of Section 4.2 then go through analogously with the only change being to the operator $\mathcal{L}_2$ defined in (4.10). With the addition of the dividend structure, it becomes

$$\mathcal{L}_2^D = \frac{\partial}{\partial t} + \frac{1}{2}f^2(y,z)x^2\frac{\partial^2}{\partial x^2} + (r - D)x\frac{\partial}{\partial x} - r = \mathcal{L}_{BS}^D(f(y)).$$

The leading-order approximation to the price $P^{\varepsilon,\delta}(t,x,y)$ of a European contract is now given by $P_0^D(t,x,z)$, the Black–Scholes price for the contract with dividends. It satisfies the problem

$$\langle \mathcal{L}_2^D \rangle P_0^D = \mathcal{L}_{BS}^D(\bar{\sigma}(z))P_0^D = 0,$$
$$P_0^D(T,x,z) = h(x).$$

The first-order perturbation due to stochastic volatility is now denoted by $P_1^{D,\varepsilon,\delta}(t,x,z)$, and, using the notation (4.61), it satisfies the source problem

$$\mathcal{L}_{BS}^D(\bar{\sigma}(z))P_1^{D,\varepsilon,\delta} = -\mathcal{H}^{\varepsilon,\delta}P_0^D,$$
$$P_1^{D,\varepsilon,\delta}(T,x,z) = 0.$$

This is simply because

$$\mathcal{L}_2^D - \langle \mathcal{L}_2^D \rangle = \mathcal{L}_2 - \langle \mathcal{L}_2 \rangle,$$

and therefore the operator $\mathcal{A}^\varepsilon$ defined in (4.32) does not depend on D. Since $\langle \mathcal{M}_1 \rangle$ defined in (4.49) does not involve D, the operator $\mathcal{H}^{\varepsilon,\delta}$ also does not depend on D.

In the case of European contracts, the first-order approximation of the price is given explicitly, as in (4.67), by

$$\tilde{P}^{D,\varepsilon,\delta} = P_0^D + P_1^{D,\varepsilon,\delta}$$
$$= P_0^D + (T - t)\left[V_0^\delta(z)\frac{\partial}{\partial \sigma} + V_1^\delta(z)x\frac{\partial}{\partial x}\left(\frac{\partial}{\partial \sigma}\right) + V_2^\varepsilon(z)x^2\frac{\partial^2}{\partial x^2} \right.$$
$$\left. + V_3^\varepsilon(z)x\frac{\partial}{\partial x}\left(x^2\frac{\partial^2}{\partial x^2}\right)\right]P_0^D,$$

where the effect of the dividend is in the leading-order term P_0^D.

9.1.2 Varying Interest Rates

In this section, we generalize the results obtained in Section 4.2 to the case where the instantaneous interest rate r is varying as a function of the two factors Y and Z, that is $r = r(Y_t, Z_t)$, where $r(y,z)$ is a bounded deterministic function. We show that the perturbation theory leads to first-order approximations which do not depend on a particular choice of r but rather on two additional group market parameters introduced below.

In this case, the price $P^{\varepsilon,\delta}$ of the European contract is given by the expected discounted payoff

$$P^{\varepsilon,\delta}(t, X_t, Y_t, Z_t) = \mathbb{E}^\star \left\{ e^{-\int_t^T r(Y_s, Z_s) ds} h(X_T) \mid X_t, Y_t, Z_t \right\},$$

which depends on the model functions $(r, f, \alpha, \beta, c, g, \Lambda_1, \Lambda_2)$ as in (4.2). The main change in the derivation of the approximation is that the operator $\mathcal{L}_2$ in (4.10) now involves the rate function $r(y,z)$. We denote it by

$$\mathcal{L}_2^r = \mathcal{L}_{BS}(f(y,z), r(y,z)),$$

where the volatility level $f(y,z)$ and the rate $r(y,z)$ are shown explicitly in the Black–Scholes operator.

Denoting by $\bar{r}(z)$ the *effective interest rate* defined, as for $\bar{\sigma}(z)$ in (4.26), by

$$\bar{r}(z) = \langle r(\cdot, z) \rangle = \int r(y,z) \Phi(dy),$$

the operator $\langle \mathcal{L}_2^r \rangle$ becomes

$$\langle \mathcal{L}_2^r \rangle = \frac{\partial}{\partial t} + \frac{1}{2} \bar{\sigma}^2(z) x^2 \frac{\partial^2}{\partial x^2} + \bar{r}(z) \left(x \frac{\partial}{\partial x} - \cdot \right)$$
$$= \mathcal{L}_{BS}(\bar{\sigma}(z), \bar{r}(z)),$$

and the leading-order term denoted by P_0^r satisfies the problem

$$\mathcal{L}_{BS}(\bar{\sigma}(z), \bar{r}(z)) P_0^r = 0, \qquad\qquad (9.1)$$
$$P_0^r(T, x, z) = h(x).$$

We then introduce a solution $\phi_r(y, z)$ of the following Poisson equation in y:

$$\mathcal{L}_0 \phi_r(y, z) = r(y, z) - \bar{r}(z),$$

so that we have

$$\mathcal{L}_0^{-1}(\mathcal{L}_2^r - \langle \mathcal{L}_2^r \rangle) = \frac{1}{2} \phi(y,z) x^2 \frac{\partial^2}{\partial x^2} + \phi_r(y,z) \left(x \frac{\partial}{\partial x} - \cdot \right).$$

The operator $\mathscr{A}^{\varepsilon}$ defined in (4.32) is now replaced by $\mathscr{A}_r^{\varepsilon}$ given by

$$\mathscr{A}_r^{\varepsilon} = \sqrt{\varepsilon}\langle\mathscr{L}_1\mathscr{L}_0^{-1}(\mathscr{L}_2^r - \langle\mathscr{L}_2^r\rangle)\rangle$$

$$= \sqrt{\varepsilon}\left[-\frac{1}{2}\left\langle\beta\Lambda_1\frac{\partial\phi}{\partial y}\right\rangle D_2 + \frac{\rho_1}{2}\left\langle\beta f\frac{\partial\phi}{\partial y}\right\rangle D_1 D_2\right.$$

$$\left.-\left\langle\beta\Lambda_1\frac{\partial\phi_r}{\partial y}\right\rangle(D_1 - D_0) + \rho_1\left\langle\beta f\frac{\partial\phi_r}{\partial y}\right\rangle D_1(D_1 - D_0)\right]. \quad (9.2)$$

Observe that the two additional terms are linear combinations of D_0, D_1, and $D_2 = D_1^2 - D_1$, where D_0 is simply the identity. Therefore, Proposition 4.6 holds for $P_{1,0}^{\varepsilon}$, denoted now by $P_{1,0}^{r,\varepsilon}$, and with P_0^r defined by (9.1) and $\mathscr{A}_r^{\varepsilon}$ given in (9.2):

$$P_{1,0}^{r,\varepsilon} = -(T-t)\mathscr{A}_r^{\varepsilon}P_0^r.$$

Introducing the new group market parameters $V_0^{r,\varepsilon}(z)$ and $V_1^{r,\varepsilon}(z)$, defined by

$$V_0^{r,\varepsilon}(z) = \sqrt{\varepsilon}\left\langle\beta\Lambda_1\frac{\partial\phi_r}{\partial y}\right\rangle, \quad (9.3)$$

$$V_1^{r,\varepsilon}(z) = -\sqrt{\varepsilon}\rho_1\left\langle\beta f\frac{\partial\phi_r}{\partial y}\right\rangle, \quad (9.4)$$

and keeping $V_2^{\varepsilon}(z)$ and $V_3^{\varepsilon}(z)$ as in (4.40) and (4.39), we can write the short time scale contribution as

$$-(T-t)\mathscr{A}_r^{\varepsilon}P_0^r = (T-t)\left[(V_3^{\varepsilon}(z)D_1 + V_2^{\varepsilon}(z))D_2\right.$$
$$\left.+ (V_1^{r,\varepsilon}(z)D_1 + V_0^{r,\varepsilon}(z))(D_1 - D_0)\right]P_0^r.$$

Since the form of the operator $\langle\mathscr{M}_1\rangle$ is not affected by the introduction of a varying rate $r(y,z)$, Proposition 4.8 holds as well with the new P_0^r. This means that the first-order long time scale contribution is only affected through P_0^r, and does not require additional group market parameters other than V_0^{δ} and V_1^{δ} already defined in (4.51) and (4.52). Introducing the generalized operator $\mathscr{H}_r^{\varepsilon,\delta}$ defined by

$$\mathscr{H}_r^{\varepsilon,\delta} = \mathscr{H}^{\varepsilon,\delta} + V_0^{r,\varepsilon}(z)(D_1 - D_0) + V_1^{r,\varepsilon}(z)D_1(D_1 - D_0)$$

$$= 2V_0^{\delta}(z)\frac{\partial}{\partial\sigma} + 2V_1^{\delta}(z)D_1\left(\frac{\partial}{\partial\sigma}\right) + V_2^{\varepsilon}(z)D_2 + V_3^{\varepsilon}(z)D_1 D_2$$

$$+ V_0^{r,\varepsilon}(z)(D_1 - D_0) + V_1^{r,\varepsilon}(z)D_1(D_1 - D_0), \quad (9.5)$$

the first-order perturbation due to stochastic volatility, denoted above by $P_1^{r,\varepsilon,\delta}(t,x,z)$, satisfies the problem

$$\mathscr{L}_{BS}(\bar{\sigma}(z),\bar{r}(z))P_1^{r,\varepsilon,\delta} = -\mathscr{H}_r^{\varepsilon,\delta}P_0^r, \qquad (9.6)$$

$$P_1^{r,\varepsilon,\delta}(T,x,z) = 0.$$

In terms of the Greeks of P_0^r evaluated at $(\bar{\sigma}(z),\bar{r}(z))$ and with a slight abuse of notation, the price approximation (4.68) becomes

$$P_0^r + (T-t)\left\{ V_0^\delta(z)\mathscr{V} + V_1^\delta(z)x\Delta\mathscr{V} + V_2^\varepsilon(z)x^2\Gamma + V_3^\varepsilon(z)x\Delta\left(x^2\Gamma\right) \right\}$$

$$+ V_0^{r,\varepsilon}(z)\mathscr{R} + V_1^{r,\varepsilon}(z)\Delta\mathscr{R}, \qquad (9.7)$$

where we have used (1.52), that is for European contracts the *Rho* is given by

$$\mathscr{R} \equiv \frac{\partial P_0^r}{\partial r} = (T-t)(D_1 - D_0)P_0^r.$$

Indeed, one can easily combine the presence of dividends as in Section 9.1.1 and varying interest rate as above. This is done by introducing the following notation for the Black–Scholes operator:

$$\mathscr{L}_{BS}^D(\sigma,r) = \frac{\partial}{\partial t} + \frac{1}{2}\sigma^2 x^2 \frac{\partial^2}{\partial x^2} + (r-D)x\frac{\partial}{\partial x} - r,$$

and defining the leading-order approximation $P_0^{D,r}(t,x,z)$ as the solution to the problem

$$\mathscr{L}_{BS}^D(\bar{\sigma}(z),\bar{r}(z))P_0^{D,r} = 0, \qquad (9.8)$$

$$P_0^{D,r}(T,x,z) = h(x).$$

The first-order perturbation due to stochastic volatility is now denoted by $P_1^{D,r,\varepsilon,\delta}(t,x,z)$, and, using the notation (9.5), it satisfies the problem

$$\mathscr{L}_{BS}^D(\bar{\sigma}(z),\bar{r}(z))P_1^{D,r,\varepsilon,\delta} = -\mathscr{H}_r^{\varepsilon,\delta}P_0^{D,r},$$

$$P_1^{D,r,\varepsilon,\delta}(T,x,z) = 0.$$

Finally, the first-order price approximation is given by (9.7), where P_0^r and its Greeks are replaced by $P_0^{D,r}$ and its corresponding Greeks. The parameter-reduction step of Section 4.3 can easily be implemented.

The generalization of the accuracy results to the case with dividends and varying interest rates follows the lines of the proofs given in Section 4.5 by simply using the modified operators and solutions introduced above.

9.2 Probabilistic Representation of the Approximate Prices

We now look at our price approximation from a different point of view. A natural question is:

"Is there a one-dimensional diffusion process $X^\star$ such that the prices $\tilde{P}^\star$ given by (4.67)

$$\tilde{P}^\star = P_{BS}^\star + (T-t)\left(V_0^\delta(z)\frac{\partial P_{BS}^\star}{\partial\sigma} + V_1^\delta(z)D_1\frac{\partial P_{BS}^\star}{\partial\sigma} + V_3^\varepsilon(z)D_1D_2P_{BS}^\star\right),$$

are the expected values of the discounted payoffs $h(X_T^\star)$?"

In fact, the answer is *no*, implying that our pricing system allows for small arbitrages of order $\mathcal{O}(\varepsilon+\delta)$. However, the prices $\tilde{P}^\star$ can be viewed as expected values of discounted *modified* payoffs as we explain now.

In Section 4.3, we saw that the leading-order price $P_{BS}^\star$ and the correction $P_1^\star$ are obtained as solutions of Black–Scholes equations with the effective volatility $\sigma^\star(z)$, and with respectively a terminal condition and a source term. From Section 1.9.3, we know that such solutions can be represented as expectations of functionals of the geometric Brownian motion $X^\star$ defined by

$$dX_t^\star = rX_t^\star\,dt + \sigma^\star(z)X_t^\star\,dW_t^\star, \tag{9.9}$$

where $W^\star$ is a standard Brownian motion under the probability $\mathbb{P}^\star$.

The leading-order term $P_{BS}^\star$ is obtained by writing the Feynman–Kac formula associated with the Black–Scholes equation (4.63) and its terminal condition $P_{BS}^\star(T,x,z) = h(x)$, so that

$$P_{BS}^\star(t,x,z) = \mathbb{E}^\star\left\{e^{-r(T-t)}h(X_T^\star) \mid X_t^\star = x\right\}, \tag{9.10}$$

where z is a fixed parameter that gives the volatility level $\sigma^\star(z)$ and $\mathbb{E}^\star$ is the expectation with respect to $\mathbb{P}^\star$.

In order to write the probabilistic representation of the first-order perturbation $P_1^\star$, it is convenient to introduce the following notation for the source term in the problem (4.64):

$$H_\star = \mathcal{H}_\star P_{BS}^\star = V_0^\delta(z)\frac{\partial P_{BS}^\star}{\partial\sigma} + V_1^\delta(z)D_1\frac{\partial P_{BS}^\star}{\partial\sigma} + V_3^\varepsilon(z)D_1D_2P_{BS}^\star. \tag{9.11}$$

The first-order perturbation $P_1^\star$ is obtained by writing the Feynman–Kac formula with this source term $H_\star$ and a zero terminal condition. We have:

$$P_1^\star(t,x,z) = \mathbb{E}^\star\left\{\int_t^T e^{-r(s-t)}H_\star(s,X_s^\star,z)ds \mid X_t^\star = x\right\}. \tag{9.12}$$

Adding (9.10) and (9.12), we get the *first-order perturbation formula*

$$\tilde{P}^\star(t,x,z) = \mathbb{E}^\star\left\{e^{-r(T-t)}h(X_T^\star) + \int_t^T e^{-r(s-t)}H_\star(s,X_s^\star,z)ds \mid X_t^\star = x\right\}.$$

$$(9.13)$$

The *path-dependent payment stream* $H_\star(t,x,z)$ (to the holder of the contract) is given by (9.11). It may be positive or negative and accounts dynamically for volatility randomness in a robust model-independent way. It also accounts for the market prices of volatility risk effectively selected by the market.

The generalizations of this probabilistic representation to cases with dividends and/or varying interest rate are straightforward by introducing the appropriate constant volatility dividend paying process and taking into account the additional corrections due to varying interest rates as in (9.7).

9.3 Second-Order Correction from Fast Scale

In this section, we derive the second correction due to the fast scale stochastic volatility factor, that is the term of order ε in the expansion presented in Chapter 4. For simplicity, we consider first models with only the fast factor and we introduce the slow factor in Section 9.4. Under the risk-neutral pricing measure, the dynamics is given by

$$\begin{cases} dX_t &= rX_t\,dt + f(Y_t)X_t\,dW_t^{(0)\star}, \\ dY_t &= \left(\frac{1}{\varepsilon}\alpha(Y_t) - \frac{1}{\sqrt{\varepsilon}}\beta(Y_t)\Lambda(Y_t)\right)dt + \frac{1}{\sqrt{\varepsilon}}\beta(Y_t)\,dW_t^{(1)\star}, \end{cases} \quad (9.14)$$

where the assumptions on the coefficients are the same as in Section 4.1.1.

Following the presentation given in Section 4.1.2, the pricing partial differential equation is given by

$$\mathcal{L}^\varepsilon P^\varepsilon = 0, \tag{9.15}$$

$$P^\varepsilon(T,x,y) = h(x), \tag{9.16}$$

with

$$\mathcal{L}^\varepsilon = \frac{1}{\varepsilon}\mathcal{L}_0 + \frac{1}{\sqrt{\varepsilon}}\mathcal{L}_1 + \mathcal{L}_2, \tag{9.17}$$

where we recall that the operators $\mathcal{L}_i$ are defined by:

$$\mathcal{L}_0 = \frac{1}{2}\beta^2(y)\frac{\partial^2}{\partial y^2} + \alpha(y)\frac{\partial}{\partial y}, \tag{9.18}$$

$$\mathcal{L}_1 = \beta(y)\left(\rho f(y)x\frac{\partial^2}{\partial x\partial y} - \Lambda(y)\frac{\partial}{\partial y}\right), \tag{9.19}$$

$$\mathcal{L}_2 = \frac{\partial}{\partial t} + \frac{1}{2}f^2(y)x^2\frac{\partial^2}{\partial x^2} + r\left(x\frac{\partial}{\partial x} - \cdot\right). \tag{9.20}$$

9.3.1 Expansion and Successive Equations

We look for an asymptotic solution of the form

$$P^\varepsilon(t,x,y) = P_0(t,x,y) + \sqrt{\varepsilon}P_1(t,x,y) + \varepsilon P_2(t,x,y) + \cdots, \qquad (9.21)$$

and the formal construction of the expansion is similar to that in Section 4.2, but now we need to compute the second term P_2. We expand $\mathscr{L}^\varepsilon P^\varepsilon$ and set successively the individual terms to zero as follows.

- Terms of order ε^{-1}:

$$\mathscr{L}_0 P_0 = 0,$$
$$P_0(t,x,y) = h(x),$$

and therefore we seek $P_0 = P_0(t,x)$ independent of y.

- Terms of order $\varepsilon^{-1/2}$:

$$\mathscr{L}_0 P_1 + \mathscr{L}_1 P_0^{0} = 0,$$
$$P_1(t,x,y) = 0,$$

and we seek $P_1 = P_1(t,x)$ independent of y.

- Terms of order one:

$$\mathscr{L}_0 P_2 + \mathscr{L}_1 P_1^{0} + \mathscr{L}_2 P_0 = 0,$$

which is a Poisson equation for P_2. The centering condition implies

$$\langle \mathscr{L}_2 \rangle P_0 = 0. \qquad (9.22)$$

Recall that $\langle \mathscr{L}_2 \rangle$ is the Black–Scholes operator with effective volatility $\bar{\sigma}$ defined in (4.26) by $\bar{\sigma}^2 = \langle f^2 \rangle$. Together with the terminal condition $P_0(T,x) = h(x)$, we see that P_0 is exactly the Black–Scholes price with volatility $\bar{\sigma}$. We also deduce that formally

$$P_2 = -\mathscr{L}_0^{-1} (\mathscr{L}_2 - \langle \mathscr{L}_2 \rangle) P_0, \qquad (9.23)$$

up to an additive constant, that is a term independent of y.

- Terms of order $\varepsilon^{1/2}$:

$$\mathscr{L}_0 P_3 + \mathscr{L}_1 P_2 + \mathscr{L}_2 P_1 = 0, \qquad (9.24)$$

which is a Poisson equation for P_3. The centering condition implies

$$\langle \mathscr{L}_2 \rangle P_1 = -\langle \mathscr{L}_1 P_2 \rangle. \qquad (9.25)$$

- Terms of order ε:

$$\mathcal{L}_0 P_4 + \mathcal{L}_1 P_3 + \mathcal{L}_2 P_2 = 0,$$

which is a Poisson equation for P_4. Its centering condition implies

$$\langle \mathcal{L}_1 P_3 + \mathcal{L}_2 P_2 \rangle = 0. \tag{9.26}$$

As before, we use the notation $D_k = x^k \frac{\partial^k}{\partial x^k}$. Summarizing the derivation of P_1 presented in Chapter 4, we have:

$$\mathcal{L}_2 - \langle \mathcal{L}_2 \rangle = \frac{1}{2} \left(f^2(y) - \langle f^2 \rangle \right) D_2 = \frac{1}{2} \mathcal{L}_0 \phi(y) D_2,$$

where ϕ is defined up to a constant with respect to y as a solution of the Poisson equation

$$\mathcal{L}_0 \phi(y) = f^2(y) - \langle f^2 \rangle.$$

Then from (9.23), we obtain:

$$P_2(t,x,y) = -\mathcal{L}_0^{-1} \left(\frac{1}{2} \mathcal{L}_0 \phi(y) \right) D_2 P_0(t,x) = -\frac{1}{2} \phi(y) D_2 P_0(t,x) + C(t,x),$$

$$(9.27)$$

where, by choosing ϕ such that $\langle \phi \rangle = 0$ and imposing $\langle P_2(T,x,y) \rangle = 0$, we must have

$$C(T,x) = 0, \tag{9.28}$$

at maturity T. This choice will be justified by the proof of accuracy of the second-order approximation given in Section 9.3.3. Note that this particular choice was not needed for the accuracy of the first-order approximation derived in Section 4.5.

Before computing $C(t,x)$, we recall how $P_1(t,x)$ is obtained. From (9.25), (9.27), and the definition (9.19) of $\mathcal{L}_1$, we get:

$$\langle \mathcal{L}_2 \rangle P_1 = -\left\langle \beta \left(\rho f D_1 \frac{\partial}{\partial y} - \Lambda \frac{\partial}{\partial y} \right) \left(-\frac{1}{2} \phi D_2 P_0 + C(t,x) \right) \right\rangle$$

$$= \frac{\rho}{2} \langle \beta f \phi' \rangle D_1 D_2 P_0 - \frac{1}{2} \langle \beta \Lambda \phi' \rangle D_2 P_0$$

$$= -V_3 D_1 D_2 P_0 - V_2 D_2 P_0, \tag{9.29}$$

where

$$V_2 = \frac{1}{2} \langle \beta \Lambda \phi' \rangle, \qquad V_3 = -\frac{\rho}{2} \langle \beta f \phi' \rangle. \tag{9.30}$$

The first correction is then given explicitly by

$$P_1 = (T - t) \left(V_2 D_2 P_0 + V_3 D_1 D_2 P_0 \right). \tag{9.31}$$

9.3.2 Computation of the Second Correction

The second correction P_2 is given by (9.27), where $C(t,x)$ remains to be computed. From the Poisson equation (9.24) and the centering condition (9.25), we deduce:

$$
\begin{aligned}
P_3 &= -\mathcal{L}_0^{-1}\left(\mathcal{L}_1 P_2 + \mathcal{L}_2 P_1 - \langle \mathcal{L}_1 P_2 + \mathcal{L}_2 P_1 \rangle\right) \\
&= -\mathcal{L}_0^{-1}\left(\mathcal{L}_1 P_2 - \langle \mathcal{L}_1 P_2 \rangle + (\mathcal{L}_2 - \langle \mathcal{L}_2 \rangle) P_1\right) \\
&= -\mathcal{L}_0^{-1}\left(-\frac{1}{2}\rho\left[\beta f\phi' - \langle \beta f\phi' \rangle\right]D_1 D_2 P_0 + \frac{1}{2}\left[\beta \Lambda \phi' - \langle \beta \Lambda \phi' \rangle\right]D_2 P_0\right. \\
&\qquad\left. + \frac{1}{2}\mathcal{L}_0 \phi D_2 P_1 \right) \\
&= \frac{1}{2}\rho \psi_1 D_1 D_2 P_0 - \frac{1}{2}\psi_2 D_2 P_0 - \frac{1}{2}\phi D_2 P_1 + F(t,x),
\end{aligned}
\tag{9.32}
$$

where $F(t,x)$ is independent of y, and $\psi_1(y)$ and $\psi_2(y)$ are solutions to the Poisson equations

$$
\mathcal{L}_0 \psi_1 = \beta f\phi' - \langle \beta f\phi' \rangle,
\tag{9.33}
$$

$$
\mathcal{L}_0 \psi_2 = \beta \Lambda \phi' - \langle \beta \Lambda \phi' \rangle.
\tag{9.34}
$$

From the centering condition (9.26) and the expression (9.27) for P_2, we get:

$$
\left\langle \mathcal{L}_2\left(-\frac{1}{2}\phi D_2 P_0\right)\right\rangle + \langle \mathcal{L}_2 \rangle C = -\langle \mathcal{L}_1 P_3 \rangle.
\tag{9.35}
$$

We then compute the first term on the left-hand side:

$$
\begin{aligned}
\left\langle \mathcal{L}_2\left(-\frac{1}{2}\phi D_2 P_0\right)\right\rangle &= -\frac{1}{2}\langle \phi \mathcal{L}_2 \rangle D_2 P_0 \\
&= -\frac{1}{2}\left[\langle \phi \rangle \langle \mathcal{L}_2 \rangle + \frac{1}{2}\left(\langle \phi f^2 \rangle - \langle \phi \rangle \langle f^2 \rangle\right)D_2\right]D_2 P_0 \\
&= -\frac{1}{2}\langle \phi \rangle \overset{\;0}{\cancel{\langle \mathcal{L}_2 \rangle D_2 P_0}} + A D_2^2 P_0,
\end{aligned}
\tag{9.36}
$$

where the first term on the right-hand side cancels by commuting $\langle \mathcal{L}_2 \rangle$ and D_2 and using (9.22), and where A is defined by

$$
A = -\frac{1}{4}\left(\langle \phi f^2 \rangle - \langle \phi \rangle \langle f^2 \rangle\right).
\tag{9.37}
$$

From (9.35), (9.36), and (9.32), it follows that:

$$\langle \mathcal{L}_2 \rangle C = -\langle \mathcal{L}_1 P_3 \rangle - A D_2^2 P_0$$

$$= -\left\langle \beta \left(\rho f D_1 \frac{\partial}{\partial y} - \Lambda \frac{\partial}{\partial y} \right) \left(\frac{1}{2} \rho \psi_1 D_1 D_2 P_0 - \frac{1}{2} \psi_2 D_2 P_0 - \frac{1}{2} \phi D_2 P_1 + F(t,x) \right) \right\rangle - A D_2^2 P_0$$

$$= -\frac{1}{2} \rho^2 \langle \beta f \psi_1' \rangle D_1^2 D_2 P_0 + \frac{\rho}{2} \left(\langle \beta f \psi_2' \rangle + \langle \beta \Lambda \psi_1' \rangle \right) D_1 D_2 P_0$$

$$- \frac{1}{2} \langle \beta \Lambda \psi_2' \rangle D_2 P_0 - V_3 D_1 D_2 P_1 - V_2 D_2 P_1 - A D_2^2 P_0$$

$$= -\left(A_2 D_1^2 D_2 + A_1 D_1 D_2 + A_0 D_2 + A D_2^2 \right) P_0$$
$$- V_3 D_1 D_2 P_1 - V_2 D_2 P_1, \tag{9.38}$$

where V_2, V_3 are given by (9.30), A is defined in (9.37), and we have introduced the new parameters:

$$A_2 = \frac{\rho^2}{2} \langle \beta f \psi_1' \rangle, \quad A_1 = -\frac{\rho}{2} \left(\langle \beta f \psi_2' \rangle + \langle \beta \Lambda \psi_1' \rangle \right), \quad A_0 = \frac{1}{2} \langle \beta \Lambda \psi_2' \rangle. \tag{9.39}$$

Using the explicit formula (9.31) for P_1, equation (9.38) is solved explicitly with the vanishing terminal condition (9.28):

$$C(t,x) = (T - t) \left[A_2 D_1^2 D_2 + A_1 D_1 D_2 + A_0 D_2 + A D_2^2 \right] P_0(t,x)$$
$$+ \frac{(T - t)^2}{2} \left[V_3^2 D_1^2 D_2^2 + 2 V_2 V_3 D_1 D_2^2 + V_2^2 D_2^2 \right] P_0(t,x), \tag{9.40}$$

and we recall that $P_2(t,x,y)$ is given by (9.27).

Observe that the parameters involved in the approximation (9.21) "absorb" the small quantities $\sqrt{\varepsilon}$ and ε as follows:

$$V_2^\varepsilon = \sqrt{\varepsilon} V_2, \ V_3^\varepsilon = \sqrt{\varepsilon} V_3, \ A_2^\varepsilon = \varepsilon A_2, \ A_1^\varepsilon = \varepsilon A_1, \ A_0^\varepsilon = \varepsilon A_0, \ A^\varepsilon = \varepsilon A, \tag{9.41}$$

and they will be the ones calibrated to the skew of implied volatilities.

9.3.3 Accuracy and Parameter Reduction

Accuracy The proof of accuracy of the first-order approximation given in Section 4.5 can be generalized to the second-order approximation as follows.

We denote

$$\tilde{P}^\varepsilon = P_0 + \sqrt{\varepsilon}P_1 + \varepsilon P_2, \tag{9.42}$$

$$\hat{P}^\varepsilon = P_0 + \sqrt{\varepsilon}P_1 + \varepsilon P_2 + \varepsilon^{3/2}P_3 + \varepsilon^2 P_4, \tag{9.43}$$

$$R^\varepsilon = P^\varepsilon - \hat{P}^\varepsilon, \tag{9.44}$$

and, from the equations satisfied by $P^\varepsilon, P_0, P_1, P_2, P_3$, and P_4 derived in Section 9.3.1, we deduce that:

$$\mathcal{L}^\varepsilon R^\varepsilon + \varepsilon^{3/2}\left(\mathcal{L}_1 P_4 + \mathcal{L}_2 P_3 + \sqrt{\varepsilon}\,\mathcal{L}_2 P_4\right) = 0. \tag{9.45}$$

Next, the terminal condition needs to be handled. As it is, we have

$$R^\varepsilon(T,x,y) = -\varepsilon P_2(T,x,y) - \varepsilon^{3/2}\left(P_3(T,x,y) + \sqrt{\varepsilon}\,P_4(T,x,y)\right).$$

As in Section 4.5, for a smooth bounded payoff with bounded derivatives, we obtain that

$$R^\varepsilon = -\varepsilon \mathbb{E}^\star\left\{ e^{-r(T-t)}P_2(T,X_T,Y_T) \mid X_t, Y_t \right\} + \mathcal{O}(\varepsilon^{3/2})$$

$$= \frac{\varepsilon}{2}\mathbb{E}^\star\left\{ e^{-r(T-t)}\phi(Y_T)D_2 P_0(T,X_T) \mid X_t, Y_t \right\} + \mathcal{O}(\varepsilon^{3/2})$$

$$= \varepsilon C \mathbb{E}^\star\left\{ \phi(Y_T) \mid Y_t \right\} + \mathcal{O}(\varepsilon^{3/2}).$$

In the penultimate step, we have used the formula (9.27) for P_2 and the imposed zero terminal condition (9.28). In the last step, we have used the fact that the process X converges in distribution as $\varepsilon \to 0$ to a geometric Brownian motion with constant volatility $\bar{\sigma}$, therefore becoming independent of the Y process, the error being pushed in the term of order $\varepsilon^{3/2}$.

If the process Y in (9.14) had $\varepsilon^{-1}\mathcal{L}_0$ as its infinitesimal generator, then $\mathbb{E}^\star\left\{ \phi(Y_T) \mid Y_t \right\}$ would converge to zero exponentially fast as $\varepsilon \to 0$ since $\langle\phi\rangle = 0$. This follows from the general theory presented in Section 3.2 for Markov processes with positive spectral gap, as the OU and CIR examples given in Sections 3.3.3 and 3.3.4.

The presence of the drift term due to the market price of volatility risk Λ implies a convergence in $\sqrt{\varepsilon}$ instead. In order to see that, we need to look more closely at $\mathbb{E}^\star\left\{ \phi(Y_T) \mid Y_t \right\}$. First, for simplicity and without loss of generality, we can assume $t = 0$. Recall that, under $\mathbb{P}^\star$, the dynamics of Y is given by the second (autonomous) equation in (9.14). By rescaling time in this equation, we need to consider $\mathbb{E}^\star\left\{ \phi(Y_{T/\varepsilon}) \right\}$ as $\varepsilon \to 0$ under the dynamics

$$dY_t = \left(\alpha(Y_t) - \sqrt{\varepsilon}\beta(Y_t)\Lambda(Y_t) \right) dt + \beta(Y_t)\,dW_t^{(1)\star}.$$

Next we use an argument similar to the one presented in Section 8.6.3. The process Y admits the invariant distribution Φ_ε given by the density

$$\Phi_\varepsilon(y) = \frac{J_\varepsilon}{\beta^2(y)} \exp\left(2 \int_0^y \frac{\alpha(z) - \sqrt{\varepsilon}\beta(z)\Lambda(z)}{\beta^2(z)} dz\right),$$

where J_ε is a normalizing constant insuring that Φ_ε is a probability density. From the ergodic property of Y (positive spectral gap), for ε small enough, one can find positive constants C and λ such that

$$\left|\mathbb{E}^*\{\phi(Y_{T/\varepsilon})\} - \langle\phi\rangle_\varepsilon\right| \leq Ce^{-\lambda T/\varepsilon},$$

where $\langle\cdot\rangle_\varepsilon$ denotes the averaging with respect to Φ_ε. Next, by expanding Φ_ε (including J_ε), we derive the expansion

$$\langle g\rangle_\varepsilon = \langle g\rangle - 2\sqrt{\varepsilon}\left\langle\left(\int_0^\cdot \frac{\Lambda(z)}{\beta(z)} dz\right)(g(\cdot) - \langle g\rangle)\right\rangle + \cdots$$

and, using the fact that ϕ is centered, $\langle\phi\rangle = 0$, we obtain

$$\left|\mathbb{E}^*\{\phi(Y_{T/\varepsilon})\}\right| \leq C'\sqrt{\varepsilon},$$

which gives the accuracy result: $|R^\varepsilon|$ is of order $\varepsilon^{3/2}$.

Parameter Reduction The parameter-reduction step presented in Section 4.3 can be generalized to the second-order approximation. It consists of absorbing in the left-hand side of (9.29) the V_2-term appearing in the right-hand side by setting

$$\sigma^* = \sqrt{\bar{\sigma}^2 + 2\sqrt{\varepsilon}V_2} = \sqrt{\bar{\sigma}^2 + 2V_2^\varepsilon},$$

and introducing P_0^*, the Black–Scholes price computed at volatility level σ^*. Then, we define P_1^* and P_2^* by

$$P_1^* = (T - t)V_3 D_1 D_2 P_0^*, \tag{9.46}$$

$$P_2^* = -\frac{1}{2}\phi(y)D_2 P_0^* + C^*, \tag{9.47}$$

$$C^* = (T - t)\left[A_2 D_1^2 D_2 + A_1 D_1 D_2 + A_0 D_2 + A D_2^2\right]P_0^*$$
$$+ \frac{(T - t)^2}{2}\left[V_3^2 D_1^2 D_2^2 + V_2 V_3 D_1 D_2^2\right]P_0^*, \tag{9.48}$$

so that

$$\mathscr{L}_{BS}(\sigma^*)P_0^* = 0,$$
$$\mathscr{L}_{BS}(\sigma^*)P_1^* = -V_3 D_1 D_2 P_0^*,$$
$$\mathscr{L}_{BS}(\sigma^*)C^* = -\left(A_2 D_1^2 D_2 + A_1 D_1 D_2 + A_0 D_2 + A D_2^2\right) P_0^*$$
$$-V_3 D_1 D_2 P_1^* - V_2 D_2 P_1^*.$$

The fact that the $\varepsilon^{3/2}$-accuracy is preserved is a simple generalization of the derivation presented in Section 4.3. One introduces the residual

$$E = \left(P_0 + \sqrt{\varepsilon}P_1 + \varepsilon P_2\right) - \left(P_0^* + \sqrt{\varepsilon}P_1^* + \varepsilon P_2^*\right),$$

and derives $\mathscr{L}_{BS}(\bar{\sigma})E = \mathcal{O}(\varepsilon^{3/2})$ by using

$$\mathscr{L}_{BS}(\sigma^*) = \mathscr{L}_{BS}(\bar{\sigma}) + V_2^\varepsilon D_2,$$

and the fact that we already know $\left(P_0 + \sqrt{\varepsilon}P_1\right) - \left(P_0^* + \sqrt{\varepsilon}P_1^*\right) = \mathcal{O}(\varepsilon)$.

9.3.4 Correction to the Skew

We now look at the effect of the second correction to the skew by expanding implied volatilities:

$$I = I_0 + \sqrt{\varepsilon}I_1 + \varepsilon I_2 + \cdots.$$

The price $P = P_0^* + \sqrt{\varepsilon}P_1^* + \varepsilon P_2^* + \cdots$ is now the price of a call option with strike price K and maturity T generated by the model (9.14). Following the method used in Chapter 5, we denote by $C_{BS}(\sigma)$ the Black–Scholes price of this option at volatility level σ, and we write:

$$P_0^* + \sqrt{\varepsilon}P_1^* + \varepsilon P_2^* + \cdots = C_{BS}(I_0 + \sqrt{\varepsilon}I_1 + \varepsilon I_2 + \cdots)$$
$$= C_{BS}(I_0) + (\sqrt{\varepsilon}I_1 + \varepsilon I_2)\partial_\sigma C_{BS}(I_0)$$
$$+ \frac{1}{2}(\sqrt{\varepsilon}I_1 + \varepsilon I_2)^2 \partial_{\sigma\sigma}^2 C_{BS}(I_0) + \cdots,$$

where for simplicity we denote partial derivatives with respect to σ by $\partial_\sigma, \partial_{\sigma\sigma}^2, \ldots$. Matching powers of ε, we get $P_0^* = C_{BS}(\sigma^*)$ with $I_0 = \sigma^*$ and:

$$I_1 = \frac{P_1^*}{\partial_\sigma P_0^*}, \tag{9.49}$$

$$I_2 = \frac{P_2^*}{\partial_\sigma P_0^*} - \frac{1}{2}I_1^2 \frac{\partial_{\sigma\sigma}^2 P_0^*}{\partial_\sigma P_0^*}. \tag{9.50}$$

We recall that for plain vanilla European options we have

$$\partial_\sigma P_0^* = \tau \sigma^* D_2 P_0^*,$$

where we denote $T - t = \tau$, and for call options:

$$\partial_\sigma P_0^\star = \frac{x\sqrt{\tau}}{\sqrt{2\pi}} e^{-d_1^2/2}, \qquad d_1 = \frac{\log(x/K) + \left(r + \frac{1}{2}\sigma^{\star 2}\right)\tau}{\sigma^\star\sqrt{\tau}}.$$

From (9.46) and (9.49) we deduce as in Chapter 5 that

$$I_1 = \frac{V_3}{\sigma^\star}\frac{D_1\partial_\sigma P_0^\star}{\partial_\sigma P_0^\star} = \frac{V_3}{\sigma^\star}\left(1 - \frac{d_1}{\sigma^\star\sqrt{\tau}}\right). \tag{9.51}$$

In order to compute I_2 from (9.50), we rewrite $P_2^\star$ given by (9.47):

$$P_2^\star = \frac{1}{\sigma^\star}\left[A_2 D_1^2\partial_\sigma + A_1 D_1\partial_\sigma + A_0\partial_\sigma + AD_2\partial_\sigma + \frac{\tau}{2}V_3^2 D_1^2 D_2\partial_\sigma\right.$$
$$\left. + \frac{\tau}{2}V_2 V_3 D_1 D_2\partial_\sigma\right] P_0^\star - \frac{1}{2\sigma^\star\tau}\phi(y)\partial_\sigma P_0^\star. \tag{9.52}$$

A direct computation of the derivatives $D_1\partial_\sigma, D_1^2\partial_\sigma, D_2\partial_\sigma, D_1 D_2\partial_\sigma, D_1^2 D_2\partial_\sigma$, and $\partial_{\sigma\sigma}^2$ of $P_0^\star$ in terms of $\partial_\sigma P_0^\star$, leads to:

$$D_1\partial_\sigma P_0^\star = \left(1 - \frac{d_1}{\sigma^\star\sqrt{\tau}}\right)\partial_\sigma P_0^\star,$$

$$D_1^2\partial_\sigma P_0^\star = \left[\left(1 - \frac{d_1}{\sigma^\star\sqrt{\tau}}\right)^2 - \frac{1}{\sigma^{\star 2}\tau}\right]\partial_\sigma P_0^\star,$$

$$D_2\partial_\sigma P_0^\star = \left(-\frac{d_1}{\sigma^\star\sqrt{\tau}} + \frac{d_1^2}{\sigma^{\star 2}\tau} - \frac{1}{\sigma^{\star 2}\tau}\right)\partial_\sigma P_0^\star,$$

$$D_1 D_2\partial_\sigma P_0^\star = \left(-\frac{d_1^3}{\sigma^{\star 3}\tau^{3/2}} + \frac{2d_1^2}{\sigma^{\star 2}\tau} + \frac{3d_1}{\sigma^{\star 3}\tau^{3/2}} - \frac{d_1}{\sigma^\star\sqrt{\tau}} - \frac{2}{\sigma^{\star 2}\tau}\right)\partial_\sigma P_0^\star,$$

$$D_1^2 D_2\partial_\sigma P_0^\star = \left(\frac{d_1^4}{\sigma^{\star 4}\tau^2} - \frac{3d_1^3}{\sigma^{\star 3}\tau^{3/2}} - \frac{6d_1^2}{\sigma^{\star 4}\tau^2} + \frac{3d_1^2}{\sigma^{\star 2}\tau} + \frac{9d_1}{\sigma^{\star 3}\tau^{3/2}} - \right.$$
$$\left. \frac{d_1}{\sigma^\star\sqrt{\tau}} + \frac{3}{\sigma^{\star 4}\tau^2} - \frac{3}{\sigma^{\star 2}\tau}\right)\partial_\sigma P_0^\star,$$

$$\partial_{\sigma\sigma}^2 P_0^\star = \frac{1}{\sigma^\star}\left(d_1^2 - \sigma^\star\sqrt{\tau}d_1\right)\partial_\sigma P_0^\star.$$

Finally, from (9.50), (9.52), and the computation above, we obtain:

$$I_2 = -\frac{1}{2\sigma^\star\tau}\phi(y)$$
$$+ \frac{1}{\sigma^\star}\left[A_2\left(\frac{d_1^2}{\sigma^{\star 2}\tau} - \frac{2d_1}{\sigma^\star\sqrt{\tau}} - \frac{1}{\sigma^{\star 2}\tau} + 1\right) + A_1\left(-\frac{d_1}{\sigma^\star\sqrt{\tau}} + 1\right) + A_0\right.$$
$$+ A\left(\frac{d_1^2}{\sigma^{\star 2}\tau} - \frac{d_1}{\sigma^\star\sqrt{\tau}} - \frac{1}{\sigma^{\star 2}\tau}\right)$$

$$+\frac{\tau V_3^2}{2}\left(\frac{d_1^4}{\sigma^{\star 4}\tau^2}-\frac{3d_1^3}{\sigma^{\star 3}\tau^{3/2}}-\frac{6d_1^2}{\sigma^{\star 4}\tau^2}+\frac{3d_1^2}{\sigma^{\star 2}\tau}+\frac{9d_1}{\sigma^{\star 3}\tau^{3/2}}\right.$$

$$\left.-\frac{d_1}{\sigma^\star\sqrt\tau}+\frac{3}{\sigma^{\star 4}\tau^2}-\frac{3}{\sigma^{\star 2}\tau}\right)$$

$$+\frac{\tau V_2 V_3}{2}\left(-\frac{d_1^3}{\sigma^{\star 3}\tau^{3/2}}+\frac{2d_1^2}{\sigma^{\star 2}\tau}+\frac{3d_1}{\sigma^{\star 3}\tau^{3/2}}-\frac{d_1}{\sigma^\star\sqrt\tau}-\frac{2}{\sigma^{\star 2}\tau}\right)\Bigg]$$

$$-\frac{1}{2}\left[\frac{V_3}{\sigma^\star}\left(1-\frac{d_1}{\sigma^\star\sqrt\tau}\right)\right]^2\left(\frac{d_1^2-\sigma^\star\sqrt\tau d_1}{\sigma^\star}\right).\tag{9.53}$$

Observe that the implied volatility becomes *quartic* in *log-moneyness* at the second order of approximation. Calibration of the parameters $(\sigma^\star, V_3^\varepsilon, A^\varepsilon, A_0^\varepsilon, A_1^\varepsilon, A_2^\varepsilon)$ to the observed implied volatility surface involves fits to a quartic in $\log(x/K)$ at fixed maturity and regressions over polynomials in τ easily derived from (9.53) (see Section 5.5.1 and Fouque *et al.* (2004b) for an example). The second-order correction is implemented numerically in the case of the Heston model in Section 10.2.2.

9.4 Second-Order Corrections from Slow and Fast Scales

In this section, we derive the additional terms in the second-order correction due to the presence of a slow factor. The expansion (to any order) when only the slow factor is present is classical and relatively easy to derive as a regular perturbation problem. Here, the additional difficulty is due to the presence of both fast and slow factors and their interaction in the second-order terms.

We consider the model (4.1) and the associated pricing partial differential equation (4.4), and use the notations (4.5)–(4.13) introduced in Chapter 4.

From the expansion (4.14), we are interested in the terms

$$P^{\varepsilon,\delta}=P_0^\varepsilon+\sqrt\delta P_1^\varepsilon+\delta P_2^\varepsilon+\cdots$$

$$=P_0+\sqrt\varepsilon P_{1,0}+\varepsilon P_{2,0}+\sqrt\delta P_{0,1}+\sqrt{\varepsilon\delta}P_{1,1}+\delta P_{0,2}+\mathcal{O}\left((\varepsilon+\delta)^{\frac{3}{2}}\right),$$

which contribute to the approximation up to the second order jointly in ε and δ. Showing explicitly the terms of order δ in (4.15), we have:

$$\left(\frac{1}{\varepsilon}\mathcal{L}_0+\frac{1}{\sqrt\varepsilon}\mathcal{L}_1+\mathcal{L}_2\right)P_0^\varepsilon$$

$$+\sqrt\delta\left\{\left(\frac{1}{\varepsilon}\mathcal{L}_0+\frac{1}{\sqrt\varepsilon}\mathcal{L}_1+\mathcal{L}_2\right)P_1^\varepsilon+\left(\mathcal{M}_1+\frac{1}{\sqrt\varepsilon}\mathcal{M}_3\right)P_0^\varepsilon\right\}$$

$$+\delta\left\{\left(\frac{1}{\varepsilon}\mathcal{L}_0+\frac{1}{\sqrt\varepsilon}\mathcal{L}_1+\mathcal{L}_2\right)P_2^\varepsilon+\left(\mathcal{M}_1+\frac{1}{\sqrt\varepsilon}\mathcal{M}_3\right)P_1^\varepsilon+\mathcal{M}_2P_0^\varepsilon\right\}$$

$$+\cdots=0.\tag{9.54}$$

The problem

$$\left(\frac{1}{\varepsilon}\mathcal{L}_0 + \frac{1}{\sqrt{\varepsilon}}\mathcal{L}_1 + \mathcal{L}_2\right)P_0^\varepsilon$$

$$= \left(\frac{1}{\varepsilon}\mathcal{L}_0 + \frac{1}{\sqrt{\varepsilon}}\mathcal{L}_1 + \mathcal{L}_2\right)\left(P_0 + \sqrt{\varepsilon}P_{1,0} + \varepsilon P_{2,0} + \cdots\right) = 0$$

is similar to the problem studied in Section 9.3. Therefore, $P_0, P_{1,0}$, and $P_{2,0}$ are given respectively by (9.22), (9.31), and (9.27) and (9.40), where the parameters $\bar{\sigma}, V_2, V_3, A, A_0, A_1, A_2$ and the function ϕ depend now on z.

The terms in $\sqrt{\delta}$ in (9.54) give

$$\left(\frac{1}{\varepsilon}\mathcal{L}_0 + \frac{1}{\sqrt{\varepsilon}}\mathcal{L}_1 + \mathcal{L}_2\right)\left(P_{0,1} + \sqrt{\varepsilon}P_{1,1} + \varepsilon P_{2,1} + \cdots\right)$$

$$+ \mathcal{M}_1\left(P_0 + \sqrt{\varepsilon}P_{1,0} + \varepsilon P_{2,0} + \cdots\right)$$

$$+ \frac{1}{\sqrt{\varepsilon}}\mathcal{M}_3\left(P_0 + \sqrt{\varepsilon}P_{1,0} + \varepsilon P_{2,0} + \cdots\right)$$

$$= 0. \qquad (9.55)$$

As in Section 4.2.4, and using the fact that $\mathcal{M}_3$ takes derivatives with respect to y, one deduces that $P_{0,1}$ and $P_{1,1}$ do not depend on y, $\sqrt{\delta}P_{0.1}$ is given explicitly by (4.57), and $P_{2,1}$ is given by (4.47). Canceling the terms of order $\sqrt{\varepsilon}$ in (9.55) gives

$$\mathcal{L}_0 P_{3,1} + \mathcal{L}_1 P_{2,1} + \mathcal{L}_2 P_{1,1} + \mathcal{M}_1 P_{1,0} + \mathcal{M}_3 P_{2,0} = 0,$$

which is a Poisson equation in $P_{3,1}$ whose centering condition characterizes $P_{1,1}$ as the solution of the following Black–Scholes equation with source and zero terminal condition:

$$\langle\mathcal{L}_2\rangle P_{1,1} = -\langle\mathcal{L}_1 P_{2,1} + \mathcal{M}_1 P_{1,0} + \mathcal{M}_3 P_{2,0}\rangle.$$

Coming back to the terms in δ in (9.54), as usual canceling the terms of order $1/\varepsilon$ and $1/\sqrt{\varepsilon}$ implies that $P_{0,2}$ and $P_{1,2}$ do not depend on y, and canceling the terms of order one gives a Poisson equation in $P_{2,2}$ whose centering condition characterizes $P_{0,2}$ as the solution of the following Black–Scholes equation with source and zero terminal condition:

$$\langle\mathcal{L}_2\rangle P_{0,2} = -\langle\mathcal{M}_1 P_{0,1} + \mathcal{M}_2 P_0\rangle.$$

Solving explicitly the previous equations characterizing $P_{1,1}$ and $P_{0,2}$ is done by computing explicitly the sources in terms of $D_n P_0$ and using the commutation property between $\langle\mathcal{L}_2\rangle$ and D_n as done previously. We omit here this tedious but straightforward computation. Similarly, parameter reduction using $\sigma^\star$ instead of $\bar{\sigma}$ can be achieved, accuracy of approximation can be derived, and formulas for second-order terms in the implied volatility due to the presence of the slow factor can be obtained by using the explicit formulas for the Greeks given in Section 9.3.4.

9.5 Periodic Day Effect

A careful analysis of high-frequency data such as presented in Section 3.5 and Fouque *et al.* (2003b) shows that volatility contains a daily periodic component which is on the same time scale as the intrinsic mean-reversion time of the fast stochastic volatility factor. It has been included in the simulations presented in Fouque *et al.* (2003b). As we shall see here, the asymptotic results for pricing and hedging are not affected by the presence of this periodic component.

For simplicity, we consider here only a fast factor. In order to model the daily effect, we replace our volatility model by $\sigma_t = f(t/\varepsilon, Y_t)$, where $f(\tau, y)$ is positive and periodic of period one in its first argument.

We consider again the European pricing problem of Chapter 4. It is convenient to introduce a new time variable $\tau = t/\varepsilon$ and consider the pricing function P^ε as a function $P^\varepsilon(t, \tau, x, y)$ and to treat τ as an independent variable. The usual time derivative is replaced by

$$\frac{\partial}{\partial t} + \frac{1}{\varepsilon}\frac{\partial}{\partial \tau}.$$

The pricing partial differential equation is now written

$$\mathscr{L}^\varepsilon P^\varepsilon = 0,$$

where analogously to the definitions in Section 4.1.2, we have

$$\mathscr{L}^\varepsilon = \frac{1}{\varepsilon}\left(\mathscr{L}_0 + \frac{\partial}{\partial \tau}\right) + \frac{1}{\sqrt{\varepsilon}}\mathscr{L}_1 + \mathscr{L}_0,$$

$$\mathscr{L}_0 = \frac{1}{2}\beta^2(y)\frac{\partial^2}{\partial y^2} + \alpha(y)\frac{\partial}{\partial y},$$

$$\mathscr{L}_1 = \beta(y)\left(\rho_1 f(\tau,y)x\frac{\partial^2}{\partial x \partial y} - \Lambda_1(\tau,y)\frac{\partial}{\partial y}\right),$$

$$\mathscr{L}_2 = \frac{\partial}{\partial t} + \frac{1}{2}f^2(\tau,y)x^2\frac{\partial^2}{\partial x^2} + r\left(x\frac{\partial}{\partial x} - \cdot\right) = \mathscr{L}_{BS}(f(\tau,y)).$$

Notice that Λ may depend on τ, but that $\mathscr{L}_0$ is the usual scaled infinitesimal generator of (Y_t). The terminal condition is again $P^\varepsilon(T, \tau, x, y) = h(x)$.

Expanding as usual,

$$P^{\varepsilon} = P_0 + \sqrt{\varepsilon}P_1 + \cdots,$$

and comparing powers of ε gives to highest order

$$\left(\mathcal{L}_0 + \frac{\partial}{\partial\tau}\right)P_0 = 0.$$

The operator $\mathcal{L}_0 + \frac{\partial}{\partial\tau}$ is the infinitesimal generator of the process (Y_t, τ_t) where the second component is simply t modulo 1 to account for the periodic component. $\mathcal{L}_0$ being independent of τ, the null space of $\mathcal{L}_0 + \frac{\partial}{\partial\tau}$ is made of the constants in (y, τ) and we choose

$$P_0 = P_0(t, x).$$

The next order gives

$$\left(\mathcal{L}_0 + \frac{\partial}{\partial\tau}\right)P_1 = 0,$$

which again leads to seeking a solution $P_1 = P_1(t, x)$. The zero-order terms yield the equation

$$\mathcal{L}_2 P_0 + \left(\mathcal{L}_0 + \frac{\partial}{\partial\tau}\right)P_2 = 0.$$

The centering condition for this Poisson equation implies that $\mathcal{L}_2 P_0$ is centered with respect to the invariant measure of the Y process *and* over one period with respect to τ. We denote here by $\langle\cdot\rangle$ the integral

$$\langle g\rangle = \int_0^1 \int_{-\infty}^{\infty} g(\tau, y)\Phi(y)\,dy\,d\tau,$$

where Φ is the density of the invariant distribution of Y.

Using the fact that P_0 does not depend on τ or y, we have

$$\langle\mathcal{L}_2\rangle P_0 = \mathcal{L}_{BS}(\bar{\sigma})P_0 = 0,$$

where $\bar{\sigma}^2 = \langle f^2\rangle$ and the average is taken in y and τ. To summarize, the theory is the same with this modified averaging.

Following the argument of Section 4.2.2, we obtain that the correction $\sqrt{\varepsilon}P_1(t, x)$ satisfies the equation

$$\mathcal{L}_{BS}(\bar{\sigma})(\sqrt{\varepsilon}P_1(t, x)) = \sqrt{\varepsilon}\left\langle\mathcal{L}_1\left(\mathcal{L}_0 + \frac{\partial}{\partial\tau}\right)^{-1}(\mathcal{L}_2 - \langle\mathcal{L}_2\rangle)\right\rangle P_0,$$

with zero terminal condition. Introducing the function $\phi(\tau, y)$, the solution of

$$\left(\mathscr{L}_0 + \frac{\partial}{\partial \tau}\right)\phi(\tau, y) = f^2(\tau, y) - \langle f^2 \rangle,$$

the source term is computed as in Section 4.2.3:

$$V_2^{\varepsilon} D_1 P_0 + V_3^{\varepsilon} D_1 D_2 P_0,$$

where V_3^{ε} and V_2^{ε} are given by the formulas (4.39) and (4.40) with the newly defined ϕ and $\langle \cdot \rangle$.

In conclusion, the use and calibration of the asymptotic theory is identical when we incorporate the day effect. Only the relations to the base model parameters are different.

9.6 Markovian Jump Volatility Models

A convenient way to introduce jumps in volatility is to consider one of the examples of jump processes given in Section 3.3.1 or 3.3.2, call it ξ_t, and model the stochastic volatility σ_t as a function

$$\sigma_t = f(Y_t, \xi_t)$$

of the diffusion processes Y and the process ξ which we assume independent. For instance, if f is only a function of ξ, the volatility is a pure jump process independent of $W^{(0)}$ and there is no leverage effect. We are mostly interested in the cases where the two components Y and ξ are present. This procedure corresponds to the addition of another *factor* which models the volatility jumps. Here, for simplicity, we ignore the slow factor Z which can easily be incorporated.

The properties of the Markov process (Y, ξ) described above can be summarized in its infinitesimal generator $\mathscr{L}_{(Y,\xi)}$ given by the sum

$$\mathscr{L}_{(Y,\xi)} = \mathscr{L}_0 + \mathscr{L}_J,$$

where $\mathscr{L}_0$ is the infinitesimal generator of the Y process and $\mathscr{L}_J$, acting on the second variable also denoted by ξ, is the infinitesimal generator of one of the jump processes described in Section 3.3.1 or 3.3.2, with an intensity normalized to one. For instance, in the case of jumps to a random point in the interval $(-1, 1)$ considered in Section 3.3.2, we have

$$\mathcal{L}_{(Y,\xi)}g(y,\xi) = \frac{1}{2}\beta^2(y)\frac{\partial^2 g}{\partial y^2}(y,\xi) + \alpha(y)\frac{\partial g}{\partial y}(y,\xi)$$

$$+\frac{1}{2}\int_{-1}^{1}(g(y,z)-g(y,\xi))\,dz \qquad (9.56)$$

for any bounded function g twice differentiable with respect to y.

An important feature of this setting is that the process (Y,ξ) is ergodic and admits the unique invariant distribution described by

$$\langle g \rangle = \langle\langle g \rangle_Y \rangle_\xi = \left\langle \langle g \rangle_\xi \right\rangle_Y, \qquad (9.57)$$

which consists of averaging in the first variable with respect to the invariant distribution Φ_Y of the process Y and in the second variable with respect to the invariant distribution of the process ξ. For instance, in the example (9.56) we have

$$\langle g \rangle = \frac{1}{2}\int_{-1}^{1}\int_{-\infty}^{+\infty} g(y,z)\Phi_Y(y)\,dy\,dz.$$

We now assume that the intensity of the jumps is large and of the same order as the rate of mean-reversion of the process Y. This is done by setting this intensity equal to the rate of mean-reversion of Y, namely $1/\varepsilon$, where we introduce the short time scale ε. The generalized Markovian model $(X^\varepsilon, Y^\varepsilon, \xi^\varepsilon)$ is then given, in the real-world measure $\mathbb{P}$, by

$$dX_t^\varepsilon = \mu X_t^\varepsilon dt + f(Y_t^\varepsilon, \xi_t^\varepsilon) X_t^\varepsilon dW_t^{(0)},$$

$$dY_t^\varepsilon = \frac{1}{\varepsilon}\alpha(Y_t^\varepsilon)dt + \frac{1}{\sqrt{\varepsilon}}\beta(Y_t^\varepsilon)\left(\rho W_t^{(0)} + \sqrt{1-\rho^2}\,dW_t^\perp\right),$$

$$\xi_t^\varepsilon = \xi_{t/\varepsilon}, \qquad (9.58)$$

where $(W^{(0)}, W^\perp, \xi)$ are independent. We assume that the factor ξ remains the same in the risk-neutral world $\mathbb{P}^\star$, so that

$$dX_t^\varepsilon = r X_t^\varepsilon dt + f(Y_t^\varepsilon, \xi_t^\varepsilon) X_t^\varepsilon dW_t^{(0)\star},$$

$$dY_t^\varepsilon = \frac{1}{\varepsilon}\alpha(Y_t^\varepsilon)dt - \frac{1}{\sqrt{\varepsilon}}\beta(Y_t^\varepsilon)\Lambda(Y_t^\varepsilon, \xi_t^\varepsilon)dt$$

$$+ \frac{1}{\sqrt{\varepsilon}}\beta(Y_t^\varepsilon)\left(\rho W_t^{(0)\star} + \sqrt{1-\rho^2}\,dW_t^{\perp\star}\right),$$

$$\xi_t^\varepsilon = \xi_{t/\varepsilon}, \qquad (9.59)$$

where $(W^{(0)\star}, W^{\perp\star}, \xi)$ are independent and Λ may also depend on ξ_t^ε.

In this setting the price of a European derivative with payoff h at maturity T is given at time $t < T$ by $P^\varepsilon(t, X_t^\varepsilon, Y_t^\varepsilon, \xi_t^\varepsilon)$, where

$$P^\varepsilon(t, x, y, \xi) = \mathbb{E}^\star\{e^{-r(T-t)}h(X_T^\varepsilon)|X_t^\varepsilon = x, Y_t^\varepsilon = y, \xi_t^\varepsilon = \xi\}$$

satisfies

$$\left(\frac{1}{\varepsilon}(\mathscr{L}_0 + \mathscr{L}_J) + \frac{1}{\sqrt{\varepsilon}}\mathscr{L}_1 + \mathscr{L}_{BS}(f(y,\xi))\right)P^\varepsilon = 0,$$

with the terminal condition $P^\varepsilon(T, x, y, \xi) = h(x)$. The operator $\mathscr{L}_1$ is given as in (4.9) and $\mathscr{L}_{BS}(f(y,\xi))$ is the Black–Scholes operator with volatility $f(y,\xi)$.

With these definitions and notations the rest of the derivation of the corrected pricing formula

$$P_0 + (T - t)(V_2^\varepsilon D_2 + V_3^\varepsilon D_1 D_2)P_0 \tag{9.60}$$

follows the lines of Section 4.2.3. The function $P_0(t, x)$ is the Black–Scholes price of the derivative, computed with the constant volatility $\bar{\sigma}$ defined according to (9.57) by

$$\bar{\sigma}^2 = \langle f^2 \rangle_{(Y,\xi)}.$$

The parameters V_2^ε and V_3^ε are obtained as in the formulas (4.39) and (4.40):

$$V_3^\varepsilon = -\frac{\rho\sqrt{\varepsilon}}{2}\left\langle \beta f \frac{\partial\phi}{\partial y} \right\rangle,$$

$$V_2^\varepsilon = \frac{\sqrt{\varepsilon}}{2}\left\langle \beta\Lambda \frac{\partial\phi}{\partial y} \right\rangle,$$

where the averages are taken with respect to the invariant distribution of (Y, ξ), and $\phi(y, \xi)$ is a solution of the Poisson equation

$$(\mathscr{L}_0 + \mathscr{L}_J)\phi(y, \xi) = f(y, \xi)^2 - \bar{\sigma}^2. \tag{9.61}$$

The main assumption on the added factor ξ is that this equation admits well-behaved solutions. This is the case for bounded processes as in the examples considered in Section 3.3.1 or 3.3.2, where a solution can be written

$$\phi(y, \xi) = -\int_0^{+\infty} \mathbb{E}\{(f(Y_t, \xi_t)^2 - \bar{\sigma}^2)|Y_0 = y, \xi_0 = \xi\}\,dt.$$

The validity, in this generalized Markovian context, of the corrected pricing formula (9.60) (along with the corresponding results in the non-Markovian case treated in Section 8.7) illustrates its **universality**. It can be computed from the Black–Scholes price, its Gamma and DeltaGamma, and calibrated on the implied volatility surface as explained in Chapter 5 without any further knowledge of the processes (Y_t) and (ξ_t) nor their current values y and ξ.

9.7 Multidimensional Models

We outline the extension of the asymptotic theory to the multidimensional case when there are N underlying assets driven by an N-dimensional Brownian motion with stochastic volatilities driven by an N-dimensional ergodic process.

Consider first the Black–Scholes case of constant volatilities using the tools introduced in Section 1.9. The asset prices $(\mathbf{X}_t)$ are geometric Brownian motions with the $N \times N$ invertible volatility matrix σ satisfying the stochastic differential equations

$$\frac{dX_i}{X_i} = \mu_i dt + \sum_{j=1}^{N} \sigma_{ij} dW_j,$$

where $(\mathbf{W}_t)$ is a standard Brownian motion in $\mathbb{R}^N$. We use subscripts now to denote components and omit the time dependence. In this complete market, there is a unique equivalent martingale measure $\mathbb{P}^\star$ under which

$$\frac{dX_i}{X_i} = rdt + \sum_{j=1}^{N} \sigma_{ij} dW_j^\star,$$

for $i = 1, \ldots, N$ and $(\mathbf{W}_t^\star)$ is a $\mathbb{P}^\star$-Brownian motion. Then a European contract with payoff function $h(\mathbf{X}_T)$ admits the *no-arbitrage* price

$$P(t, \mathbf{x}) = \mathbb{E}^\star \left\{ e^{-r(T-t)} h(\mathbf{X}_T) | \mathbf{X}_t = \mathbf{x} \right\}.$$

The pricing function $P(t, \mathbf{x})$ satisfies the partial differential equation

$$\frac{\partial P}{\partial t} + \frac{1}{2} \sum_{i,j=1}^{N} v_{ij} x_i x_j \frac{\partial^2 P}{\partial x_i \partial x_j} + r \left(\sum_{j=1}^{N} x_j \frac{\partial P}{\partial x_j} - P \right) = 0,$$

where $\{v_{ij}\}$ is the symmetric diffusion matrix $\sigma \sigma^T$:

$$v_{ij} = \sum_{k=1}^{N} \sigma_{ik} \sigma_{jk}.$$

The terminal condition is $P(T, \mathbf{x}) = h(\mathbf{x})$.

In other words, $P(t, \mathbf{x})$ is a function of the $N(N+1)/2$ independent entries of the matrix v, and consequently an implied volatility matrix is identified by $N(N+1)/2$ "independent" observed option prices.

Now consider the stochastic volatility extension:

$$\frac{dX_i}{X_i} = \mu_i dt + \sum_{j=1}^{N} f_{ij}(\mathbf{Y}) dW_j, \tag{9.62}$$

$$dY_k = \alpha_k(m_k - Y_k) dt + \beta_k d\hat{Z}_k, \tag{9.63}$$

where $i,k = 1,\ldots,N$ and, for simplicity, we assume that the Y-processes are OUs and we ignore the slow factors. The process $(\hat{\mathbf{Z}}_t)$ is a standard Brownian motion in $\mathbb{R}^N$ that is correlated to $(\mathbf{W}_t)$ by

$$\mathbb{E}\{dW_j d\hat{Z}_k\} = d\langle W_j, \hat{Z}_k \rangle_t = \rho_{jk} dt.$$

Alternatively, we could write

$$\hat{Z}_k = \sum_{j=1}^{N} \rho_{jk} W_j + \left(1 - \sum_{j=1}^{N} \rho_{jk}^2\right)^{1/2} Z_k,$$

where $(\mathbf{W}_t, \mathbf{Z}_t)$ is a standard Brownian motion in $\mathbb{R}^{2N}$. Note that each volatility $f_{ij}(\mathbf{Y})$ is driven by the whole N-dimensional OU process $(\mathbf{Y}_t)$, so there is no need to incorporate further correlations between the components of $(\hat{\mathbf{Z}}_t)$.

Under an equivalent martingale measure $\mathbb{P}^\star$, we have

$$\frac{dX_i}{X_i} = rdt + \sum_{j=1}^{N} f_{ij}(\mathbf{Y}) dW_j^\star, \tag{9.64}$$

$$dY_k = \left[\alpha_k(m_k - Y_k) - \beta_k \Lambda_k(\mathbf{Y})\right] dt + \beta_k d\hat{Z}_k^\star, \tag{9.65}$$

for some volatility risk premium $\Lambda(\mathbf{Y})$ chosen by the market and assumed to be a function of the OU processes. A European contract admits the price given by

$$P(t,\mathbf{x},\mathbf{y}) = \mathbb{E}^\star \left\{ e^{-r(T-t)} h(\mathbf{X}_T) | \mathbf{X}_t = \mathbf{x}, \mathbf{Y}_t = \mathbf{y} \right\},$$

and the pricing function $P(t,\mathbf{x},\mathbf{y})$ satisfies

$$\frac{\partial P}{\partial t} + \frac{1}{2} \sum_{i,j=1}^{N} v_{ij}(\mathbf{y}) x_i x_j \frac{\partial^2 P}{\partial x_i \partial x_j} + \sum_{i,j,k=1}^{N} \rho_{jk} \beta_k f_{ij} x_i \frac{\partial^2 P}{\partial x_i \partial y_k} + \frac{1}{2} \sum_{k=1}^{N} \beta_k^2 \frac{\partial^2 P}{\partial y_k^2}$$

$$+ \frac{1}{2} \sum_{k \neq l} \beta_k \beta_l \left(\sum_j \rho_{jk} \rho_{jl} \right) \frac{\partial^2 P}{\partial y_k \partial y_l} + \sum_{k=1}^{N} [\alpha_k(m_k - y_k) - \beta_k \Lambda_k] \frac{\partial P}{\partial y_k}$$

$$+ r \left(\sum_{j=1}^{N} x_j \frac{\partial P}{\partial x_j} - P \right) = 0,$$

with terminal condition $P(T, \mathbf{x}, \mathbf{y}) = h(\mathbf{x})$.

Introducing the usual scaling modeling fast mean-reversion in the volatilities, we write

$$\alpha_k = c_k/\varepsilon,$$
$$\beta_k^2/2\alpha_k = v_k^2,$$

so that $\beta_k = v_k\sqrt{2c_k/\varepsilon}$. Note that each component of the OU process can mean-revert at a different rate α_k, but that with c_k fixed $\mathcal{O}(1)$ constants, the rates are assumed of the same order. The c_k are defined up to a multiplicative constant which we can normalize by choosing $c_1 = 1$, for instance. It is possible to introduce phenomena on different scales by allowing the c_k to be powers of ε, but this will lead to a more involved theory than we present here.

We now write our pricing problems as

$$\left(\frac{1}{\varepsilon}\mathcal{L}_0 + \frac{1}{\sqrt{\varepsilon}}\mathcal{L}_1 + \mathcal{L}_2\right) P^\varepsilon = 0,$$

where

$$\mathcal{L}_0 = \sum_k c_k \left(v_k^2 \frac{\partial^2}{\partial y_k^2} + (m_k - y_k)\frac{\partial}{\partial y_k}\right)$$
$$+ \sum_{k \neq l} v_k v_l \sqrt{c_k c_l} \left(\sum_j \rho_{jk}\rho_{jl}\right)\frac{\partial^2}{\partial y_k \partial y_l}, \tag{9.66}$$

$$\mathcal{L}_1 = \sum_{i,j,k} v_k\sqrt{2c_k}\,\rho_{jk}f_{ij}x_i\frac{\partial^2}{\partial x_i \partial y_k} - \sum_k v_k\sqrt{2c_k}\,\Lambda_k\frac{\partial}{\partial y_k}, \tag{9.67}$$

$$\mathcal{L}_2 = \frac{\partial}{\partial t} + \frac{1}{2}\sum_{i,j=1}^N v_{ij}(\mathbf{y})x_i x_j\frac{\partial^2}{\partial x_i \partial x_j} + r\left(\sum_{j=1}^N x_j\frac{\partial}{\partial x_j} - \bullet\right). \tag{9.68}$$

Notice that $\mathcal{L}_0$ is the infinitesimal generator of the N-dimensional OU process with the $\mathcal{N}(\mathbf{m}, v^2)$-invariant distribution where $\mathbf{m} \in \mathbb{R}^N$ has components m_k and v^2 is the covariance matrix with diagonal entries v_k^2 and off-diagonal entries

$$\text{cov}\,(Y_k, Y_l) = \frac{\beta_k \beta_l}{\alpha_k + \alpha_l}\sum_j \rho_{jk}\rho_{jl}.$$

We know from Chapter 4 that the key to constructing an asymptotic expansion

$$P^\varepsilon = P_0 + \sqrt{\varepsilon}P_1 + \varepsilon P_2 + \cdots$$

is the Poisson equation

$$\mathcal{L}_0 \chi(\mathbf{y}) + g(\mathbf{y}) = 0.$$

This equation has well-behaved solutions only if g is centered with respect to the invariant distribution of the N-dimensional OU process:

$$\langle g \rangle = \int_{\mathbb{R}^N} \Phi(\mathbf{y}) g(\mathbf{y}) d\mathbf{y} = 0,$$

where $\Phi(\mathbf{y})$ is the density function for the N-dimensional normal distribution $\mathcal{N}(\mathbf{m}, v^2)$ satisfying

$$\mathcal{L}_0^\star \Phi = 0.$$

Notice that, when $g = 0$, the solutions are constants in $\mathbf{y}$.

Then the arguments of Chapter 4 go through analogously and P_0 is a function of $(t, \mathbf{x})$ only which satisfies

$$\mathcal{L}_{BS}^N(\bar{v}) P_0 = 0,$$

with $P(T, \mathbf{x}) = h(\mathbf{x})$ and where we define $\mathcal{L}_{BS}^N(v)$ by

$$\mathcal{L}_{BS}^N = \frac{\partial}{\partial t} + \frac{1}{2} \sum_{i,j=1}^N v_{ij} x_i x_j \frac{\partial^2}{\partial x_i \partial x_j} + r \left(\sum_{j=1}^N x_j \frac{\partial}{\partial x_j} - \bullet \right),$$

a function of the symmetric matrix v. In this case an *effective volatility square matrix* $\bar{\sigma}$ satisfies $\bar{\sigma}\bar{\sigma}^T = \bar{v}$ with the average diffusion matrix given by

$$\bar{v}_{ij} = \sum_k \langle f_{ik} f_{jk} \rangle.$$

Carrying on, $P_1^\varepsilon = \sqrt{\varepsilon} P_1$ is also a function of $(t, \mathbf{x})$ only which satisfies

$$\mathcal{L}_{BS}^N(\bar{v}) P_1^\varepsilon = \sqrt{\varepsilon} \langle \mathcal{L}_1 \mathcal{L}_0^{-1} (\mathcal{L}_2 - \langle \mathcal{L}_2 \rangle) \rangle P_0 = \mathcal{A}^\varepsilon P_0,$$

with zero terminal condition. It remains to compute the operator $\mathcal{A}^\varepsilon$ on the right-hand side.

Introducing the symmetric matrix $\phi(\mathbf{Y})$ satisfying the Poisson equations

$$\mathcal{L}_0 \phi_{ij} = \sum_k f_{ik} f_{jk} - \bar{v}_{ij},$$

we generalize (4.38) as follows:

$$\mathcal{A}^\varepsilon = - \sum_{i,l,m} V_{3,ilm}^\varepsilon x_m \frac{\partial}{\partial x_m} \left(x_i x_l \frac{\partial^2}{\partial x_i \partial x_l} \right) - \sum_{i,m} V_{2,im}^\varepsilon x_i x_m \frac{\partial^2}{\partial x_i \partial x_m}.$$

Analogously to (4.39) and (4.40), the small parameters $V^\varepsilon_{3,ilm}$ and $V^\varepsilon_{2,im}$ are given by

$$V^\varepsilon_{3,ilm} = -\sum_k \frac{v_k\sqrt{\varepsilon c_k}}{\sqrt{2}} \sum_j \rho_{jk} \left\langle f_{ij}\frac{\partial\phi_{lm}}{\partial y_k}\right\rangle, \tag{9.69}$$

$$V^\varepsilon_{2,im} = \sum_k \frac{v_k\sqrt{\varepsilon c_k}}{\sqrt{2}} \left\langle \Lambda_k\frac{\partial\phi_{im}}{\partial y_k}\right\rangle. \tag{9.70}$$

Notice that $V^\varepsilon_{3,ilm}$ is symmetric in the last two indices so the parameter reduction is from N^2 unspecified functions f_{ij} and N volatility risk premia Λ_j plus $N(N+1)/2$ entries ρ_{ij} and $3N$ **Y** parameters $(\alpha, \mathbf{m}, \beta)$ to $N^2(N+1)/2$ V^ε_3 parameters plus N^2 V^ε_2 parameters and $N(N+1)/2$ $\bar{v}$ entries.

Notes

The cases with dividends and varying interest rate appear here for the first time. The derivation of the second-order corrections with the probabilistic "boundary-layer" argument appears here for the first time. A matched asymptotic expansion method is presented in Howison (2005). Averaging simultaneously with respect to random and periodic components has been studied extensively in the context of waves in random media, we refer to Fouque *et al.* (2007) and references cited therein.

The multidimensional extension presented in Section 9.7 will be used in Chapters 13 and 14 in the context of multiname credit derivatives.

10

Around the Heston Model

Since its publication in 1993, the Heston model (Heston, 1993) has received considerable attention from academics and practitioners alike. The Heston model belongs to the class of stochastic volatility models described in Chapter 2. Among stochastic volatility models, the Heston model enjoys wide popularity because it provides an explicit, easy-to-compute, integral formula for calculating European option prices. In terms of the computational resources needed to calibrate a model to market data, the existence of such a formula makes the Heston model extremely efficient compared to models that rely on partial differential equations techniques or Monte Carlo simulations for computation and calibration.

In this chapter, we show that our asymptotic analysis can be applied to Heston's model in several ways. First, we approximate the Heston model in the two regimes where the CIR volatility factor is fast mean-reverting or slowly varying, and in each regime we derive formulas for the V parameters arising in the analysis presented in Chapter 4. The main advantages of this approximation technique are that, as for general stochastic volatility models treated before, the computation and calibration steps simplify vastly, which in turn enables consistent pricing of more complex derivatives.

As has been explained in previous chapters, a single factor of stochastic volatility is not enough to capture the main features of the implied volatility surface. Any stochastic volatility model in which the volatility is modeled as a one-factor diffusion (as is the case in the Heston model) has trouble fitting implied volatility levels across all strikes and maturities, see for instance Gatheral (2006). Despite its success, the Heston model has a number of documented shortcomings. For example, it has been statistically verified that the model misprices far in-the-money and out-of-the-money European options (Fiorentini *et al.*, 2002; Zhang and Shu, 2003). In particular, the

Heston model has difficulty fitting implied volatility levels for options with short expirations (Gatheral, 2006).

In Section 10.3, following Fouque and Lorig (2009), we propose one way to bring the Heston model into the realm of multiscale stochastic volatility models without sacrificing analytic tractability. Specifically, we add a fast mean-reverting component of volatility on top of the CIR process that drives the volatility in the Heston model. Using this multiscale model, we perform a singular perturbation expansion around the Heston model, in order to obtain a correction to the Heston price of a European option. This correction is easy to implement, as it has an integral representation that is quite similar to that of the European option pricing formula produced by the Heston model.

Regarding short maturities, we present in Section 10.4 results derived in Feng *et al.* (2009) for a fast mean-reverting Heston model using an explicit computation of the moment generating function in order to derive a large deviation principle and its applications to pricing and implied volatilities.

10.1 The Heston Model

We start by summarizing the Heston model and the associated formulas.

10.1.1 European Derivatives under the Heston Model

There are a number of excellent sources where one can read about the Heston stochastic volatility model – so many, in fact, that a detailed review of the model would seem superfluous. However, in order to establish some notation, we will briefly review the dynamics of the Heston model here, as well as a preferred method for solving the corresponding European option pricing problem. We closely follow Shaw (2006).

Let X_t be the price of a stock, and denote by r the risk-free rate. Then, under the pricing risk-neutral probability measure $\mathbb{P}^\star$, the Heston model takes the following form:

$$
\begin{aligned}
dX_t &= rX_t dt + \sqrt{Z_t}\, X_t dW_t^x, \\
dZ_t &= \kappa\,(m - Z_t)\,dt + \sigma\sqrt{Z_t}\, dW_t^z, \\
d\,\langle W^x, W^z \rangle_t &= \rho dt,
\end{aligned}
\tag{10.1}
$$

where here, W_t^x and W_t^z are one-dimensional standard Brownian motions with correlation ρ, such that $|\rho| \le 1$. The CIR process, Z_t, is the stochastic variance of the stock and $\sqrt{Z_t}$ is its volatility. The positive constants κ, m, and σ satisfy $\sigma^2 \leq 2\kappa m$, ensuring that Z_t remains positive for all t as seen in Section 3.3.4.

Note that the model is given directly under a risk-neutral measure. In fact, from the general presentation given in Chapter 2, this means that the market price of volatility risk γ_t is chosen such that the drift of Z_t remains affine, that is γ_t is proportional to $\sqrt{Z_t}$, and that we assume a bound on the stochastic Sharpe ratio $\frac{\mu(Z_t)-r}{\sqrt{Z_t}}$ in order to use Girsanov's theorem.

We denote by P_H the price of a European option, as calculated under the Heston model:

$$P_H(t,x,z) = \mathbb{E}^\star \left\{ e^{-r(T-t)} h(X_T) \middle| X_t = x, Z_t = z \right\},$$

where $P_H(t,x,z)$ satisfies the following partial differential equation and terminal condition:

$$\mathcal{L}_H P_H(t,x,z) = 0, \tag{10.2}$$
$$P_H(T,x,z) = h(x), \tag{10.3}$$

where

$$\mathcal{L}_H = \frac{\partial}{\partial t} - r + rx\frac{\partial}{\partial x} + \frac{1}{2}zx^2\frac{\partial^2}{\partial x^2} + \kappa(m-z)\frac{\partial}{\partial z}$$
$$+ \frac{1}{2}\sigma^2 z\frac{\partial^2}{\partial z^2} + \rho\sigma zx\frac{\partial^2}{\partial x\partial z}. \tag{10.4}$$

In order to find a solution for $P_H(t,x,z)$, it is convenient to transform variables as follows:

$$\tau(t) = T - t,$$
$$q(t,x) = r(T-t) + \log x,$$
$$P_H(t,x,z) = P'_H(\tau(t),q(t,x),z)e^{-r\tau(t)},$$

so that:

$$\mathcal{L}'_H P'_H(\tau,q,z) = 0,$$
$$P'_H(0,q,z) = h(e^q),$$

with

$$\mathcal{L}'_H = -\frac{\partial}{\partial \tau} + \frac{1}{2}z\left(\frac{\partial^2}{\partial q^2} - \frac{\partial}{\partial q}\right) + \kappa(m-z)\frac{\partial}{\partial z} +$$
$$+ \frac{1}{2}\sigma^2 z\frac{\partial^2}{\partial z^2} + \rho\sigma z\frac{\partial^2}{\partial q\partial z}. \tag{10.5}$$

Denoting by $G(\tau,q,z)$ the Green's function, solution to the following problem:

$$\mathcal{L}'_H G(\tau,q,z) = 0, \tag{10.6}$$
$$G(0,q,z) = \delta(q), \tag{10.7}$$

then

$$P'_H(\tau,q,z) = \int_{\mathbb{R}} G(\tau,q-p,z)h(e^p)dp. \tag{10.8}$$

In the Fourier domain with respect to q, we have

$$
\begin{aligned}
P'_H(\tau,q,z) &= \frac{1}{2\pi}\int_{\mathbb{R}} e^{-ikq}\widehat{P}'_H(\tau,k,z)dk \\
&= \frac{1}{2\pi}\int_{\mathbb{R}} e^{-ikq}\widehat{G}(\tau,k,z)\widehat{h}(k)dk,
\end{aligned}
$$

where $\widehat{P}'_H(\tau,k,z)$, $\widehat{G}(\tau,k,z)$, and $\widehat{h}(k)$ denote the Fourier transforms of $P'_H(\tau,q,z)$, $G(\tau,q,z)$, and $h(e^q)$ respectively. Multiplying equations (10.6) and (10.7) by $e^{ikq'}$ and integrating over $\mathbb{R}$ in q', we find that $\widehat{G}(\tau,k,z)$ satisfies the following initial value problem:

$$\widehat{\mathcal{L}}'_H \widehat{G}(\tau,k,z) = 0, \tag{10.9}$$
$$\widehat{G}(0,k,z) = 1, \tag{10.10}$$

with

$$\widehat{\mathcal{L}}'_H = -\frac{\partial}{\partial\tau} + \frac{1}{2}z\left(-k^2+ik\right) + \frac{1}{2}\sigma^2 z\frac{\partial^2}{\partial z^2} + (\kappa m - (\kappa+\rho\sigma ik)z)\frac{\partial}{\partial z}.$$

We now seek a solution $\widehat{G}(\tau,k,z)$ of the exponential affine form

$$\widehat{G}(\tau,k,z) = e^{C(\tau,k)+zD(\tau,k)}. \tag{10.11}$$

Substituting (10.11) into (10.9) and (10.10), and collecting terms of like-powers of z, we find that $C(\tau,k)$ and $D(\tau,k)$ must satisfy the following ordinary differential equations:

$$\frac{dC}{d\tau}(\tau,k) = \kappa m D(\tau,k), \quad C(0,k) = 0, \tag{10.12}$$

$$
\begin{aligned}
\frac{dD}{d\tau}(\tau,k) &= \frac{1}{2}\sigma^2 D^2(\tau,k) - (\kappa+\rho\sigma ik)D(\tau,k) \\
&\quad + \frac{1}{2}\left(-k^2+ik\right), \quad D(0,k) = 0. \tag{10.13}
\end{aligned}
$$

Equations (10.12) and (10.13) can be solved analytically:

$$C(\tau,k) = \frac{\kappa m}{\sigma^2}\left((\kappa+\rho ik\sigma+d(k))\,\tau - 2\log\left(\frac{1-g(k)e^{\tau d(k)}}{1-g(k)}\right)\right), \quad (10.14)$$

$$D(\tau,k) = \frac{\kappa+\rho ik\sigma+d(k)}{\sigma^2}\left(\frac{1-e^{\tau d(k)}}{1-g(k)e^{\tau d(k)}}\right), \quad (10.15)$$

$$d(k) = \sqrt{\sigma^2(k^2-ik)+(\kappa+\rho ik\sigma)^2}, \quad (10.16)$$

$$g(k) = \frac{\kappa+\rho ik\sigma+d(k)}{\kappa+\rho ik\sigma-d(k)}. \quad (10.17)$$

The solution to the European option pricing problem in the Heston framework, is then given by:

$$P_H(t,x,z) = \frac{e^{-r\tau}}{2\pi}\int e^{-ikq}e^{C(\tau,k)+zD(\tau,k)}\widehat{h}(k)dk, \quad (10.18)$$

where we recall $\tau = T-t$, $q = r\tau+\log x$, and $P_H(t,x,z) = P'_H(\tau,q,z)e^{-r\tau}$.

10.1.2 Numerical Evaluation of Call Options

In the case of a European call option, we have:

$$\widehat{h}(k) = \int_{\mathbb{R}}e^{ikq}(e^q-K)^+dq = \frac{K^{1+ik}}{ik-k^2}. \quad (10.19)$$

We note that the integral in (10.19) will not converge unless the imaginary part of k is greater than 1. Thus, we decompose k into its real and imaginary parts, and impose the following condition on the latter:

$$k = k_r+ik_i,$$
$$k_i > 1. \quad (10.20)$$

When we integrate over k in (10.18), we hold $k_i > 1$ fixed, and integrate k_r over $\mathbb{R}$, so that

$$P_H(t,x,z) = \frac{e^{-r\tau}}{2\pi}\int_{\mathbb{R}}e^{-ikq}e^{C(\tau,k)+zD(\tau,k)}\frac{K^{1+ik}}{ik-k^2}dk_r. \quad (10.21)$$

In order for any numerical integration scheme to work, we must verify the continuity of the integrand in (10.21). First, by (10.20), the poles at $k=0$ and $k=i$ are avoided. The only other worrisome term in the integrand of (10.21) is $e^{C(\tau,k)}$, which may be discontinuous due to the presence of the log in $C(\tau,k)$ given by (10.14).

We recall that any $\zeta \in \mathbb{C}$ can be represented in polar notation as $\zeta = r\exp(im)$, where $m \in [-\pi, \pi)$. In this notation, $\log \zeta = \log r + im$. Now, suppose we have a map $\zeta(k_r) : \mathbb{R} \to \mathbb{C}$. We see that whenever $\zeta(k_r)$ crosses the negative real axis, $\log \zeta(k_r)$ will be discontinuous (due to m jumping from $-\pi$ to π or from π to $-\pi$). Thus, in order for $\log \zeta(k_r)$ to be continuous, we must ensure that $\zeta(k_r)$ does not cross the negative real axis.

We now return our attention to $C(\tau, k)$ given by (10.14). We note that $C(\tau, k)$ has two algebraically equivalent representations, (10.14) and the following representation:

$$C(\tau, k) = \frac{\kappa m}{\sigma^2} \left((\kappa + \rho i k \sigma - d(k)) \tau - 2 \log \zeta(\tau, k) \right), \qquad (10.22)$$

$$\zeta(\tau, k) := \frac{e^{-\tau d(k)}/g(k) - 1}{1/g(k) - 1}. \qquad (10.23)$$

It turns out that, under most reasonable conditions, $\zeta(\tau, k)$ does not cross the negative real axis (Lord and Kahl, 2006). As such, as one integrates over k_r, no discontinuities will arise from the $\log \zeta(\tau, k)$ term which appears in (10.22). Therefore, if we use expression (10.22) when evaluating (10.21), the integrand will be continuous.

Aside from using equations (10.22) for $C(\tau, k)$, there are a few other tricks one can use to facilitate the numerical evaluation of (10.21). Denote by $I(k)$ the integrand appearing in (10.21), so that:

$$P_H = \int_{\mathbb{R}} I(k) dk_r.$$

First, we note that the real and imaginary parts of $I(k)$ are even and odd functions of k_r, respectively. As such, instead of integrating in k_r over $\mathbb{R}$, one can integrate in k_r over $\mathbb{R}_+$, drop the imaginary part, and multiply the result by 2.

Second, numerically integrating in k_r over $\mathbb{R}_+$ requires that one arbitrarily truncates the integral at some k_{cutoff}. Rather than doing this, one can make the following variable transformation, suggested by Jackel and Kahl (2005):

$$k_r = \frac{-\log u}{C_\infty},$$

$$C_\infty := \frac{\sqrt{1 - \rho^2}}{\sigma}(z + \kappa m \tau). \qquad (10.24)$$

Then, for some arbitrary $I(k)$, we have

$$\int_0^\infty I(k) dk_r = \int_0^1 I\left(\frac{-\log u}{C_\infty} + ik_i\right) \frac{1}{uC_\infty} du.$$

Thus, we avoid having to establish a cutoff value, k_{cutoff} (and avoid the error that comes along with doing so).

10.2 Approximations to the Heston Model

In this section we approximate the Heston model in the two regimes where the CIR volatility factor Z_t is fast mean-reverting or slowly varying, and in each regime we derive formulas for the V parameters arising in the analysis presented in Chapter 4. In Section 10.2.2, we implement the second-order correction derived in Section 9.3.1.

10.2.1 Fast Mean-Reverting Heston Model

In the regime of fast mean-reverting volatility, the Heston model (10.1) can be reparametrized by setting $\kappa = \frac{1}{\varepsilon}$ and $\sigma = \frac{v\sqrt{2}}{\sqrt{\varepsilon}}$, so that

$$
\begin{aligned}
dX_t &= rX_t dt + \sqrt{Z_t}\, X_t dW_t^x, \\
dZ_t &= \frac{1}{\varepsilon}(m - Z_t)\, dt + \frac{v}{\sqrt{\varepsilon}}\sqrt{2Z_t}\, dW_t^z, \\
d\langle W^x, W^z \rangle_t &= \rho dt,
\end{aligned}
\tag{10.25}
$$

where the condition $v^2 < m$, independent of the small time scale parameter ε, ensures that Z_t stays positive at all times. From Section 3.3.4, we know that the invariant distribution of Z is a Gamma distribution with mean m and parameter $\theta = m/v^2 \geq 1$. This is a fast mean-reverting one-factor stochastic volatility model which falls in the class of models treated by a singular perturbation approach in Chapter 4. Here, Z denotes the fast factor and the volatility function is given by $f(z) = \sqrt{z}$. Since there is no term of order $1/\sqrt{\varepsilon}$ in the drift of Z_t, one has that V_2^ε given by (4.40) is zero, and consequently no parameter-reduction step is needed ($\bar{\sigma} = \sigma^\star$). In order to identify the approximation to derivative prices, one needs the other two group parameters, $\bar{\sigma}$ and V_3^ε defined respectively by (4.26) and (4.39), given here by:

$$
\begin{aligned}
\bar{\sigma}^2 &= \int_0^\infty f^2(z)\Phi(z)dz \\
&= \frac{1}{v^{2\theta}\Gamma(\theta)}\int_0^\infty z^\theta e^{-z/v^2}dz \\
&= m,
\end{aligned}
\tag{10.26}
$$

$$V_3^\varepsilon = -\frac{\rho\sqrt{\varepsilon}}{2}\int_0^\infty v\sqrt{2z}f(z)\phi'(z)\Phi(z)dz$$

$$= \frac{\rho v\sqrt{\varepsilon}}{\sqrt{2}}\frac{1}{v^{2\theta}\Gamma(\theta)}\int_0^\infty z^\theta e^{-z/v^2}dz$$

$$= \frac{\rho m v\sqrt{\varepsilon}}{\sqrt{2}}, \tag{10.27}$$

$$\theta = \frac{m}{v^2} \geq 1, \tag{10.28}$$

where we have used the fact that, in this case, the solution to the Poisson equation $\mathcal{L}_0\phi = f^2 - \bar{\sigma}^2$ defining ϕ (see (4.36)) satisfies $\phi'(z) = -1$ from the results derived in Section 3.3.4. Note that, as expected in general for an increasing function f, V_3^ε has the sign of ρ which has the sign of the skew of implied volatility as follows from (5.21). It is also useful to write the group parameters $\bar{\sigma}$ and V_3^ε in terms of the original parameters m, κ, and σ in (10.1). From (10.26) we have $\bar{\sigma}^2 = m$, and from $\kappa = \frac{1}{\varepsilon}$, $\sigma = \frac{v\sqrt{2}}{\sqrt{\varepsilon}}$, and (10.27) we deduce:

$$V_3^\varepsilon = \frac{\rho m \sigma}{2\kappa},$$

where κ is large and $\frac{2\kappa m}{\sigma^2} \geq 1$ just of order one (not large). Finally, in this fast mean-reverting regime, the Heston approximated price of a European derivative is given by

$$P_H(t,x,z) \sim P_{BS}(t,x) + (T-t)V_3^\varepsilon x\frac{\partial}{\partial x}\left(x^2\frac{\partial^2 P_{BS}}{\partial x^2}\right),$$

where $P_{BS}(t,x)$ is the price of this same derivative computed with the constant volatility $\bar{\sigma} = \sqrt{m}$. Note again that this approximation does not depend on the present volatility level z, and that the parameters $\bar{\sigma}$ and V_3^ε are calibrated to the observed implied volatility skew as explained in Chapter 5.

Recall that from (5.18)–(5.19) with $V_0^\delta = V_1^\delta = V_2^\varepsilon = 0$, the implied volatility I_H generated by this fast mean-reverting Heston model (10.25) is approximated by

$$I_H \sim \bar{\sigma} + \frac{V_3^\varepsilon}{2\bar{\sigma}}\left(1 - \frac{2r}{\bar{\sigma}^2}\right) + \frac{V_3^\varepsilon}{\bar{\sigma}^3}\left(\frac{\log(K/x)}{T-t}\right), \tag{10.29}$$

with, here, $\bar{\sigma} = \sqrt{m}$ and $V_3^\varepsilon = (\rho m \sigma)/(2\kappa)$.

In Figure 10.1, we show examples of approximated Heston skews in the regime of fast mean-reverting volatility. Indeed, the affine approximation cannot account for the turn of the skew (the smile) generated by the Heston model at shorter maturity (left plots). For this choice of parameters, the

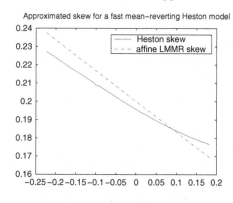
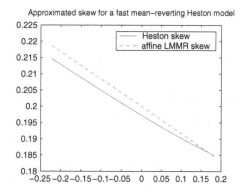

Figure 10.1 Implied volatility skews, as functions of moneyness K/x, generated by the Heston formula (10.21) and approximated by the affine LMMR formula (10.29) in the regime of fast mean-reverting volatility. The parameters are as follows: $x = 100$, $z = 0.04$, $r = 0.02$, $\kappa = 10$, $m = 0.04$, $\sigma = 0.15\sqrt{20}$, $\rho = -0.5$. The *left plots* are for $T = 0.5$ and the *right plots* for $T = 1$.

approximation error around the money is on the order of half a percent with the model-independent formula (10.29). There are several ways to improve the quality of this first-order approximation. One of them, which is discussed in the following Section 10.2.2, is indeed to take into account the next term in the expansion. At shorter maturities, one can also use *large deviation* techniques, as we will discuss briefly in the context of Heston's model in Section 10.4. Then there is also the possibility of building a correction around the Heston model instead of the Black–Scholes model. This last direction is the topic of Section 10.3.

10.2.2 Second-Order Approximation for a Fast Mean-Reverting Heston Model

In Section 9.3.1 we derived the formula (9.53) for the second term in the expansion of the implied volatility. We now implement it on the fast mean-reverting Heston model (10.25) considered in the previous section. In this model, we have set $\Lambda = 0$ so that formula (9.53) simplifies since in this case $V_2 = 0$, and therefore, $\bar{\sigma} = \sigma^\star$. We also have that ψ_2 introduced in (9.34) is zero and consequently $A_1 = A_0 = 0$ from their definition (9.39).

As we have noted in the previous subsection, $\phi' = -1$, so we choose $\phi(z) = m - z$ to satisfy the condition $\langle \phi \rangle = 0$. Using $\beta(z) = v\sqrt{2z}$, an easy calculation shows that $\psi_1(z)$ introduced in (9.33) is given by $\psi_1(z) = -v\sqrt{2}\phi(z)$, where the variable y is denoted by z here. It follows that A_2 defined by (9.39) is now given by $A_2 = v^2\rho^2\bar{\sigma}^2$.

The new quantity A defined in (9.37) is easily computed:

$$A = -\frac{1}{4}\left(\langle\phi f^2\rangle - \langle\phi\rangle\langle f^2\rangle\right) = \frac{1}{4}\left(\langle z^2\rangle - \langle z\rangle^2\right) = \frac{mv^2}{4},$$

where we have used the variance of the invariant distribution of the square-root process Z.

Finally, using again $\bar{\sigma}^2 = m$, $\kappa = \frac{1}{\varepsilon}$, and $\sigma = \frac{v\sqrt{2}}{\sqrt{\varepsilon}}$, we deduce:

$$V_3^{\varepsilon} = \sqrt{\varepsilon}V_3 = \frac{\rho m\sigma}{2\kappa}, \quad A_2^{\varepsilon} = \varepsilon A_2 = \frac{2(V_3^{\varepsilon})^2}{m}, \quad A^{\varepsilon} = \varepsilon A = \frac{m\sigma^2}{8\kappa^2},$$

where κ is large and $\frac{2\kappa m}{\sigma^2} \geq 1$ just of order one.

Substituting these coefficients in (9.53), we obtain the second-order approximation formula for the Heston's implied volatility I_H:

$$
\begin{aligned}
I_H \sim \bar{\sigma} &+ \frac{V_3^{\varepsilon}}{\bar{\sigma}}\left(1 - \frac{d_1}{\bar{\sigma}\sqrt{\tau}}\right) + \frac{z-m}{2\bar{\sigma}\tau\kappa} \\
&+ \frac{A_2^{\varepsilon}}{\bar{\sigma}}\left(\frac{d_1^2}{\bar{\sigma}^2\tau} - \frac{2d_1}{\bar{\sigma}\sqrt{\tau}} - \frac{1}{\bar{\sigma}^2\tau} + 1\right) + \frac{A^{\varepsilon}}{\bar{\sigma}}\left(\frac{d_1^2}{\bar{\sigma}^2\tau} - \frac{d_1}{\bar{\sigma}\sqrt{\tau}} - \frac{1}{\bar{\sigma}^2\tau}\right) \\
&+ \frac{\tau(V_3^{\varepsilon})^2}{2\bar{\sigma}}\left(\frac{d_1^4}{\bar{\sigma}^4\tau^2} - \frac{3d_1^3}{\bar{\sigma}^3\tau^{3/2}} - \frac{6d_1^2}{\bar{\sigma}^4\tau^2} + \frac{3d_1^2}{\bar{\sigma}^2\tau} + \frac{9d_1}{\bar{\sigma}^3\tau^{3/2}}\right. \\
&\qquad\qquad\left. - \frac{d_1}{\bar{\sigma}\sqrt{\tau}} + \frac{3}{\bar{\sigma}^4\tau^2} - \frac{3}{\bar{\sigma}^2\tau}\right) \\
&- \frac{(V_3^{\varepsilon})^2}{2\bar{\sigma}^3}\left(1 - \frac{d_1}{\sigma^{\star}\sqrt{\tau}}\right)^2\left(d_1^2 - \sigma^{\star}\sqrt{\tau}d_1\right).
\end{aligned}
\tag{10.30}
$$

In Figure 10.2, we show examples of the effect of the second correction on the quality of approximation to the Heston price. As expected, this quality improves as κ increases, as can be seen in the bottom plots. From our numerical experiments, in this range of parameters, we observe that the main effect of the second correction comes from the A-terms in (10.30). In fact, V_3^{ε} is of order 10^{-4} (first correction), and A_2^{ε} and A^{ε} are of order 10^{-5}, so that the terms in $(V_3^{\varepsilon})^2$ are negligible and the implied volatility is effectively *quadratic* in log-moneyness.

10.2.3 Slowly Varying Heston Model

In the regime of slow varying volatility, the Heston model (10.1) can be reparametrized by setting $\kappa = \delta$ and $\sigma = v\sqrt{2\delta}$, so that

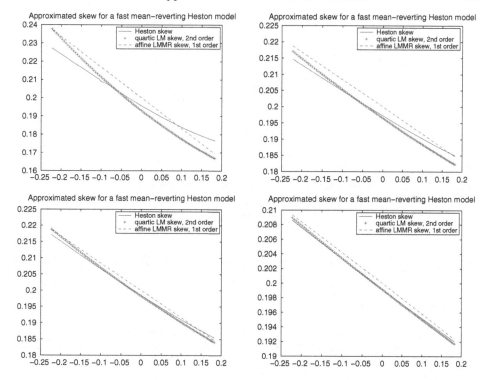

Figure 10.2   Implied volatility skews, as functions of moneyness K/x, generated by the Heston formula (10.21) and approximated by the affine LMMR formula (10.29) and the quartic LM formula (10.30) in the regime of fast mean-reverting volatility. The parameters are as follows. Top plots: $x = 100$, $z = 0.04$, $r = 0.02$, $\kappa = 10$, $m = 0.04$, $\sigma = 0.15\sqrt{20}$, $\rho = -0.5$. In the *bottom plots*, $\kappa = 10 \times 4$, $\sigma = 0.15\sqrt{20} \times 2$ and the other parameters remain the same. The *left plots* are for $T = 0.5$ and the *right plots* for $T = 1$.

$$dX_t = rX_t dt + \sqrt{Z_t}\, X_t dW_t^x,$$
$$dZ_t = \delta\,(m - Z_t)\,dt + v\sqrt{2\delta Z_t}\, dW_t^z, \qquad (10.31)$$
$$d\,\langle W^x, W^z \rangle_t = \rho dt,$$

where δ is a small positive parameter. This is a slowly varying one-factor stochastic volatility model which falls in the class of models treated by a regular perturbation approach in Chapter 4. Here, Z denotes the slow factor and the volatility function is given by $f(z) = \sqrt{z}$. Again, there should be a constant market price of risk term of order $\sqrt{\delta}$ in the equation for Z, which we will simply ignore so that V_0^δ given by (4.51) is zero.

In this case, since there is no fast mean-reverting factor, the effective volatility $\bar{\sigma}(z)$ is simply the volatility frozen at level $\sqrt{z}$; that is,

$$\bar{\sigma}^2(z) = z. \tag{10.32}$$

The group parameter V_1^δ defined in (4.52) is given by

$$\begin{aligned}
V_1^\delta &= \frac{\rho v \sqrt{2\delta z}}{2} \sqrt{z}(\sqrt{z})' \\
&= \frac{\rho v \sqrt{\delta z}}{2\sqrt{2}}. \tag{10.33}
\end{aligned}$$

In terms of the original parameters in (10.1), we have

$$V_1^\delta = \frac{\rho \sigma \sqrt{z}}{4},$$

where, as expected in this regime, the drift of Z does not play a role. Finally, in this slowly varying regime, the Heston approximated price of a European derivative is given by

$$P_H(t,x,z) \sim P_{BS}(t,x) + (T-t)V_1^\delta x \frac{\partial}{\partial x}\left(\frac{\partial P_{BS}}{\partial \sigma}\right),$$

where $P_{BS}(t,x)$ is the price of this same derivative computed with the constant volatility $\bar{\sigma} = \sqrt{z}$, and the parameters $\bar{\sigma}$ and V_1^δ are calibrated to the observed implied volatility skew as explained in Chapter 5.

Recall that from (5.18), (5.20) with $V_0^\delta = V_2^\varepsilon = V_3^\varepsilon = 0$, the implied volatility I_H generated by this slowly varying Heston model (10.31) is approximated by

$$I_H \sim \bar{\sigma} + \frac{V_1^\delta (T-t)}{2}\left(1 - \frac{2r}{\bar{\sigma}^2}\right) + \frac{V_1^\delta (T-t)}{\bar{\sigma}^2}\left(\frac{\log(K/x)}{T-t}\right), \tag{10.34}$$

with, here, $\bar{\sigma} = \sqrt{z}$ and $V_1^\delta = (\rho \sigma \sqrt{z})/4$.

In Figure 10.3, we show examples of approximated Heston skews in the regime of slowly varying volatility. For this choice of parameters, the approximation error around-the-money is on the order of a tenth of a percent with the model-independent formula (10.2.3).

Indeed, a single-factor Heston model does not allow for modeling several volatility time scales which we know are needed to fit the implied volatility surface. It is then natural to consider multifactor versions of this model. This can be achieved in many ways, for instance by modeling the mean m of Z as a slowly varying process, or, as we do next, building a fast mean-reverting factor around the original Heston model. The challenge here is to keep the computational tractability of the Heston model.

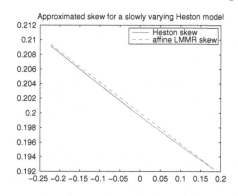
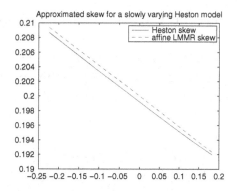

Figure 10.3   Implied volatility skews, as functions of moneyness K/x, generated by the Heston formula (10.21) and approximated by the affine LMMR formula (10.2.3) in the regime of slowly varying volatility. The parameters are as follows: $x = 100, z = 0.04, r = 0.02, \kappa = 0.1, m = 0.04, \sigma = 0.15\sqrt{0.2}, rho = -0.5$. The *left plots* are for $T = 0.5$ and the *right plots* for $T = 1$.

10.3 A Fast Mean-Reverting Correction to the Heston Model

The results of this section are based on Fouque and Lorig (2009). The idea here is to add a fast mean-reverting factor to the original Heston's model. By using a singular perturbation around the original Heston's model, we will be able to keep the computational tractability of this model, and at the same time bring additional flexibility through the correction due to the fast factor.

10.3.1 A Multiscale Generalized Heston Model

The two-factor stochastic volatility model we consider is as follows:

$$dX_t = rX_tdt + f(Y_t)\sqrt{Z_t}X_tdW_t^x, \tag{10.35}$$
$$dZ_t = \kappa(\theta - Z_t)dt + \sigma\sqrt{Z_t}dW_t^z, \tag{10.36}$$
$$dY_t = \frac{Z_t}{\varepsilon}(m - Y_t)dt + v\sqrt{\frac{2Z_t}{\varepsilon}}dW_t^y. \tag{10.37}$$

Here, W_t^x, W_t^y, and W_t^z are one-dimensional standard Brownian motions with the correlation structure

$$d\langle W^x, W^y\rangle_t = \rho_{xy}dt, \tag{10.38}$$
$$d\langle W^x, W^z\rangle_t = \rho_{xz}dt, \tag{10.39}$$
$$d\langle W^y, W^z\rangle_t = \rho_{yz}dt, \tag{10.40}$$

where the correlation coefficients ρ_{xy}, ρ_{xz}, and ρ_{yz} are constants satisfying $\rho_{xy}^2 < 1, \rho_{xz}^2 < 1, \rho_{yz}^2 < 1$, and $\rho_{xy}^2 + \rho_{xz}^2 + \rho_{yz}^2 - 2\rho_{xy}\rho_{xz}\rho_{yz} < 1$.

As it should be, in (10.35) the stock price discounted by the risk-free rate r is a martingale under the pricing risk-neutral measure. Its volatility is driven by two processes Y and Z, through the product $f(Y_t)\sqrt{Z_t}$. The process Z is a CIR process with long-run mean θ, rate of mean-reversion κ, and "CIR-volatility" σ. We assume that κ, θ, and σ are positive, and that $2\kappa\theta > \sigma^2$, which ensures that Z_t remains positive at all times.

Note that given Z, the process Y in (10.37) appears as an OU process evolving on the time scale ε/Z_t. This way of "modulating" the time scale of the process Y by Z_t has also been used in Cotton *et al.* (2004) in the context of interest rate modeling. Multiple time scales are incorporated in this model through the parameter $\varepsilon > 0$, which is intended to be small, so that Y is fast-reverting.

We do not specify the precise form of $f(y)$, which will not play an essential role in the asymptotic results derived in this section. However, in order to ensure that the volatility of the stock has the same behavior at zero and infinity as in the case of a pure Heston model, we assume there exist constants c_1 and c_2 such that $0 < c_1 \leq f(y) \leq c_2 < \infty$ for all $y \in \mathbb{R}$.

We note that if one chooses $f(y) = 1$, the multiscale model becomes ε-independent and reduces to the pure Heston model (10.1). Thus, the multiscale model (10.35)–(10.37) can be thought of as a Heston-like model with a fast-varying factor of volatility, $f(Y_t)$, built on top of the CIR process Z, which drives the volatility in the Heston model.

10.3.2 Pricing Equation

We consider a European option expiring at time $T > t$ with payoff $h(X_T)$. As the dynamics of the stock in the multiscale model (10.35)–(10.37) is specified under a risk-neutral measure, the price of the option, denoted by P_t, is expressed as an expectation of the option payoff, discounted at the risk-free rate:

$$P_t = \mathbb{E}^\star \left\{ e^{-r(T-t)} h(X_T) \mid X_t, Y_t, Z_t \right\} = P^\varepsilon(t, X_t, Y_t, Z_t),$$

where we have used the Markov property of (X_t, Y_t, Z_t), and defined the pricing function $P^\varepsilon(t, x, y, z)$, the superscript ε denoting the dependence on the small parameter ε. Using the Feynman–Kac formula, $P^\varepsilon(t, x, y, z)$ satisfies the following partial differential equation and boundary condition:

$$\mathcal{L}^\varepsilon P^\varepsilon = 0, \tag{10.41}$$

$$P^\varepsilon(T, x, y, z) = h(x), \tag{10.42}$$

where

$$\mathcal{L}^{\varepsilon} = \frac{z}{\varepsilon}\mathcal{L}_0 + \frac{z}{\sqrt{\varepsilon}}\mathcal{L}_1 + \mathcal{L}_2, \tag{10.43}$$

$$\mathcal{L}_0 = v^2 \frac{\partial^2}{\partial y^2} + (m-y)\frac{\partial}{\partial y}, \tag{10.44}$$

$$\mathcal{L}_1 = \rho_{yz}\sigma v\sqrt{2}\frac{\partial^2}{\partial y\partial z} + \rho_{xy}v\sqrt{2}f(y)x\frac{\partial^2}{\partial x\partial y}, \tag{10.45}$$

$$\mathcal{L}_2 = \frac{\partial}{\partial t} + \frac{1}{2}f^2(y)zx^2\frac{\partial^2}{\partial x^2} + r\left(x\frac{\partial}{\partial x} - \cdot\right)$$
$$+ \frac{1}{2}\sigma^2 z\frac{\partial^2}{\partial z^2} + \kappa(\theta-z)\frac{\partial}{\partial z} + \rho_{xz}\sigma f(y)zx\frac{\partial^2}{\partial x\partial z}. \tag{10.46}$$

Note that $\mathcal{L}_0$ is the infinitesimal generator of an OU process with unit rate of mean-reversion, and $\mathcal{L}_2$ is the pricing operator of the Heston model volatility and correlation modulated by $f(y)$.

10.3.3 Asymptotic Analysis

For a general function f, there is no analytic solution to the terminal value problem (10.41)–(10.42). Thus, we proceed with an asymptotic analysis as developed in Chapter 4. Specifically, we perform a singular perturbation with respect to the small parameter ε, expanding our solution in powers of $\sqrt{\varepsilon}$:

$$P^{\varepsilon} = P_0 + \sqrt{\varepsilon}P_1 + \varepsilon P_2 + \cdots. \tag{10.47}$$

We then plug (10.47) and (10.43) into (10.41) and (10.42), and collect terms of equal powers of $\sqrt{\varepsilon}$.

By choosing $P_0 = P_0(t,x,z)$ and $P_1 = P_1(t,x,z)$ independent of y, the terms of order $1/\varepsilon$ and $1/\sqrt{\varepsilon}$ vanish. Matching terms of order 1 leads to the Poisson equation

$$z\mathcal{L}_0 P_2 + \mathcal{L}_2 P_0 = 0, \tag{10.48}$$

which requires the solvability condition

$$\langle \mathcal{L}_2 P_0 \rangle = \langle \mathcal{L}_2 \rangle P_0 = 0, \tag{10.49}$$

where the brackets stand for averaging with respect to the invariant distribution $\mathcal{N}(m, v^2)$ associated with the infinitesimal generator $\mathcal{L}_0$. The solution $P_0(t,x,z)$ of (10.49) must satisfy the terminal condition $P_0(T,x,z) = h(x)$. From (10.46) we deduce

$$\langle \mathcal{L}_2 \rangle = \frac{\partial}{\partial t} + \frac{1}{2} \langle f^2 \rangle zx^2 \frac{\partial^2}{\partial x^2} + r \left(x \frac{\partial}{\partial x} - \cdot \right)$$

$$+ \frac{1}{2} \sigma^2 z \frac{\partial^2}{\partial z^2} + \kappa(\theta - z) \frac{\partial}{\partial z} + \rho_{xz} \langle f \rangle \sigma zx \frac{\partial^2}{\partial x \partial z}, \qquad (10.50)$$

which is nothing else than the Heston operator $\mathcal{L}_H$ defined in (10.4) when we impose the normalization condition $\langle f^2 \rangle = 1$ and set $\rho = \rho_{xz} \langle f \rangle$ as an effective correlation coefficient. Therefore, $P_0 = P_H$ is given semi-analytically by (10.18).

Using equation (10.48) and the centering condition (10.49), we deduce:

$$P_2 = -\frac{1}{z} \mathcal{L}_0^{-1} (\mathcal{L}_2 - \langle \mathcal{L}_2 \rangle) P_0. \qquad (10.51)$$

Canceling the terms of order $\sqrt{\varepsilon}$ leads to the following Poisson equation in P_3:

$$z \mathcal{L}_0 P_3 + z \mathcal{L}_1 P_2 + \mathcal{L}_2 P_1 = 0, \qquad (10.52)$$

which requires the solvability condition

$$\langle \mathcal{L}_2 \rangle P_1 = -z \langle \mathcal{L}_1 P_2 \rangle, \qquad (10.53)$$

defining P_1 as the solution of a Heston partial differential equation with source and zero terminal condition $P_1(t,x,z) = 0$.

Plugging (10.51) into (10.53), using the definition (10.45) of $\mathcal{L}_1$, and introducing the Poisson equations

$$\mathcal{L}_0 \phi = \frac{1}{2} (f^2 - \langle f^2 \rangle), \qquad (10.54)$$

$$\mathcal{L}_0 \psi = f - \langle f \rangle, \qquad (10.55)$$

one easily deduces

$$\langle \mathcal{L}_2 \rangle (\sqrt{\varepsilon} P_1) = \mathcal{A}^\varepsilon P_0, \qquad (10.56)$$

$$P_1(T,x,z) = 0, \qquad (10.57)$$

where here

$$\mathcal{A}^\varepsilon = H_1^\varepsilon zx^2 \frac{\partial^3}{\partial z \partial x^2} + H_2^\varepsilon zx \frac{\partial^3}{\partial z^2 \partial x}$$

$$+ H_3^\varepsilon zx \frac{\partial}{\partial x} \left(x^2 \frac{\partial^2}{\partial x^2} \right) + H_4^\varepsilon z \frac{\partial}{\partial z} \left(x \frac{\partial}{\partial x} \right)^2, \qquad (10.58)$$

$$H_1^\varepsilon = \rho_{yz} \sigma v \sqrt{2} \langle \phi' \rangle \sqrt{\varepsilon}, \qquad (10.59)$$

$$H_2^\varepsilon = \rho_{xz}\rho_{yz}\sigma^2 v v \sqrt{2}\langle \psi' \rangle \sqrt{\varepsilon}, \tag{10.60}$$

$$H_3^\varepsilon = \rho_{xy}v v \sqrt{2}\langle f\phi' \rangle \sqrt{\varepsilon}, \tag{10.61}$$

$$H_4^\varepsilon = \rho_{xy}\rho_{xz}\sigma v v \sqrt{2}\langle f\psi' \rangle \sqrt{\varepsilon}. \tag{10.62}$$

What is interesting here is that a careful analysis in the Fourier domain of the problem (10.56)–(10.57) reveals that the correction $\sqrt{\varepsilon}P_1(t,x,z)$ can also be computed semi-analytically using the following formulas (we refer to Fouque and Lorig (2009) for the details of the derivation):

$$\sqrt{\varepsilon}P_1(t,x,z)$$
$$= \frac{e^{-r\tau}}{2\pi}\int_{\mathbb{R}} e^{-ikq}\left(\kappa\theta\widehat{f_0}(\tau,k) + z\widehat{f_1}(\tau,k)\right)\widehat{G}(\tau,k,z)\widehat{h}(k)dk, \tag{10.63}$$

where $\widehat{G}(\tau,k,z)$ is given by (10.11), and $\widehat{f_0}$ and $\widehat{f_1}$ are computed as follows:

$$\widehat{f_0}(\tau,k) = \int_0^\tau \widehat{f_1}(t,k)dt, \tag{10.64}$$

$$\widehat{f_1}(\tau,k) = \int_0^\tau b(s,k)e^{A(\tau,k,s)}ds, \tag{10.65}$$

$$A(\tau,k,s) = (\kappa + \rho\sigma ik + d(k))\frac{1-g(k)}{d(k)g(k)}\log\left(\frac{g(k)e^{\tau d(k)} - 1}{g(k)e^{sd(k)} - 1}\right)$$
$$+d(k)(\tau - s), \tag{10.66}$$

$$d(k) = \sqrt{\sigma^2(k^2 - ik) + (\kappa + \rho ik\sigma)^2},$$

$$g(k) = \frac{\kappa + \rho ik\sigma + d(k)}{\kappa + \rho ik\sigma - d(k)},$$

$$b(\tau,k) = -[H_1^\varepsilon D(\tau,k)\left(-k^2 + ik\right) + H_2^\varepsilon D^2(\tau,k)\left(-ik\right)$$
$$+ H_3^\varepsilon\left(ik^3 + k^2\right) + H_4^\varepsilon D(\tau,k)\left(-k^2\right)], \tag{10.67}$$

where $D(\tau,k)$ is given by (10.15), and the four group parameters $(H_j^\varepsilon, j = 1,\ldots,4)$ given by (10.59)–(10.62) can be calibrated from observed implied volatility skews.

10.3.4 The Multiscale Implied Volatility Surface

In this subsection, we show how the implied volatility surface produced by our multiscale model compares to that produced by the Heston model. Let us recall that the price P^ε of a European option in the multiscale model (10.35)–(10.37) is approximated by

$$P^\varepsilon \sim P_H + \sqrt{\varepsilon}P_1, \tag{10.68}$$

where the Heston price P_H is given by (10.18), and the correction $\sqrt{\varepsilon}P_1$ is given by (10.63).

Interestingly, the accuracy of this approximation, even in the smooth case (smooth payoff $h(x)$), is not the same as the one obtained in Chapter 4, Theorem 4.10. This is due to the fact that the moments of Y given by (10.37) are not bounded uniformly with respect to ε. Instead, as proved in Appendix C of Fouque and Lorig (2009), for every $\alpha < 1$, the first moment for instance can be bounded by $C\varepsilon^{\alpha-1}$ for some constant C. Therefore, following the proof of accuracy given in Fouque and Lorig (2009), one obtains an error of order ε^α (instead of ε in the case of a singular perturbation around Black–Scholes).

It is important to note that, although adding a fast mean-reverting factor of volatility on top of the Heston model introduces five new parameters (v, m, ε, ρ_{xy}, ρ_{yz}) plus an unknown function f to the dynamics of the stock (see (10.35) and (10.37)), neither knowledge of the values of these five parameters, nor the specific form of the function f, is required to price options using our approximation. The effect of adding a fast mean-reverting factor of volatility on top of the Heston model is *entirely* captured by the four group parameters H_j^ε, which are constants that can be obtained by calibrating the multiscale model to option prices on the market.

By setting $H_j^\varepsilon = 0$ for $j = 1,\ldots,4$, we see that $P_1 = 0$, $P^\varepsilon = P_H$, and the resulting implied volatility surface, obtained by inverting the Black–Scholes formula, corresponds to the implied volatility surface produced by the Heston model. If we then vary a single H_j^ε while holding $H_l^\varepsilon = 0$ for $l \neq j$, we can see exactly how the multiscale implied volatility surface changes as a function of each of the H_j^ε. The results of this procedure are plotted in Figure 10.4.

Because they are on the order of $\sqrt{\varepsilon}$, typical values of the H_j^ε are quite small. However, in order to highlight their effect on the implied volatility surface, the range of values plotted for the H_j^ε in Figure 10.4 was intentionally chosen to be large. It is clear from Figure 10.4 that each H_j^ε has a distinct effect on the implied volatility surface. Thus, the multiscale model provides considerable flexibility when it comes to calibrating the model to the implied volatily surface produced by options on the market.

10.4 Large Deviations and Short Maturity Asymptotics

Large deviations theory provides a natural framework for approximating the exponentially small probabilities associated with the behavior of a diffusion process over a small time interval (we refer to Dembo and Zeitouni (1998)

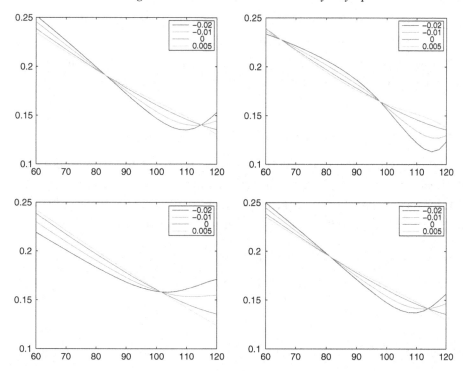

Figure 10.4 Implied volatility curves plotted as a function of the strike price for European calls in the multiscale model. In this example, the initial stock price is $x = 100$. The Heston parameters are set to $z = 0.04$, $m = 0.024$, $\kappa = 3.4$, $\sigma = 0.39$, $\rho_{xz} = -0.64$, and $r = 0.0$. In the upper-left figure we vary only H_1^ε, fixing $H_j^\varepsilon = 0$ for $j \neq 1$. Likewise, in the upper-right, lower-left, and lower-right figures, we vary only H_2^ε, H_3^ε, and H_4^ε, respectively, fixing all other $H_j^\varepsilon = 0$. In all the four plots, $H_j^\varepsilon = 0$ corresponds to the implied volatility curve of the Heston model.

for general background). In the context of financial mathematics, large deviations theory arises in the computation of small-maturity, out-of-the-money call or put option prices, or the probability of reaching a default level in a small time period. The theory of large deviations has been applied to local and stochastic volatility models (see references in the Notes), and has given very interesting results on the behavior of implied volatilities near maturity. In the context of stochastic volatility models, the rate function involved in the large deviation estimates is given in terms of a distance function, which in general cannot be calculated in closed-form.

In this section, we consider the particular case of the Heston model in the regime in which the maturity is small but large compared to the mean-reversion time of the stochastic volatility factor. This is a realistic situation where, for instance, the maturity is one month and the volatility mean-reversion time is on the order of a few days. We present the results obtained in Feng *et al.* (2009), which provides an explicit formula for the asymptotic smile/skew. The analysis is based on a precise study of the moment-generating function and its asymptotic.

10.4.1 Fast Mean-Reversion and Short Maturity Scaling

In what follows, ε will denote a typical short maturity and we will consider the Heston model (10.1) when the rate of mean-reversion κ is of order ε^2. In order to achieve this scaling, we replace κ by κ/ε^2, replace σ by v/ε, and finally rescale time: $t \mapsto \varepsilon t$ to obtain in the distribution

$$dX_t = \varepsilon r X_t dt + \sqrt{\varepsilon Z_t}\, X_t dW_t^x,$$
$$dZ_t = \frac{\kappa}{\varepsilon}(m - Z_t)\,dt + \frac{v}{\sqrt{\varepsilon}}\sqrt{Z_t}\,dW_t^z, \qquad (10.69)$$
$$d\langle W^x, W^z\rangle_t = \rho dt.$$

Note that the log-price is given by

$$\log X_t =: R_t^\varepsilon = \log x + \varepsilon r t - \frac{\varepsilon}{2}\int_0^t Z_s^2 ds + \sqrt{\varepsilon}\int_0^t \sqrt{Z_s}dW_s^x, \quad (10.70)$$

where we show explicitly the dependence in the small parameter ε.

10.4.2 Moment-Generating Function and its Asymptotic

The analysis of this particular model relies on an explicit calculation of a moment-generating function, and evaluating its limit. First we define the rescaled moment-generating function of X:

$$\Lambda_\varepsilon(p) = \Lambda_\varepsilon(p;x,z,t) = \varepsilon \log \mathbb{E}^\star\left\{X_t^{\frac{p}{\varepsilon}}|X_0 = x, Z_0 = z\right\}$$
$$= \varepsilon \log \mathbb{E}^\star\left\{e^{\frac{p}{\varepsilon}R_t^\varepsilon}|R_0^\varepsilon = \log x, Z_0 = z\right\}. \quad (10.71)$$

Using (10.70), introducing independent Brownian motions, and using Girsanov's theorem leads to

$$\mathbb{E}^\star\left\{X_t^{\frac{p}{\varepsilon}}|X_0 = x, Z_0 = z\right\} = e^{prt}x^{\frac{p}{\varepsilon}}\mathbb{E}^Q\left\{e^{\frac{p(p-\varepsilon)}{2\varepsilon}\int_0^t \tilde{Z}_s ds}|\tilde{Z}_0 = z\right\}, \quad (10.72)$$

where, under the measure Q, the process $\tilde{Z}$ satisfies the equation

$$d\tilde{Z}_t = \frac{1}{\varepsilon}\left(\kappa m - (\kappa - v\rho p)\tilde{Z}_t\right)dt + \frac{v}{\sqrt{\varepsilon}}\sqrt{\tilde{Z}_t}dW_t^Q, \quad (10.73)$$

driven by a Brownian motion W^Q. The result (10.72)–(10.73) is given in Andersen and Piterbarg (2007), and, as observed in Feng *et al.* (2009), the proof allows "$+\infty = +\infty$" and therefore (10.72) holds for $p \in \mathbb{R}$.

The moments (10.72) can then be computed explicitly, as for instance in Hurd and Kuznetsov (2008). A careful analysis of the limit of $\Lambda_\varepsilon(p)$ given in Feng *et al.* (2009) leads to a *large deviation principle* for the family $\{X_t, \varepsilon > 0\}$ or equivalently $\{R_t^\varepsilon, \varepsilon > 0\}$, that is

$$\lim_{\varepsilon \to 0^+} \varepsilon \log \mathbb{P}^\star \{R_t^\varepsilon - \log x > q\} = -I(q; \log x, t),$$

for $q > 0$ and a *rate function* I given below, and with a similar limit for $q < 0$.

Theorem 10.1 *Assume $R_0^\varepsilon = \log x$. For each $t > 0$, the family $\{R_t^\varepsilon, \varepsilon > 0\}$ satisfies the large deviation principle with good rate function*

$$I(q; \log x, t) = \Lambda^*(q - \log x; 0, t),$$

where Λ^* is the Legendre transform of Λ

$$\Lambda^*(q; \log x, t) \equiv \sup_{p \in \mathbb{R}} \{qp - \Lambda(p; \log x, t)\},$$

and $\Lambda(p; \log x, t) : \mathbb{R} \times \mathbb{R} \times \mathbb{R}_+ \mapsto \mathbb{R} \cup \{+\infty\}$ is given explicitly by:

$$\Lambda(p; \log x, t) = p \log x + \frac{\kappa m t}{v^2} \left((\kappa - v\rho p) - \sqrt{(\kappa - \rho v p)^2 - v^2 p^2} \right) \quad (10.74)$$

$$\text{for } -\frac{\kappa}{v(1-\rho)} \le p \le \frac{\kappa}{v(1+\rho)},$$

$$= +\infty \quad \text{otherwise.}$$

The function $\Lambda(p)$, and the rate function $\Lambda^*(q)$ given below, are plotted in Figure 10.5 in the three cases, $\rho > 0$, $\rho = 0$, and $\rho < 0$.

Lemma 10.2 *The rate function Λ^* is given explicitly by*

$$\Lambda^*(q; 0, t) = qp(q; t) - \Lambda(p(q; t); 0, t),$$

where $p(q; t)$ is defined by

$$p(q; t) = \frac{\kappa}{v(1 - \rho^2)} \left(-\rho + \frac{qv + \kappa m t \rho}{\sqrt{(qv + \kappa m t \rho)^2 + (1 - \rho^2)\kappa^2 m^2 t^2}} \right) \quad (10.75)$$

$$\in \text{int}(\text{Dom}(\Lambda)) = \left(-\frac{\kappa}{v(1-\rho)}, \frac{\kappa}{v(1+\rho)} \right) :$$

$\Lambda^*(q; 0, t)$ *is finite for all $q \in \mathbb{R}$; it is strictly increasing for $q > 0$ and strictly decreasing for $q < 0$; it is continuous in $(q, t) \in \mathbb{R} \times \mathbb{R}_+$; and $\Lambda^*(0; 0, t) = 0$.*

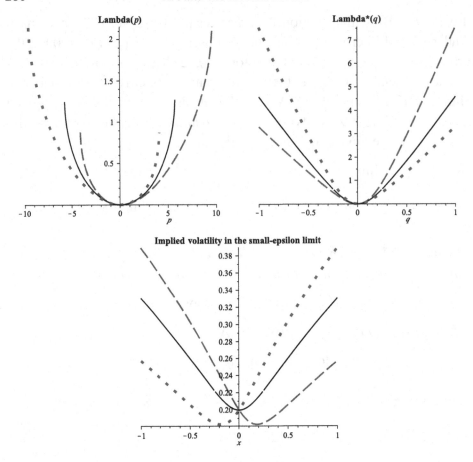

Figure 10.5 Here we have plotted Λ, Λ^*, and the implied volatility in the small-ε limit as a function of the log-moneyness. The parameters are $t = 1$, *ergodic mean* $m = 0.04$, *convexity* $v/\kappa = 1.74$ ($\kappa = 1.15$, $v = 0.2$), and *skew* $\rho = -0.4$ (dashed), $\rho = 0$ (solid), $\rho = +0.4$ (dotted).

Remarks

(i) Note that $\Lambda^*(q; \log x, t) = \Lambda^*(q - \log x; 0, t)$, since the only $\log x$ dependence in Λ is the linear term $p \log x$.

(ii) Note also the scaling property $\Lambda(p; \log x, t) = t \Lambda(p; \frac{\log x}{t}, 1)$.

(iii) In this asymptotic regime, the limiting quantities Λ and Λ^* do not depend on the starting level of volatility $\sqrt{z}$, and they depend on the κ (mean-reversion rate) and v (volatility-of-volatility) parameters only through their ratio v/κ.

(iv) The previous remark will also apply to asymptotic option prices and implied volatilities described below. In this regime, therefore, the relevant features of the Heston model are captured by just three parameters: the ergodic mean m, the correlation ρ, and the ratio v/κ. They control, respectively, the implied volatility skew's *level, slope,* and *convexity.*

We refer to Feng *et al.* (2009) for the proofs of this result and the consequences below.

10.4.3 Applications to Pricing and Implied Volatilities

A practical application of this result is the following rare event estimate for pricing out-of-the-money options of small maturity.

Corollary 10.3 *Suppose that log-moneyness is positive,* $\log(\frac{K}{x}) > 0$, *and* $t > 0$ *fixed. Then*

$$\lim_{\varepsilon \to 0^+} \varepsilon \log \mathbb{E}^\star \left\{ e^{-\varepsilon rt} (X_t - K)^+ | X_0 = x, Z_0 = z \right\} = -\Lambda^* \left(\log \left(\frac{K}{x} \right); 0, t \right),$$

independently of the initial square volatility level z. Note that the maturity of the option is $T = \varepsilon t$, *which goes to zero in the limit. The discounting factor* $e^{-\varepsilon rt}$ *plays no role in this asymptotic result.*

Moreover, the asymptotic implied volatility can be computed. Let $\sigma_\varepsilon(t, L)$ denote the Black–Scholes implied volatility for the European call option with strike price K, out-of-the-money so that $L = \log(K/x) > 0$, with short maturity $T = \varepsilon t$ for $t > 0$ fixed, and computed under the dynamics given by (10.69). Then we have:

Corollary 10.4

$$\lim_{\varepsilon \to 0^+} \sigma_\varepsilon^2(t, L) = \frac{L^2}{2\Lambda^*(L; 0, t)t}, \quad L = \log \left(\frac{K}{x} \right) > 0.$$

Similarly, by considering out-of-the-money put options, one obtains the same formula for $L < 0$. The at-the-money volatility is obtained by taking the limit $L \to 0$, which gives $\sqrt{m}$ coinciding with the effective volatility found in (10.26).

In fact, the results in Corollaries 10.3 and 10.4 hold for any fast mean-reverting stochastic volatility model (other than Heston's) which satisfies a large deviation principle, as in Theorem 10.1, provided the asymptotic rate function satisfies: $\Lambda^*(q; 0, t)$ *is finite for all* $q \in \mathbb{R}$; *it is strictly increasing for* $q > 0$ *and strictly decreasing for* $q < 0$; *and* $\Lambda^*(0; 0, t) = 0$.

In Figure 10.5 we show plots of the functions Λ and Λ^*, and of the implied volatility smile/skew obtained in the limit $\varepsilon \to 0^+$.

Notes

References for computational issues with the Heston model are given in the text in Section 10.1. The approximation results given in Section 10.2 are direct applications of the approach developed in the previous chapters. The approximation presented in Section 10.3 is based on Fouque and Lorig (2009). Multidimensional Heston models based on Wishart processes are studied in Benabid *et al.* (2009) using perturbation methods.

The large deviation theory has been used for local and stochastic volatility models in Zhu and Avellaneda (1997), Avellaneda *et al.* (2002), Berestycki *et al.* (2004), and Henry-Labordère (2005). The results presented in Section 10.4 are based on Feng *et al.* (2009).

11

Other Applications

In Section 11.1, we present two variance-reduction techniques in the context of multiscale stochastic volatility models. In the same context, we apply in Section 11.2 the perturbation method to Merton's problem of finding an optimal portfolio allocation using power utility functions. We present in Section 11.3 an application to *forward-looking* estimation of stock *betas* using skews of implied volatilities.

11.1 Application to Variance Reduction in Monte Carlo Computations

Monte Carlo methods are natural and essential tools in computational finance. Examples include pricing and hedging financial instruments with complex structure or high dimensionality (Glasserman, 2003). Variance-reduction techniques play a crucial role in making Monte Carlo simulations practical in terms of computational time. There are many different such techniques, and here we concentrate on two of them: importance sampling and control variate, applied to simulations of multiscale stochastic volatility dynamics presented in this book. The objective is to sample from the full stochastic volatility model using the approximations derived in the previous chapters as tools to speed up the convergence of the Monte Carlo estimates. We give a brief description of the methods, and refer to the two papers of Fouque and Han (2004b, 2007) for more details and numerical illustrations.

11.1.1 Importance Sampling

Our starting point is the stochastic volatility dynamics given under a risk-neutral pricing measure $\mathbb{P}^\star$ by (4.1), and the quantity of interest is the price of a European option given by the expectation (4.2). In this application, we suppose that all the parameters are known. To simplify the notation, we write the stochastic volatility model in (4.1) in vector form as follows:

$$dV_t = b(t, V_t)dt + a(t, V_t)d\eta_t, \tag{11.1}$$

where we set

$$v = \begin{pmatrix} x \\ y \\ z \end{pmatrix}, \qquad V_t = \begin{pmatrix} X_t \\ Y_t \\ Z_t \end{pmatrix}, \qquad \eta_t = \begin{pmatrix} \eta_t^{(0)} \\ \eta_t^{(1)} \\ \eta_t^{(2)} \end{pmatrix},$$

with $(\eta^{(0)}, \eta^{(1)}, \eta^{(2)})$ being independent standard Brownian motions, and where we define the drift

$$b(t, v) = \begin{pmatrix} rx \\ \frac{1}{\varepsilon}\alpha(y) - \frac{1}{\sqrt{\varepsilon}}\beta(y)\Lambda_1(y, z) \\ \delta c(z) - \sqrt{\delta}\, g(z)\Lambda_2(y, z) \end{pmatrix},$$

and the diffusion matrix

$$a(t, v) = \begin{pmatrix} f(y, z)x & 0 & 0 \\ \frac{1}{\sqrt{\varepsilon}}\beta(y)\rho_1 & \frac{1}{\sqrt{\varepsilon}}\beta(y)\sqrt{1 - \rho_1^2} & 0 \\ \sqrt{\delta}\, g(z)\rho_2 & \sqrt{\delta}\, g(z)\rho_{12} & \sqrt{\delta}\, g(z)\sqrt{1 - \rho_2^2 - \rho_{12}^2} \end{pmatrix}.$$

The price $P(t, x, y, z)$ of a European option with maturity T, payoff function $H(x)$, and computed at time $t < T$ is given by

$$P(t, v) = \mathbb{E}^\star \left\{ e^{-r(T-t)} H(X_T) \mid V_t = v \right\}. \tag{11.2}$$

A basic Monte Carlo approximation for the price (11.2) is based on calculating the sample mean

$$P(t, x, y, z) \approx \frac{1}{N} \sum_{k=1}^{N} e^{-r(T-t)} H(X_T^{(k)}), \tag{11.3}$$

where N is the total number of independent realizations of the process, and $X_T^{(k)}$ denotes the terminal value of the stock price in the kth trajectory. Importance sampling techniques consist of changing the weights of these realizations in order to reduce the variance of the estimator (11.3).

Under classical integrability conditions on the function $h(t, v)$, we introduce the process

$$Q_t = \exp\left(\int_0^t h(s, V_s)d\eta_s + \frac{1}{2}\int_0^t \|h(s, V_s)\|^2 ds \right)$$

so that the Radon–Nikodym derivative

$$\frac{d\tilde{\mathbb{P}}}{d\mathbb{P}^\star} = (Q_T)^{-1}$$

defines a new probability $\tilde{\mathbb{P}}$ equivalent to $\mathbb{P}^\star$. By Girsanov's theorem, under this new measure $\tilde{\mathbb{P}}$, the process $(\tilde{\eta}_t)$ defined by

$$\tilde{\eta}_t = \eta_t + \int_0^t h(s, V_s) ds$$

is a standard Brownian motion. The option price P at time $t = 0$ can be written under $\tilde{\mathbb{P}}$ as

$$P(0, v) = \tilde{\mathbb{E}}\left\{ e^{-rT} H(X_T) Q_T \mid V_0 = v \right\}, \qquad (11.4)$$

where

$$Q_T = \exp\left(\int_0^T h(s, V_s) d\tilde{\eta}_s - \frac{1}{2} \int_0^T ||h(s, V_s)||^2 ds \right), \qquad (11.5)$$

and the dynamics of our model becomes

$$dV_t = (b(t, V_t) - a(t, V_t) h(t, V_t)) dt + a(t, V_t) d\tilde{\eta}_t. \qquad (11.6)$$

Applying Itô's formula to $P(t, V_t) Q_t$ is a straightforward computation which gives

$$H(V_T) Q_T = P(0, v) + \int_0^T Q_s \left(a' \nabla P + P h \right) (s, V_s) \cdot d\tilde{\eta}_s,$$

where a' denotes the transpose of a and the gradient is with respect to the variable v. Therefore, the variance of the payoff $H(V_T) Q_T$ in (11.4) is simply

$$\mathrm{Var}_{\tilde{\mathbb{P}}}(H(V_T) Q_T) = \tilde{\mathbb{E}}\left\{ \int_0^T Q_s^2 ||a' \nabla P + P h||^2 ds \right\}.$$

Indeed, if the quantity P to be computed was known, one could obtain a zero variance by choosing

$$h = -\frac{1}{P} \left(a' \nabla P \right). \qquad (11.7)$$

Our strategy is to use in (11.7), known approximations to the exact value P. Then, the Monte Carlo simulations are done under the new measure $\tilde{\mathbb{P}}$:

$$P(t, x, y, z) \approx \frac{1}{N} \sum_{k=1}^{N} e^{-r(T-t)} H(X_T^{(k)}) Q_T^{(k)}, \qquad (11.8)$$

where N is the total number of simulations, and $X_T^{(k)}$ and $Q_T^{(k)}$ denote the final value of the kth realized trajectory (11.6) and weight (11.5), respectively.

As an approximation to the exact value P, we use the leading-order Black–Scholes price $P_0(t, x, z) = P_{BS}(t, x; \bar{\sigma}(z))$ given in Definition 4.3.

Numerical simulations presented in Fouque and Han (2004b) show a variance reduction by a factor of ten in the regime where $\varepsilon = 0.05$ and $\delta = 0.1$. Indeed, the results are even better when ε and δ become smaller, and as expected they are not as spectacular when these parameters are of order 1. It is also shown that the introduction of the first correction, that is using the approximation $P_0 + P_1$ in (4.67), significantly further improves the variance reduction. The case of Asian options is also considered in Fouque and Han (2004b).

11.1.2 Control Variates

We briefly present a control variate technique proposed in Fouque and Han (2007) and we refer to this paper for more details, numerical results, and applications to path-dependent contracts.

Our starting point is again the stochastic volatility dynamics given under a risk-neutral pricing measure $\mathbb{P}^\star$ by (4.1), and the quantity of interest is the price of a European option given by the expectation (4.2).

The key (and obvious) observation is the following martingale representation of the option price:

$$P^{\varepsilon,\delta}(0,X_0,Y_0,Z_0) = e^{-rT}H(X_T) - M_0(P^{\varepsilon,\delta}) - \frac{1}{\sqrt{\varepsilon}}M_1(P^{\varepsilon,\delta}) - \sqrt{\delta}M_2(P^{\varepsilon,\delta}),$$

$$(11.9)$$

where the centered martingales (M_0, M_1, M_2) are defined by

$$M_0(P^{\varepsilon,\delta}) = \int_0^T e^{-rs}\frac{\partial P^{\varepsilon,\delta}}{\partial x}(s,X_s,Y_s,Z_s)f(Y_s,Z_s)X_s dW_s^{(0)*}, \quad (11.10)$$

$$M_1(P^{\varepsilon,\delta}) = \int_0^T e^{-rs}\frac{\partial P^{\varepsilon,\delta}}{\partial y}(s,X_s,Y_s,Z_s)\beta(Y_s)dW_s^{(1)*}, \qquad (11.11)$$

$$M_2(P^{\varepsilon,\delta}) = \int_0^T e^{-rs}\frac{\partial P^{\varepsilon,\delta}}{\partial z}(s,X_s,Y_s,Z_s)g(Z_s)dW_s^{(2)*}. \qquad (11.12)$$

These martingales play the role of "perfect" control variates for Monte Carlo simulations and their integrands would be the perfect Delta hedges if $P^{\varepsilon,\delta}$ were known and volatility factors traded. Unfortunately, neither the option price $P^{\varepsilon,\delta}(s,X_s,Y_s,Z_s)$ nor its gradient at any time $0 \le s \le T$ are in any analytic form, even though all the parameters of the model have been calibrated as we suppose here.

One can choose an approximate option price to substitute $P^{\varepsilon,\delta}$ used in the martingales (11.10), (11.11), (11.12) and still retain martingale properties. When time scales ε and $1/\delta$ are well separated, namely

$0 < \varepsilon \ll 1 \ll 1/\delta$, we can use, as an approximation, the Black–Scholes price $P_0(t,x,z) = P_{BS}(t,x;\bar{\sigma}(z))$ given in Definition 4.3. A martingale control variate estimator is formulated as

$$\frac{1}{N}\sum_{i=k}^{N}\left[e^{-rT}H(X_T^{(k)}) - M_0^{(k)}(P_{BS}) - \sqrt{\delta}M_2^{(k)}(P_{BS})\right]. \quad (11.13)$$

Since the control variates $\mathcal{M}_0$ and $\mathcal{M}_2$ are martingales, we simply call them *martingale control variates*. Note that there is no M_1 martingale term since the approximation P_{BS} does not depend on y and the y-derivative vanishes in (11.11) with $P^{\varepsilon,\delta}$ replaced by P_{BS}.

We refer to Fouque and Han (2007) for a detailed analysis of the variance of this estimator in the European case, comparison with important sampling presented in the previous section, and applications to barrier and American options.

11.2 Portfolio Optimization under Stochastic Volatility

We look at the optimal asset allocation problem facing an investor who has to decide how much of his or her wealth to invest in a risky asset (X_t) and how much in the riskless bond to maximize expected utility of wealth at the final time T. First, we shall briefly review the case of constant volatility, when the optimal strategy is known as the Merton solution. Then we look at the same problem when the price of the risky asset has randomly changing fast mean-reverting volatility. The asymptotic analysis reveals a perturbation around a constant volatility Merton problem, with an effective volatility different from $\bar{\sigma}$ estimated for pricing problems. However, this solution requires us to track the unobservable volatility and we present a suboptimal, but more practical solution in Section 11.2.3. Indeed, a slow volatility factor can be incorporated in the model, but, for simplicity, we consider here only the fast factor.

11.2.1 Constant Volatility Merton Problem

We suppose the lognormal model

$$dX_t = \mu X_t dt + \sigma X_t dW_t$$

for the stock price (X_t) and a constant interest rate r. An investor has wealth $\mathcal{W}_t$ at time t made of a_t stocks and b_t bonds

$$\mathcal{W}_t = a_t X_t + b_t e^{rt}.$$

He trades in a self-financing manner so that his wealth process satisfies

$$d\mathscr{W}_t = a_t dX_t + b_t d(e^{rt}).$$

Denoting by u_t the fraction of wealth in the stock and $1 - u_t$ in the bond at time t, we have $a_t = u_t \mathscr{W}_t / X_t$ and $b_t = (1 - u_t)\mathscr{W}_t / e^{rt}$, so that

$$d\mathscr{W}_t = u_t \mathscr{W}_t \frac{dX_t}{X_t} + (1 - u_t)r\mathscr{W}_t dt.$$

Using the geometric Brownian structure of the stock price, $(\mathscr{W}_t)$ satisfies the equation

$$d\mathscr{W}_t = \mathscr{W}_t[(r + (\mu - r)u_t)dt + \sigma u_t dW_t],$$

with initial wealth $\mathscr{W}_0 = w$. Notice that this is a closed equation because X does not appear, and in particular this implies that $(\mathscr{W}_t)$ is a Markov process by itself if the control u_t is chosen through a Markov control policy, that is, u_t is a function of $(t, \mathscr{W}_t)$.

The goal of the investor is to choose the strategy (u_t) to maximize his expected utility at some given finite terminal time T

$$\mathbb{E}\{U(\mathscr{W}_T)\},$$

where we shall restrict ourselves to the example of a so-called HARA (hyperbolic absolute risk averse) utility function

$$U(w) = w^p/p$$

for some $0 < p < 1$. The expectation is with respect to the subjective probability. In the spirit of dynamic programming, we define

$$V(t, w) = \sup_u \mathbb{E}\left\{\frac{\mathscr{W}_T^p}{p} \,\middle|\, \mathscr{W}_t = w\right\},$$

the maximum expected utility as a function of the starting time t and starting wealth w.

By the Bellman principle, for which we give references in the Notes, $V(t, x)$ satisfies the nonlinear Hamilton–Jacobi–Bellman (HJB) partial differential equation

$$\frac{\partial V}{\partial t} + \sup_u \left\{\frac{1}{2}\sigma^2 u^2 w^2 \frac{\partial^2 V}{\partial w^2} + (r + (\mu - r)u)w\frac{\partial V}{\partial w}\right\} = 0,$$

with terminal condition $V(T, w) = w^p/p$ and where, in this Markovian framework, the *control* turns out to be $u = u(t, w)$, a function of time and wealth. When the HJB equation has a solution to which Itô's formula can be applied, the verification that it is the optimal expected utility follows readily.

The special form of the chosen utility function motivates the transformation

$$V(t,w) = \frac{w^p}{p} v(t).$$

This leads to the linear ordinary differential equation

$$\frac{dv}{dt} = pv \sup_u \left\{ \frac{1}{2} \sigma^2 u^2 (p-1) + (\mu - r)u + r \right\}, \tag{11.14}$$

with $v(T) = 1$. The supremum is attained at

$$u^\star = \frac{\mu - r}{\sigma^2 (1-p)}. \tag{11.15}$$

Notice that $u^\star$ is simply a constant which is positive when $\mu > r$. This implies that the optimal strategy does not depend on the wealth, which allows *mutual funds* where the portfolio allocation is independent of the wealth of the participant. Moreover, the fraction invested in the stock is constant in time and the portfolio needs to be continuously rebalanced.

The corresponding maximum expected utility is given by

$$V(t,w) = \frac{w^p}{p} e^{\left(r + \frac{(\mu - r)^2}{2\sigma^2 (1-p)}\right) p(T-t)}. \tag{11.16}$$

11.2.2 Stochastic Volatility Merton Problem

We now tackle the same problem when the market model incorporates random volatility:

$$dX_t = \mu X_t dt + f(Y_t) X_t dW_t,$$

$$dY_t = \alpha(m - Y_t)dt + \beta(\rho dW_t + \sqrt{1 - \rho^2} dZ_t),$$

with a short time scale modeled as

$$\alpha = 1/\varepsilon \text{ and } \beta = \sqrt{2}v/\sqrt{\varepsilon}.$$

For clarity of exposition, we explicitly write the driving process (Y_t) as an OU process but, as in previous chapters, this is not essential to the asymptotic analysis.

Now, the wealth process $(\mathcal{W}_t)$ satisfies

$$d\mathcal{W}_t = \mathcal{W}_t \left[(r + (\mu - r)u_t)dt + u_t f(Y_t)dW_t\right],$$

where again u_t denotes the fraction of wealth in the stock. Assuming first that both wealth and volatility are observable, we are in a Markovian case where u_t will be of the form $u = u(t, w, y)$. This is an ideal but not realistic scenario since we know that in general volatility is not directly observable. The maximum expected HARA utility is now a function $V(t, w, y)$ defined by

$$V(t, w, y) = \sup_u \mathbb{E}\left\{\frac{\mathcal{W}_T^p}{p} \,\middle|\, \mathcal{W}_t = w, Y_t = y\right\}.$$

The HJB equation becomes

$$\frac{\partial V}{\partial t} + \sup_u \left\{\frac{1}{2}f(y)^2 u^2 w^2 \frac{\partial^2 V}{\partial w^2} + \beta\rho f(y)uw\frac{\partial^2 V}{\partial w \partial y} + (r + (\mu - r)u)w\frac{\partial V}{\partial w}\right\}$$
$$+ \frac{1}{\varepsilon}\mathcal{L}_0 V = 0,$$

with terminal condition $V(T, w, y) = w^p/p$ and where $\mathcal{L}_0$ is the infinitesimal generator of the normalized (rate of mean-reversion equal to one) OU process (Y_t).

Using the transformation $V(t, w, y) = \frac{w^p}{p}v(t, y)$, we obtain the following equation for v:

$$\frac{\partial v}{\partial t} + p\sup_u\left\{\frac{1}{2}(p-1)f(y)^2 u^2 v + u\left[\beta\rho f(y)\frac{\partial v}{\partial y} + (\mu - r)v\right] + rv\right\}$$
$$+ \frac{1}{\varepsilon}\mathcal{L}_0 v = 0, \tag{11.17}$$

with $v(T, y) = 1$. The supremum is attained at

$$u^\star(t, y) = \frac{\beta\rho f(y)\frac{\partial v}{\partial y} + (\mu - r)v}{(1-p)f(y)^2 v}, \tag{11.18}$$

and we can rewrite the equation for v as

$$\frac{\partial v}{\partial t} + p\frac{\left(\beta\rho f(y)\frac{\partial v}{\partial y} + (\mu - r)v\right)^2}{2(1-p)f(y)^2 v} + \frac{1}{\varepsilon}\mathcal{L}_0 v + rpv = 0.$$

As expected, the optimal strategy requires knowledge of y.

Rearranging terms in powers of ε, this equation becomes:

$$\frac{\partial v}{\partial t} + \frac{1}{\varepsilon}\left[\mathcal{L}_0 v + \frac{pv^2\rho^2}{(1-p)v}\left(\frac{\partial v}{\partial y}\right)^2\right] + \frac{\sqrt{2}\rho vp(\mu - r)}{\sqrt{\varepsilon}(1-p)f(y)}\left(\frac{\partial v}{\partial y}\right)$$
$$+ \frac{p(\mu - r)^2 + 2p(1-p)rf(y)^2}{2(1-p)f(y)^2}v = 0. \tag{11.19}$$

Expanding v as

$$v = v_0 + \sqrt{\varepsilon}v_1 + \varepsilon v_2 + \varepsilon^{3/2}v_3 + \cdots,$$

and comparing powers of ε, the term in $1/\varepsilon$ gives

$$\mathscr{L}_0 v_0 + \frac{pv^2\rho^2}{(1-p)v_0}\left(\frac{\partial v_0}{\partial y}\right)^2 = 0.$$

This is a differential equation in the y variable which reduces to

$$v^2\frac{v_0''}{v_0'} + (m-y) + \frac{pv^2\rho^2}{(1-p)}\frac{v_0'}{v_0} = 0,$$

where partial derivatives with respect to y are denoted by v_0' and v_0''. Integrating this gives

$$v_0^{1+c} = c_1(t)\int_0^y e^{(m-z)^2/2v^2}dz + c_2(t),$$

where $c = \rho^2 p/(1-p)$ is positive for our assumed $0 < p < 1$, and c_1 and c_2 are some functions of time. These solutions do not belong to any reasonable space where the HJB equation is well-posed, unless $c_1(t) = 0$ and consequently v_0 does not depend on y. In other words, we must have $v_0 = v_0(t)$.

Comparing $1/\sqrt{\varepsilon}$ terms in the equation gives

$$\mathscr{L}_0 v_1 = 0,$$

because v_0 does not depend on y. This implies that $v_1 = v_1(t)$ is independent of y as well. This is an important observation because it implies that the leading-order term in (11.18),

$$u^\star(t,y) = \frac{\frac{\sqrt{2}v}{\sqrt{\varepsilon}}\rho f(y)\frac{\partial}{\partial y}(v_0 + \sqrt{\varepsilon}v_1 + \cdots) + (\mu - r)(v_0 + \sqrt{\varepsilon}v_1 + \cdots)}{(1-p)f(y)^2(v_0 + \sqrt{\varepsilon}v_1 + \cdots)},$$

is simply given by

$$u_0^\star(y) = \frac{\mu - r}{(1-p)f(y)^2}, \tag{11.20}$$

which is independent of time and is Merton's solution (11.15) where σ^2 is replaced by $f(y)^2$.

We can compute the leading term $v_0(t)$ of the maximum expected utility which does not depend on y, even though the optimal strategy does. Looking at the order-one terms in (11.19) gives

$$\frac{\partial v_0}{\partial t} + \mathscr{L}_0 v_2 + \left(\frac{(\mu - r)^2}{2(1-p)f(y)^2} + r\right)pv_0 = 0, \tag{11.21}$$

which is a Poisson equation for v_2 with respect to $\mathscr{L}_0$. Its centering condition gives

$$\frac{\partial v_0}{\partial t} + \left(\frac{(\mu - r)^2}{2(1-p)} \left\langle \frac{1}{f^2} \right\rangle + r \right) p v_0 = 0,$$

where as usual, $\langle \cdot \rangle$ denotes the integral with respect to the invariant density of the OU process Y.

Setting

$$\sigma^\star = \frac{1}{\sqrt{\langle 1/f^2 \rangle}}, \tag{11.22}$$

the leading-order maximum expected utility is exactly the Merton formula (11.16) with σ replaced by $\sigma^\star$:

$$V(t, w, y) = \frac{w^p}{p} e^{\left(r + \frac{(\mu - r)^2}{2(\sigma^\star)^2 (1-p)} \right) p (T-t)} + \mathcal{O}(\sqrt{\varepsilon}).$$

We have seen that v_1 is also independent of y, and it can be computed from the order-$\sqrt{\varepsilon}$ terms in (11.19), leading to

$$\mathscr{L}_0 v_3 + \frac{\partial v_1}{\partial t} + \left(\frac{(\mu - r)^2}{2(1-p) f(y)^2} + r \right) p v_1 + \frac{2 p v \rho (\mu - r)}{(1-p) f(y)} \frac{\partial v_2}{\partial y} = 0, \tag{11.23}$$

which is a Poisson equation in v_3. Note that from (11.21) we know that

$$v_2 = -\frac{(\mu - r)^2 p}{2(1-p)} (\phi^\star(y) + k(t)) v_0,$$

where $k(t)$ is some constant in y and $\phi^\star$ satisfies the Poisson equation

$$\mathscr{L}_0 \phi^\star = \frac{1}{f^2(y)} - \frac{1}{(\sigma^\star)^2}.$$

The centering condition for (11.23) gives the following ordinary differential equation for $v_1^\varepsilon = \sqrt{\varepsilon} v_1$:

$$\frac{\partial v_1^\varepsilon}{\partial t} + \left(\frac{(\mu - r)^2}{2(1-p)(\sigma^\star)^2} + r \right) p v_1^\varepsilon + \frac{p^2 (\mu - r)^3}{(1-p)^2} A^{\star, \varepsilon} v_0 = 0,$$

where we define

$$A^{\star, \varepsilon} = -\rho v \sqrt{\varepsilon} \langle (\phi^\star)' / f \rangle,$$

which describes the correcting effect due to stochastic volatility. With the terminal condition $v_1^\varepsilon(T) = 0$, this gives the corrected maximum expected utility

$$V(t,w,y) = \frac{w^p}{p}\left(1 + \frac{p^2(\mu-r)^3 A^{\star,\varepsilon}}{(1-p)^2}(T-t)\right)e^{\left(r+\frac{(\mu-r)^2}{2(\sigma^\star)^2(1-p)}\right)p(T-t)} + \mathscr{O}(\varepsilon).$$

Observe that in this ideal scenario the analysis identifies the market parameters $\sigma^\star$ and $A^{\star,\varepsilon}$ as important statistics of the volatility.

We also note that the transformation

$$v(t,y) = q^\Delta(t,y)$$

with

$$\Delta = \frac{1-p}{1-p+\rho^2 p}$$

leads to a *linear* parabolic partial differential equation for $q(t,y)$:

$$\frac{\partial q}{\partial t} + \frac{p(\mu-r)\beta\rho}{(1-p)f(y)}\left(\frac{\partial q}{\partial y}\right) + \frac{p(1-p+\rho^2 p)}{1-p}\left[r + \frac{(\mu-r)^2}{2f(y)^2(1-p)}\right]q$$
$$+ \frac{1}{\varepsilon}\mathscr{L}_0 q = 0.$$

Here Δ is known as the distortion power. The asymptotic analysis can alternatively be performed on this equation analogously to Chapter 4. We give a reference for this transformation in the Notes at the end of this chapter.

In practice, the volatility level, or equivalently Y_t, is not observable. The natural thing to do would be to filter Y_t from discrete observations of the price process X_t. This would lead to non-Markovian controls u depending on the past of X_t. This *filtering* problem is extremely complex, since Y_t appears in the "noise term" of the observation X_t. We will not pursue that route in this book and we give some references in the Notes. Instead, we look at an intermediate solution which consists of restricting our set of strategies to those for which the leading-order term does not depend on volatility.

11.2.3 A Practical Solution

We now go back to the HJB equation (11.17) and restrict ourselves to strategies u that do not depend on the unobserved y: $u = u(t)$. As a result, the supremum in equation (11.17) is not well-defined unless the quantity to be maximized does not depend on y.

Expanding v as

$$v = v_0 + \sqrt{\varepsilon}v_1 + \cdots,$$

and the control u as

$$u = u_0 + \sqrt{\varepsilon} u_1 + \cdots,$$

and collecting terms in (11.17), the $1/\varepsilon$ term implies

$$\mathcal{L}_0 v_0 = 0,$$

and so $v_0 = v_0(t)$. Now looking at the $1/\sqrt{\varepsilon}$ terms

$$\mathcal{L}_0 v_1 = 0,$$

because v_0 does not depend on y. We deduce that $v_1 = v_1(t)$ also.

The order-one terms in the equation give

$$\sup_{u_0} \left\{ \frac{\partial v_0}{\partial t} + \left(\frac{1}{2} p(p-1) f(y)^2 u_0^2 + p(r + (\mu - r)u_0) \right) v_0 + \mathcal{L}_0 v_2 \right\} = 0.$$

Since we are maximizing over u_0 independent of y, we have to choose v_2 so that the argument of the sup does not depend on y. If we define

$$A_{(f(y),u_0)} = \frac{1}{2} p(p-1) f(y)^2 u_0^2 + p(r + (\mu - r)u_0),$$

the only way that the argument

$$\frac{\partial v_0}{\partial t} + A_{(f(y),u_0)} v_0 + \mathcal{L}_0 v_2$$

can be independent of y is by choosing v_2 to be a solution of the Poisson equation

$$\mathcal{L}_0 v_2 + (A_{(f(y),u_0)} - \langle A_{(f(y),u_0)} \rangle) v_0 = 0.$$

Observe that in terms of the function $\phi(y)$, the solution of the Poisson equation

$$\mathcal{L}_0 \phi = f^2(y) - \langle f^2 \rangle$$

introduced in Section 4.2.2, v_2 is given by

$$v_2 = \frac{1}{2} p(1-p) u_0^2 v_0 \phi.$$

Consequently, the order-one equation becomes

$$\frac{\partial v_0}{\partial t} + \sup_{u_0} \left\{ \langle A_{(f(y),u_0)} \rangle) v_0 \right\} = 0,$$

with $v_0(T) = 1$, where

$$\langle A_{(f(y),u_0)} \rangle = A_{(\bar{\sigma},u_0)} = \frac{1}{2} p(p-1) \bar{\sigma}^2 u_0^2 + p(r + (\mu - r)u_0),$$

and $\bar{\sigma}^2 = \langle f^2 \rangle$, the historical square volatility.

Notice that this is exactly the equation (11.14) with volatility $\bar{\sigma}$, which leads to the Merton solution

$$u_0^\star = \frac{\mu - r}{\bar{\sigma}^2(1 - p)},$$

and the corresponding maximum expected utility

$$V(t,w,y) = \frac{w^p}{p} e^{\left(r + \frac{(\mu - r)^2}{2\bar{\sigma}^2(1-p)}\right)p(T-t)} + \mathcal{O}(\sqrt{\varepsilon}).$$

From the inequality

$$\langle f^2 \rangle \left\langle \frac{1}{f^2} \right\rangle \geq \left\langle f \times \frac{1}{f} \right\rangle^2 = 1,$$

it follows that $\bar{\sigma} \geq \sigma^\star$, where $\sigma^\star$ was defined in (11.22). In other words, as expected, the leading-order maximum expected utility would be greater if we could observe the volatility.

We now compute the correction v_1 by collecting the $\sqrt{\varepsilon}$ terms in (11.17). By the same argument, we can choose v_3 to make the quantity that is to be maximized at this order independent of y. Following steps analogous to the computation of v_1 in the previous section, leads to an equation for the correction $v_1^\varepsilon = \sqrt{\varepsilon} v_1$:

$$\frac{\partial v_1^\varepsilon}{\partial t} + \left(\frac{(\mu - r)^2}{2\bar{\sigma}^2(1 - p)} + r\right) p v_1^\varepsilon + \left(\frac{p}{1-p}\right)^2 \left(\frac{\mu - r}{\bar{\sigma}^2}\right)^3 V_3^\varepsilon v_0 = 0,$$

where we have used

$$A_{(\bar{\sigma},u_0^\star)} = \left(\frac{(\mu - r)^2}{2\bar{\sigma}^2(1 - p)} + r\right) p$$

and the small constant V_3^ε:

$$V_3^\varepsilon = -\frac{\rho v \sqrt{\varepsilon}}{\sqrt{2}} \langle f \phi' \rangle,$$

which is exactly the quantity in (4.39) in the OU case and without the slow factor as considered here. The estimation of this parameter from the observed smile is discussed in Section 5.1. It is given by $V_3^\varepsilon = a^\varepsilon \bar{\sigma}^3$ as in (5.21), where a^ε is the slope of the fit to the smile.

We notice also that the first correction to the strategy u_1 does not affect the first correction to the maximum expected utility.

Using the formula for v_0, we get the corrected maximum expected utility

$$V(t,w,y) =$$

$$\frac{w^p}{p}\left[1+\left(\frac{p}{1-p}\right)^2\left(\frac{\mu-r}{\bar{\sigma}^2}\right)^3 V_3^\varepsilon(T-t)\right]e^{\left(r+\frac{(\mu-r)^2}{2\bar{\sigma}^2(1-p)}\right)p(T-t)}+\mathcal{O}(\varepsilon).$$

The correction coefficient V_3^ε has the sign of the correlation ρ, which is typically negative. This shows that negative correlation diminishes the investor's maximum expected utility.

References for further studies of portfolio optimization problems are given in the Notes.

11.3 Application to CAPM Forward-Looking Beta Estimation

In this application, originally developed in Fouque and Kollman (2011), we consider call option price approximations for both the market index and an individual asset in the context of a continuous time capital asset pricing model (CAPM) in a stochastic volatility environment. These approximations show the role played by the asset's beta parameter as a component of the parameters of the call option price of the asset. They also show how these parameters, in combination with the parameters of the call option price for the market, can be used to extract the beta parameter. A calibration technique for the beta parameter is derived using the estimated option price parameters of both the asset and market index. The resulting estimator of the beta parameter is not only simple to implement but has the advantage of being *forward-looking* as it is calibrated from skews of implied volatilities.

11.3.1 Discrete-Time CAPM and Forward-Looking Betas

The concept of stock betas was developed in the context of the CAPM of Sharpe (1996) and was based on previous portfolio theory in Markowitz (1952). The beta of a stock represents the scale of the risk of the asset relative to the systematic risk of the market and is critical in the development and performance of stock portfolios. We examine the role of a stocks beta parameter in option prices on the stock in the presence of stochastic volatility and develop a calibration technique for the beta parameter using the option prices, or equivalently, implied volatilities.

The estimation of the beta parameter is an important issue in financial practices that deal with CAPM and is used, amongst other things, for portfolio construction and performance measurement. The original discrete-time

CAPM defined the log-price return on individual asset R_a as a linear function of the risk-free interest rate R_f, the log-return of the market R_M, and a Gaussian error term ε_a:

$$R_a = R_f + \beta_a(R_M - R_f) + \varepsilon_a. \tag{11.24}$$

The beta coefficient is usually estimated using historical returns on the asset and market index. The classic approach used a simple linear regression of asset returns on market returns as implied by (11.24). This regression approach leads to a simple estimation of beta as the ratio of the covariance of historical market and asset returns to the variance of historical market returns. Other approaches have accounted for the fact that the beta parameter may not be constant in time. To this end, Scholes and Williams (1977) provided an approach to estimating the beta using historical non-synchronous data. However, a fundamental flaw in estimating the beta parameter using historical data is that it is inherently backward-looking, which can be a major drawback for the use of betas in forward-looking portfolio construction. As such, many studies on beta estimation, such as French *et al.* (1983) and Siegel (1995), and more recently Christoffersen *et al.* (2008a), have attempted to extract the parameter from option prices on the underlying market and asset processes. In fact, by taking first and second risk-neutral moments in (11.24), the following formula is derived in Christoffersen *et al.* (2008a):

$$\beta_a = \left(\frac{\text{SKEW}_a}{\text{SKEW}_M}\right)^{\frac{1}{3}} \left(\frac{\text{VAR}_a}{\text{VAR}_M}\right)^{\frac{1}{2}}, \tag{11.25}$$

where VAR_a (resp. VAR_M) and SKEW_a (respectively SKEW_M) are the variance and the risk-neutral skewness of returns of the asset (respectively of the market). Then, they use results from Carr and Madan (2001) which relate these moments to option prices (*Quad* and *Cubic*) on the asset (respectively on the market). These results are direct consequences of the fact that the second derivative of a call payoff with respect to the strike price is a δ-function at the asset price, so that by integration by parts one has:

$$e^{rT} \int_0^\infty 2C(0,x;K,T)dK = \mathbb{E}^\star \left\{ X_T^2 \mid X_0 = x \right\},$$

$$e^{rT} \int_0^\infty 6KC(0,x;K,T)dK = \mathbb{E}^\star \left\{ X_T^3 \mid X_0 = x \right\}.$$

The advantage of this approach is that option prices are inherently forward-looking on the underlying price process. However, it requires call option prices at all strikes.

Our main result, formula (11.44), obtained by using the theory developed in this book, can be viewed as a simplified version of (11.25) allowing for a direct calibration to the skews of implied volatilities.

11.3.2 Continuous-Time CAPM with Stochastic Volatility

A simple continuous-time CAPM for market price M_t and asset price X_t evolves as follows:

$$\frac{dM_t}{M_t} = \mu_m dt + \sigma_m dW_t^{(1)},$$

$$\frac{dX_t}{X_t} = rdt + \beta\left(\frac{dM_t}{M_t} - rdt\right) + \sigma dW_t^{(2)},$$

where μ_m is the rate of return of the market, σ_m and σ are positive constant volatilities, and $W^{(1)}$, $W^{(2)}$ are independent Brownian motions

$$d\langle W^{(1)}, W^{(2)}\rangle_t = 0.$$

This model is consistent with (11.24) in that the excess return of the asset $\frac{dX_t}{X_t} - rdt$ is an affine function of the excess return of the market $\frac{dM_t}{M_t} - rdt$ through the β coefficient and a Brownian-driven noise process. Most importantly, the process preserves the definition of the β coefficient as the covariance of the asset and market returns divided by the market variance: that is, formally,

$$\frac{\langle\frac{dX_t}{X_t}, \frac{dM_t}{M_t}\rangle}{\langle\frac{dM_t}{M_t}, \frac{dM_t}{M_t}\rangle} = \frac{\langle rdt + \beta\left(\frac{dM_t}{M_t} - rdt\right) + \sigma dW_t^{(2)}, \frac{dM_t}{M_t}\rangle}{\langle\frac{dM_t}{M_t}, \frac{dM_t}{M_t}\rangle}$$

$$= \frac{\langle\beta\frac{dM_t}{M_t}, \frac{dM_t}{M_t}\rangle}{\langle\frac{dM_t}{M_t}, \frac{dM_t}{M_t}\rangle} = \beta, \tag{11.26}$$

where the second equality holds due to the independence of M_t and $W_t^{(2)}$. Observe that the evolution of X_t is given by

$$\frac{dX_t}{X_t} = (r + \beta(\mu_m - r))dt + \beta\sigma_m dW_t^{(1)} + \sigma dW_t^{(2)},$$

that is a geometric Brownian motion with volatility $\sqrt{\beta^2\sigma_m^2 + \sigma^2}$. Even if this quantity is known, along with the volatility σ_m of the market process, one cannot disentangle β and σ.

Moreover, the assumption of constant volatility in an asset market is not satisfactory as we have seen in Chapter 2. Following the modeling approach developed in previous chapters, we introduce a stochastic volatility component to the market price process, that is we replace σ_m by a stochastic process $\sigma_t = f(Y_t)$:

$$\frac{dM_t}{M_t} = \mu_m dt + f(Y_t)dW_t^{(1)},\tag{11.27}$$

$$\frac{dX_t}{X_t} = r dt + \beta \left(\frac{dM_t}{M_t} - r dt\right) + \sigma dW_t^{(2)},\tag{11.28}$$

$$dY_t = \frac{1}{\varepsilon}(m - Y_t)dt + \frac{v\sqrt{2}}{\sqrt{\varepsilon}}dZ_t.\tag{11.29}$$

For simplicity, we only consider a fast volatility factor driven by a mean-reverting OU process Y_t with a large mean-reversion rate $1/\varepsilon$ and the invariant (long-run) distribution $\mathcal{N}(m, v^2)$. This model implies stochastic volatility in the asset price through its dependence on the market return, and assumes correlation between the Brownian motions driving the market returns and volatility process:

$$d\langle W^{(1)}, Z\rangle_t = \rho \, dt.$$

In fact, in what follows we will assume that $\rho \neq 0$, however, we continue to assume independence between $W_t^{(2)}$ and the other two Brownian motions $W_t^{(1)}$ and Z_t in order to preserve the interpretation of β in (11.26). We could suppose that the volatility σ is also fluctuating with Y_t, so that $\sigma = \sigma(Y_t)$. The results presented below would remain essentially the same. The introduction of another independent fast mean-reverting volatility factor is more involved but can be treated as well.

Note that, in this section, we do not try to work with the most elaborate model, but rather to present the main idea, that is calibration of β, in the context of the simplest model which allows it. In particular, the asymptotic analysis that we will use does not depend crucially on the particular choice of an OU process for Y. Another choice could be a CIR process combined with a function f of the form $f(y) = \sqrt{y}$. In that case, (M, Y) would simply be a Heston model – discussed in detail in Chapter 10. More importantly, we are assuming ε to be small, that is the volatility is fast mean-reverting. A generalization to the case of fast and slow time scales could easily be derived along the lines of Chapter 4.

The first step is to rewrite the dynamics of the market and of the asset under a pricing risk-neutral measure.

11.3.3 Pricing Risk-Neutral Measure

The market (or index) and the asset being both tradable, their discounted prices need to be martingales under a pricing risk-neutral measure. In order to achieve that, we first write

$$Z_t = \rho \, dW_t^{(1)} + \sqrt{1-\rho^2} \, dW_t^{(3)},$$

with now $(W_t^{(1)}, W_t^{(2)}, W_t^{(3)})$ being three independent standard Brownian motions, and then we rewrite the system (11.27), (11.28), (11.29) as:

$$\frac{dM_t}{M_t} = rdt + f(Y_t)\left(dW_t^{(1)} + \frac{\mu_m - r}{f(Y_t)} dt \right),$$

$$\frac{dX_t}{X_t} = rdt + \beta f(Y_t)\left(dW_t^{(1)} + \frac{\mu_m - r}{f(Y_t)} dt \right) + \sigma dW_t^{(2)},$$

$$dY_t = \frac{1}{\varepsilon}(m - Y_t)dt - \frac{v\sqrt{2}}{\sqrt{\varepsilon}} \Lambda(Y_t)dt$$
$$+ \frac{v\sqrt{2}}{\sqrt{\varepsilon}} \left[\rho \left(dW_t^{(1)} + \frac{\mu - r}{f(Y_t)} dt \right) + \sqrt{1-\rho^2} \left(dW_t^{(3)} + \gamma(Y_t)dt \right) \right],$$

where $\gamma(Y_t)$ is a market price of volatility risk, which we suppose to depend on Y_t only, and we define

$$\Lambda(Y_t) = \rho \frac{\mu_m - r}{f(Y_t)} + \sqrt{1-\rho^2} \, \gamma(Y_t).$$

Setting

$$dW_t^{(1)\star} = dW_t^{(1)} + \frac{\mu_m - r}{f(Y_t)} dt,$$

$$dW_t^{(2)\star} = dW_t^{(2)},$$

$$dW_t^{(3)\star} = dW_t^{(3)} + \gamma(Y_t)dt,$$

by Girsanov's theorem, there is an equivalent probability $\mathbb{P}^{\star(\gamma)}$ such that $(W_t^{(1)\star}, W_t^{(2)\star}, W_t^{(3)\star})$ are independent standard Brownian motions under $\mathbb{P}^{\star(\gamma)}$, called the pricing equivalent martingale measure and determined by the market price of volatility risk γ. We assume here that the Sharpe ratio $\frac{\mu_m - r}{f(Y_t)}$ and the volatility premium $\gamma(Y_t)$ are bounded, which, depending on the choice of function f, may require that μ_m depends on Y_t. Finally, under $\mathbb{P}^{\star(\gamma)}$, the dynamics (11.27)–(11.29) becomes:

$$\frac{dM_t}{M_t} = rdt + f(Y_t)dW_t^{(1)\star}, \tag{11.30}$$

$$\frac{dX_t}{X_t} = rdt + \beta f(Y_t)dW_t^{(1)\star} + \sigma dW_t^{(2)\star}, \tag{11.31}$$

$$dY_t = \frac{1}{\varepsilon}(m - Y_t)dt - \frac{v\sqrt{2}}{\sqrt{\varepsilon}}\Lambda(Y_t)dt + \frac{v\sqrt{2}}{\sqrt{\varepsilon}}dZ_t^\star, \tag{11.32}$$

$$Z_t^\star = \rho W_t^{(1)\star} + \sqrt{1 - \rho^2}\, W_t^{(3)\star}.$$

As before, we take the point of view that by pricing options on the index M and on the particular asset X, the market is "completing" itself and indirectly choosing the market price of volatility risk γ.

11.3.4 Market and Asset Option Prices

In looking first at option prices on the market index we will only focus on the autonomous evolution of (M_t, Y_t) described by equations (11.30) and (11.32) under the risk-neutral pricing measure. We use the singular perturbation technique developed in Chapter 4, including the parameter-reduction step presented in Section 4.3. Let $P^{M,\varepsilon}(t,\xi,y)$ denote the price of a European call option written on the market index M, with maturity T and strike K, evaluated at time $t < T$ with current values $M_t = \xi$ and $Y_t = y$, where we explicitly show the dependence on the small volatility mean-reversion time ε. The function $P^{M,\varepsilon}(t,\xi,y)$ satisfies the partial differential equation:

$$\mathcal{L}^\varepsilon P^{M,\varepsilon} = 0,$$
$$P^{M,\varepsilon}(T,\xi,y) = (\xi - K)^+,$$

where

$$\mathcal{L}^\varepsilon = \frac{1}{\varepsilon}\mathcal{L}_0 + \frac{1}{\sqrt{\varepsilon}}\mathcal{L}_1 + \mathcal{L}_2,$$

$$\mathcal{L}_0 = v^2\frac{\partial^2}{\partial y^2} + (m - y)\frac{\partial}{\partial y},$$

$$\mathcal{L}_1 = \rho v\sqrt{2}f(y)\xi\frac{\partial^2}{\partial\xi\partial y} - v\sqrt{2}\Lambda(y)\frac{\partial}{\partial y},$$

$$\mathcal{L}_2 = \frac{\partial}{\partial t} + \frac{1}{2}f(y)^2\xi^2\frac{\partial^2}{\partial\xi^2} + r(\xi\frac{\partial}{\partial\xi} - \cdot) = \mathcal{L}_{BS}(f(y)).$$

Then, the singular perturbation technique presented in Chapter 4 gives the following approximation:

$$P^{M,\varepsilon} \sim P^{M\star} + (T-t)V_3^{M,\varepsilon} \xi \frac{\partial}{\partial \xi}\left(\xi^2 \frac{\partial^2 P^{M\star}}{\partial \xi^2}\right), \tag{11.33}$$

where $P^{M\star}$ is the corresponding Black–Scholes call price with constant volatility equal to the *adjusted effective volatility* $\sigma^{M\star}$ given as in (4.62) by

$$\sigma^{M\star} = \sqrt{\bar{\sigma}^2 + 2V_2^{M,\varepsilon}}, \tag{11.34}$$

and $\bar{\sigma}$ is the *effective volatility* defined as in (4.26) by

$$\bar{\sigma}^2 = \langle f^2 \rangle \equiv \int f(y)^2 \frac{1}{\sqrt{2\pi v}} e^{-\frac{(y-m)^2}{2v^2}} dy, \tag{11.35}$$

with the average being taken with respect to the invariant distribution of the OU process Y. The small parameters $V_2^{M,\varepsilon}$ and $V_3^{M,\varepsilon}$ are given as in (4.40) and (4.39) by

$$V_2^{M,\varepsilon} = \frac{\sqrt{\varepsilon}v}{\sqrt{2}} \langle \phi' \Lambda \rangle, \tag{11.36}$$

$$V_3^{M,\varepsilon} = -\frac{\sqrt{\varepsilon}\rho v}{\sqrt{2}} \langle \phi' f \rangle, \tag{11.37}$$

where ϕ is a solution of the Poisson equation $\mathcal{L}_0 \phi = f^2 - \langle f^2 \rangle$.

Now, we briefly show how to adapt the derivation of the approximation for the market options to the case for options written on the individual asset. Let $P^{a,\varepsilon}$ denote the price of a European option written on the asset X, with maturity T and payoff h, evaluated at time $t < T$ with current values $X_t = x$ and $Y_t = y$, where as before we explicitly show the dependence on the small volatility mean-reversion time ε. Then, we have

$$P^{a,\varepsilon} = \mathbb{E}^{\star(\gamma)}\left\{e^{-r(T-t)}h(X_T) \mid \mathscr{F}_t\right\} = P^{a,\varepsilon}(t,X_t,Y_t),$$

since (X_t,Y_t) is Markovian as can be seen from (11.31)–(11.32), and where the function $P^{a,\varepsilon}(t,x,y)$ satisfies the partial differential equation

$$\mathcal{L}^{a,\varepsilon} P^{a,\varepsilon} = 0,$$
$$P^{a,\varepsilon}(T,x,y) = h(x),$$

where

$$\mathscr{L}^{a,\varepsilon} = \frac{1}{\varepsilon}\mathscr{L}_0 + \frac{1}{\sqrt{\varepsilon}}\mathscr{L}_1^a + \mathscr{L}_2^a,$$

$$\mathscr{L}_0 = v^2\frac{\partial^2}{\partial y^2} + (m-y)\frac{\partial}{\partial y},$$

$$\mathscr{L}_1^a = \rho v\sqrt{2}\beta f(y)x\frac{\partial^2}{\partial x\partial y} - v\sqrt{2}\Lambda(y)\frac{\partial}{\partial y},$$

$$\mathscr{L}_2^a = \frac{\partial}{\partial t} + \frac{1}{2}\left(\beta^2 f(y)^2 + \sigma^2\right)x^2\frac{\partial^2}{\partial x^2} + r\left(x\frac{\partial}{\partial x} - \cdot\right)$$

$$= \mathscr{L}_{BS}(\sqrt{\beta^2 f(y)^2 + \sigma^2}).$$

Observe that the only differences with options on the market index are the factor β in $\mathscr{L}_1^a$, and the modified square volatility $\beta^2 f(y)^2 + \sigma^2$ in $\mathscr{L}_2^a$. Going line by line in the derivation of the approximation, it is easy to see that the only modifications are:

(i) $\bar{\sigma}^2$ in (11.35) is replaced by

$$\bar{\sigma}_a^2 = \langle\beta^2 f^2 + \sigma^2\rangle = \beta^2\bar{\sigma}^2 + \sigma^2. \tag{11.38}$$

(ii) $V_2^{M,\varepsilon}$ in (11.36) is replaced by $V_2^{a,\varepsilon} = \beta^2 V_2^{M,\varepsilon}$:

$$V_2^{a,\varepsilon} = \frac{\beta^2\sqrt{\varepsilon}v}{\sqrt{2}}\langle\phi'\Lambda\rangle. \tag{11.39}$$

(iii) $V_3^{M,\varepsilon}$ in (11.37) is replaced by $V_3^{a,\varepsilon} = \beta^3 V_3^{M,\varepsilon}$:

$$V_3^{a,\varepsilon} = -\frac{\beta^3\sqrt{\varepsilon}\rho v}{\sqrt{2}}\langle\phi'f\rangle. \tag{11.40}$$

(iv) $\sigma^{M\star}$ in (11.34) is replaced by $\sigma^{a\star}$, given by

$$\sigma_a^\star = \sqrt{\bar{\sigma}_a^2 + 2V_2^{a,\varepsilon}} = \sqrt{\beta^2\bar{\sigma}^2 + \sigma^2 + 2V_2^{a,\varepsilon}}. \tag{11.41}$$

(v) The option price approximation becomes

$$P^{a,\varepsilon} \sim P_0^{a\star} + (T-t)V_3^{a,\varepsilon}x\frac{\partial}{\partial x}\left(x^2\frac{\partial^2 P_0^{a\star}}{\partial x^2}\right), \tag{11.42}$$

where $P_0^{a\star}$ is the Black–Scholes price with volatility $\sigma^{a\star}$ given by (11.41).

11.3.5 Beta Estimation

From the expressions for $V_3^{M,\varepsilon}$ and $V_3^{a,\varepsilon}$ given respectively in (11.37) and (11.40), one deduces that if $\rho \neq 0$, then $V_3^{M,\varepsilon} \neq 0$ and $V_3^{a,\varepsilon} \neq 0$, and $V_3^{a,\varepsilon} = \beta^3 V_3^{M,\varepsilon}$, so that β is given by:

$$\beta = \left(\frac{V_3^{a,\varepsilon}}{V_3^{M,\varepsilon}}\right)^{\frac{1}{3}}. \tag{11.43}$$

Therefore in order to estimate β in a forward-looking fashion using the implied skew parameters from option prices, we need to calibrate our two parameters $V_3^{a,\varepsilon}$ and $V_3^{M,\varepsilon}$.

Using the results of Section 5.1 with the fast factor only, we simply regress linearly on LMMR for implied volatilities of the market and for implied volatilities of the particular asset:

$$I^M \approx b^{M\star} + a^{M,\varepsilon} \text{LMMR}^{(M)},$$
$$I^a \approx b^{a\star} + a^{a,\varepsilon} \text{LMMR}^{(a)}.$$

Then, using the calibration formula (5.21), we deduce

$$V_3^{M,\varepsilon} \approx a^{M,\varepsilon}(b^{M\star})^3,$$
$$V_3^{a,\varepsilon} \approx a^{a,\varepsilon}(b^{a\star})^3,$$

and therefore, in view of (11.43), we propose the following estimator:

$$\hat{\beta} = \left(\frac{a^{a,\varepsilon}}{a^{M,\varepsilon}}\right)^{1/3} \left(\frac{b^{a\star}}{b^{M\star}}\right). \tag{11.44}$$

Observe the similarity between formula (11.25) and our formula (11.44) where $a^{a,\varepsilon}, a^{M,\varepsilon}$ are skews and $b^{a\star}, b^{M\star}$ are *at-the-money* volatilities.

11.3.6 Examples

We give here two examples of beta fits. The first one is for Amgen on February 18, 2009 and the second one is for Goldman Sachs on February 19, 2009.

In Figure 11.1(a), we present the implied volatility skews and their affine LMMR fits, for the S&P 500 index and Amgen (AMGN) on the particular day of February 18, 2009 (around-the-money options with LMMR values between -1 and 1 are used in the fits). The beta estimate for Amgen is found to be 1.03 on that particular day, which was slightly above the historical estimate.

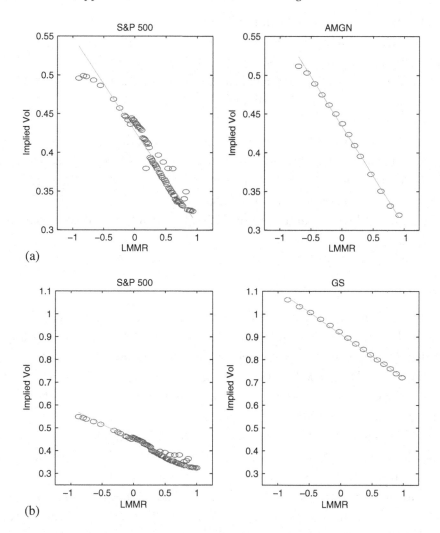

Figure 11.1   (a) Implied volatility (y-axis) of June/July 2009 maturity options for the S&P 500 and Amgen, plotted against the option's LMMR. These are for February 18, 2009 option prices. The parameters fit for each series are: S&P 500: $a^{M,\varepsilon} = -0.121$ and $b^{M\star} = 0.428 \Rightarrow V_3^{M,\varepsilon} = -0.0095$; Amgen: $a^{a,\varepsilon} = -0.0128$ and $b^{a\star} = 0.434 \Rightarrow V_3^{a,\varepsilon} = -0.0105$. From (11.44), the beta estimate for Amgen is 1.03.

(b) Implied volatility (y-axis) of June/July 2009 maturity options for the S&P 500 and Goldman Sachs, plotted against the option's LMMR. These are for February 19, 2009 option prices. The parameters fit for each series are: S&P 500: $a^{M,\varepsilon} = -0.137$ and $b^{M\star} = 0.448 \Rightarrow V_3^{M,\varepsilon} = -0.0123$; GS: $a^{a,\varepsilon} = -0.191$ and $b^{a\star} = 0.912 \Rightarrow V_3^{a,\varepsilon} = -0.1450$. From (11.44), the beta estimate for Goldman Sachs is 2.28.

In Figure 11.1(b), we present the implied volatility skews and their affine LMMR fits, for the S&P 500 index and Goldman Sachs (GS) on the particular day of February 19, 2009 (around-the-money options with LMMR values between -1 and 1 are used in the fits). The beta estimate for Goldman Sachs is found to be 2.28 on that particular day, which was at the level of the historical estimate.

In both examples, it is interesting to note the quality of the fits, and also the fact that the coefficient V_3s are small, which justifies *a posteriori* the validity of our model with fast mean-reverting stochastic volatility. Note also the high levels of volatility on these dates.

We refer to Fouque and Kollman (2011) for additional examples and more details on the stability of this estimator.

Notes

Regarding Section 11.1, for background material on Monte Carlo simulations and variance-reduction techniques, we refer to Glasserman (2003). For details on the results presented in this section we refer to the articles of Fouque and Han (2004b, 2007). We also refer to Carmona *et al.* (2009) for a variance-reduction technique using interacting particle methods for computing tail probabilities with applications to credit derivatives.

The analysis of the constant volatility optimal asset allocation problem in Section 11.2.1 appears in Merton (1969, 1971). Much work has been done on these problems, and we refer for instance to Duffie (2001) and Karatzas and Shreve (1998). Details about HJB equations and stochastic control can be found in Fleming and Soner (1993), among other sources. The distortion power transformation discussed at the end of Section 11.2.2 appears in Zariphopoulou (2001).

Extensions of the fast mean-reverting portfolio asymptotics to partial hedging and other related state-dependent utility problems are discussed in Jonsson and Sircar (2002a,b). Asymptotic approximations for utility indifference prices under stochastic volatility are constructed in Sircar and Zariphopoulou (2005).

For filtering volatility from observed returns, we refer for instance to Cvitanic and Rozovskii (2006) and Batalova *et al.* (2006).

References regarding Section 11.3 on CAPM are given in the text. Additionally, a model with a nonlinear beta is introduced in Fouque and Tashman (2011a) and applied to portfolio optimization in Fouque and Tashman (2011b).

12

Interest Rate Models

In this chapter we illustrate the role of stochastic volatility in the case of interest rate products traded in *fixed income markets*. Our main example is pricing of bonds when the interest rate is defined in terms of a Vasicek model. As in the previous chapters, we use a two-factor stochastic volatility model and show how one can derive bond price approximations in the regime of separation of time scales. Market bond pricing data are often quoted in terms of the yield curve corresponding to the effective or continously compounded interest rate for the bond as a function of time to maturity. We show how the bond price approximation gives a flexible way of parameterizing this yield curve, also called the term structure of interest rates.

The stochastic volatility Vasicek model that we consider here is introduced in Section 12.1 and we carry out the asymptotic expansion for the associated bond price in Section 12.2. The bond price approximation leads to a particular form for the yield curve and we discuss this and calibration issues in Section 12.2.8. The Vasicek example illustrates how our singular and regular perturbation approach easily generalizes to typical problems in the fixed income market. There are many other interest rate products and also interest rate models that can be analyzed in our framework, and we comment on some of these. In Section 12.4 we use the CIR model and in Section 12.3 a quadratic model for the interest rate. We then consider the pricing problem for bond options in Section 12.5. In the Notes we give references to extensive works on classic models of interest rate and interest rate products.

12.1 The Vasicek Model

The basic interest rate product that we consider is a zero-coupon bond with a face value one and maturity T. This is a security which pays one (dollar) for sure at time T and nothing at all at other times. Coupon bonds, on the other hand, give the owner a payment stream during the interval $(0,T)$. These instruments have the common property that they provide the owner with a deterministic cash flow, and for this reason they are also known as fixed income instruments. In the case with a constant interest rate r, the bond price at time t, $P(t,r;T)$, would simply be $P(t,r;T) = \exp(-(T-t)r)$. In the general case we model the short rate r as a stochastic process

$$dr_t = \mu(r_t,t)dt + \sigma(r_t,t)dW_t, \tag{12.1}$$

this quantity is sometimes also refered to as the spot or short rate. The no-arbitrage price at time t of a zero-coupon bond maturing at time T is then given by

$$P(t,x;T) = \mathbb{E}^\star \left\{ e^{-\int_t^T r_s ds} \mid r_t = x \right\}, \tag{12.2}$$

with the expectation being with respect to the pricing measure and x being the current level of the interest rate.

The Vasicek model for the short rate, r, under the subjective probability measure $\mathbb{P}$ is given by

$$dr_t = a(r_\infty - r_t)dt + \sigma dW_t, \tag{12.3}$$

where (W_t) is a standard $\mathbb{P}$-Brownian motion, σ the constant volatility, a is the rate of mean reversion, and r_∞ the long-run mean interest level.

The model under an equivalent martingale (pricing) measure $\mathbb{P}^\star$ becomes

$$dr_t = a(\bar{r} - r_t)dt + \sigma dW_t^\star, \tag{12.4}$$

where $(W_t^\star)$ is a standard $\mathbb{P}^\star$-Brownian motion. Assuming a constant market price of interest rate risk, λ, the risk-adjusted mean-reversion level for the interest rate is

$$\bar{r} = r_\infty - \frac{\lambda\sigma}{a},$$

where the notation reflects the fact that in the stochastic volatility case, discussed below, the analog of this quantity will be a homogenized value. Thus, in the risk-neutral world $\mathbb{P}^\star$, the short rate (r_t) is an OU process as introduced in Section 1.1.6. Here, the long-run mean level is $\bar{r}$ and the rate of mean-reversion is a. Indeed, in this model the short rate can (and will) become negative, though, with a very small probability in a finite time, starting sufficiently above zero (see the Notes for a reference where this issue is also discussed). Here we adopt the point of view that the computational advantage with explicit solutions is worth this drawback associated with the Gaussian formulation.

12.1.1 Bond Pricing in the Vasicek Model

The pricing partial differential equation again follows from the Feynman–Kac formula (1.80). We apply it with $X_t = r_t$, the discount factor $e^{-\int_t^T r_s ds}$, and the payoff $h = 1$, to find that $P(t,x;T)$ solves

$$\frac{\partial P}{\partial t} + \frac{1}{2}\sigma^2\frac{\partial^2 P}{\partial x^2} + a(\bar{r}-x)\frac{\partial P}{\partial x} - xP = 0,$$

with the terminal condition $P(T,x;T) = 1$. We write this as

$$\mathscr{L}_V P = 0, \tag{12.5}$$
$$P(T,x;T) = 1, \tag{12.6}$$

with $\mathscr{L}_V = \mathscr{L}_V(\sigma,\bar{r},a)$ being the Vasicek operator

$$\mathscr{L}_V = \frac{\partial}{\partial t} + \frac{1}{2}\sigma^2\frac{\partial^2}{\partial x^2} + a(\bar{r}-x)\frac{\partial}{\partial x} - x\cdot = 0, \tag{12.7}$$

at volatility level σ, mean-reversion level $\bar{r}$, and rate of mean-reversion a.

This problem has nonconstant coefficient, but we can still find the solution explicitly. Consider first the special case $\sigma = \bar{r} = 0$. Then (12.5) becomes

$$\frac{\partial P}{\partial t} - x\left(a\frac{\partial P}{\partial x} + P\right) = 0,$$

with $P(T,x;T) = 1$, which admits the solution

$$P(t,x;T) = e^{-B(T-t)x},$$

where

$$\frac{dB(\tau)}{d\tau} = -aB(\tau) + 1, \quad B(0) = 0, \tag{12.8}$$

so that

$$B(\tau) = \frac{1 - e^{-a\tau}}{a}, \tag{12.9}$$

using the notation $\tau = T - t$ for the *time to maturity*. In the general case, we write

$$P(T-\tau,x;T) = A(\tau)e^{-B(\tau)x}, \tag{12.10}$$

and find from (12.5) that

$$\frac{A'}{A} = \frac{1}{2}\sigma^2 B^2 - a\bar{r}B. \tag{12.11}$$

Using (12.8) one readily obtains

$$A(\tau) = \exp\left(-\left[R_\infty(\tau - B(\tau)) + \frac{\sigma^2}{4a}B^2(\tau)\right]\right), \tag{12.12}$$

where we have defined

$$R_\infty = \bar{r} - \frac{\sigma^2}{2a^2}. \tag{12.13}$$

The explicit formula for the zero-coupon bond price is therefore

$$P(t,r_t;T) = \exp\left(-\left[R_\infty(T-t) + (r_t - R_\infty)\frac{1 - e^{-a(T-t)}}{a}\right.\right.$$
$$\left.\left. + \frac{\sigma^2}{4a^3}\left(1 - e^{-a(T-t)}\right)^2\right]\right). \tag{12.14}$$

The ansatz (12.10) leads more generally to a problem that is separable in A and B for short rate models of the form (12.1) if μ and σ^2 are given, possibly time-dependent, *affine* functions in r. The Vasicek model is the simplest in this larger class of such affine models, for which the computation of bond prices and derivatives is relatively easy.

In Figure 12.1 we show the zero-coupon bond price as a function of time to maturity. We used the parameters $\bar{r} = 0.06$, $\sigma = 0.05$, and $a = 0.3$ in nondimensionalized units. The top solid line corresponds to the current value for the short rate being $r_t = 0.02$ and the bottom solid line to $r_t = 0.10$. The two dashed lines are the analogous bond prices when we increase the volatility to $\sigma = 0.10$, which gives a smaller (risk-adjusted) return. Observe that the bond price approaches its face value as the time to maturity approaches zero (pull-to-par). Notice also that if $\tau = T - t \ll 1/a$ then

$$P(t,r_t;t+\tau) \sim \exp(-r_t\tau),$$

as if the interest rate was frozen at the present level. If $\tau \gg 1/a$ then

$$-\frac{1}{\tau}\log P(t,r_t;t+\tau) \sim R_\infty,$$

where R_∞ is the long-maturity yield. The time $1/a$ corresponds to the correlation time for the process r_t, that is, $(r_t, r_{t+\Delta})$ are approximately uncorrelated if $\Delta > 1/a$.

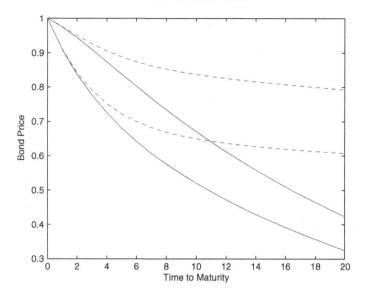

Figure 12.1   The bond price as a function of time to maturity τ. We use $\bar{r} = 0.06$ and $a = 0.3$ in nondimensionalized units. The solid lines correspond to $\sigma = 0.05$ and the dashed lines to $\sigma = 0.10$. The current short rate is $r_t = 0.10$ for the lower curves and $r_t = 0.02$ for the upper curves.

12.1.2 Yield Curve

It is convenient to represent the bond price term structure in terms of the *yield curve* which measures the effective interest rate as a function of time to maturity. This quantity is also referred to as the continuously compounded spot rate over the interval (t, T) and is defined by

$$R(t, \tau; r_t) = -\frac{1}{\tau} \log\left(P(t, r_t; t + \tau)\right).$$

In the deterministic case ($\sigma = 0$), R is simply

$$R(t, \tau; r_t) = \frac{1}{\tau} \int_t^{t+\tau} r_s \, ds.$$

In the general Vasicek case we find explicitly

$$R(t, \tau; r_t) = R_\infty + (r_t - R_\infty) \frac{B(\tau)}{\tau} + \frac{\sigma^2}{4a} \frac{B^2(\tau)}{\tau}, \qquad (12.15)$$

which shows that $R(t, \tau; r_t)$ is an affine function of the current rate r_t and justifies the name *affine model of term structure*. As already observed at the end of the previous section, for all t, $R(t, \tau; r_t)$ converges to $R_\infty = \bar{r} - \sigma^2/(2a^2)$ as $\tau \to +\infty$ and to r_t as $\tau \to 0$.

In Figure 12.2, we show with the solid and dashed lines the yield curves associated with the bond prices in Figure 12.1, the top curves now

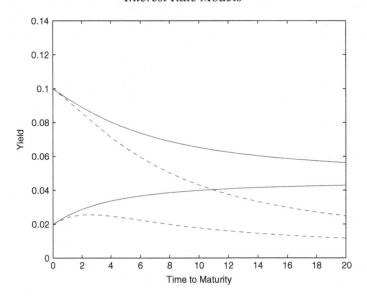

Figure 12.2   The yields as functions of time to maturity τ. We use $\bar{r} = 0.06$, $\sigma = 0.05$, and $a = 0.3$ in nondimensionalized units. The current short rate is $r_t = 0.10$ for the top solid line and $r_t = 0.02$ for the bottom. The two dashed lines are the corresponding yield curves when we increase the volatility level to $\sigma = 0.10$.

correspond to the higher current short rate. An increase in the volatility risk, corresponding to the dashed lines, reduces the overall yield with the long-term yield being determined by $R_\infty = \bar{r} - \frac{\sigma^2}{2a^2}$.

Above, we found expressions for bond prices and yields by solving the pricing partial differential equation. Alternatively, formulas for bond prices and yields can be derived probabilistically as follows. Under the pricing measure, the short rate at a future time $s > t$ is

$$r_s = \bar{r} + (r_t - \bar{r})e^{-a(s-t)} + \sigma \int_t^s e^{-a(s-v)} dW_v^\star.$$

Conditioned on the current rate r_t, r_s is normally distributed with mean $\bar{r} + (r_t - \bar{r}) \exp(-a(s-t))$ and variance $\sigma^2(1 - \exp(-2a(s-t)))/2a$. Thus, in the above affine model the price at some future time s is lognormally distributed and the yield is normally distributed. The integral

$$\int_t^T r(s)\,ds = \bar{r}(T - t) + (r_t - \bar{r})\frac{1 - e^{-a(T-t)}}{a} + \sigma \int_t^T \frac{1 - e^{-a(T-s)}}{a} dW_s^\star,$$

conditioned on r_t, is normally distributed with mean

$$m(t,T;r_t) = \bar{r}(T - t) + (r_t - \bar{r})\frac{1 - e^{-a(T-t)}}{a}$$

and variance

$$v(t,T) = \sigma^2 \int_t^T \left(\frac{1 - e^{-a(T-s)}}{a} \right)^2 ds.$$

Therefore, using the formula for the mean of a lognormal random variable, the bond price and the yield can be expressed as

$$P(t, r_t; T) = \mathbb{E}^\star \left\{ e^{-\int_t^T r_s ds} \mid r_t \right\} = e^{-m(t,T;r_t) + v(t,T)/2},$$

$$R(t, T - t; r_t) = -\frac{1}{\tau} \log\left(P(t, r_t; T) \right) = \frac{m(t,T;r_t) - v(t,T)/2}{\tau},$$

and it can easily be checked that these expressions coincide with the ones in (12.14) and (12.15).

12.1.3 Bond Volatility

The bond price $P(t, r_t; T)$ is a function of the stochastic process r_t, and is thus itself a stochastic process which solves a stochastic differential equation that follows from applying Itô's formula. We now write this evolution equation for the bond price $P(t, r_t; T)$ in order to identify its volatility. We apply Itô's formula to (12.14) and obtain

$$dP(t, r_t; T) = P(t, r_t; T) \left(r_t dt + \Sigma(t, T) d\left(-W_t^\star\right) \right), \qquad (12.16)$$

where the bond *volatility term structure* $\Sigma(t, T)$ is given by

$$\Sigma(t, T) = \sigma B(T - t) = \sigma \left(\frac{1 - e^{-a(T-t)}}{a} \right). \qquad (12.17)$$

The form of the drift term shows that, as it should be, the discounted bond price is a martingale under the risk-neutral pricing measure. In the Vasicek model the bond volatility is a time-dependent function going exponentially to zero as time to maturity goes to zero when the pull-to-par phenomenon sets in.

12.1.4 Forward Rates

Another natural way to look at interest rates and bonds, is to introduce the *forward rates* $f(t, T)$ defined here by

$$f(t, T) = -\frac{\partial}{\partial T} \log P(t, r_t; T), \qquad (12.18)$$

so that

$$P(t, r_t; T) = e^{-\int_t^T f(t,u)du}.$$ (12.19)

Using the explicit form (12.14) of the bond price, one obtains the following stochastic differential equation with respect to t for the forward rate:

$$df(t, T) = \sigma(t, T)\Sigma(t, T)dt + \sigma(t, T)dW_t^{\star},$$ (12.20)

where the *forward rates volatility term structure* $\sigma(t, T)$ is given by

$$\sigma(t, T) = \sigma e^{-a(T-t)},$$ (12.21)

and where the relation

$$\Sigma(t, T) = \sigma B(T - t) = \int_t^T \sigma(t, u)du$$ (12.22)

is satisfied, as can be seen from (12.17) and (12.21). This is the classical no-arbitrage condition for HJM models of forward rates. In Figure 12.3, we show the forward rates when using the same parameters as in Figure 12.2. The limits of the forward rate for small and large times to maturity are r_t and R_∞ respectively, as was the case for the yield.

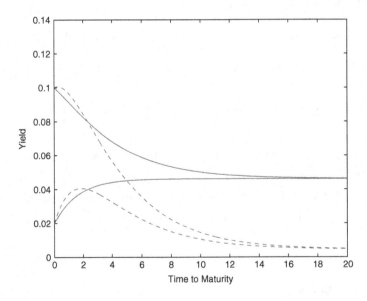

Figure 12.3 The forward rates as a function of time to maturity τ. We use $\bar{r} = 0.06$, $\sigma = 0.05$, and $a = 0.3$ in nondimensionalized units. The current short rate is $r_t = 0.10$ for the top solid line and $r_t = 0.02$ for the bottom. The two dashed lines are the corresponding forward rates when we increase the volatility level to $\sigma = 0.10$.

12.2 The Bond Price and its Expansion

The standard Vasicek model (12.4) has a constant volatility σ. Here we will analyze the situation where the volatility is stochastic. The stochastic volatility is modeled as in Chapter 4, that is, it is driven by two volatility factors, one fast factor (Y_t) and one slow factor (Z_t). The former evolves on the short time scale ε, it has a mean-reversion time of order ε, the latter evolves on the long time scale $1/\delta$.

With a stochastic volatility model as in (4.1), the generalized Vasicek model becomes, under the physical measure,

$$dr_t = a(r_\infty - r_t)dt + f(Y_t, Z_t)dW_t^{(0)}, \qquad (12.23)$$

and under an equivalent martingale (pricing) measure

$$dr_t = (a(r_\infty - r_t) - \lambda(Y_t, Z_t)f(Y_t, Z_t))dt + f(Y_t, Z_t)dW_t^{(0)\star}, \quad (12.24)$$
$$dY_t = \left(\frac{1}{\varepsilon}\alpha(Y_t) - \frac{1}{\sqrt{\varepsilon}}\beta(Y_t)\Lambda_1(Y_t, Z_t)\right)dt + \frac{1}{\sqrt{\varepsilon}}\beta(Y_t)dW_t^{(1)\star},$$
$$dZ_t = \left(\delta\,c(Z_t) - \sqrt{\delta}\,g(Z_t)\Lambda_2(Y_t, Z_t)\right)dt + \sqrt{\delta}\,g(Z_t)dW_t^{(2)\star}.$$

The volatility $f(Y_t, Z_t)$ and the correlation structure of the driving Brownian motions are modeled exactly as described in Section 4.1.1.

Observe that (12.24) can be solved so that the short rate r_t is given by

$$r_t = r_0 + \int_0^t e^{-a(t-s)}(ar_\infty - \lambda(Y_s, Z_s)f(Y_s, Z_s))\,ds$$
$$+ \int_0^t e^{-a(t-s)}f(Y_s, Z_s)\,dW_s^{(0)\star}.$$

Since the process (r_t, Y_t, Z_t) is Markovian under the risk-neutral probability measure $\mathbb{P}^\star$, chosen by the market, the price of the bond is a function of the present values of these processes and of the present time $t < T$. The bond price is

$$P^{\varepsilon,\delta}(t, r_t, Y_t, Z_t; T) = \mathbb{E}^\star\left\{e^{-\int_t^T r_s\,ds} \mid r_t, Y_t, Z_t\right\}.$$

We need to specify all the parameters in the above model in order to compute this expected value. As before, our perturbation approach will simplify enormously this problem by approximating $P^{\varepsilon,\delta}$ by a quantity which depends only on a few *group market parameters*.

12.2.1 The Full Bond-Pricing Problem

The pricing partial differential equation for the bond price again follows by
an application of the multidimensional Feynman–Kac formula presented in
Section 1.9.3. The problem that determines $P^{\varepsilon,\delta}$ is

$$\mathcal{L}^{\varepsilon,\delta} P^{\varepsilon,\delta} = \frac{\partial P^{\varepsilon,\delta}}{\partial t} + \mathcal{L}_{(r,Y,Z)} P^{\varepsilon,\delta} - x P^{\varepsilon,\delta} = 0, \qquad (12.25)$$

$$P^{\varepsilon,\delta}(T,x,y,z;T) = 1,$$

with $\mathcal{L}_{(r,Y,Z)}$ now denoting the infinitesimal generator of the Markovian process (r_t, Y_t, Z_t) defined in (12.24) and with x denoting the value for the short
rate r_t. As in Section 4.1.2, we decompose the pricing operator according
to the small parameters ε and δ:

$$\mathcal{L}^{\varepsilon,\delta} = \frac{1}{\varepsilon}\mathcal{L}_0 + \frac{1}{\sqrt{\varepsilon}}\mathcal{L}_1 + \mathcal{L}_2 + \sqrt{\delta}\,\widetilde{\mathcal{M}}_1 + \delta\mathcal{M}_2 + \sqrt{\frac{\delta}{\varepsilon}}\mathcal{M}_3. \quad (12.26)$$

The operators $\mathcal{L}_0$, $\mathcal{M}_2$, and $\mathcal{M}_3$ are defined as in (4.8)–(4.13), the operator
$\mathcal{L}_2$ has been replaced by the Vasicek operator $\widetilde{\mathcal{L}}_2$ evaluated at the volatility
$\sigma = f(y,z)$ and the market price of interest rate risk $\lambda = \lambda(y,z)$, moreover,
the operators $\mathcal{L}_1$ and $\mathcal{M}_1$ are modified according to:

$$\widetilde{\mathcal{L}}_1 = \beta(y)\left(\rho_1 f(y,z)\frac{\partial^2}{\partial x \partial y} - \Lambda_1(y,z)\frac{\partial}{\partial y}\right),$$

$$\widetilde{\mathcal{L}}_2 = \frac{\partial}{\partial t} + \frac{1}{2}f^2(y,z)\frac{\partial^2}{\partial x^2} + (a(r_\infty - x) - f(y,z)\lambda(y,z))\frac{\partial}{\partial x} - x\cdot,$$

$$\widetilde{\mathcal{M}}_1 = g(z)\left(\rho_2 f(y,z)\frac{\partial^2}{\partial x \partial z} - \Lambda_2(y,z)\frac{\partial}{\partial z}\right).$$

Recall that the small parameter ε gives rise to a singular perturbation
problem. The terms associated with only the small parameter δ give rise
to a *regular* perturbation problem about the Vasicek operator $\widetilde{\mathcal{L}}_2$. In the
next section we carry out the combined regular and singular perturbation
expansion.

It turns out that in the stochastic volatility case the yield is affine in the
short rate r_t, it has however a general dependence on the volatility factors
Y_t and Z_t. Writing the bond price in the form

$$P^{\varepsilon,\delta}(t,x,y,z;T) = \tilde{A}(T-t,y,z)e^{-B(T-t)x}, \qquad (12.27)$$

with B defined in (12.9), we find that $\tilde{A} = \tilde{A}(\tau, y, z)$ solves the problem

$$\frac{\partial \tilde{A}}{\partial \tau} = \left(\frac{1}{\varepsilon}\mathcal{L}_0 + \frac{1}{\sqrt{\varepsilon}}b + c + \sqrt{\delta}d + \delta\mathcal{M}_2 + \sqrt{\frac{\delta}{\varepsilon}}\mathcal{M}_3 \right) \tilde{A}, \quad (12.28)$$

$$\tilde{A}(0, y, z) = 1,$$

where

$$b(\tau, y, z) = -\beta(y) \left(\rho_1 f(y, z)B(\tau)\frac{\partial}{\partial y} + \Lambda_1(y, z)\frac{\partial}{\partial y} \right),$$

$$c(\tau, y, z) = \frac{1}{2}f^2(y, z)B^2(\tau) - (ar_\infty - f(y, z)\lambda(y, z))B(\tau),$$

$$d(\tau, y, z) = -g(z) \left(\rho_2 f(y, z)B(\tau)\frac{\partial}{\partial z} + \Lambda_2(y, z)\frac{\partial}{\partial z} \right).$$

The asymptotic analysis can be carried out on the x-independent differential equation (12.28), however, we continue with the partial differential equation formulation (12.25) since it allows us to more directly use the results from the analysis developed in Chapter 4.

12.2.2 Bond Price Expansion

The equation (12.25) is analogous to (4.5), however, now the Black–Scholes operator $\mathcal{L}_2$ has been replaced by the Vasicek operator $\widetilde{\mathcal{L}_2}$ and xf has been replaced by f to give the modified operators $\widetilde{\mathcal{L}_1}$ and $\widetilde{\mathcal{M}_1}$. We exploit the techniques derived in Section 4.2 to obtain the expression for the Vasicek bond price approximation.

The first-order approximation of the bond price expanded in the small parameters ε and δ is given by

$$P^{\varepsilon, \delta} \approx \tilde{P}^{\varepsilon, \delta} = P_0 + \sqrt{\varepsilon}P_{1,0} + \sqrt{\delta}P_{0,1},$$

where $P_{1,0}$ and $P_{0,1}$ are the first corrections due to fast and slow scales respectively.

We introduce the operator $\mathcal{L}_V(\bar{\sigma}(z), \bar{r}(z), a)$ that is obtained from averaging with respect to the invariant distribution for the fast scale variable y:

$$\mathcal{L}_V(\bar{\sigma}(z), \bar{r}(z), a) = \langle \widetilde{\mathcal{L}_2} \rangle, \quad (12.29)$$

which is the Vasicek operator at volatility $\bar{\sigma}$ and a risk-adjusted short rate $\bar{r}$ defined by

$$\bar{\sigma}^2(z) = \langle f^2(\cdot,z) \rangle = \int f^2(y,z)\Phi(dy), \qquad (12.30)$$

$$\bar{r}(z) = r_\infty - \frac{\langle f(\cdot,z)\lambda(\cdot,z) \rangle}{a}. \qquad (12.31)$$

A modification of Definition 4.3 now gives the leading-order term $P_0(t,x;T,z)$:

Definition 12.1 The leading-order term P_0 is the Vasicek price which solves

$$\mathscr{L}_V(\bar{\sigma}(z),\bar{r}(z),a)P_0 = 0,$$
$$P_0(T,x;T,z) = 1.$$

This is the constant volatility Vasicek price evaluated at effective parameters which results from averaging with respect to the fast variable and "freezing" the slow factor at its current level z. The explicit form for P_0 is given in (12.10); using the notation $\tau = T - t$ for the time to maturity, we can write

$$P_0(t,x;t+\tau,z) = A(\tau;z)e^{-B(\tau)x}, \qquad (12.32)$$

and as in (12.12), we obtain

$$A(\tau;z) = \exp\left(-\left[R_\infty(z)(\tau - B(\tau)) + \frac{\bar{\sigma}^2(z)}{4a}B^2(\tau)\right]\right), \qquad (12.33)$$

where we defined the z-dependent long-maturity yield

$$R_\infty(z) = \bar{r}(z) - \frac{\bar{\sigma}^2(z)}{2a^2}. \qquad (12.34)$$

Next, we obtain $P_{1,0}$ from a modification of Definition 4.4. We start by introducing the operator

$$\mathscr{A}_V = \left\langle \widetilde{\mathscr{L}}_1 \mathscr{L}_0^{-1}\left(\widetilde{\mathscr{L}}_2 - \langle \widetilde{\mathscr{L}}_2 \rangle\right) \right\rangle, \qquad (12.35)$$

which will be computed explicitly below.

Definition 12.2 The function $P_{1,0}(t,x,z)$ is the solution of the inhomogeneous problem

$$\mathscr{L}_V(\bar{\sigma}(z),\bar{r}(z),a)P_{1,0} = \mathscr{A}_V P_0, \qquad (12.36)$$
$$P_{1,0}(T,x;T,z) = 0. \qquad (12.37)$$

Thus, the correction $P_{1,0}$ is the solution of a Vasicek equation with a source term $\mathscr{A}_V P_0$ and a zero terminal condition.

We consider next the correction $P_{0,1}$ due to the slow volatility factor. Following Definition 4.7 in Chapter 4, we define:

Definition 12.3 The function $P_{0,1}(t,x,z)$ is the unique solution of the problem

$$\mathscr{L}_V(\bar{\sigma}(z),\bar{r}(z),a)P_{0,1} = -\langle\widetilde{\mathscr{M}_1}\rangle P_0, \tag{12.38}$$

$$P_{0,1}(T,x;T,z) = 0. \tag{12.39}$$

In Sections 12.2.3 and 12.2.4 below, the operators $\mathscr{A}_V$ and $\langle\widetilde{\mathscr{M}_1}\rangle$, and the associated price corrections, are computed explicitly with the resulting price approximation being summarized in Section 12.2.5.

12.2.3 The Fast Scale Correction

In order to obtain an expression for the operator $\mathscr{A}_V$ in (12.35) we introduce $\phi(y,z)$ and $\psi(y,z)$, solutions of the following Poisson equations with respect to the variable y:

$$\mathscr{L}_0\phi(y,z) = f^2(y,z) - \bar{\sigma}^2(z), \tag{12.40}$$

$$\mathscr{L}_0\psi(y,z) = \lambda(y,z)f(y,z) - \langle\lambda(\cdot,z)f(\cdot,z)\rangle. \tag{12.41}$$

Again, these functions are defined up to additive functions that do not depend on the variable y and which will not affect $\mathscr{A}_V$ since the operator $\widetilde{\mathscr{L}_1}$ takes derivatives with respect to y. We then have

$$\mathscr{L}_0^{-1}(\widetilde{\mathscr{L}_2} - \langle\widetilde{\mathscr{L}_2}\rangle) = \frac{1}{2}\phi(y,z)\frac{\partial^2}{\partial x^2} - \psi(y,z)\frac{\partial}{\partial x}, \tag{12.42}$$

and therefore

$$\widetilde{\mathscr{L}_1}\mathscr{L}_0^{-1}(\widetilde{\mathscr{L}_2} - \langle\widetilde{\mathscr{L}_2}\rangle)$$

$$= \beta(y)\left(\rho_1 f(y,z)\frac{\partial^2}{\partial x\partial y} - \Lambda_1(y,z)\frac{\partial}{\partial y}\right)\left(\frac{1}{2}\phi(y,z)\frac{\partial^2}{\partial x^2} - \psi(y,z)\frac{\partial}{\partial x}\right)$$

$$= \beta(y)\Lambda_1(y,z)\frac{\partial\psi}{\partial y}\frac{\partial}{\partial x}$$

$$- \left(\beta(y)\rho_1 f(y,z)\frac{\partial\psi}{\partial y} + \frac{1}{2}\beta(y)\Lambda_1(y,z)\frac{\partial\phi}{\partial y}\right)\frac{\partial^2}{\partial x^2} + \frac{1}{2}\rho_1\beta(y)f(y,z)\frac{\partial\phi}{\partial y}\frac{\partial^3}{\partial x^3}.$$

By averaging this expression in y with respect to the invariant distribution Φ and incorporating the small multiplicative factor $\sqrt{\varepsilon}$, we obtain from (12.35)

$$-\sqrt{\varepsilon}\,\mathscr{A}_V = V_1^{\varepsilon}(z)\frac{\partial}{\partial x} + V_2^{\varepsilon}(z)\frac{\partial^2}{\partial x^2} + V_3^{\varepsilon}(z)\frac{\partial^3}{\partial x^3}, \tag{12.43}$$

where we introduce the new small quantities

$$V_1^\varepsilon(z) = -\sqrt{\varepsilon}\left\langle \beta\Lambda_1 \frac{\partial\psi}{\partial y}\right\rangle,$$ (12.44)

$$V_2^\varepsilon(z) = \rho_1\sqrt{\varepsilon}\left\langle \beta f \frac{\partial\psi}{\partial y}\right\rangle + \frac{\sqrt{\varepsilon}}{2}\left\langle \beta\Lambda_1 \frac{\partial\phi}{\partial y}\right\rangle,$$ (12.45)

$$V_3^\varepsilon(z) = -\frac{\rho_1\sqrt{\varepsilon}}{2}\left\langle \beta f \frac{\partial\phi}{\partial y}\right\rangle.$$ (12.46)

Recall that here the averaging is with respect to the variable y for fixed z, note moreover that we suppress the z-dependence for simplicity of notation. The problem for $\sqrt{\varepsilon}P_{1,0}$ then becomes

$$\mathcal{L}_V(\bar\sigma(z),\bar r(z),a)(\sqrt{\varepsilon}P_{1,0}) = -\sum_{k=1}^{3} V_k^\varepsilon(z)\frac{\partial^k}{\partial x^k}P_0,$$

$$\sqrt{\varepsilon}P_{1,0}(T,x;T,z) = 0.$$

Before we give the explicit solution we shall combine the source operator with the one associated with the slow scale in order to get the total correction in a convenient form.

12.2.4 The Slow Scale Correction

The correction due to the slow scale is defined in (12.38). The source term for this problem involves the operator

$$\langle\widetilde{\mathcal{M}_1}\rangle = \rho_2 g\langle f\rangle\frac{\partial^2}{\partial x\partial z} - g\langle\Lambda_2\rangle\frac{\partial}{\partial z}.$$

In fact, this operator will act on functions that depend on z only through the parameters $\bar\sigma$ and $\bar r$. It is therefore natural to express the z-derivatives in terms of σ- and r-derivatives:

$$\sqrt{\delta}\langle\widetilde{\mathcal{M}_1}\rangle = V_{0,\sigma}^\delta(z)\frac{\partial}{\partial\sigma} + V_{0,r}^\delta(z)\frac{\partial}{\partial r} + V_{1,\sigma}^\delta(z)\frac{\partial^2}{\partial x\partial\sigma} + V_{1,r}^\delta(z)\frac{\partial^2}{\partial x\partial r},$$

where

$$V_{0,\sigma}^\delta(z) = -g(z)\sqrt{\delta}\,\langle\Lambda_2(\cdot,z)\rangle\bar\sigma'(z),$$ (12.47)

$$V_{0,r}^\delta(z) = -g(z)\sqrt{\delta}\,\langle\Lambda_2(\cdot,z)\rangle\bar r'(z),$$ (12.48)

$$V_{1,\sigma}^\delta(z) = \rho_2 g(z)\sqrt{\delta}\,\langle f(\cdot,z)\rangle\bar\sigma'(z),$$ (12.49)

$$V_{1,r}^\delta(z) = \rho_2 g(z)\sqrt{\delta}\,\langle f(\cdot,z)\rangle\bar r'(z).$$ (12.50)

In the next section we will combine the slow scale source term with the fast one and obtain an explicit expression for the combined correction, after introducing a convenient parameter-reduction step.

12.2.5 Combined Corrections

It follows from the above analysis that the leading-order bond price P_0 depends only on the three parameters

$$(\bar{\sigma}(z), \bar{r}(z), a), \tag{12.51}$$

which define the Vasicek operator (12.7). The price P_0 solves the problem

$$\mathscr{L}_V(\bar{\sigma}(z), \bar{r}(z), a)P_0 = 0,$$
$$P_0(T, x; T, z) = 1,$$

and can be written explicitly as

$$P_0(t, x; T, z) = A(\tau; z)e^{-B(\tau)x},$$

where $B(\tau) = (1 - \exp(-a\tau))/a$, and $A(\tau; z)$ is defined by (12.33) and depends on all the three parameters in (12.51). The corrected bond price is given by

$$\widetilde{P^{\varepsilon, \delta}} = P_0 + \sqrt{\varepsilon}P_{1,0} + \sqrt{\delta}P_{0,1},$$

with $P_{1,0}$ and $P_{0,1}$ given in (12.37) and (12.39), respectively. We use the notation

$$P_1^{\varepsilon, \delta} = \sqrt{\varepsilon}P_{1,0} + \sqrt{\delta}P_{0,1},$$

for the combined correction. The problem characterizing $P_1^{\varepsilon, \delta}$ can be written:

$$\mathscr{L}_V(\bar{\sigma}(z), \bar{r}(z), a)P_1^{\varepsilon, \delta} = -\mathscr{H}_V^{\varepsilon, \delta}P_0, \tag{12.52}$$
$$P_1^{\varepsilon, \delta}(T, x, z) = 0,$$

where, in the source term, we have used the operator notation in terms of "Greeks" as

$$\mathscr{H}_V^{\varepsilon, \delta} = \sum_{k=1}^{3} V_k^{\varepsilon}\frac{\partial^k}{\partial x^k} + \left(V_{0,r}^{\delta}\frac{\partial}{\partial r} + V_{1,r}^{\delta}\frac{\partial^2}{\partial r \partial x}\right)$$
$$+ \left(V_{0,\sigma}^{\delta}\frac{\partial}{\partial \sigma} + V_{1,\sigma}^{\delta}\frac{\partial^2}{\partial \sigma \partial x}\right). \tag{12.53}$$

In conclusion, the source term depends on the seven parameters

$$V_1^\varepsilon, V_2^\varepsilon, V_3^\varepsilon, V_{0,\sigma}^\delta, V_{0,r}^\delta, V_{1,\sigma}^\delta, V_{1,r}^\delta, \tag{12.54}$$

which depend on z and are given by (12.44)–(12.46), (12.47)–(12.50) respectively.

Observe again that the price approximation does not depend on y and that z is a parameter. Furthermore, the parameters in (12.54), which gives the modulation in the term structure, are those that determine the operator $\mathcal{H}_V^{\varepsilon,\delta}$. Therefore, if we can calibrate the model parameters with respect to observations of the yield curve for liquid contracts, these would determine the source operator $\mathcal{H}_V^{\varepsilon,\delta}$. The calibrated source operator can then be used to price less liquid contracts, such as contracts with conversion features that in general lead to boundary value problems, however, with a similar form for the source operator.

To the order of our approximation we can now proceed with a reduction of parameters step analogously to what was done in the equity case in Section 4.3.

12.2.6 Group Parameter Reduction

In Section 4.3 we considered European options and found that we could reduce the set of pricing parameters by adjusting $\bar{\sigma}$ with V_2^ε. The situation is analogous in the bond pricing case. In fact, in this case the parameter V_2^ε can be absorbed by the volatility, and the parameter V_1^ε by the interest rate level $\bar{r}$. The parameter V_2^ε scales a second derivative term and can be absorbed by the diffusion, whereas the coefficient V_1^ε scales a first derivative and can be absorbed by the drift term. In order to show this, we note first that the corrected price $\widetilde{P^{\varepsilon,\delta}}$ solves

$$\mathcal{L}_V(\bar{\sigma}(z), \bar{r}(z), a)\widetilde{P^{\varepsilon,\delta}} = -\mathcal{H}_V^{\varepsilon,\delta} P_0,$$
$$\widetilde{P^{\varepsilon,\delta}}(T, x, z) = 1,$$

with the source operator defined in (12.53). If we define

$$\sigma^\star(z) = \sqrt{\bar{\sigma}^2(z) + 2V_2^\varepsilon(z)}, \tag{12.55}$$
$$r^\star(z) = \bar{r}(z) + V_1^\varepsilon(z)/a, \tag{12.56}$$

then, dropping the z-dependence, we find

$$\frac{1}{2}\bar{\sigma}^2\frac{\partial^2\widetilde{P^{\varepsilon,\delta}}}{\partial x^2} + V_2^\varepsilon\frac{\partial^2 P_0}{\partial x^2} = \frac{1}{2}\sigma^{\star 2}\frac{\partial^2\widetilde{P^{\varepsilon,\delta}}}{\partial x^2} + \mathcal{O}(\varepsilon+\delta),$$

$$\bar{ar}\frac{\partial\widetilde{P^{\varepsilon,\delta}}}{\partial x} + V_1^\varepsilon\frac{\partial P_0}{\partial x} = ar^\star\frac{\partial\widetilde{P^{\varepsilon,\delta}}}{\partial x} + \mathcal{O}(\varepsilon+\delta).$$

We can then modify the argument presented in Section 4.3 to find that, to the same order of accuracy, we can replace $\widetilde{P^{\varepsilon,\delta}}$ by $P_0^\star + P_1^\star$, solving

$$\mathscr{L}_V(\sigma^\star,r^\star,a)P_0^\star = 0,$$
$$P_0^\star(T,x) = 1,$$

and

$$\mathscr{L}_V(\sigma^\star,r^\star,a)P_1^\star = -\mathscr{H}_V^\star P_0^\star, \tag{12.57}$$
$$P_1^\star(T,x) = 0,$$

for the reduced source operator

$$\mathscr{H}_V^\star = V_3^\varepsilon\frac{\partial^3}{\partial x^3} + \left(V_{0,r}^\delta\frac{\partial}{\partial r} + V_{1,r}^\delta\frac{\partial^2}{\partial r\partial x}\right) + \left(V_{0,\sigma}^\delta\frac{\partial}{\partial\sigma} + V_{1,\sigma}^\delta\frac{\partial^2}{\partial\sigma\partial x}\right). \tag{12.58}$$

The form of the source operator obtained in this section is valid for general contracts (with more general boundary conditions) in the stochastic volatility Vasicek short rate modeling approach and the associated calibration parameters are in general

$$a,\sigma^\star,r^\star,V_3^\varepsilon,V_{0,r}^\delta,V_{1,r}^\delta,V_{0,\sigma}^\delta,V_{1,\sigma}^\delta. \tag{12.59}$$

In the next subsection we exploit the special form of the solution for bonds and derive explicit formulas that can be used for calibration purposes.

12.2.7 Bond Approximation Formulas

The explicit form for $P_0^\star$ is given as in (12.10), so that we can write

$$P_0^\star(t,x;t+\tau,z) = A^\star(\tau;\sigma^\star,r^\star)e^{-B(\tau)x},$$

where

$$A^\star(\tau;\sigma,r) = \exp\left[-\left(r(\tau-B(\tau)) + \frac{\sigma^2}{4a^2}(aB^2(\tau) - 2\tau + 2B(\tau))\right)\right]. \tag{12.60}$$

We then find

$$
\mathcal{H}_V^\star P_0^\star(t,x;t+\tau,z) =
$$

$$
-\left[B^3(\tau)\left(-V_3^\varepsilon + \frac{\bar\sigma V_{1,\sigma}^\delta}{2a} \right) + B^2(\tau)\left(-V_{1,r}^\delta - \frac{\bar\sigma V_{0,\sigma}^\delta}{2a} + \frac{\bar\sigma V_{1,\sigma}^\delta}{a^2} \right) \right.
$$

$$
+ B(\tau)\left(V_{0,r}^\delta + \frac{\bar\sigma V_{0,\sigma}^\delta}{a^2} \right) + \tau\left(-V_{0,r}^\delta + \frac{\bar\sigma V_{0,\sigma}^\delta}{a^2} \right)
$$

$$
\left. + \tau B(\tau)\left(V_{1,r}^\delta - \frac{\bar\sigma V_{1,\sigma}^\delta}{a^2} \right) \right] P_0^\star(t,x;t+\tau,z). \tag{12.61}
$$

In order to carry out a further parameter-reduction step we write this in the form

$$
\mathcal{H}_V^\star P_0^\star(t,x;t+\tau,z) = \tag{12.62}
$$

$$
-\left(\tilde{V}_1^\delta \frac{\partial}{\partial x} + \tilde{V}_2^\delta \frac{\partial^2}{\partial x^2} + \tilde{V}_3^{\varepsilon,\delta} \frac{\partial^3}{\partial x^3} + \tau \tilde{V}_{0,r,\sigma} + \tau B(\tau)\tilde{V}_{1,r,\sigma} \right) P_0^\star(t,x;t+\tau,z),
$$

with

$$
\tilde{V}_1^\delta = -V_{0,r}^\delta - \frac{\bar\sigma V_{0,\sigma}^\delta}{a^2}, \quad \tilde{V}_2^\delta = -V_{1,r}^\delta - \frac{\bar\sigma V_{0,\sigma}^\delta}{2a} + \frac{\bar\sigma V_{1,\sigma}^\delta}{a^2},
$$

$$
\tilde{V}_3^{\varepsilon,\delta} = V_3^\varepsilon - \frac{\bar\sigma V_{1,\sigma}^\delta}{2a}, \quad \tilde{V}_{0,r,\sigma} = -V_{0,r}^\delta + \frac{\bar\sigma V_{0,\sigma}^\delta}{a^2}, \quad \tilde{V}_{1,r,\sigma} = V_{1,r}^\delta - \frac{\bar\sigma V_{1,\sigma}^\delta}{a^2}.
$$

In fact, we can now repeat the parameter reduction of the previous section. The particular form of the bond price leading to the representation (12.62) for the source means that now also the parameter $\tilde{V}_2$ can be absorbed by the volatility, and the parameter $\tilde{V}_1$ by the interest rate level $\bar r$. Therefore, introducing

$$
\tilde\sigma^\star = \sqrt{\bar\sigma^2 + 2(V_2^\varepsilon + \tilde{V}_2^\delta)}, \tag{12.63}
$$

$$
\tilde r^\star = \bar r + (V_1^\varepsilon + \tilde{V}_1^\delta)/a, \tag{12.64}
$$

we can write $\widetilde{P^{\varepsilon,\delta}} \sim \tilde{P}_0^\star + \tilde{P}_1^\star$ for

$$
\tilde{P}_0^\star(t,x;t+\tau,z) = A^\star(\tau;\tilde\sigma^\star,\tilde r^\star)e^{-B(\tau)x}, \tag{12.65}
$$

and for $\tilde{P}_1^\star$ solving

$$
\mathcal{L}_V(\sigma^\star,r^\star,a)\tilde{P}_1^\star = -\tilde{\mathcal{H}}_V^\star \tilde{P}_0^\star, \tag{12.66}
$$

$$
\tilde{P}_1^\star(T,x) = 0,
$$

with the further reduced source operator

$$\tilde{\mathscr{H}}_V^\star = \tilde{V}_3^{\varepsilon,\delta}\frac{\partial^3}{\partial x^3} + \tau \tilde{V}_{0,r,\sigma}^\delta + \tau B(\tau)\tilde{V}_{1,r,\sigma}^\delta.$$

We can solve for $\tilde{P}_1^\star$ by writing the ansatz

$$\tilde{P}_1^\star(t,x;T,z) = \tilde{D}(\tau,z)\tilde{P}_0^\star(t,x;T,z) = \tilde{D}(\tau,z)A^\star(\tau;\tilde{\sigma}^\star,\tilde{r}^\star)e^{-B(\tau)x}, \quad (12.67)$$

where $A^\star(\tau;\sigma,r)$ is defined in (12.60). Then we find

$$\frac{\partial \tilde{D}}{\partial \tau} = -(\tau \tilde{V}_{0,r,\sigma}^\delta + \tau B\tilde{V}_{1,r,\sigma}^\delta - B^3\tilde{V}_3^{\varepsilon,\delta}),$$

$$\tilde{D}(0,z) = 0.$$

Using repeatedly that $B = (1 - B')/a$, one readily obtains

$$\tilde{D}(\tau,z) = -\tilde{V}_{0,r,\sigma}^\delta\left(\frac{\tau^2}{2}\right) - \tilde{V}_{1,r,\sigma}^\delta\left(\frac{\tau^2}{2a} - \frac{1}{a^2}(B(\tau)(a\tau+1) - \tau)\right)$$

$$+ \tilde{V}_3^{\varepsilon,\delta}\left(\frac{\tau}{a^3} - \frac{B(\tau)}{a^3} - \frac{B^2(\tau)}{2a^2} - \frac{B^3(\tau)}{3a}\right). \quad (12.68)$$

The short rate reverts to its mean after epochs of characteristic duration $1/a$. In the case that the mean-reversion time is large relative to the time to maturity, that is $\tau \ll 1/a$, we find

$$\tilde{D}(\tau,z) \sim -(a\tau)^2\left(\frac{\tilde{V}_{0,r,\sigma}^\delta}{2a^2}\right) + (a\tau)^3\left(\frac{\tilde{V}_{1,r,\sigma}^\delta}{3a^3}\right) + (a\tau)^4\left(\frac{\tilde{V}_3^{\varepsilon,\delta}}{3a^4}\right), \quad (12.69)$$

whereas in the regime $\tau \gg 1/a$,

$$\tilde{D}(\tau,z) \sim -(a\tau)^2\left(\frac{\tilde{V}_{0,r,\sigma}^\delta}{2a^2}\right) + \left(\frac{(a\tau)^2}{2} - 1\right)\left(\frac{\tilde{V}_{1,r,\sigma}^\delta}{3a^3}\right)$$

$$+ \left(a\tau - \frac{11}{6}\right)\left(\frac{\tilde{V}_3^{\varepsilon,\delta}}{3a^4}\right). \quad (12.70)$$

In the context of calibration the data is typically provided in the form of bond yield curves. Next, we discuss the characteristic form for the yield that derives from the Vasicek stochastic volatility modeling.

12.2.8 Yield Correction

From (12.67) it follows that we can write

$$\widetilde{P^{\varepsilon,\delta}} \sim (1 + \tilde{D})\tilde{P}_0^\star, \quad (12.71)$$

where $\tilde{D}$ is given in (12.68).

The corresponding yield curve $\widetilde{R^{\varepsilon,\delta}} = -\frac{1}{\tau}\log\widetilde{P^{\varepsilon,\delta}}$ satisfies

$$\widetilde{R^{\varepsilon,\delta}} \sim \tilde{R}_0^\star - \frac{1}{\tau}\log\left(1+\tilde{D}\right) \sim \tilde{R}_0^\star - \frac{\tilde{D}}{\tau}, \tag{12.72}$$

where we have used the fact that the correction $\tilde{D}$ is small. Moreover, $\tilde{R}_0^\star = -\frac{1}{\tau}\log\tilde{P}_0^\star$ is the yield corresponding to the constant volatility bond price $\tilde{P}_0^\star$. Using (12.65) and (12.60), we obtain

$$\tilde{R}_0^\star = \tilde{R}_\infty^\star + (x - \tilde{R}_\infty^\star)\frac{B(\tau)}{\tau} + \frac{\tilde{\sigma}^{\star 2}}{4a}\frac{B^2(\tau)}{\tau},$$

where we defined

$$\tilde{R}_\infty^\star = \tilde{r}^\star - \frac{\tilde{\sigma}^{\star 2}}{2a^2}.$$

From (12.68) we have

$$\frac{\tilde{D}}{\tau} = -a^2\tau\left(\frac{\tilde{V}_{0,r,\sigma}^\delta}{2a^2}\right) + \left(1 + \frac{a\tau}{2} - aB(\tau) - \frac{B(\tau)}{\tau}\right)\left(\frac{\tilde{V}_{1,r,\sigma}^\delta}{a^2}\right)$$
$$+ \left(1 - \frac{B(\tau)}{\tau} - \frac{aB^2(\tau)}{2\tau} - \frac{a^2B^3(\tau)}{3\tau}\right)\left(\frac{\tilde{V}_3^{\varepsilon,\delta}}{a^3}\right).$$

Note that the asymptotics for relatively small and large maturities follow from (12.69) and (12.70).

In terms of calibration and application in the context of multiscale Vasicek bonds we have thus identified the six calibration parameters

$$a, \tilde{\sigma}^\star, \tilde{r}^\star, \tilde{V}_{0,r,\sigma}^\delta, \tilde{V}_{1,r,\sigma}^\delta, \tilde{V}_3^{\varepsilon,\delta}. \tag{12.73}$$

There are many ways of constructing estimators for these parameters from (12.72) and the observed yield curve. The practical implementation, which depends on the structure of the data (the maturities in particular), is not addressed here.

Recall that from (12.59) the parameters needed to price other instruments are

$$a, \sigma^\star, r^\star, V_3^\varepsilon, V_{0,r}^\delta, V_{1,r}^\delta, V_{0,\sigma}^\delta, V_{1,\sigma}^\delta.$$

To complete the calibration of these parameters, additional traded instruments can be used. However, from (12.31) and (12.47)–(12.50), we see that if $\lambda(\cdot,z) = f(\cdot,z)$ or $\lambda(\cdot,z) = 1/f(\cdot,z)$ then there are only two, rather than four, independent parameters in (12.47)–(12.50) and the calibration of the yield fully specifies the source operator in (12.58).

12.3 The Quadratic Model

In the Vasicek model the interest rate may become negative. An important model with a non-negative interest rate is the so-called quadratic Gaussian model. We discuss it here briefly in the context of multiscale stochastic volatility. This example also serves to illustrate how the multiscale framework can easily be adapted to various interest rate models.

In the quadratic model we start by letting the short rate r_t be defined by the square

$$r_t = X_t^2,$$

with X_t being an Ornstein–Uhlenbeck process where for simplicity we assume here that X_t has zero long-run mean. More generally, we can choose r to be a second-order polynomial in X. We first consider the case with constant volatility σ, so that

$$dX_t = -\kappa X_t dt + \sigma dW_t^\star. \tag{12.74}$$

The bond price is given by

$$P(t,x;T) = \mathbb{E}^\star \left\{ e^{-\int_t^T X_s^2 ds} \mid X_t = x \right\}.$$

Applying the Feynman–Kac formula (1.80) we find that $P(t,x;T)$ satisfies the pricing equation

$$\mathcal{L}_Q P = 0, \tag{12.75}$$
$$P(T,x;T) = 1,$$

with $\mathcal{L}_Q = \mathcal{L}_Q(\sigma,\kappa)$ being the (centered) quadratic model operator:

$$\mathcal{L}_Q = \frac{\partial}{\partial t} + \frac{1}{2}\sigma^2 \frac{\partial^2}{\partial x^2} - \kappa x \frac{\partial}{\partial x} - x^2 \cdot . \tag{12.76}$$

We make the ansatz

$$P(T - \tau, x; T) = A(\tau) e^{-B(\tau)x^2}$$

for the solution of (12.75). This produces the following ordinary differential equations:

$$\frac{dB}{d\tau} = -2\sigma^2 B^2 - 2\kappa B + 1,$$
$$\frac{dA}{d\tau} = -\sigma^2 BA,$$

with $A(0) = 1, B(0) = 0$. The former is a Riccati equation for the quadratic coefficient. By introducing the following parameters:

$$f = \sqrt{1 + 2\sigma^2/\kappa^2}, \quad \gamma = \frac{f-1}{f+1}, \quad \tau_0 = \frac{1}{2\kappa f}, \quad a = \frac{1}{\kappa(f+1)}, \quad (12.77)$$

we find explicitly

$$\log A(\tau) = -\sigma^2 \left(a\tau + \frac{a(\gamma+1)\tau_0}{\gamma} \log \left(\frac{1 + \gamma \exp(-\tau/\tau_0)}{1+\gamma} \right) \right),$$

$$B(\tau) = a \left(\frac{1 - e^{-\tau/\tau_0}}{1 + \gamma e^{-\tau/\tau_0}} \right),$$

so that the constant volatility bond price takes the form

$$P(T - \tau, x; T) = \left[\frac{1+\gamma}{1 + \gamma e^{-\tau/\tau_0}} \right]^{a\sigma^2 \tau_0(1+1/\gamma)} e^{-B(\tau)x^2 - a\sigma^2 \tau}. \quad (12.78)$$

We remark that the associated yield satisfies the following asymptotics at the two ends of the curve:

$$\lim_{\tau \to 0^+} -\frac{1}{\tau} \log P = \frac{ax^2}{\tau_0(1+\gamma)},$$

$$\lim_{\tau \to \infty} -\frac{1}{\tau} \log P = a\sigma^2 = \frac{\sigma^2}{\kappa(1+f)}.$$

We turn now to the stochastic volatility case where the dynamics of X_t is given by

$$dX_t = -\kappa X_t dt + f(Y_t, Z_t) dW_t^{(0)\star}, \quad (12.79)$$

$$dY_t = \frac{1}{\varepsilon} \alpha(Y_t) + \frac{1}{\sqrt{\varepsilon}} \beta(Y_t) dW_t^{(1)\star},$$

$$dZ_t = \delta c(Z_t) + \sqrt{\delta} g(Z_t) dW_t^{(2)\star},$$

and where we again state the model directly in terms of the pricing measure so that the $W^{(i)\star}$'s are Brownian motion with respect to the equivalent martingale measure and we do not explicitly introduce market price of risk factors.

In the stochastic volatility case the bond pricing function solves (12.25) with $\mathscr{L}^{\varepsilon,\delta}$ having the form

$$\mathscr{L}^{\varepsilon,\delta} = \frac{1}{\varepsilon}\mathscr{L}_0 + \frac{1}{\sqrt{\varepsilon}}\mathscr{L}_1 + \mathscr{L}_2 + \sqrt{\delta}\mathscr{M}_1 + \delta\mathscr{M}_2 + \sqrt{\frac{\delta}{\varepsilon}}\mathscr{M}_3. \quad (12.80)$$

These operators are now defined according to the quadratic model (12.79) and read

$$\mathcal{L}_0 = \frac{1}{2}\beta^2(y)\frac{\partial^2}{\partial y^2} + \alpha(y)\frac{\partial}{\partial y},$$

$$\mathcal{L}_1 = \beta(y)\rho_1 f(y,z)x\frac{\partial^2}{\partial x\partial y},$$

$$\mathcal{L}_2 = \frac{\partial}{\partial t} + \frac{1}{2}f^2(y,z)x\frac{\partial^2}{\partial x^2} - \kappa x\frac{\partial}{\partial x} - x^2\cdot,$$

$$\mathcal{M}_1 = xg(z)\rho_2 f(y,z)x\frac{\partial^2}{\partial x\partial z},$$

$$\mathcal{M}_2 = \frac{1}{2}g^2(z)\frac{\partial^2}{\partial z^2} + xc(z)\frac{\partial}{\partial z},$$

$$\mathcal{M}_3 = x\beta(y)\rho_{12}g(z)\frac{\partial^2}{\partial y\partial z}.$$

The leading-order price P_0 is given by (12.78) with σ replaced by $\bar{\sigma}(z)$. The source operators that define the price corrections are

$$\sqrt{\varepsilon}\mathcal{A}_Q = \left\langle \mathcal{L}_1\mathcal{L}_0^{-1}\left(\mathcal{L}_2 - \langle\mathcal{L}_2\rangle\right)\right\rangle$$

$$= \frac{\rho_1\sqrt{\varepsilon}}{2}\left\langle \beta f(\cdot,z)\frac{\partial\phi}{\partial y}\right\rangle\frac{\partial}{\partial x^3} \equiv -V_3^\varepsilon\frac{\partial}{\partial x^3},$$

$$-\sqrt{\delta}\langle\mathcal{M}_1\rangle = -\rho_2 g(z)\langle f(\cdot,z)\rangle\frac{\partial^2}{\partial x\partial z} \equiv -V_1^\delta\frac{\partial^2}{\partial x\partial z},$$

so that the correction due to the fast scale solves explicitly

$$\mathcal{L}_Q(\bar{\sigma}(z),\kappa)(\sqrt{\varepsilon}P_{1,0}) = -V_3^\varepsilon(z)\frac{\partial^3 P_0}{\partial x^3}, \tag{12.81}$$

$$P_{0,1}(T,x;T,z) = 0, \tag{12.82}$$

while the correction due to the slow scale is determined by

$$\mathcal{L}_Q(\bar{\sigma}(z),\kappa)(\sqrt{\delta}P_{0,1}) = -V_1^\delta(z)\frac{\partial^2 P_0}{\partial z\partial x}, \tag{12.83}$$

$$P_{0,1}(T,x;T,z) = 0. \tag{12.84}$$

The fast scale correction follows by making the ansatz

$$\sqrt{\varepsilon}P_{1,0} = (xD_1^\varepsilon + x^3 D_3^\varepsilon)P_0,$$

where here the D_i^ε are small maturity-dependent parameters that in view of (12.81) solve

$$\frac{dD_1^\varepsilon}{d\tau} = (-2\bar\sigma^2 B - \kappa)D_1^\varepsilon + 3\bar\sigma^2 D_3^\varepsilon + 12V_3^\varepsilon B^2, \qquad (12.85)$$

$$\frac{dD_3^\varepsilon}{d\tau} = 3(-2\bar\sigma^2 B - \kappa)D_3^\varepsilon - 8V_3^\varepsilon B^3, \qquad (12.86)$$

with $D_1^\varepsilon(0) = D_3^\varepsilon(0) = 0$. Similarily, we find that the ansatz

$$\sqrt{\varepsilon}P_{0,1} = (xD_1^\delta + x^3 D_3^\delta)P_0$$

and (12.83) give

$$\frac{dD_1^\delta}{d\tau} = (-2\bar\sigma^2 B - \kappa)D_1^\delta + 3\bar\sigma^2 D_3^\delta + \tilde V_1^\delta\left(-2\frac{\partial B}{\partial\sigma} - BC_\sigma\right), \quad (12.87)$$

$$\frac{dD_3^\delta}{d\tau} = 3(-2\bar\sigma^2 B - \kappa)D_3^\delta + \tilde V_1^\delta\left(2A\frac{\partial A}{\partial\sigma}\right), \qquad (12.88)$$

for

$$\tilde V_1^\delta = \bar\sigma' V_1^\delta, \quad D_1^\delta(0) = D_3^\delta(0) = 0.$$

This system of first-order ordinary differential equations can be solved analytically or simply via numerical quadrature.

In Figure 12.4 we show the fast term factors D_1^ε and D_3^ε in the top plot and the slow correction factors D_1^δ and D_3^δ in the bottom plot and how all these approach their asymptotic value. In the example we used $\bar\sigma = 0.2$. We also used $\kappa = \tilde V_1^\delta = V_3^\varepsilon = 1$, which corresponds to a time scale and magnitude normalization.

12.4 The CIR Model

In this section we introduce and discuss a third model for interest that we refer to as the CIR (Cox–Ingersoll–Ross) model. This serves to further illustrate the flexibility of our stochastic volatility perturbation approach framework.

We recall first that the CIR model, also known as the square root process, in standard form can be written as

$$dr_t = \kappa(r^\star - r_t)dt + \sqrt{r_t}\sigma dW_t^{(0)\star}, \qquad (12.89)$$

with $\kappa, r^\star, \sigma$ strictly positive parameters. Both the Vasicek and the CIR models are examples of affine models for which the yield is affine in the

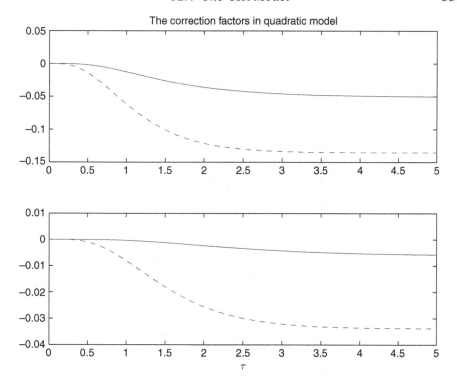

Figure 12.4 The bond price (fast) correction term structure factors D_1^ε and D_3^ε are shown in the top plot by the solid and dashed lines, respectively, correspondingly D_1^δ and D_3^δ are shown in the bottom plot. We use the parameters $\kappa = \tilde{V}_1^\delta = V_3^\varepsilon = 1$ and $\bar{\sigma} = 0.25$. The factors are plotted as a function of time to maturity for the bond.

state variable, they are also particular examples of the larger Hull–White class of models with time-dependent parameters.

The bond price is again given by the expectation

$$P(t,x;T) = \mathbb{E}^\star \left\{ e^{-\int_t^T r_s ds} \mid r_t = x \right\}, \tag{12.90}$$

with x being the current level of the interest rate. By applying the Feynman–Kac formula (1.80) we find that in the *constant volatility* case, $P(t,x;T)$ satsifies the pricing equation

$$\mathscr{L}_{CIR} P = 0, \tag{12.91}$$
$$P(T,x;T) = 1,$$

where $\mathscr{L}_{CIR} = \mathscr{L}_{CIR}(\sigma, r^\star, \kappa)$ is the CIR operator:

$$\mathscr{L}_{CIR} = \frac{\partial}{\partial t} + \frac{1}{2}\sigma^2 x \frac{\partial^2}{\partial x^2} + \kappa(r^\star - x)\frac{\partial}{\partial x} - x \cdot. \tag{12.92}$$

The solution of (12.91) is given by the one-factor affine CIR formula

$$P(T - \tau, x; T) = A(\tau)e^{-B(\tau)x},$$

where here the functions $A(\tau)$ and $B(\tau)$ are solutions of

$$\frac{dA(\tau)}{d\tau} = -\kappa r^* A(\tau)B(\tau),$$

$$\frac{dB(\tau)}{d\tau} = -\frac{\sigma^2}{2}B^2(\tau) - \kappa B(\tau) + 1.$$

These functions are explicitly given by

$$A(\tau) = \left[\frac{2\gamma e^{\frac{1}{2}(\gamma+\kappa)\tau}}{(\gamma+\kappa)(e^{\gamma\tau}-1)+2\gamma}\right]^{2\kappa r^*/\sigma^2}, \qquad (12.93)$$

$$B(\tau) = \left[\frac{2(e^{\gamma\tau}-1)}{(\gamma+\kappa)(e^{\gamma\tau}-1)+2\gamma}\right], \qquad (12.94)$$

$$\gamma = \sqrt{\kappa^2 + 2\sigma^2}.$$

The following stochastic volatility extension of the CIR model will lead to closed expressions for the yield curve. We model the interest rate under the equivalent martingale (pricing) measure by

$$dr_t = \kappa(r^* - r_t))dt + \sqrt{r_t}f(Y_t, Z_t)dW_t^{(0)\star}, \qquad (12.95)$$

$$dY_t = \frac{r_t}{\varepsilon}\alpha(Y_t) + \sqrt{\frac{r_t}{\varepsilon}}\beta(Y_t)\,dW_t^{(1)\star},$$

$$dZ_t = r_t\delta\,c(Z_t) + \sqrt{r_t\delta}\,g(Z_t)\,dW_t^{(2)\star},$$

and we do not articulate explicitly the relation between physical and pricing measures for this model in order to work with simple expressions capturing the main aspects of the CIR dynamics. Note that the equations for the volatility factors are not autonomous but depend on r_t; this modification has been introduced in order to stay within the affine yield framework, as we will see below. This modeling corresponds to a "local time scaling" by the positive level of r_t, that is, the local scaling of the quadratic variation of the driving process $\int_0^t \sqrt{r_s}dW_s^{(0)\star}$ so that the parameters ε and $1/\delta$ retain their interpretation as time scale parameters. Here we assume that the function $f(y, z)$ and the coefficients of the equations for Y, Z are such that a unique strong solution exists and r_t stays positive for all time.

In the stochastic volatility case the price solves (12.25), with $\mathscr{L}^{\varepsilon,\delta}$ having the form

$$\mathscr{L}^{\varepsilon,\delta} = \frac{1}{\varepsilon}\mathscr{L}_0 + \frac{1}{\sqrt{\varepsilon}}\mathscr{L}_1 + \mathscr{L}_2 + \sqrt{\delta}\,\mathscr{M}_1 + \delta\mathscr{M}_2 + \sqrt{\frac{\delta}{\varepsilon}}\mathscr{M}_3, \quad (12.96)$$

where the operators are now defined according to the CIR model (12.95):

$$\mathscr{L}_0 = x\left(\frac{1}{2}\beta^2(y)\frac{\partial^2}{\partial y^2} + \alpha(y)\frac{\partial}{\partial y}\right),$$

$$\mathscr{L}_1 = \beta(y)\rho_1 f(y,z)x\frac{\partial^2}{\partial x\partial y},$$

$$\mathscr{L}_2 = \frac{\partial}{\partial t} + \frac{1}{2}f^2(y,z)x\frac{\partial^2}{\partial x^2} + \kappa(r^\star - x)\frac{\partial}{\partial x} - x\cdot,$$

$$\mathscr{M}_1 = xg(z)\rho_2 f(y,z)x\frac{\partial^2}{\partial x\partial z},$$

$$\mathscr{M}_2 = \frac{1}{2}xg^2(z)\frac{\partial^2}{\partial z^2} + xc(z)\frac{\partial}{\partial z},$$

$$\mathscr{M}_3 = x\beta(y)\rho_{12}g(z)\frac{\partial^2}{\partial y\partial z}.$$

By carrying out the regular singular expansion method introduced in Chapter 4 with these modified operators, we find that the leading-order price solves the CIR constant volatility problem (12.91) evaluated at the effective volatility level determined as in (12.30), so that

$$P_0(T - \tau, x; T) = A(\tau)e^{-B(\tau)x}, \quad (12.97)$$

where $A(\tau)$ and $B(\tau)$ are given by (12.93) and (12.94) with σ replaced by $\bar{\sigma}(z)$. The correction due to the fast scale solves

$$\mathscr{L}_{CIR}(\bar{\sigma}(z), r^\star(z), a)P_{1,0} = -xV_3^\varepsilon(z)\frac{\partial^3 P_0}{\partial x^3},$$

$$P_{0,1}(T, x; T, z) = 0,$$

while the correction due to the slow scale is determined by

$$\mathscr{L}_{CIR}(\bar{\sigma}(z), r^\star(z), a)P_{0,1} = -xV_1^\delta(z)\frac{\partial^2 P_0}{\partial z\partial x},$$

$$P_{0,1}(T, x; T, z) = 0.$$

Here we have introduced the new z-dependent small parameters

$$V_1^\delta(z) = \sqrt{\delta}\rho_2 g(z)\langle f \rangle,$$

$$V_3^\varepsilon(z) = -\frac{\sqrt{\varepsilon}\rho_1}{2}\left\langle \beta f \frac{\partial \phi}{\partial y} \right\rangle,$$

with ϕ solving the same Poisson equation as in (12.40) and the rectangular brackets representing averaging with respect to the invariant distribution associated with the Poisson equation.

We now make the "educated" ansatz

$$\widetilde{P^{\varepsilon,\delta}} = (1 + D_0^{\varepsilon,\delta} + D_1^{\varepsilon,\delta}x + D_2^\delta x^2)P_0 \qquad (12.98)$$

for the price approximation. It then follows that the correction factors solve

$$\frac{\partial D_0^{\varepsilon,\delta}}{\partial \tau} = \kappa r^\star D_1^{\varepsilon,\delta},$$

$$\frac{\partial D_1^{\varepsilon,\delta}}{\partial \tau} = -(\kappa + B\bar{\sigma}^2)D_1^{\varepsilon,\delta} + (2\kappa r^\star + \bar{\sigma}^2)D_2^\delta - \left(V_3^\varepsilon B^3 + \tilde{V}_1^\delta B\left(\frac{\partial A/\partial \sigma}{A}\right)\right),$$

$$\frac{\partial D_2^\delta}{\partial \tau} = -2(\kappa + B\bar{\sigma}^2)D_2^\delta + \tilde{V}_1^\delta B\frac{\partial B}{\partial \sigma},$$

with zero initial conditions and where the functions A and B are given by (12.93) and (12.94) and $\tilde{V}_1^\delta = \bar{\sigma}'V_1^\delta$.

In Figure 12.5 we show the factors $D_0^{\varepsilon,\delta}$, $D_1^{\varepsilon,\delta}$, and D_2^δ by respectively dotted, dashed, and solid lines. In the example we used $\bar{\sigma} = 0.2$, $\kappa = 1$, $r^\star = 0.05$ and $\tilde{V}_1^\delta = V_3^\varepsilon = 0.1$ and the factors have been computed numerically from the system of ordinary differential equations that they satisfy.

We end this section with a remark regarding the connection between the quadratic model of the previous section and the CIR model introduced here. Let x_t be the Ornstein–Uhlenbeck process

$$dx_t = -\kappa x_t dt + \sigma dW_t,$$

then by an application of Itô's formula we find that $r_t = \frac{1}{2}x_t^2$ solves

$$dr_t = \kappa\left(\frac{\sigma^2}{2\kappa} - r_t\right)dt + \sigma\sqrt{r_t}dW_t,$$

which is a CIR process (12.89) with the parameter constraint $r^\star = \sigma^2/(2\kappa)$. Note however that, in the context of stochastic volatility, the drift coefficient will become stochastic as well.

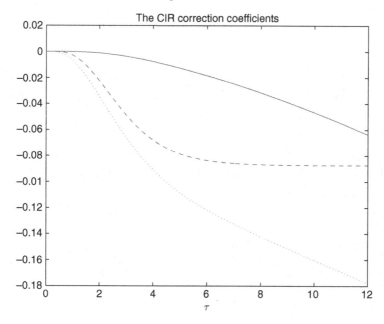

Figure 12.5   The CIR price correction term structure factors $D_0^{\varepsilon,\delta}$, $D_1^{\varepsilon,\delta}$, and D_2^{δ} are shown by respectively dotted, dashed, and solid lines as functions of time to maturity. In this example, we used $\bar{\sigma} = 0.2$, $\kappa = 1$, $r^{\star} = 0.05$ and $\tilde{V}_1^{\delta} = V_3^{\varepsilon} = 0.1$.

12.5 Options on Bonds

We have discussed pricing of zero-coupon bonds which pay a unit amount at maturity. The bond price is itself a stochastic process and we now consider the situation when it plays the role of the underlying tradable security. We may then want to buy, for instance, a call option on this security which gives us the right to buy the bond at a given strike price K and a specific maturity time T. We shall denote the maturity of the *bond itself* by T_0. Note that for the call option we will have $T < T_0$ and $K < 1$. More generally, we could consider a bond option with a payoff function h at maturity T.

12.5.1 *Pricing via Partial Differential Equations*

Given a Markovian model for the short rate, with constant volatility for instance, the bond option price denoted by $Q(t,x;T)$ is given by

$$Q(t,x;T) = \mathbb{E}^{\star}\left\{e^{-\int_t^T r_s ds} h(P(T,r_T;T_0)) \mid r_t = x\right\}, \qquad (12.99)$$

where $P(T,r_T;T_0)$ denotes the price of the bond with maturity T_0, at time T and short rate r_T. Applying the Feynman–Kac formula we find that Q is also characterized as the solution of the following partial differential equation:

$$\left(\frac{\partial}{\partial t} + \mathscr{L} - x\right)Q = 0, \tag{12.100}$$

$$Q(T,x;T_0) = h(P(T,x;T_0)), \tag{12.101}$$

where $\mathscr{L}$ is the infinitesimal generator of the process (r_t). Note that the terminal condition is itself given as the solution of a partial differential equation. Alternatively, in order to compute the expectation in (12.99), we need to use the joint distribution of $(\int_t^T r_s ds, r_T)$ under the pricing measure. A convenient alternative approach to this problem is to introduce the forward measure, which we do next.

12.5.2 *Pricing via Forward Measures*

For simplicity of notation, in what follows, we use the notation

$$P(u,T) \equiv P(u,r_u;T),$$

and similarly for the bond option price Q. Without loss of generality, we set $t = 0$, so that the price of a call on the bond is given by

$$Q(0,T) = \mathbb{E}^\star\left\{e^{-\int_0^T r_s ds}\left(P(T,T_0) - K\right)^+\right\}$$

$$= \mathbb{E}^\star\left\{e^{-\int_0^T r_s ds}P(T,T_0)\mathbf{1}_{\{P(T,T_0)>K\}}\right\}$$

$$- K\mathbb{E}^\star\left\{e^{-\int_0^T r_s ds}\mathbf{1}_{\{P(T,T_0)>K\}}\right\}. \tag{12.102}$$

We rewrite the first expectation in the form

$$\mathbb{E}^\star\left\{e^{-\int_0^T r_s ds}P(T,T_0)\mathbf{1}_{\{P(T,T_0)>K\}}\right\}$$

$$= P(0,T)\mathbb{E}^\star\left\{\frac{e^{-\int_0^T r_s ds}}{P(0,T)}P(T,T_0)\mathbf{1}_{\{P(T,T_0)>K\}}\right\},$$

where $P(0,T)$ is the price at $t = 0$ (the present time) of a bond maturing at T, which is given by the market. Similarly, we rewrite the second expectation in (12.102) in the form

$$\mathbb{E}^\star\left\{e^{-\int_0^T r_s ds}\mathbf{1}_{\{P(T,T_0)>K\}}\right\} = P(0,T)\mathbb{E}^\star\left\{\frac{e^{-\int_0^T r_s ds}}{P(0,T)}\mathbf{1}_{\{P(T,T_0)>K\}}\right\}.$$

Now we define the forward measure $\mathbb{P}^T$ by its Radon–Nikodym derivative

$$\xi_T \equiv \frac{d\mathbb{P}^T}{d\mathbb{P}^\star} = \frac{e^{-\int_0^T r_s ds}}{P(0,T)}, \tag{12.103}$$

where by definition

$$\mathbb{E}^{\star}\left\{e^{-\int_0^T r_s ds}\right\} = P(0,T),$$

so that $\mathbb{P}^T$ is a probability measure. Finally, we can write (12.102) as

$$Q(0,T) = P(0,T)\left(\mathbb{E}^T\left\{P(T,T_0)\mathbf{1}_{\{P(T,T_0)>K\}}\right\}\right.$$
$$\left. -K\mathbb{P}^T\left\{P(T,T_0) > K\right\}\right). \tag{12.104}$$

When the dynamics of the bond price $P(t,T)$ is governed by

$$dP(t,T) = P(t,T)\left(r_t dt - \Sigma(t,T)dW_t^{\star}\right), \tag{12.105}$$

with a deterministic volatility $\Sigma(t,T)$, as for instance in the Vasicek model (12.17), the density (12.103) restricted to $\mathcal{F}_t$ for $t \in [0,T]$ is given by

$$\xi_t \equiv \left.\frac{d\mathbb{P}^T}{d\mathbb{P}^{\star}}\right|_{\mathcal{F}_t} \equiv \mathbb{E}^{\star}\{\xi_T \mid \mathcal{F}_t\} = \frac{e^{-\int_0^t r_s ds}P(t,T)}{P(0,T)}. \tag{12.106}$$

In fact, what we have done is related to a *change of numeraire*, meaning that tradable quantities are measured in the unit of the bond maturing at time T. For instance, the value at time $t \le T$ of \$1 put in the savings account at time zero, is $e^{\int_0^t r_s ds}$, which in units of the bond becomes $e^{\int_0^t r_s ds}/P(t,T)$. Similarly, if there is another traded asset in the market, say a stock with price S_t, then its value at time t in units of the bond will be $S_t/P(t,T)$. Indeed, the bond itself in units of the bond is $P(t,T)/P(t,T) = 1$. Moreover, it is a general result that traded quantities measured in units of the bond are martingales with respect to the forward measure $\mathbb{P}^T$. Using the conditional change of measure formula (1.59) for $s \le t \le T$ and (12.106), we have for the savings account

$$\mathbb{E}^T\left\{\frac{e^{\int_0^t r_u du}}{P(t,T)} \mid \mathcal{F}_s\right\} = \frac{1}{\xi_s}\mathbb{E}^{\star}\left\{\frac{e^{\int_0^t r_u du}}{P(t,T)}\xi_t \mid \mathcal{F}_s\right\}$$

$$= \frac{1}{\xi_s}\mathbb{E}^{\star}\left\{\frac{e^{\int_0^t r_u du}}{P(t,T)}\frac{e^{-\int_0^t r_u du}P(t,T)}{P(0,T)} \mid \mathcal{F}_s\right\}$$

$$= \frac{1}{\xi_s P(0,T)} = \frac{e^{\int_0^s r_u du}}{P(s,T)}.$$

In fact, for this martingale property to hold, the important property of the savings account is that when discounted (by itself) it is a $\mathbb{P}^{\star}$-martingale. This is clearly seen for a general traded asset S_t:

$$\mathbb{E}^T\left\{\frac{S_t}{P(t,T)}\mid\mathscr{F}_s\right\} = \frac{1}{\xi_s}\mathbb{E}^\star\left\{\frac{S_t}{P(t,T)}\,\xi_t\mid\mathscr{F}_s\right\}$$

$$= \frac{1}{\xi_s}\mathbb{E}^\star\left\{\frac{S_t}{P(t,T)}\,\frac{e^{-\int_0^t r_u du}P(t,T)}{P(0,T)}\mid\mathscr{F}_s\right\}$$

$$= \frac{1}{\xi_s P(0,T)}\mathbb{E}^\star\left\{S_t\,e^{-\int_0^t r_u du}\mid\mathscr{F}_s\right\}$$

$$= \frac{S_s\,e^{-\int_0^s r_u du}}{\xi_s P(0,T)} = \frac{S_s}{P(s,T)}.$$

We now derive the dynamics of the short rate process (r_t) under $\mathbb{P}^T$ for $t \le T$. From (12.105) we have

$$P(t,T) = P(0,T)\,e^{\int_0^t r_s ds - \frac{1}{2}\int_0^t \Sigma(s,T)^2 ds - \int_0^t \Sigma(s,T)dW_s^\star}.$$

Using (12.106) we deduce

$$\xi_t = \frac{e^{-\int_0^t r_s ds}}{P(0,T)}P(0,T)\,e^{\int_0^t r_s ds - \frac{1}{2}\int_0^t \Sigma(s,T)^2 ds - \int_0^t \Sigma(s,T)dW_s^\star}$$

$$= e^{-\frac{1}{2}\int_0^t \Sigma(s,T)^2 ds - \int_0^t \Sigma(s,T)dW_s^\star}.$$

Here we recognize a Girsanov density (see Section 1.9.2). Therefore, the process W^T defined by

$$W_t^T \equiv W_t^\star - \int_0^t \Sigma(s,T)ds$$

is a Brownian motion under $\mathbb{P}^T$. If the dynamics of the short rate (r_t) is given under $\mathbb{P}^\star$ by

$$dr_t = a(r_t)dt + \sigma(r_t)dW_t^\star,$$

then it is given under $\mathbb{P}^T$ by

$$dr_t = (a(r_t) + \sigma(r_t)\Sigma(t,T))\,dt + \sigma(r_t)dW_t^T.$$

Combined with the representation (12.104) of the bond option, we have arrived at an efficient way for computing the bond option price as illustrated in the case of the Vasicek model given in the following section.

12.5.3 Bond Option Pricing under the Vasicek Model

Recall from Section 12.1 that under $\mathbb{P}^\star$,

$$dr_t = a(\bar{r} - r_t)dt + \sigma dW_t^\star,$$

and that the bond volatility $\Sigma(t,T)$ is given by (12.105):

$$\Sigma(t,T) = \sigma \left(\frac{1-e^{-a(T-t)}}{a} \right).$$

Therefore, from the previous section, the dynamics under $\mathbb{P}^T$ is given by

$$dr_t = a \left(\bar{r} + \frac{\sigma^2}{a^2}(1-e^{-a(T-t)}) - r_t \right) dt + \sigma dW_t^T. \qquad (12.107)$$

The bond price $P(T,T_0)$ is given by $P(T,T_0) = A(T_0-T)e^{-B(T_0-T)r_T}$, where B and A are given by (12.9) and (12.12), respectively. Therefore,

$$\{P(T,T_0) > K\} = \{r_T < R\},$$

where we defined

$$R \equiv \frac{\log K - \log A(T_0 - T)}{B(T_0 - T)},$$

and the bond option price (12.104) can be rewritten as

$$Q(0,T) = P(0,T) \left(\mathbb{E}^T \left\{ A(T_0 - T)e^{-B(T_0-T)r_T} \mathbf{1}_{\{r_T < R\}} \right\} - K\mathbb{P}^T \left\{ r_T < R \right\} \right).$$

From (12.107) we deduce that, under $\mathbb{P}^T$, r_T is Gaussian with mean and variance given by

$$\mathbb{E}^T\{r_T\} = r_0 + a(\bar{r} - r_0)B(T) + \frac{\sigma^2}{2}B(T)^2,$$

$$\mathrm{Var}^T\{r_T\} = \frac{\sigma^2}{2}B(2T).$$

Finally, a direct Gaussian calculation gives

$$Q(0,T) = P(0,T)\left(N(d_1) - KN(d_2)\right), \qquad (12.108)$$

where N is the $\mathcal{N}(0,1)$-cdf, and d_1 and d_2 are easily computed as in the derivation of the Black–Scholes formula.

12.5.4 Stochastic Volatility Correction Factors

In this section we analyze how the bond option price is modified when we let the volatility be stochastic as in Section 12.2. As in the case with bond price approximation, the correction terms due to the fast and slow scale perturbations in the volatility will solve the problems in Definitions 12.2 and 12.3 respectively, however, now with P_0 replaced by the leading-order or constant volatility bond option price Q that we considered in the previous section and that we shall denote by Q_0. We denote the combined first-order correction terms by $Q_1^{\varepsilon,\delta}$, that is

$$Q_1^{\varepsilon,\delta} = \sqrt{\varepsilon}\,Q_{1,0} + \sqrt{\delta}\,Q_{0,1}.$$

One then finds that the problem characterizing $Q_1^{\varepsilon,\delta}$ can be written:

$$\mathscr{L}_V(\bar{\sigma},\bar{r},a)Q_1^{\varepsilon,\delta} = -\mathscr{H}_V^{\varepsilon,\delta}Q_0, \tag{12.109}$$

$$Q_1^{\varepsilon,\delta}(T,x,z) = \left(\sqrt{\varepsilon}\,P_{1,0}(T,T_0) + \sqrt{\delta}\,P_{0,1}(T,T_0)\right)h'(P_0(T,T_0)),$$

where, in the source term, the operator $\mathscr{H}_V^{\varepsilon,\delta}$ is given by (12.53), Q_0 is the bond option price evaluated at $\sigma = \bar{\sigma}$, and the terminal condition has been obtained by first-order Taylor expansions of the payoff $h(P^{\varepsilon,\delta}(T,T_0))$. Here we assume that h is smooth, but this first-order expansion for a call payoff remains valid, in which case $h'(x) = \mathbf{1}_{\{x>K\}}$. The leading-order term $P_0(T,T_0)$ is the bond price at time T maturing at T_0, computed with constant volatility $\bar{\sigma}$. The bond price corrections $P_{1,0}(T,T_0)$ and $P_{0,1}(T,T_0)$ are given in Definitions 12.2 and 12.3, respectively.

Without entering into the details, the parameter-reduction step presented for the bond price corrections in Section 12.2.6 can be carried out in a similar way for the leading-order bond option price Q_0 and the combined bond option price correction $Q_1^{\varepsilon,\delta}$. In particular, the constant volatility $\bar{\sigma}$ is replaced by the adjusted effective volatility $\tilde{\sigma}^\star$ given by (12.63), and $\bar{r}$ is replaced by $\tilde{r}^\star$ given by (12.64). The explicit formulas for the bond price approximation are given in Section 12.2.7.

The combined bond option price correction $Q_1^{\varepsilon,\delta}$ solves the partial differential equation problem (12.109) where the parameters $(\bar{\sigma},\bar{r})$ have been replaced by $(\tilde{\sigma}^\star,\tilde{r}^\star)$. We make the following ansatz:

$$Q_1^{\varepsilon,\delta}(t) = \sum_{k=0}^{3} D_k(t)\frac{\partial^k Q_0}{\partial x^k} + D_4(t)\frac{\partial Q_0}{\partial z} + + D_5(t)\frac{\partial^2 Q_0}{\partial z\partial x} + Q_T^{\varepsilon,\delta}(t), \tag{12.110}$$

where $Q_T^{\varepsilon,\delta}(t)$ denotes the contribution from the terminal condition and the other terms come from the source term in (12.109).

Specializing to a call option, Q_0 is given explicitly in the Section 12.5.3 with the parameters $(\tilde{\sigma}^\star,\tilde{r}^\star)$. The contribution from the terminal condition is given by:

$$Q_T^{\varepsilon,\delta}(t) = \mathbb{E}^\star \left\{ e^{-\int_t^T r_s ds} P_1^{\varepsilon,\delta}(T,r_T;T_0) \mathbf{1}_{\{P_0(T,r_T;T_0)>K\}} \mid r_t \right\}$$
$$= \tilde{D}(T_0 - T)\mathbb{E}^\star \left\{ e^{-\int_t^T r_s ds} P_0(T,r_T;T_0) \mathbf{1}_{\{P_0(T,r_T;T_0)>K\}} \mid r_t \right\},$$

where we have used the form of $P_1^{\varepsilon,\delta}$ in (12.67) and where the factor $\tilde{D}$ is given in (12.68). The dynamics of (r_t) follows an OU process with parameters $(a, \tilde{r}^\star, \tilde{\sigma}^\star)$. We observe that the expectation above is identical (but evaluated at time t) to the first term appearing in the bond option price (12.102). Therefore, using the explicit formula (12.108), we deduce

$$Q_T^{\varepsilon,\delta}(t) = \tilde{D}(T_0 - T)P_0(t,T)N(d_1).$$

The contribution from the source is obtained by computing the $D_i(t)$ which solve a linear system of differential equations that follows from plugging the ansatz (12.110) into equation (12.109). We omit the details here.

In conclusion, the bond option price correction $Q_1^{\varepsilon,\delta}(t)$ given in (12.110) can be computed using only the six parameters $(a, \tilde{\sigma}^\star, \tilde{r}^\star, V_0^\delta, V_1^\delta, V_3^\varepsilon)$, which are precisely the parameters in (12.73) calibrated from the yield curve.

Notes

For a general background on interest rates modeling we refer for instance to Musiela and Rutkowski (2002) and Björk (2004), and for an infinite-dimensional perspective to Carmona and Tehranchi (2006). Regarding the issue of negative interest rate with the Vasicek model, we also refer to Rogers (1995). Multifactor quadratic models using spectral decompositions have been analyzed in Boyarchenko and Levendorskii (2007). Short rate models with fast mean-reverting stochastic volatility have been studied using asymptotic methods and calibrated to data in Cotton *et al.* (2004). Bond pricing with fast and slow factors is studied in DeSantiago *et al.* (2008).

13

Credit Risk I: Structural Models with Stochastic Volatility

Defaultable instruments, or credit-linked derivatives, are financial securities that pay their holders amounts that are contingent on the occurrence (or not) of one or more *default events* such as the bankruptcy of a firm or nonrepayment of a loan. There are primarily two approaches to modeling default risk:

- Structural models, which we consider in this chapter.
- Intensity-based models, presented in the following chapter.

These two chapters can be read in any order. In each of these, we study the valuation of single-name credit derivatives, that is those which depend on the credit risk of a single firm, and multiname credit derivatives that depend on the credit risk of many firms. References for background on credit modeling are given in the Notes at the end of Chapter 1. The single-name material presented in Section 13.1 is from Fouque *et al.* (2006) and the multiname material in Section 13.2 is from Fouque *et al.* (2008).

13.1 Single-Name Credit Derivatives

We concentrate here on defaultable bonds already introduced in Section 1.8 using structural models built on geometric Brownian motion. We will introduce the other important single-name credit default swap (CDS) contract, in Section 14.1.5.

In structural models default occurs when the value of the firm's assets drops below some debt threshold. Such models have the benefit of providing economic intuition for the cause of default, but, in fact, they predict essentially zero-yield spreads at short maturities, because defaults are "predictable" in the sense that we can see them coming.

We recall the Black–Cox first-passage model that was discussed in the constant volatility case in Section 1.8.2, with the defaultable bond price $P^B(t,T)$ defined in (1.106) given by (1.110). The *yield spread* $Y(0,T)$ at time zero is defined by

$$e^{-Y(0,T)T} = \frac{P^B(0,T)}{P(0,T)}, \qquad (13.1)$$

where $P(0,T)$ is the default-free zero-coupon bond price given, in the case of constant interest rate r, by $P(0,T) = e^{-rT}$. In other words, $r + Y(0,T)$ is the effective rate of return over the period $(0,T)$, where the spread $Y(0,T)$ is due to the default risk. The price of the defaultable bond is given by $P^B(0,T) = u(0,x)$ obtained explicitly in (1.110), leading to the explicit formula for the yield spread

$$Y(0,T) = -\frac{1}{T} \log\left(N(d_2(T)) - \left(\frac{x}{B}\right)^{1-k} N(d_2^-(T)) \right). \qquad (13.2)$$

In Figure 13.1 we show the yield spread curve $Y(0,T)$ as a function of maturity T for some typical values of the constant volatility, the other parameters are the constant interest rate r and the ratio of initial value to default level x/B. As is well documented in the literature, in this first-passage model the likelihood of default is essentially zero for short maturities even for highly levered firms, corresponding to x/B close to one, as illustrated in Figure 13.2. As stated by Eom *et al.* (2004), one of the challenges for theoretical pricing models is to raise the average predicted spread relative to crude models such as the constant volatility model presented in this section, without overstating the risks associated with volatility or leverage. Several approaches have been proposed that aim at the modeling in this regard. These include the introduction of jumps (Zhou, 2001b; Hilberink and Rogers, 2002), stochastic interest rate (Longstaff and Schwartz, 1995), or imperfect information on X or B (Duffie and Lando, 2001; Duffie and Singleton, 2003). Based on a comprehensive empirical analysis (Eom *et al.*, 2004), empirical studies show that the generalizations proposed in the literature (Longstaff and Schwartz, 1995; Collin-Dufresne and Goldstein, 2001) still have difficulties in predicting realistic credit spreads.

In this chapter, we introduce multiscale stochastic volatility into the firm's value process and demonstrate significant improvement in the ability to calibrate market-observed yield spreads. A popular alternative to structural models is to model default events as pure surprises using a jump process. These are studied in Chapter 14.

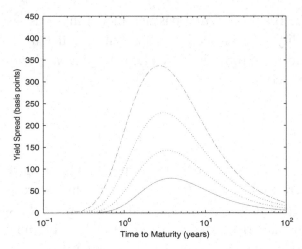

Figure 13.1   The figure shows the sensitivity of the yield spread curve to the volatility level. The ratio of the initial value to the default level x/B is set to 1.3, the interest rate r is 6% and the curves increase with the values of σ: 10%, 11%, 12%, and 13%. Note that the time to maturity is in units of years and plotted on the log scale and the yield spread is quoted in basis points.

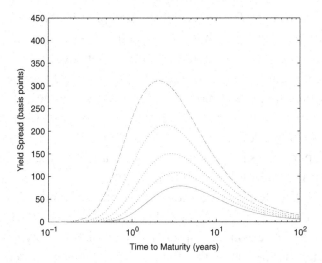

Figure 13.2   This figure shows the sensitivity of the yield spread to the leverage level. The volatility level is set to 10%, the interest rate is 6%. The curves increase with the decreasing ratios x/B: $(1.3, 1.275, 1.25, 1.225, 1.2)$.

13.1.1 Black–Cox Model for Defaultable Bond Price with Stochastic Volatility

Recall that in applying the Merton or Black–Cox structural framework, we essentially consider the firm's stock price X as a proxy for the value of its assets. In the constant volatility case, one may try to calibrate the distance-to-default, defined in (1.105), from the traded prices of credit derivatives, or one might take the volatility σ to be that estimated from the firm's stock price returns, and use credit derivative prices to imply the leverage B/x of the firm. Here, we shall study the effects of multiscale stochastic volatility in the firm's asset value process and look at the ability to capture yield spreads. We therefore work in the stochastic volatility environment described by the processes (X, Y, Z) evolving according to (4.1) under the risk-neutral pricing measure $\mathbb{P}^\star$.

Technically, our problem is to construct an approximation for the price of a zero-coupon defaultable bond, which is viewed in the Black–Cox approach as a down-and-out barrier binary option, with the barrier level and the strike equal to B, the debt level. Let T denote the expiration date of the bond and define

$$\tau_t = \inf\{s \geq t, X_s \leq B\}.$$

Then the pricing function $P(t, x, y, z)$ of the barrier binary option, given survival up to time t, is defined by

$$P(t, x, y, z) = e^{-r(T-t)} \mathbb{P}^\star \{\tau_t > T \mid X_t = x, Y_t = y, Z_t = z\},$$

and satisfies the partial differential equation (4.5) in the domain $\{x > B\}$, with terminal condition $P(T, x, y, z) = \mathbf{1}_{\{x > B\}}$, and boundary condition at $x = B$, $P(t, B, y, z) = 0$. As with the barrier call option in Section 6.2, the asymptotic calculations of Chapter 4 are not affected by the smaller fixed domain $\{x > B\}$ of this problem, as long as we keep track of the boundary condition.

Our first-order approximation is

$$\widetilde{P}(t, x, z) = P_0(t, x, z) + P_1(t, x, z),$$

where $P_0(t, x, z) = P^\star_{BS}(t, x)$ is the Black–Scholes barrier binary price with volatility parameter $\sigma^\star$ which depends on z. For simplicity of notation, we do not show explicitly the dependence of z inherited from $(\sigma^\star, V_0^\delta, V_1^\delta, V_3^\varepsilon)$, in the various quantities that follow.

As discussed in Section 1.8.2, the Black–Scholes price $P^\star_{BS}$ can be obtained by the method of images and is given by

$$P_{BS}^\star(t,x) = e^{-r(T-t)} \left(N(d_2^+ (T-t)) - \left(\frac{x}{B}\right)^{p^\star} N(d_2^- (T-t)) \right), \quad (13.3)$$

where we define

$$p^\star = 1 - \frac{2r}{(\sigma^\star)^2}, \quad (13.4)$$

and we denote

$$d_2^\pm(s) = \frac{\pm \log(x/B) + \left(r - (\sigma^\star)^2/2\right) s}{\sigma^\star \sqrt{s}}. \quad (13.5)$$

Greeks of the Black–Cox Formula For convenience, we give formulas for some of the Greeks of (13.3) that will be used to express the stochastic volatility correction. Using the relations

$$\frac{\partial d_2^\pm}{\partial x} = \frac{\pm 1}{x\sigma\sqrt{T-t}}, \quad N''(z) = -zN'(z),$$

and defining

$$F_2(t,x) = x^2 \frac{\partial^2 P_{BS}^\star}{\partial x^2}(t,x), \qquad F_3(t,x) = x \frac{\partial}{\partial x} \left(x^2 \frac{\partial^2 P_{BS}^\star}{\partial x^2}(t,x) \right),$$

one obtains successively

$$e^{r(T-t)} F_2(t,x) = N'(d_2^+) \left[-\frac{d_2^+}{(\sigma^\star \sqrt{T-t})^2} - \frac{1}{\sigma^\star \sqrt{T-t}} \right]$$

$$+ N'(d_2^-) \left[\frac{d_2^-}{(\sigma^\star \sqrt{T-t})^2} + \frac{2p-1}{\sigma^\star \sqrt{T-t}} \right] \left(\frac{x}{B}\right)^{p^\star}$$

$$+ N(d_2^-)\left[(1-p^\star)p^\star\right] \left(\frac{x}{B}\right)^{p^\star}, \quad (13.6)$$

and

$$e^{r(T-t)} F_3(t,x) =$$

$$N'(d_2^+) \left[\frac{(d_2^+)^2 - 1}{(\sigma^\star \sqrt{T-t})^3} + \frac{d_2^+}{(\sigma^\star \sqrt{T-t})^2} \right]$$

$$+ N'(d_2^-) \left[\frac{(d_2^-)^2 - 1}{(\sigma^\star \sqrt{T-t})^3} + \frac{(3p^\star - 1)d_2^-}{(\sigma^\star \sqrt{T-t})^2} + \frac{p^\star(3p^\star - 2)}{\sigma^\star \sqrt{T-t}} \right] \left(\frac{x}{B}\right)^{p^\star}$$

$$+ N(d_2^-) \left[(1-p^\star)(p^\star)^2 \right] \left(\frac{x}{B}\right)^{p^\star}. \quad (13.7)$$

The Vega and Delta-Vega defined by

$$F_0(t,x) = \frac{\partial P_{BS}^\star}{\partial \sigma^\star}, \qquad F_1(t,x) = x \frac{\partial}{\partial x} \left(\frac{\partial P_{BS}^\star}{\partial \sigma^\star} \right)$$

are given explicitly by

$$e^{r(T-t)}F_0(t,x) = N'(d_2^+)\left[-\frac{d_2^+}{\sigma^\star} - \sqrt{T-t}\right]$$

$$+ N'(d_2^-)\left[\frac{d_2^-}{\sigma^\star} + \sqrt{T-t}\right]\left(\frac{x}{B}\right)^{p^\star}$$

$$+ N(d_2^-)\left[\frac{2}{\sigma^\star}(p^\star - 1)\log\left(\frac{x}{B}\right)\right]\left(\frac{x}{B}\right)^{p^\star}, \quad (13.8)$$

and

$$e^{r(T-t)}F_1(t,x) = N'(d_2^+)\left[\frac{(d_2^+)^2 - 1}{(\sigma^\star)^2\sqrt{T-t}} + \frac{d_2^+}{\sigma^\star}\right]$$

$$+ N'(d_2^-)\left[\frac{(d_2^-)^2 - 1}{(\sigma^\star)^2\sqrt{T-t}} + (1+p^\star)\frac{d_2^-}{\sigma} + p^\star\sqrt{T-t}\right.$$

$$\left. + \frac{2(1-p^\star)}{(\sigma^\star)^2\sqrt{T-t}}\log\left(\frac{x}{B}\right)\right]\left(\frac{x}{B}\right)^{p^\star}$$

$$+ N(d_2^-)\left[\frac{2}{\sigma^\star}(p^\star - 1)\left(1 + p^\star\log\left(\frac{x}{B}\right)\right)\right]\left(\frac{x}{B}\right)^{p^\star}. \quad (13.9)$$

First Correction The stochastic volatility correction $P_1(t,x,z)$ satisfies the partial differential equation

$$\mathcal{L}_{BS}(\sigma^\star)P_1 = -\left(2V_0^\delta\frac{\partial}{\partial\sigma} + 2V_1^\delta D_1\left(\frac{\partial}{\partial\sigma}\right) + V_3^\varepsilon D_1 D_2\right)P_{BS}^\star$$

$$\text{in } x > B, t < T, \quad (13.10)$$

$$P_1(t,B,z) = 0, \quad (13.11)$$

$$P_1(T,x,z) = 0, \quad (13.12)$$

where we have used the notation $D_n = x^n\partial^n/\partial x^n$.

As in the case of the barrier option in Section 6.2, we can transform the current source with zero boundary condition problem into a *boundary value problem with no source* from where it can be reduced to a one-dimensional integral, which is much simpler to compute numerically than solving the full source problem (13.10). The calculations are similar to that case, and we do not repeat them explicitly here. Separating the fast and slow scale components, we first write $P_1(t,x,z)$ as

$$P_1(t,x,z) = P_1^{(\varepsilon)}(t,x,z) + P_1^{(\delta)}(t,x,z).$$

The fast scale part is given by

$$P_1^{(\varepsilon)}(t,x,z) = (T-t)V_3^\varepsilon F_3(t,x) + v_1^\star(t,x), \quad (13.13)$$

where F_3 is given in (13.7) and

$$v_1^\star(t,x) = \frac{\left(\frac{x}{B}\right)^{\frac{p^\star}{2}}}{\sigma^\star\sqrt{2\pi}} \int_t^T \frac{\log(x/B)}{(s-t)^{3/2}} e^{-\frac{(\log(x/B))^2}{2(\sigma^\star)^2(s-t)}} e^{-\left(r+(\sigma^\star p^\star)^2/8\right)(s-t)} g(s)\, ds,$$

$$g(t) = V_3^\varepsilon e^{-r(T-t)} \left[\frac{1}{(\sigma^\star)^3} \left(\frac{2}{\sqrt{T-t}} + 4p^\star r\sqrt{T-t} \right) N'(d) \right.$$

$$\left. + (T-t)(p^\star - 1)(p^\star)^2 N(d) \right],$$

$$d = -\frac{p\sigma^\star\sqrt{T-t}}{2}.$$

Note that g is proportional to the small parameter V_3^ε.

The slow scale contribution $P_1^{(\delta)}(t,x,z)$ is the sum of three components

$$P_1^{(\delta)}(t,x,z) = u_{1a}(t,x) + u_{1c}(t,x) + u_{1d}(t,x), \qquad (13.14)$$

given respectively by

$$u_{1a}(t,x) = 2(T-t)\left(V_0^\delta \frac{\partial P_{BS}^\star}{\partial\sigma^\star} + V_1^\delta x\frac{\partial}{\partial x}\left(\frac{\partial P_{BS}^\star}{\partial\sigma^\star}\right)\right),$$

$$u_{1c}(t,x) = -(T-t)^2\sigma^\star\left(V_0^\delta x^2\frac{\partial^2 P_{BS}^\star}{\partial x^2} + V_1^\delta x\frac{\partial}{\partial x}\left(x^2\frac{\partial^2 P_{BS}^\star}{\partial x^2}\right)\right),$$

$$u_{1d}(t,x) = \frac{\left(\frac{x}{B}\right)^{\frac{p^\star}{2}}}{\sigma^\star\sqrt{2\pi}} \int_t^T \frac{\log(x/B)}{(s-t)^{3/2}} \exp\left(-\frac{(\log(x/B))^2}{2\sigma^{\star 2}(s-t)} \right.$$

$$\left. - \left[\frac{(\sigma^\star p^\star)^2}{8} + r\right](s-t) \right) g_d(s)\, ds,$$

$$g_d(t) = -2(T-t)\left(V_0^\delta F_0(t,B) + V_1^\delta F_1(t,B)\right)$$

$$+ (T-t)^2\sigma^\star\left(V_0^\delta F_2(t,B) + V_1^\delta F_3(t,B)\right),$$

with F_0, F_1, and F_2 given in (13.8), (13.9), and (13.6), respectively. Observe that u_{1d} is also a sum of terms proportional to either V_0^δ or V_1^δ, which are the two small parameters to be fitted in order to capture the yield spread behavior at longer maturities.

13.1.2 Effect of Stochastic Volatility on Yield Spreads

In terms of the yield spreads $Y(0,T)$, we obtain the following approximation:

$$r + Y(0,T)$$

$$= -\frac{1}{T} \log(P(0,x,y,z))$$

$$\approx -\frac{1}{T} \log \left(P_{BS}^{\star}(0,x) + P_1^{(\varepsilon)}(0,x) + P_1^{(\delta)}(0,x) \right)$$

$$\approx -\frac{1}{T} \log \left(P_{BS}^{\star}(0,x) \right) - \frac{1}{T} \left(\frac{P_1^{(\varepsilon)}(0,x)}{P_{BS}^{\star}(0,x)} \right) - \frac{1}{T} \left(\frac{P_1^{(\delta)}(0,x)}{P_{BS}^{\star}(0,x)} \right). \quad (13.15)$$

- The first term is the yield spread produced by the constant volatility model discussed in Section 1.8.2, evaluated at the volatility level $\sigma^{\star}$. Therefore $\sigma^{\star}$ is the parameter which controls the yield curve for intermediate maturities (say one to ten years).

- The second term is the correction scaled by the small parameter V_3^{ε}. In Figure 13.3, the yield corresponding to this price approximation (13.13) is represented by the dashed line, and the yield corresponding to the constant volatility price $P_{BS}^{\star}(t,x)$ is represented by the solid line. The stochastic volatility strongly affects the yields for short maturities and

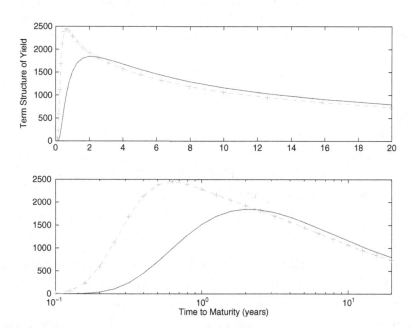

Figure 13.3 Effect of the fast volatility factor. The approximated yield for $\sigma^{\star} = 0.12, r = 0.0, V_3^{\varepsilon} = -0.0003, x/B = 1.2$, and $V_0^{\delta} = V_1^{\delta} = 0$. The solid line corresponds to the constant volatility leading-order term. The crossed dashed line incorporates the fast factor stochastic volatility correction. The top plot is on the linear scale and the bottom plot is on the log maturity scale.

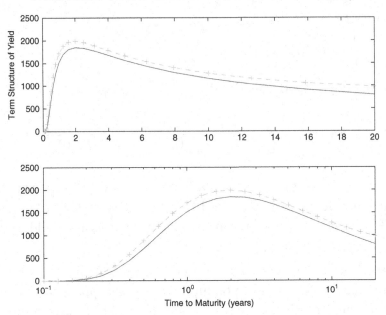

Figure 13.4   Effect of the slow volatility factor. The approximated yield for $\sigma^\star = 0.12, r = 0.0, V_0^\delta = 0.0003, V_1^\delta = -0.0005, x/B = 1.2$, and $V_3^\varepsilon = 0$. The solid line corresponds to the constant volatility leading-order term. The dashed line incorporates the slow varying stochastic volatility correction. The bottom plot is in the log scale and the top plot is in the original scale.

the effect is very different from that obtained if only the volatility level is changed.

- The third term is the correction scaled by the small parameters V_0^δ and V_1^δ. In Figure 13.4, the yield corresponding to this price approximation (13.14) is represented by the dashed line, and the yield corresponding to the constant volatility price $P_{BS}^\star$ is represented by the solid line. The stochastic volatility affects the yields for longer maturities with small effects on short maturities.

In Figure 13.5 we show the yield corresponding to the price approximation (13.15) that includes the correction terms from both the fast and the slow scales by the dashed line, and the yield corresponding to the constant volatility price $P_{BS}^\star(t,x)$ by the solid line. The multiscale stochastic volatility affects the yields significantly for all maturities.

13.1.3 Accuracy of Approximation

The accuracy of the approximation (13.15) is obtained by generalizing the case of a digital option, discussed in Section 6.1.2, to the case of a digital

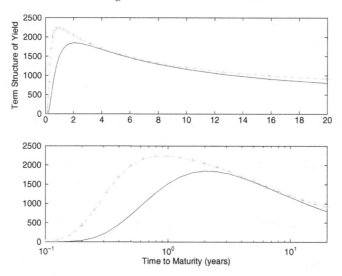

Figure 13.5   The approximated yield for $\sigma^\star = 0.12, r = 0.0, V_0^\delta = 0.0003, V_1^\delta = -0.0005, V_3^\varepsilon = -0.0003, x/B = 1.2$. The solid line corresponds to the constant volatility leading-order term. The dashed line incorporates the slow and fast varying stochastic volatility corrections. The bottom plot is in the log scale and the top plot is in the original scale.

barrier option. The first step is to regularize the terminal payoff by replacing it by the Black–Scholes price of the barrier option, with a small time-to-maturity η. As in the proof of convergence in the European case, the argument consists of controlling the blow-up of the successive x-derivatives at maturity and at the barrier, but now of the Black–Scholes price of the contract. By the method of images, this is reduced to the analysis of the Black–Scholes digital option formula and its derivatives. Consequently, there is a discontinuity in the payoff at the corner $(t = T, x = B)$ as in the case of a European binary option. Therefore, the order of accuracy in this case is as with the digital, namely $\mathcal{O}(\varepsilon^{2/3}\log|\varepsilon| + \delta)$.

13.1.4 Calibration

In this subsection, we discuss calibration of the asymptotic approximation obtained in the previous section from yield spread data. The parameters of interest are $(\sigma^\star, V_0^\delta, V_1^\delta, V_3^\varepsilon)$. We demonstrate the versatility of the stochastic volatility models through the approximation formulas above, by manually fitting them to some market data.

The specific components of the model, namely the base Black–Cox model enhanced with fast and slow stochastic volatility factors, have natural effects on the yield spreads produced: the base volatility $\sigma^\star$ and the leverage

B/x entering the Black–Cox formula set the basic level of the curve; the fast factor, whose effect is described through the parameter V_3^ε, influences the slope of the short end of the curve; and the parameters V_0^δ and V_1^δ associated with the slow factor impact the level and slope, respectively, of the long end of the curve. Of course, the effects of each parameter are not entirely independent, but the physical interpretation of their roles makes it natural to employ a visual fitting as a starting point for an automated procedure.

We take yield spreads from market prices of corporate bonds for two firms: Ford on December 9, 2004, when it was rated BBB; and IBM, a firm rated A or higher, on December 1, 2004. The spreads are obtained from bondpage.com. For simplicity, we assume a constant interest rate $r = 0.025$ throughout.

We first fit the Black–Cox yield spread by varying the volatility $\bar{\sigma}$ and the leverage B/x. This is shown by the solid line in Figures 13.6 and 13.7. As expected and well documented, the shape generated by this model doesn't capture the data well, especially at shorter maturities.

We next exploit the roles of the parameters $(V_3^\varepsilon, V_0^\delta, V_1^\delta, V_2^\delta)$ in adjusting the yield spread for stochastic volatility. This is illustrated in Figure 13.8 for the Ford data.

The fitted parameters $(V_0^\delta, V_1^\delta, V_2^\delta, V_3^\varepsilon)$ (reported in the figure captions) are small, validating the use of the asymptotic approximation. Our corrections enable us to match yield spreads for maturities one year and above,

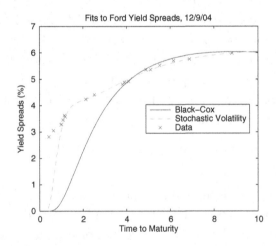

Figure 13.6 Black–Cox and two-factor stochastic volatility fits to Ford yield spread data. The short rate is fixed at $r = 0.025$. The fitted Black–Cox parameters are $\bar{\sigma} = 0.35$ and $x/B = 2.875$. The fitted stochastic volatility parameters are $\sigma^\star = 0.385$, corresponding to $V_2^\delta = 0.0129$, $V_3^\varepsilon = -0.012$, $V_1^\delta = 0.016$, and $V_0^\delta = -0.008$.

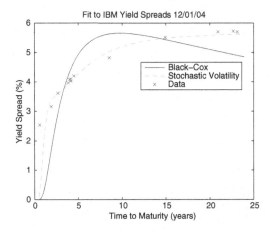

Figure 13.7   Black–Cox and two-factor stochastic volatility fits to IBM yield spread data. The short rate is fixed at $r = 0.025$. The fitted Black–Cox parameters are $\bar{\sigma} = 0.35$ and $x/B = 3$. The fitted stochastic volatility parameters are $\sigma^\star = 0.36$, corresponding to $V_2^\delta = 0.00355$, $V_3^\varepsilon = -0.0112$, $V_1^\delta = 0.013$, and $V_0^\delta = -0.0045$.

compared with only four years and above with the simpler Black–Cox model.

13.2 Multiname Credit Derivatives

Default dependency structure is a crucial issue in pricing multiname credit derivatives as well as in credit risk management. For a multiname credit derivative, the default dependency structure among the underlying portfolio of reference entities is as important as, and in many cases even more important than, the individual term structures of default probabilities.

In this section, we extend the first-passage model to model default dependency in two directions: by extending to multidimensional models and by incorporating stochastic volatility. We derive approximations for the joint survival probabilities and subsequently for the distribution of number of defaults in a basket of names.

13.2.1 Model Setup

We consider a pool of n defaultable bonds whose underlying firms' value processes $\{X_t^{(i)}\}_{i=1}^n$ exhibit the following *multifactor stochastic volatility* dynamics under the physical probability measure $\mathbb{P}$:

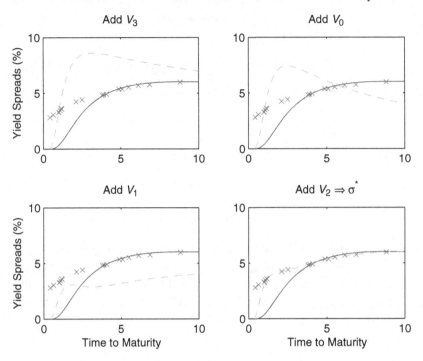

Figure 13.8   The effect of introducing the correction parameters successively. The solid curves in each plot show the Black–Cox fit to the
Ford data with $\bar{\sigma} = 0.35$ and $x/B = 2.875$. The top left plot shows that
the effect of adding the *fast factor skew correction* parameterized by
$V_3 = V_3^\varepsilon = -0.012$ (with $V_0^\delta = V_1^\delta = V_2^\delta = 0$) is to get closer to the
short maturity yields. Then the level of the curve at longer maturities is
adjusted by bringing in the *slow factor level correction* parameterized by
$V_0 = V_0^\delta = -0.008$. Next, introducing the *slow factor skew correction*
parameterized by $V_1 = V_1^\delta = 0.016$ twists the curve to match the slope at
medium to long maturities. Finally, the level of the whole curve is adjusted
by $V_2 = V_2^\delta$ (through σ^*), as shown in the bottom right plot.

$$dX_t^{(1)} = \mu_1 X_t^{(1)} dt + f_1(Y_t, Z_t) X_t^{(1)} dW_t^{(1)},$$
$$dX_t^{(2)} = \mu_2 X_t^{(2)} dt + f_2(Y_t, Z_t) X_t^{(2)} dW_t^{(2)},$$
$$\vdots$$
$$dX_t^{(n)} = \mu_n X_t^{(n)} dt + f_n(Y_t, Z_t) X_t^{(n)} dW_t^{(n)},$$
$$dY_t = \frac{1}{\varepsilon}\alpha(Y_t)\,dt + \frac{1}{\sqrt{\varepsilon}}\beta(Y_t)\,dW_t^{(Y)},$$
$$dZ_t = \delta c(Z_t)\,dt + \sqrt{\delta}\,g(Z_t)\,dW_t^{(Z)},$$

where the $W^{(i)}$'s are standard Brownian motions and we consider first the
uncorrelated case corresponding to $d\langle W^{(i)}, W^{(j)}\rangle_t \equiv \rho_{ij}dt = 0$ for $i \neq j$.
The stochastic volatility correlation structure is given by:

$$d\langle W^{(Y)}, W^{(i)}\rangle_t = \rho_{iY}\, dt, \quad d\langle W^{(Z)}, W^{(i)}\rangle_t = \rho_{iZ}\, dt,$$
$$d\langle W^{(Y)}, W^{(Z)}\rangle_t = \rho_{YZ}\, dt,$$

with all the ρ being constant numbers between -1 and 1. Note that $\sum_{i=1}^{n}\rho_{iY}^2 \le 1$ and $\sum_{i=1}^{n}\rho_{iZ}^2 \le 1$ must hold if the Brownian motions $W^{(i)}$ are to be independent. The f_i are positive functions, smooth with respect to the slow variable z, and are assumed here, for instance, to be bounded above and below away from zero. If the stochastic volatility is turned off by choosing f_i constant, $i = 1, \ldots, n$, then the defaults become independent and the default of a given firm follows the model developed by Black and Cox (1976).

We start by assuming independence among the Brownian motions $W^{(i)}$ for $i = 1, 2, \ldots, n$ for two reasons. Firstly, we try to avoid the intractability caused by the interdependence, as can be seen in Zhou (2001a) where results on the joint distribution of two hitting times are derived. This dependency will be restored in Section 13.2.4 and made tractable by using a perturbation argument around the uncorrelated case. Secondly, we argue that the dependence among the defaultable names introduced through stochastic volatilities is as important as the dependence generated by the interdependence among the Brownian motions driving them. This important issue will be discussed further in Section 13.2.4.

Under the risk-neutral probability measure $\mathbb{P}^\star$, chosen by the market through derivatives trading, the dynamics becomes

$$dX_t^{(1)} = rX_t^{(1)}dt + f_1(Y_t, Z_t)X_t^{(1)}dW_t^{(1)\star},$$
$$dX_t^{(2)} = rX_t^{(2)}dt + f_2(Y_t, Z_t)X_t^{(2)}dW_t^{(2)\star},$$
$$\vdots$$
$$dX_t^{(n)} = rX_t^{(n)}dt + f_n(Y_t, Z_t)X_t^{(n)}dW_t^{(n)\star},$$
$$dY_t = \left(\frac{1}{\varepsilon}\alpha(Y_t) - \frac{1}{\sqrt{\varepsilon}}\beta(Y_t)\Lambda_Y(Y_t, Z_t)\right) dt + \frac{1}{\sqrt{\varepsilon}}\beta(Y_t)dW_t^{(Y)\star},$$
$$dZ_t = \left(\delta c(Z_t) - \sqrt{\delta}g(Z_t)\Lambda_Z(Y_t, Z_t)\right) dt + \sqrt{\delta}\, g(Z_t)dW_t^{(Z)\star},$$

where r is the risk-free interest rate (assumed constant here), the $W^{(i)\star}$ are standard Brownian motions with $d\langle W^{(i)\star}, W^{(j)\star}\rangle_t = 0$ for $i \ne j$, and

$$d\langle W^{(Y)\star}, W^{(i)\star}\rangle_t = \rho_{iY}\, dt, \quad d\langle W^{(Z)\star}, W^{(i)\star}\rangle_t = \rho_{iZ}\, dt,$$
$$d\langle W^{(Y)\star}, W^{(Z)\star}\rangle_t = \rho_{YZ}\, dt.$$

We now study the joint survival probability of n given firms. These will be used in computing the probability distribution of the number of defaults among N names, as explained in Section 13.2.3.

For fixed time $T > 0$, our objective is to find the joint (risk-neutral) survival probability

$$u^{\varepsilon,\delta}(t,\mathbf{x},y,z) \equiv \mathbb{P}^{\star}\left\{\tau_t^{(1)} > T, \ldots, \tau_t^{(n)} > T \mid \mathbf{X}_t = \mathbf{x}, Y_t = y, Z_t = z\right\}, \quad (13.16)$$

where $t < T$, $\mathbf{X}_t \equiv (X_t^{(1)}, \ldots, X_t^{(n)})$, $\mathbf{x} \equiv (x_1, \ldots, x_n)$, and $\tau_t^{(i)}$ is the default time of firm i, defined as follows:

$$\tau_t^{(i)} = \inf\left\{s \geq t, X_s^{(i)} \leq B_i(s)\right\},$$

where $B_i(t)$ is the exogenously prespecified default threshold at time t for firm i. Here we follow Black and Cox (1976) and we assume that

$$B_i(t) = K_i e^{\eta_i t},$$

with $K_i > 0$ and $\eta_i \geq 0$, all being constant numbers. It is very common that credit derivatives have long maturities. Therefore it is more realistic to assume time-varying default thresholds (exponentially growing in our case) than constant ones. Observe that $u^{\varepsilon,\delta}$ is zero whenever $x_i \leq B_i(t)$ for some i, and therefore we only need to focus on the case where $x_i > B_i(t)$ for all $i = 1, 2, \ldots, n$.

13.2.2 Approximated Joint Survival Probabilities

As usual, $u^{\varepsilon,\delta}$ defined in (13.16) is the solution to the following boundary value problem:

$$\mathcal{L}^{\varepsilon,\delta} u^{\varepsilon,\delta} = 0, \quad x_i > B_i(t), \text{ for all } i, t < T,$$
$$u^{\varepsilon,\delta}(t,x_1,x_2,\ldots,x_n,y,z) = 0, \quad \exists i \in \{1,\ldots,n\}, x_i = B_i(t), t \leq T,$$
$$u^{\varepsilon,\delta}(T,x_1,x_2,\ldots,x_n,y,z) = 1, \quad x_i > B_i(T), \text{ for all } i,$$

where

$$\mathcal{L}^{\varepsilon,\delta} = \frac{1}{\varepsilon}\mathcal{L}_0 + \frac{1}{\sqrt{\varepsilon}}\mathcal{L}_1 + \mathcal{L}_2 + \sqrt{\delta}\mathcal{M}_1 + \delta\mathcal{M}_2 + \sqrt{\frac{\delta}{\varepsilon}}\mathcal{M}_3, \quad (13.17)$$

with the notations

$$\mathcal{L}_0 = \alpha(y)\frac{\partial}{\partial y} + \frac{1}{2}\beta^2(y)\frac{\partial^2}{\partial y^2},$$

$$\mathcal{L}_1 = \beta(y)\left(\sum_{i=1}^{n} \rho_{iY} f_i(y,z) x_i \frac{\partial^2}{\partial x_i \partial y} - \Lambda_Y(y,z)\frac{\partial}{\partial y}\right),$$

$$\mathcal{L}_2 = \frac{\partial}{\partial t} + \sum_{i=1}^{n}\left(\frac{1}{2}f_i^2(y,z) x_i^2 \frac{\partial^2}{\partial x_i^2} + r x_i \frac{\partial}{\partial x_i}\right), \quad (13.18)$$

$$\mathcal{M}_1 = g(z)\left(\sum_{i=1}^{n} \rho_{iz} f_i(y,z) x_i \frac{\partial^2}{\partial x_i \partial z} - \Lambda_Z(y,z) \frac{\partial}{\partial z}\right),$$

$$\mathcal{M}_2 = c(z) \frac{\partial}{\partial z} + \frac{1}{2} g^2(z) \frac{\partial^2}{\partial z^2},$$

$$\mathcal{M}_3 = \rho_{YZ} \beta(y) g(z) \frac{\partial^2}{\partial y \partial z}.$$

We expand $u^{\varepsilon,\delta}$ in terms of powers of $\sqrt{\varepsilon}$ and $\sqrt{\delta}$:

$$u^{\varepsilon,\delta} = u_0 + \sqrt{\varepsilon}\, u_{1,0} + \sqrt{\delta}\, u_{0,1} + \varepsilon\, u_{2,0} + \sqrt{\varepsilon\delta}\, u_{1,1} + \delta\, u_{0,2} + \cdots, \quad (13.19)$$

and retain

$$\tilde{u} \equiv u_0 + \sqrt{\varepsilon}\, u_{1,0} + \sqrt{\delta}\, u_{0,1} \qquad (13.20)$$

as our approximation for $u^{\varepsilon,\delta}$.

Leading-Order Term u_0

The leading-order term u_0 is independent of y and is determined by the following PDE system with respect to the variables (t,x), the variable z being simply a parameter:

$$\langle \mathcal{L}_2 \rangle u_0 = 0, \quad x_i > B_i(t), \text{ for all } i, t < T, \qquad (13.21)$$
$$u_0(t,x_1,x_2,\ldots,x_n) = 0, \quad \exists i \in \{1,\ldots,n\}, x_i = B_i(t), t \le T,$$
$$u_0(T,x_1,x_2,\ldots,x_n) = 1, \quad x_i > B_i(T), \text{ for all } i,$$

where, from the definition (13.18) of $\mathcal{L}_2$, we have

$$\langle \mathcal{L}_2 \rangle = \frac{\partial}{\partial t} + \sum_{i=1}^{n} \left(\frac{1}{2} \langle f_i^2(\cdot,z) \rangle x_i^2 \frac{\partial^2}{\partial x_i^2} + r x_i \frac{\partial}{\partial x_i}\right), \qquad (13.22)$$

where the averaging is with respect to the invariant distribution of the process Y_t under the real-world measure $\mathbb{P}$.

Proposition 13.1 *The leading-order term u_0 in the approximation (13.20) is given by:*

$$u_0 = \prod_{i=1}^{n} Q_i \equiv \prod_{i=1}^{n} \left[N\left(d_{2(i)}^{+}\right) - \left(\frac{x_i}{B_i(t)}\right)^{p_i} N\left(d_{2(i)}^{-}\right)\right], \qquad (13.23)$$

where $N(\cdot)$ is the standard cumulative normal distribution function, and

$$d^{\pm}_{2(i)} \equiv \frac{\pm \ln \frac{x_i}{B_i(t)} + \left(r - \eta_i - \frac{\sigma_i^2(z)}{2}\right)(T-t)}{\sigma_i(z)\sqrt{T-t}},$$

$$\sigma_i(z) \equiv \sqrt{\langle f_i^2(\cdot,z)\rangle} \quad \textit{(effective volatility of firm i)},$$

$$p_i \equiv 1 - \frac{2(r - \eta_i)}{\sigma_i^2(z)}.$$

Proof This is simply the product of one-dimensional survival probabilities, appropriately modified to take into account the exponentially
time-varying barrier.   □

Correction Term $\sqrt{\varepsilon}\,u_{1,0}$ A straightforward generalization of the fast scale
asymptotics of Section 4.2.2 to the case of multidimensional x shows that
the term $u_{1,0}$ is determined by the following PDE system:

$$\langle \mathscr{L}_2\rangle u_{1,0} = \mathscr{A}_n u_0, \quad x_i > B_i(t), \text{ for all } i, t < T, \quad (13.24)$$

$$u_{1,0}(t,x_1,x_2,\dots,x_n) = 0, \quad \exists\, i \in \{1,\dots,n\}, x_i = B_i(t), t \le T,$$

$$u_{1,0}(T,x_1,x_2,\dots,x_n) = 0, \quad x_i > B_i(t), \text{ for all } i,$$

where the operator $\mathscr{A}_n$ is given by

$$\mathscr{A}_n \equiv \langle \mathscr{L}_1 \mathscr{L}_0^{-1}(\mathscr{L}_2 - \langle \mathscr{L}_2\rangle)\rangle$$

$$= \frac{v_Y}{\sqrt{2}}\left[\sum_{i=1}^{n}\sum_{j=1}^{n}\rho_{iY}\left\langle f_i\frac{\partial\phi_j}{\partial y}\right\rangle x_i\frac{\partial}{\partial x_i}\left(x_j^2\frac{\partial^2}{\partial x_j^2}\right) - \sum_{j=1}^{n}\left\langle \Lambda_1\frac{\partial\phi_j}{\partial y}\right\rangle x_j^2\frac{\partial^2}{\partial x_j^2}\right]$$

$$= -\frac{v_Y}{\sqrt{2}}\sum_{i=1}^{n}\left\langle \Lambda_1\frac{\partial\phi_i}{\partial y}\right\rangle x_i^2\frac{\partial^2}{\partial x_i^2} + \frac{v_Y}{\sqrt{2}}\sum_{i=1}^{n}\rho_{iY}\left\langle f_i\frac{\partial\phi_i}{\partial y}\right\rangle x_i\frac{\partial}{\partial x_i}\left(x_i^2\frac{\partial^2}{\partial x_i^2}\right)$$

$$+ \frac{v_Y}{\sqrt{2}}\sum_{\substack{i,j=1\\i\ne j}}^{n}\rho_{iY}\left\langle f_i\frac{\partial\phi_j}{\partial y}\right\rangle x_i\frac{\partial}{\partial x_i}\left(x_j^2\frac{\partial^2}{\partial x_j^2}\right). \quad (13.25)$$

Here $\phi_i(y,z)$, for $i = 1,2,\dots,n$, denote solutions to the following Poisson
equations with respect to the variable y:

$$\mathscr{L}_0\phi_i(y,z) = f_i^2(y,z) - \langle f_i^2(\cdot,z)\rangle, \quad (13.26)$$

and $\langle\cdot\rangle$ denotes the average with respect to the invariant distribution of Y.

The following result shows that the n-dimensional problem (13.24) can
be reduced to many uncoupled one- and two-dimensional problems.

Proposition 13.2 *The correction term $\sqrt{\varepsilon}\,u_{1,0}$ is given by*

$$\sqrt{\varepsilon}\,u_{1,0} = \sum_{i=1}^{n} R_i^{(2)} w_i^{(2)} \prod_{\substack{j=1 \\ j\neq i}}^{n} Q_j + \sum_{i=1}^{n} R_i^{(3)} w_i^{(3)} \prod_{\substack{j=1 \\ j\neq i}}^{n} Q_j$$

$$+ \sum_{\substack{i,j=1 \\ i\neq j}}^{n} R_{ij}^{(3)} w_{ij}^{(3)} \prod_{\substack{k=1 \\ k\neq i,j}}^{n} Q_k, \qquad (13.27)$$

where the coefficients $R_i^{(2)}, R_i^{(3)}, R_{ij}^{(3)}$ depend on the parameter z and are given by

$$R_i^{(2)} = -\frac{\sqrt{\varepsilon}}{2} \left\langle \beta(\cdot)\Lambda_Y(\cdot,z)\frac{\partial \phi_i}{\partial y}(\cdot,z) \right\rangle, \qquad (13.28)$$

$$R_i^{(3)} = \frac{\sqrt{\varepsilon}}{2} \rho_{iY} \left\langle \beta(\cdot)f_i(\cdot,z)\frac{\partial \phi_i}{\partial y}(\cdot,z) \right\rangle, \qquad (13.29)$$

$$R_{ij}^{(3)} = \frac{\sqrt{\varepsilon}}{2} \rho_{iY} \left\langle \beta(\cdot)f_i(\cdot,z)\frac{\partial \phi_j}{\partial y}(\cdot,z) \right\rangle, \quad i\neq j, \qquad (13.30)$$

with ϕ_i given by (13.26) and Q_i given in Proposition 13.1, and where the functions $w_i^{(2)}(t,x_i)$, $w_i^{(3)}(t,x_i)$, and $w_{ij}^{(3)}(t,x_i,x_j)$ depend on the parameter z and are given by the following problems:

$$\left[\frac{\partial}{\partial t} + \frac{1}{2}\sigma_i^2(z)x_i^2\frac{\partial^2}{\partial x_i^2} + rx_i\frac{\partial}{\partial x_i}\right] w_i^{(2)} = x_i^2\frac{\partial^2 Q_i}{\partial x_i^2}, \quad x_i > B_i(t), t < T, \quad (13.31)$$

$$w_i^{(2)}(t,B_i(t)) = 0, \quad t \leq T,$$
$$w_i^{(2)}(T,x_i) = 0, \quad x_i > B_i(t),$$

$$\left[\frac{\partial}{\partial t} + \frac{1}{2}\sigma_i^2(z)x_i^2\frac{\partial^2}{\partial x_i^2} + rx_i\frac{\partial}{\partial x_i}\right] w_i^{(3)} = x_i\frac{\partial}{\partial x_i}\left(x_i^2\frac{\partial^2 Q_i}{\partial x_i^2}\right),$$
$$x_i > B_i(t), t < T, \qquad (13.32)$$

$$w_i^{(3)}(t,B_i(t)) = 0, \quad t \leq T,$$
$$w_i^{(3)}(T,x_i) = 0, \quad x_i > B_i(t),$$

$$\left[\frac{\partial}{\partial t} + \frac{1}{2}\sigma_i^2(z)x_i^2\frac{\partial^2}{\partial x_i^2} + \frac{1}{2}\sigma_j^2(z)x_j^2\frac{\partial^2}{\partial x_j^2} + rx_i\frac{\partial}{\partial x_i} + rx_j\frac{\partial}{\partial x_j}\right] w_{ij}^{(3)} \quad (13.33)$$

$$= \left(x_i\frac{\partial Q_i}{\partial x_i}\right)\left(x_j^2\frac{\partial^2 Q_j}{\partial x_j^2}\right), \quad x_i > B_i(t), x_j > B_j(t), t < T,$$

$$w_{ij}^{(3)}(t,x_i,x_j) = 0, \quad \text{if} \quad x_i = B_i(t) \quad \text{or} \quad x_j = B_j(t), t \le T,$$

$$w_{ij}^{(3)}(T,x_i,x_j) = 0, \quad x_i > B_i(t), x_j > B_j(t),$$

with $\sigma_i(z)$ given in Proposition 13.1.

Proof Note that with the form (13.25) of $\mathscr{A}_n$ and the definitions of (13.28), (13.29), (13.30), we have:

$$\sqrt{\varepsilon}\mathscr{A}_n = \sum_{i=1}^{n} R_i^{(2)} x_i^2 \frac{\partial^2}{\partial x_i^2} + \sum_{i=1}^{n} R_i^{(3)} x_i \frac{\partial}{\partial x_i}\left(x_i^2 \frac{\partial^2}{\partial x_i^2}\right)$$

$$+ \sum_{\substack{i,j=1 \\ i \ne j}}^{n} R_{ij}^{(3)} x_i \frac{\partial}{\partial x_i}\left(x_j^2 \frac{\partial^2}{\partial x_j^2}\right).$$

By linearity of (13.24), it is enough to check that

$$\langle \mathscr{L}_2 \rangle \left(w_i^{(2)} \prod_{j=1,j \ne i}^{n} Q_j \right) = x_i^2 \frac{\partial^2 u_0}{\partial x_i^2}, \tag{13.34}$$

$$\langle \mathscr{L}_2 \rangle \left(w_i^{(3)} \prod_{j=1,j \ne i}^{n} Q_j \right) = x_i \frac{\partial}{\partial x_i}\left(x_i^2 \frac{\partial^2 u_0}{\partial x_i^2} \right), \tag{13.35}$$

$$\langle \mathscr{L}_2 \rangle \left(w_{ij}^{(3)} \prod_{k=1,k \ne i,j}^{n} Q_k \right) = x_i \frac{\partial}{\partial x_i}\left(x_j^2 \frac{\partial^2 u_0}{\partial x_j^2} \right). \tag{13.36}$$

Using the form (13.22) of $\langle \mathscr{L}_2 \rangle$ and $u_0 = \prod_{i=1}^{n} Q_i$, one can easily check that (13.34), (13.35), and (13.36) are satisfied. The boundary and terminal conditions for the correction $\sqrt{\varepsilon}u_{1,0}$ are directly inherited from the boundary and terminal conditions for the functions $w_i^{(2)}$, $w_i^{(3)}$, $w_{ij}^{(3)}$, and Q_i. □

The problems (13.31), (13.32), and (13.33) are boundary value problems with sources. Generalizing the one-dimensional case presented in Section 13.1.1, one can write explicit formulas for the solutions up to single and double integrals with respect to explicitly known hitting time distributions. These formulas are quite lengthy and can be found in Fouque *et al.* (2008).

Correction Term $\sqrt{\delta}u_{0,1}$ A straightforward generalization of the slow scale asymptotics of Section 4.2.4 to the case of multidimensional x shows that the term $u_{0,1}$ is determined by the following PDE system:

$$\langle \mathscr{L}_2 \rangle u_{0,1} = -\langle \mathscr{M}_1 \rangle u_0, \quad x_i > B_i(t), \text{for all } i, t < T, \quad (13.37)$$
$$u_{0,1}(t, x_1, x_2, \ldots, x_n) = 0, \quad \exists i \in \{1, \ldots, n\}, x_i = B_i(t), t \le T,$$
$$u_{0,1}(T, x_1, x_2, \ldots, x_n) = 0, \quad x_i > B_i(t), \text{ for all } i,$$

where the operator $\langle \mathscr{M}_1 \rangle$ is given by:

$$\langle \mathscr{M}_1 \rangle = g(z) \left[\sum_{i=1}^{n} \rho_{iZ} \langle f_i(\cdot, z) \rangle x_i \frac{\partial^2}{\partial x_i \partial z} - \langle \Lambda_Z(\cdot, z) \rangle \frac{\partial}{\partial z} \right].$$

The following result shows that, as for $u_{1,0}$, the multidimensional problem (13.37) for $u_{0,1}$ can be reduced to many uncoupled one- and two-dimensional problems.

Proposition 13.3 *The correction term $\sqrt{\delta} u_{0,1}$ is given by*

$$\sqrt{\delta} u_{0,1} = \sum_{i=1}^{n} R_i^{(0)} w_i^{(0)} \prod_{\substack{j=1 \\ j \ne i}}^{n} Q_j + \sum_{i=1}^{n} R_i^{(1)} w_i^{(1)} \prod_{\substack{j=1 \\ j \ne i}}^{n} Q_j$$
$$+ \sum_{\substack{i,j=1 \\ i \ne j}}^{n} R_{ij}^{(1)} w_{ij}^{(1)} \prod_{\substack{k=1 \\ k \ne i,j}}^{n} Q_k, \quad (13.38)$$

where the coefficients $R_i^{(0)}, R_i^{(1)}, R_{ij}^{(1)}$ depend on the parameter z and are given by

$$R_i^{(0)} = -\sqrt{\delta} g(z) \langle \Lambda_Z(\cdot, z) \rangle \sigma_i'(z), \quad (13.39)$$
$$R_i^{(1)} = \sqrt{\delta} g(z) \rho_{iZ} \langle f_i(\cdot, z) \rangle \sigma_i'(z), \quad (13.40)$$
$$R_{ij}^{(1)} = \sqrt{\delta} g(z) \rho_{iZ} \langle f_i(\cdot, z) \rangle \sigma_j'(z), \quad i \ne j, \quad (13.41)$$

the $\sigma_i(z)$ and Q_i are given in Proposition 13.1, $\sigma_i' = d\sigma_i/dz$, and where the functions $w_i^{(0)}(t, x_i; z)$, $w_i^{(1)}(t, x_i; z)$, and $w_{ij}^{(1)}(t, x_i, x_j; z)$ depend on the parameter z and are given by the following problems:

$$\left[\frac{\partial}{\partial t} + \frac{1}{2} \sigma_i^2(z) x_i^2 \frac{\partial^2}{\partial x_i^2} + r x_i \frac{\partial}{\partial x_i} \right] w_i^{(0)} = -\frac{\partial Q_i}{\partial \sigma_i}, \quad x_i > B_i(t), t < T, \quad (13.42)$$
$$w_i^{(0)}(t, B_i(t)) = 0, \quad t \le T,$$
$$w_i^{(0)}(T, x_i) = 0, \quad x_i > B_i(t),$$

$$\left[\frac{\partial}{\partial t}+\frac{1}{2}\sigma_i^2(z)x_i^2\frac{\partial^2}{\partial x_i^2}+rx_i\frac{\partial}{\partial x_i}\right]w_i^{(1)}=-x_i\frac{\partial}{\partial x_i}\left(\frac{\partial Q_i}{\partial \sigma_i}\right),$$

$$x_i>B_i(t),t<T,\quad(13.43)$$

$$w_i^{(1)}(t,B_i(t))=0,\quad t\leq T,$$

$$w_i^{(1)}(T,x_i)=0,\quad x_i>B_i(t),$$

$$\left[\frac{\partial}{\partial t}+\frac{1}{2}\sigma_i^2(z)x_i^2\frac{\partial^2}{\partial x_i^2}+\frac{1}{2}\sigma_j^2(z)x_j^2\frac{\partial^2}{\partial x_j^2}+rx_i\frac{\partial}{\partial x_i}+rx_j\frac{\partial}{\partial x_j}\right]w_{ij}^{(1)}\quad(13.44)$$

$$=-\left(x_i\frac{\partial Q_i}{\partial x_i}\right)\left(\frac{\partial Q_j}{\partial \sigma_j}\right),\quad x_i>B_i(t),x_j>B_j(t),t<T,$$

$$w_{ij}^{(1)}(t,x_i,x_j)=0,\quad\text{if}\quad x_i=B_i(t)\quad\text{or}\quad x_j=B_j(t),t\leq T,$$

$$w_{ij}^{(1)}(T,x_i,x_j)=0,\quad x_i>B_i(t),x_j>B_j(t).$$

Proof The proof is very similar to that of Proposition 13.2. Since u_0 depends on z only through $\sigma_i(z)$, we have

$$\frac{\partial u_0}{\partial z}=\sum_{j=1}^{n}\sigma_j'(z)\frac{\partial u_0}{\partial \sigma_j},$$

and therefore

$$\sqrt{\delta}\langle\mathscr{M}_1\rangle u_0=\sqrt{\delta}\,g(z)\left[\sum_{i=1}^{n}\rho_{iz}\langle f_i(\cdot,z)\rangle x_i\frac{\partial}{\partial x_i}\left(\sum_{j=1}^{n}\sigma_j'(z)\frac{\partial u_0}{\partial \sigma_j}\right)\right.$$

$$\left.-\langle\Lambda_z(\cdot,z)\rangle\sum_{j=1}^{n}\sigma_j'(z)\frac{\partial u_0}{\partial \sigma_j}\right]$$

$$=\sqrt{\delta}\,g(z)\left[\sum_{i=1}^{n}\sum_{j=1}^{n}\rho_{iz}\langle f_i(\cdot,z)\rangle\sigma_j'(z)x_i\frac{\partial}{\partial x_i}\left(\frac{\partial u_0}{\partial \sigma_j}\right)\right.$$

$$\left.-\langle\Lambda_z(\cdot,z)\rangle\sum_{i=1}^{n}\sigma_i'(z)\frac{\partial u_0}{\partial \sigma_i}\right]$$

$$=\sum_{i=1}^{n}R_i^{(0)}\frac{\partial u_0}{\partial \sigma_i}+\sum_{i=1}^{n}R_i^{(1)}x_i\frac{\partial}{\partial x_i}\left(\frac{\partial u_0}{\partial \sigma_i}\right)+\sum_{\substack{i,j=1\\i\neq j}}^{n}R_{ij}^{(1)}x_i\frac{\partial}{\partial x_i}\left(\frac{\partial u_0}{\partial \sigma_j}\right),$$

where we have used the definitions (13.39), (13.40), (13.41) of $(R_i^{(0)},R_i^{(1)},R_{ij}^{(1)})$. By linearity of (13.37), it is enough to check that

$$\langle \mathcal{L}_2 \rangle \left(w_i^{(1)} \prod_{j=1, j\neq i}^n Q_j \right) = -\frac{\partial u_0}{\partial \sigma_i}, \tag{13.45}$$

$$\langle \mathcal{L}_2 \rangle \left(w_i^{(1)} \prod_{j=1, j\neq i}^n Q_j \right) = -x_i \frac{\partial}{\partial x_i} \left(\frac{\partial u_0}{\partial \sigma_i} \right), \tag{13.46}$$

$$\langle \mathcal{L}_2 \rangle \left(w_{ij}^{(1)} \prod_{k=1, k\neq i,j}^n Q_k \right) = -x_i \frac{\partial}{\partial x_i} \left(\frac{\partial u_0}{\partial \sigma_j} \right). \tag{13.47}$$

Using the form (13.22) of $\langle \mathcal{L}_2 \rangle$ and $u_0 = \prod_{i=1}^n Q_i$, one can easily check that (13.45), (13.46), and (13.47) are satisfied. The boundary and terminal conditions for the correction $\sqrt{\delta}\, u_{0,1}$ are directly inherited from the boundary and terminal conditions for the functions $w_i^{(0)}$, $w_i^{(1)}$, $w_{ij}^{(1)}$, and Q_i. □

The problems (13.42), (13.43), and (13.44) are boundary value problems with sources. Again, one can write explicit formulas for the solutions up to single and double integrals with respect to explicitly known hitting time distributions. These formulas can be found in Fouque *et al.* (2008).

Summary of the Approximation Combining the results of Propositions 13.1, 13.2, and 13.3, we get that the approximation $\tilde{u}$ in (13.20) is given by

$$\tilde{u} = \prod_{i=1}^n Q_i + \sum_{i=1}^n R_i^{(2)} w_i^{(2)} \prod_{\substack{j=1 \\ j\neq i}}^n Q_j + \sum_{i=1}^n R_i^{(3)} w_i^{(3)} \prod_{\substack{j=1 \\ j\neq i}}^n Q_j + \sum_{\substack{i,j=1 \\ i\neq j}}^n R_{ij}^{(3)} w_{ij}^{(3)} \prod_{\substack{k=1 \\ k\neq i,j}}^n Q_k$$

$$+ \sum_{i=1}^n R_i^{(0)} w_i^{(0)} \prod_{\substack{j=1 \\ j\neq i}}^n Q_j + \sum_{i=1}^n R_i^{(1)} w_i^{(1)} \prod_{\substack{j=1 \\ j\neq i}}^n Q_j + \sum_{\substack{i,j=1 \\ i\neq j}}^n R_{ij}^{(1)} w_{ij}^{(1)} \prod_{\substack{k=1 \\ k\neq i,j}}^n Q_k, \tag{13.48}$$

where $(R_i^{(2)}, R_i^{(3)}, R_{ij}^{(3)})$ are small of order $\sqrt{\varepsilon}$, $(R_i^{(0)}, R_i^{(1)}, R_{ij}^{(1)})$ are small of order $\sqrt{\delta}$, and they all depend on the parameter z. The functions Q_i, $w_i^{(2)}$, $w_i^{(3)}$, $w_i^{(0)}$, and $w_i^{(1)}$ depend on the variable x_i, the functions $w_{ij}^{(3)}$ and $w_{ij}^{(1)}$ depend on the variables (x_i, x_j), and they all depend on the parameter z.

Numerical Illustration of the Accuracy of Approximation In order to illustrate the quality of the approximation of the joint survival probability given by (13.48), we have conducted the following numerical experiments. For $n = 10$ names (Table 13.1), and for $n = 25$ names (Table 13.2), we compute the zero-order approximation u_0 given by (13.23), and the first-order approximation $\tilde{u}$ given by (13.48) with the explicit formulas given in the Appendix of Fouque *et al.* (2008). We present the results for four sets of

Table 13.1: *Joint survival probability for ten firms ($n = 10$)*

ε	δ	u_0	$\tilde{u}$	u_{MC}	Absolute (relative) error
1/100	1/50	0.740389	0.75079	0.7502	0.0006(0.08%)
1/50	1/20	0.740389	0.756015	0.7529	0.003(0.4%)
1/20	1/10	0.740389	0.763647	0.7567	0.007(0.9%)
1	1	0.740389	0.82833	.7653	0.063(8.2%)

Table 13.2: *Joint survival probability for twenty five firms ($n = 25$)*

ε	δ	u_0	$\tilde{u}$	u_{MC}	Absolute (relative) error
1/100	1/50	0.471683	0.481506	0.4789	0.003 (0.5%)
1/50	1/20	0.471683	0.486892	0.4803	0.006 (1.4%)
1/20	1/20	0.471683	0.488478	0.4854	0.003 (0.6%)
1/20	1/10	0.471683	0.493648	0.4843	0.009 (1.9%)

values of the small parameters ε and δ. Since there is no explicit formula for the true value, we obtain it by Monte Carlo simulations with a very large number of realizations (10^5) and using an Euler scheme with a very small time step (10^{-4}) in order to ensure accuracy of the true value proxy denoted by u_{MC}. The absolute and relative errors are shown in the last columns. In all cases we have used the following parameter values:

$$X_0^{(i)} = 20, r = 0.05, \eta_i = 0.06, K_i = 10,$$
$$\rho_{ij} = 0, \rho_{iY} = \rho_{iZ} = 1/(2\sqrt{n}), \rho_{YZ} = 0,$$
$$\alpha(y) = 0.3 - y, \quad c(z) = 0.3 - z, \quad \beta(y) = 0.14, \quad g(z) = 0.14,$$
$$Y_0 = Z_0 = 0.3, \Lambda_Y = \Lambda_Z = 0,$$
$$f_i(y,z) = 0.3 \exp(y+z)/\exp(0.64), T = 1.$$

As expected, the first-order approximation $\tilde{u}$ converges to the (simulated) true value u_{MC} as (ε, δ) goes to $(0,0)$. In fact, as often observed in homogenization, the approximation remains very accurate even in regimes where these parameters are not so small. We also observe that, in the present case with $\rho_{ij} = 0$, the volatility–name correlations ρ_{iY}, ρ_{iZ} are small (since $\sum_{i=1}^n \rho_{iY}^2 \le 1$ and $\sum_{i=1}^n \rho_{iZ}^2 \le 1$), and the effect of the correction due to stochastic volatility is also relatively small (u_0 is already close to u_{MC}). This will not be the case with name–name correlations $\rho_{ij} \ne 0$, as shown in Section 13.2.4.

13.2.3 Loss Distribution

We consider now a portfolio consisting of N defaultable bonds and we denote by a_i the number of bond i in this portfolio. Assuming zero recovery rate from default for each bond, the loss at time t of this portfolio is given by the random variable

$$L(t) \equiv \sum_{i=1}^{N} a_i \chi_i(t),$$

where $\chi_i(t)$ takes on 1 if bond i defaults before t and 0 otherwise. Our objective is to study the distribution of $L(T)$ for a maturity T smaller than all the bond maturities.

General Case Since the portfolio consists only of a finite number of bonds, L must have a discrete distribution function. It hence suffices to find the probability value that a subset of names in the portfolio default. It turns out that this can be done recursively. For example,

$$\mathbb{P}^{\star}\{\text{only bond 1 defaults in the portfolio before time } T\}$$
$$= \mathbb{P}^{\star}\{\tau_1 \leq T, \tau_2 > T, \ldots, \tau_N > T\}$$
$$= \mathbb{P}^{\star}\{\tau_2 > T, \ldots, \tau_N > T\} - \mathbb{P}^{\star}\{\tau_1 > T, \tau_2 > T, \ldots, \tau_N > T\},$$

while τ_i is the default time of bond i. Both terms on the right-hand side of the last equality are known (approximately) by formula (13.48). In general, if we denote

$$I \equiv \{i_1, i_2, \ldots, i_n\} \subset J \equiv \{j_1, j_2, \ldots, j_m\} \subset \bar{N} \equiv \{1, 2, \ldots, N\},$$
$$D_{|J|}^{|I|}(I; J) \equiv \mathbb{P}^{\star}\{(\text{bond } i \text{ defaults}, i \in I) \cap (\text{bond } j \text{ survives}, j \in J \backslash I)\},$$

then

$$D_m^n(\{i_1, \ldots, i_n\}; J)$$
$$= \mathbb{P}^{\star}\left\{(\tau_i \leq T, i \in \{i_1, \ldots, i_{n-1}, i_n\}) \cap (\tau_j > T, j \in J \backslash I)\right\}$$
$$= \mathbb{P}^{\star}\left\{(\tau_i \leq T, i \in \{i_1, \ldots, i_{n-1}\}) \cap (\tau_j > T, j \in J \backslash I)\right\}$$
$$\quad - \mathbb{P}^{\star}\left\{(\tau_i \leq T, i \in \{i_1, \ldots, i_{n-1}\}) \cap (\tau_j > T, j \in J \backslash \{i_1, \ldots, i_{n-1}\})\right\}.$$

Therefore,

$$D_m^n(\{i_1, \ldots, i_n\}; \{j_1, \ldots, j_m\})$$
$$= D_{m-1}^{n-1}(\{i_1, \ldots, i_{n-1}\}; \{j_1, \ldots, j_m\} \backslash \{i_n\})$$
$$\quad - D_m^{n-1}(\{i_1, \ldots, i_{n-1}\}; \{j_1, \ldots, j_m\}). \tag{13.49}$$

Formula (13.49) is recursive and by implementing it once, one can reduce by one the superscript n (= the number of defaults) on D. This can be done repeatedly until one reduces the superscript to zero, implying a total survival of an appropriate subset of names in the portfolio. Finally, the probability of total survival is (approximately) $\tilde{u}$ given by (13.48), with names/indices $\{1,2,\ldots,n\}$ replaced by the appropriate set of indices/names.

It can be shown that in order to compute $D_m^n(\cdot,\cdot)$, one has to evaluate $\tilde{u}$-like forms 2^n times. This may be computationally expensive for a portfolio of big size, say 50 names or more. For a smaller size portfolio, say $N = 20$, however, the computational cost is acceptable to find for instance the probability of 50% loss, namely $D_{20}^{10}(\cdot,\cdot)$. To be more precise, the computation of $D_{20}^{10}(\cdot,\cdot)$ would require computing $2^{10} = 1024$ $\tilde{u}$-like forms given by (13.48). However, every item in equation (13.48) is in closed-form (or up to double integrals), and therefore can be computed fairly fast when an appropriate programming language is chosen (say, C/C++). This can be done within 0.05 to 0.5 seconds on a 2GHz CPU with 2GB RAM PC. Therefore the whole computation of $D_{20}^{10}(\cdot,\cdot)$ can be done within 51.2 to 512 seconds, which is what we mean by acceptable.

Special Case: Homogeneous Portfolio We now consider a fully homogeneous portfolio, that is

$$f_i(y,z) = f(y,z), \; \rho_{iY} = \rho_Y, \; \rho_{iZ} = \rho_Z, \; a_i = a, \; X_0^{(i)} = x_i = x, \quad (13.50)$$

for all $i = 1,\ldots,N$. We first recall a classical result.

Lemma 13.4 *Let τ_i be the default time of bond i, and suppose that under a probability $\mathbb{P}$*

$$\mathbb{P}(\tau_i > T, i \in \{i_1, i_2, \ldots, i_m\}) = \mathbb{P}(\tau_j > T, j \in \{j_1, j_2, \ldots, j_m\}),$$

for any two equal-sized sets of indices $\{i_1, i_2, \ldots, i_m\}$ and $\{j_1, j_2, \ldots, j_m\}$ chosen from $\{1,2,\ldots,N\}$ with $1 \le m \le N$. Define

$$S_m \equiv \mathbb{P}(\tau_i > T, i \in \{1,2,\ldots,m\}), \quad 1 \le m \le N. \quad (13.51)$$

Then, for $0 \le k < N$, the probability that exactly k, among the N, bonds default before T, is given by

$$F_k^{(N)} \equiv \mathbb{P}\left(\sum_{i=1}^N \chi_i(T) = k \right) = \binom{N}{k} \sum_{j=0}^k \binom{k}{j} (-1)^j S_{N+j-k}. \quad (13.52)$$

Proof We provide here a short proof which consists of computing the moment-generating function of the number of survivals. Denote by $\chi_i' \equiv 1 - \chi_i$ the event "firm i survives after T". On the one hand, we have

$$\Phi(z) \equiv \mathbb{E}\left(z^{(\sum_{i=1}^{N}\chi_i')}\right) = \mathbb{E}\left(z^{(N-\sum_{i=1}^{N}\chi_i)}\right) = \sum_{k=0}^{N}z^{(N-k)}F_k^{(N)}$$

$$= \sum_{k=0}^{N}z^k F_{N-k}^{(N)}. \tag{13.53}$$

On the other, we have

$$\Phi(z)$$

$$= \mathbb{E}\left(\prod_{i=1}^{N}z^{\chi_i'}\right)$$

$$= \mathbb{E}\left(\prod_{i=1}^{N}(1+(z-1)\chi_i')\right) = \mathbb{E}\left(\sum_{(i_1,\ldots,i_n),n=0,\ldots,N}(z-1)^n\chi_{i_1}'\cdots\chi_{i_n}'\right)$$

$$= \sum_{n=0}^{N}(z-1)^n\sum_{(i_1,\ldots,i_n)}\mathbb{P}(\tau_i > T, i \in \{i_1,i_2,\ldots,i_n\}) = \sum_{n=0}^{N}(z-1)^n\binom{N}{n}S_n$$

$$= \sum_{n=0}^{N}\left(\sum_{k=0}^{n}(-1)^{n-k}\binom{n}{k}z^k\right)\binom{N}{n}S_n$$

$$= \sum_{k=0}^{N}z^k\left(\sum_{n=k}^{N}(-1)^{n-k}\binom{n}{k}\binom{N}{n}S_n\right). \tag{13.54}$$

Since the function $\Phi(z)$ is the polynomial in z given by (13.53) and (13.54), we deduce that

$$F_{N-k}^{(N)} = \sum_{n=k}^{N}(-1)^{n-k}\binom{n}{k}\binom{N}{n}S_n$$

$$= \binom{N}{N-k}\sum_{n=k}^{N}(-1)^{n-k}\binom{N-k}{n-k}S_n.$$

We finally obtain (13.52):

$$F_k^{(N)} = \binom{N}{k}\sum_{n=N-k}^{N}(-1)^{n-N+k}\binom{k}{n-N+k}S_n$$

$$= \binom{N}{k}\sum_{j=0}^{k}(-1)^j\binom{k}{j}S_{N-k+j},$$

by setting $j = n - N + k$. $\qquad\square$

With the homogeneity assumption, we set $q \equiv Q_1(t,x)$ for the survival probability after t of one given bond, to obtain from the results in Section 13.2.2:

$$u_0 = \prod_{i=1}^{n} Q_i = q^n,$$

$$\sqrt{\varepsilon}\, u_{1,0} = \sum_{i=1}^{n} R_i^{(2)} w_i^{(2)} \prod_{\substack{j=1\\j\neq i}}^{n} Q_j + \sum_{i=1}^{n} R_i^{(3)} w_i^{(3)} \prod_{\substack{j=1\\j\neq i}}^{n} Q_j + \sum_{\substack{i,j=1\\i\neq j}}^{n} R_{ij}^{(3)} w_{ij}^{(3)} \prod_{\substack{k=1\\k\neq i,j}}^{n} Q_k$$

$$= nR_1^{(2)} w_1^{(2)}(t,x) q^{n-1} + nR_1^{(3)} w_1(t,x)^{(3)} q^{n-1} + n(n-1)R_{12}^{(3)} w_{12}^{(3)}(t,x,x) q^{n-2},$$

$$\sqrt{\delta}\, u_{0,1} = \sum_{i=1}^{n} R_i^{(0)} w_i^{(0)} \prod_{\substack{j=1\\j\neq i}}^{n} Q_j + \sum_{i=1}^{n} R_i^{(1)} w_i^{(1)} \prod_{\substack{j=1\\j\neq i}}^{n} Q_j + \sum_{\substack{i,j=1\\i\neq j}}^{n} R_{ij}^{(1)} w_{ij}^{(1)} \prod_{\substack{k=1\\k\neq i,j}}^{n} Q_k$$

$$= nR_1^{(0)} w_1^{(0)}(t,x) q^{n-1} + nR_1^{(1)} w_1^{(1)}(t,x) q^{n-1} + n(n-1)R_{12}^{(1)} w_{12}^{(1)}(t,x,x) q^{n-2},$$

where we have used the fact that the R and w do not depend on a particular choice of names (i,j). We define the quantities

$$A \equiv \sum_{k=0}^{3} R_1^{(k)} w_1^{(k)}(t,x), \tag{13.55}$$

$$B \equiv R_{12}^{(1)} w_{12}^{(1)}(t,x,x) + R_{12}^{(3)} w_{12}^{(3)}(t,x,x), \tag{13.56}$$

which also depend on the parameter z, and we rewrite the joint survival probabilities (13.51) as

$$S_n \approx \tilde{u} \equiv u_0 + \sqrt{\varepsilon}\, u_{1,0} + \sqrt{\delta}\, u_{0,1} = q^n + Anq^{n-1} + Bn(n-1)q^{n-2}, \quad n \geq 2.$$

Note that $S_1 = q + A$, and hence the previous formula for S_n is actually valid for all $n \geq 1$, as well as for $n = 0$ with $S_0 \equiv 1$.

The approximated loss distribution is now given by

$$\mathbb{P}^\star(L = k)$$

$$= \binom{N}{k} \sum_{j=0}^{k} \binom{k}{j} (-1)^j S_{N+j-k} = \binom{N}{k} \sum_{i=0}^{k} \binom{k}{i} (-1)^{k-i} S_{N-i}$$

$$\approx \binom{N}{k} \sum_{i=0}^{k} \binom{k}{i} (-1)^{k-i} \left[q^{N-i} + A(N-i)q^{N-i-1} \right.$$

$$\left. + B(N-i)(N-i-1)q^{N-i-2} \right]$$

$$= \binom{N}{k} \sum_{i=0}^{k} \binom{k}{i} (-1)^{k-i} q^{N-i} + A \binom{N}{k} \sum_{i=0}^{k} \binom{k}{i} (-1)^{k-i} (N-i) q^{N-i-1}$$

$$+B\binom{N}{k}\sum_{i=0}^{k}\binom{k}{i}(-1)^{k-i}(N-i)(N-i-1)q^{N-i-2}$$

$$\equiv I_0 + AI_1 + BI_2,$$

where I_0, I_1, and I_2 are obtained by straightforward calculation:

$$I_0 = \binom{N}{k}\sum_{i=0}^{k}\binom{k}{i}(-1)^{k-i}q^{N-i} = \binom{N}{k}(1-q)^k q^{N-k}, \qquad (13.57)$$

$$I_1 = \binom{N}{k}\left[\sum_{i=0}^{k}\binom{k}{i}(-1)^{k-i}s^{N-i}\right]'_{s=q} = \binom{N}{k}\left[s^{N-k}(1-s)^k\right]'_{s=q}$$

$$= \binom{N}{k}\left[(N-k)q^{N-k-1}(1-q)^k - kq^{N-k}(1-q)^{k-1}\right]$$

$$= \left[\frac{N-k}{q} - \frac{k}{1-q}\right]I_0, \qquad (13.58)$$

$$I_2 = \binom{N}{k}\left[\sum_{i=0}^{k}\binom{k}{i}(-1)^{k-i}s^{N-i}\right]''_{s=q} = \binom{N}{k}\left[s^{N-k}(1-s)^k\right]''_{s=q}$$

$$= \binom{N}{k}\left[(N-k)(N-k-1)q^{N-k-2}(1-q)^k\right.$$

$$\left. -2k(N-k)q^{N-k-1}(1-q)^{k-1} + k(k-1)q^{N-k}(1-q)^{k-2}\right]$$

$$= \left[\frac{(N-k)(N-k-1)}{q^2} - \frac{2k(N-k)}{q(1-q)} + \frac{k(k-1)}{(1-q)^2}\right]I_0. \qquad (13.59)$$

To summarize, for $0 \le k < N$,

$$\mathbb{P}^*(L = k) \approx I_0 + AI_1 + BI_2, \qquad (13.60)$$

where I_0, I_1, and I_2 are explicitly given by (13.57), (13.58), (13.59), and A and B defined in (13.55), (13.56) are small of order $\mathcal{O}(\sqrt{\varepsilon} + \sqrt{\delta})$.

Note that I_0 corresponds to the case where all the underlying assets are mutually independent, which gives rise to the classic binomial distribution. Since the method used here is perturbation, we call (13.60) the *perturbed binomial formula*.

Numerical Illustration In the homogeneous portfolio case discussed above, we implemented some numerical computation to illustrate the effect of introducing stochastic volatilities. We compute the approximated loss distribution given by (13.60) where only four parameters (N, q, A, B) are

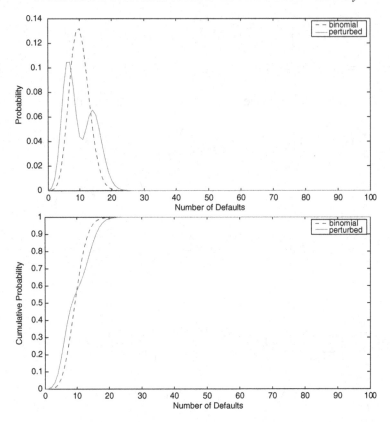

Figure 13.9   Perturbed binomial loss distribution.

needed. The purpose of this computation is to show stylized features of the loss distribution generated by the correlation of defaults due to the presence of stochastic volatility.

The parameters used in Figure 13.9 are as follows:

$$N = 100, \quad q = 0.9, \quad A = 0.00, \quad B = 0.0006.$$

The upper graph presents the probability mass function and the lower one presents the cumulative distribution function for the loss. Note that in order to make comparison between our result and the classic binomial distribution, we chose $A = 0$ because a nonzero A would make single-name survival probabilities distinct under these two scenarios. By choosing $A = 0$, the single-name survival probability is $q = 0.9 = 90\%$. Note also that B is chosen very small to ensure that the approximation is in its range of validity.

It can be observed from Figure 13.9 that:

- The probability, with stochastic volatilities, that the total portfolio loss is less than 3% is bigger than that from the classic binomial distribution.

- The probability, with stochastic volatilities, that the total portfolio loss is bigger than 15% is bigger than that from the classic binomial distribution.

In terms of CDO tranches (we refer to Section 14.2 for an introduction), this implies that the expected loss and hence the fair spread of the equity tranche become less, while the expected loss and hence the fair spread of the senior tranche become bigger, than in the classic binomial case. This phenomenon is consistent with what has been documented so far in the CDO literature if one increases the default correlation between the underlying bonds and keeps other things unchanged. For example, see Duffie and Gârleanu (2001). Note that a positive number B indicates positive correlation between the underlying bonds. Here we assume that the equity tranche absorbs 0–3% loss and the senior absorbs 15–100% loss.

Furthermore, the binomial probability mass function achieves its maximum 0.1319 at $k = 11$, while the probability mass function with stochastic volatilities achieves its maximum 0.1047 at $k = 8$.

13.2.4 Models with Name–Name Correlation

We now consider the *correlated case*, namely where the Brownian motions $W^{(i)}$ driving the names are correlated. Using the notation of Section 13.2.1, this means taking $d\langle W^{(i)\star}, W^{(j)\star}\rangle_t = \rho_{ij}dt$ for $i \neq j$ under the risk-neutral probability measure $\mathbb{P}^\star$ with $|\rho_{ij}| < 1$. Observe that for nonzero ρ_{ij}'s the conditions $\sum_{i=1}^{n}\rho_{iY}^2 \leq 1$ and $\sum_{i=1}^{n}\rho_{iZ}^2 \leq 1$ are no longer needed. We denote by $u^{\varepsilon,\delta,\rho}$ the joint survival probability defined by

$$u^{\varepsilon,\delta,\rho}(t,\mathbf{x},y,z)$$
$$\equiv \mathbb{P}^\star\{\tau_t^{(1)} > T, \ldots, \tau_t^{(n)} > T \mid \mathbf{X}_t = \mathbf{x}, Y_t = y, Z_t = z\}. \quad (13.61)$$

Then $u^{\varepsilon,\delta,\rho}$ is the solution to the boundary value problem

$$\mathcal{L}^{\varepsilon,\delta,\rho}u^{\varepsilon,\delta,\rho} = 0, \quad x_i > B_i(t), \text{ for all } i, t < T,$$
$$u^{\varepsilon,\delta,\rho}(t,x_1,x_2,\ldots,x_n,y,z) = 0, \quad \exists i \in \{1,\ldots,n\}, x_i = B_i(t), t \leq T,$$
$$u^{\varepsilon,\delta,\rho}(T,x_1,x_2,\ldots,x_n,y,z) = 1, \quad x_i > B_i(t), \text{ for all } i,$$

where the operator $\mathcal{L}^{\varepsilon,\delta,\rho}$ can be written as

$$\mathcal{L}^{\varepsilon,\delta,\rho} = \mathcal{L}^{\varepsilon,\delta} + \sum_{i<j}^{n} \rho_{ij}\mathcal{L}_\rho^{(ij)}, \quad (13.62)$$

where we have used $\rho_{ij} = \rho_{ji}$, the definition (13.17) of the operator $\mathcal{L}^{\varepsilon,\delta}$, and the notation

$$\mathscr{L}_\rho^{(ij)} = f_i(y,z)f_j(y,z)x_i x_j \frac{\partial^2}{\partial x_i \partial x_j}. \qquad (13.63)$$

The leading-order term in the small ε and small δ expansion carried out in the previous sections would correspond to the correlated multiname case with constant volatility. As explained in the introduction, this is not a tractable model because of the lack of simple formulas for the joint distribution of hitting times. This leads us to perform an additional expansion around the independent case where the ρ_{ij} are zero. We therefore consider the case where the ρ_{ij} are small and of the same order.

Expanding $u^{\varepsilon,\delta,\rho}$ in powers of $\sqrt{\varepsilon}$, $\sqrt{\delta}$, and ρ_{ij}, we get:

$$u^{\varepsilon,\delta,\rho} = u^{\varepsilon,\delta} + \sum_{i<j}^{n} \rho_{ij}\left(u_{0,0,1}^{(ij)} + \sqrt{\varepsilon}u_{1,0,1}^{(ij)} + \sqrt{\delta}u_{0,1,1}^{(ij)} + \cdots\right) + \cdots, \qquad (13.64)$$

and retain the lowest-order terms:

$$\tilde{u} \equiv u_0 + \sqrt{\varepsilon}\,u_{1,0} + \sqrt{\delta}\,u_{0,1} + \sum_{i<j}^{n} \rho_{ij}u_{0,0,1}^{(ij)}, \qquad (13.65)$$

as our approximation for $u^{\varepsilon,\delta,\rho}$ where the first three terms have been computed in Section 13.2.2.

Correction Terms $\rho_{ij}u_{0,0,1}^{(ij)}$ By now routine computation, the term $u_{0,0,1}^{(ij)}$ is determined by the following PDE system:

$$\langle\mathscr{L}_2\rangle u_{0,0,1}^{(ij)} = -\langle\mathscr{L}_\rho^{(ij)}\rangle u_0, \quad x_l > B_l(t), \text{ for all } l, t < T, \quad (13.66)$$

$$u_{0,0,1}^{(ij)}(t,x_1,x_2,\ldots,x_n) = 0, \quad \exists\, l \in \{1,\ldots,n\}, x_l = B_l(t), t \le T,$$

$$u_{0,0,1}^{(ij)}(T,x_1,x_2,\ldots,x_n) = 0, \quad x_l > B_l(t), \text{ for all } l,$$

where the operator $\langle\mathscr{L}_\rho^{(ij)}\rangle$ is given by

$$\langle\mathscr{L}_\rho^{(ij)}\rangle = \langle f_i(\cdot,z)f_j(\cdot,z)\rangle x_i x_j \frac{\partial^2}{\partial x_i \partial x_j}.$$

The following result shows that each $u_{0,0,1}^{(ij)}$ corresponds to a single two-dimensional problem.

Proposition 13.5   *The correction term* $\rho_{ij}u_{0,0,1}^{(ij)}$ *is given by*

$$\rho_{ij}u_{0,0,1}^{(ij)} = R_{ij}^{(4)}w_{ij}^{(4)}\prod_{\substack{k=1\\k\ne i,j}}^{n} Q_k, \qquad (13.67)$$

where the coefficient $R_{ij}^{(4)}$ depends on the parameter z and is given by

$$R_{ij}^{(4)} = \rho_{ij}\langle f_i(\cdot,z)f_j(\cdot,z)\rangle, \quad i \neq j, \tag{13.68}$$

and where the function $w_{ij}^{(4)}(t,x_i,x_j)$ depends on the parameter z and is given by the following problem:

$$\left[\frac{\partial}{\partial t} + \frac{1}{2}\sigma_i^2(z)x_i^2\frac{\partial^2}{\partial x_i^2} + \frac{1}{2}\sigma_j^2(z)x_j^2\frac{\partial^2}{\partial x_j^2} + rx_i\frac{\partial}{\partial x_i} + rx_j\frac{\partial}{\partial x_j}\right]w_{ij}^{(4)} \tag{13.69}$$

$$= -\left(x_i\frac{\partial Q_i}{\partial x_i}\right)\left(x_j\frac{\partial Q_j}{\partial x_j}\right), \quad x_i > B_i(t), x_j > B_j(t), t < T,$$

$$w_{ij}^{(4)}(t,x_i,x_j) = 0, \quad \text{if} \quad x_i = B_i(t) \quad \text{or} \quad x_j = B_j(t), t \leq T,$$

$$w_{ij}^{(4)}(T,x_i,x_j) = 0, \quad x_i > B_i(t), x_j > B_j(t),$$

with $\sigma_i(z)$ given in Proposition 13.1.

Proof The proof is very similar to that of Proposition 13.2. With the definition (13.68) of $R_{ij}^{(4)}$, we have

$$\rho_{ij}\langle \mathcal{L}_\rho^{(ij)}\rangle u_0 = R_{ij}^{(4)}x_i\frac{\partial}{\partial x_i}\left(x_j\frac{\partial u_0}{\partial x_j}\right). \tag{13.70}$$

It is therefore enough to check that

$$\langle \mathcal{L}_2\rangle\left(w_{ij}^{(4)}\prod_{k=1,k\neq i,j}^n Q_k\right) = -x_i\frac{\partial}{\partial x_i}\left(x_j\frac{\partial u_0}{\partial x_j}\right). \tag{13.71}$$

Using the form (13.22) of $\langle\mathcal{L}_2\rangle$ and $u_0 = \prod_{i=1}^n Q_i$, one can easily check that (13.71) is satisfied. The boundary and terminal conditions for the correction $\rho_{ij}u_{0,0,1}^{(ij)}$ are directly inherited from the boundary and terminal conditions for the functions $w_{ij}^{(4)}$, and Q_i. $\quad\square$

Formulas for the solutions $w_{ij}^{(4)}$ of (13.69) can be found in Fouque *et al.* (2008).

Summary of Approximation with Name–Name Correlation Combining the results of Propositions (13.1)–(13.5), we get that the approximation $\tilde{u}$ in (13.65) is given by

$$\tilde{u} = \prod_{i=1}^{n} Q_i + \sum_{i=1}^{n} R_i^{(2)} w_i^{(2)} \prod_{\substack{j=1 \\ j \neq i}}^{n} Q_j + \sum_{i=1}^{n} R_i^{(3)} w_i^{(3)} \prod_{\substack{j=1 \\ j \neq i}}^{n} Q_j + \sum_{\substack{i,j=1 \\ i \neq j}}^{n} R_{ij}^{(3)} w_{ij}^{(3)} \prod_{\substack{k=1 \\ k \neq i,j}}^{n} Q_k$$

$$+ \sum_{i=1}^{n} R_i^{(0)} w_i^{(0)} \prod_{\substack{j=1 \\ j \neq i}}^{n} Q_j + \sum_{i=1}^{n} R_i^{(1)} w_i^{(1)} \prod_{\substack{j=1 \\ j \neq i}}^{n} Q_j + \sum_{\substack{i,j=1 \\ i \neq j}}^{n} R_{ij}^{(1)} w_{ij}^{(1)} \prod_{\substack{k=1 \\ k \neq i,j}}^{n} Q_k$$

$$+ \sum_{i<j}^{n} R_{ij}^{(4)} w_{ij}^{(4)} \prod_{\substack{k=1 \\ k \neq i,j}}^{n} Q_k, \tag{13.72}$$

where $R_{ij}^{(4)}$ are small of order ρ_{ij} and depend on the parameter z, and the functions $w_{ij}^{(4)}$ depend on the variables (x_i, x_j) and the parameter z.

Loss Distribution (Homogeneous Portfolio) For simplicity we only consider the homogeneous portfolio case. In addition to the homogeneity conditions (13.50), we also assume that $\rho_{ij} = \rho$ for all (i,j) and $|\rho| \ll 1$.

Using the same notation as in Section 13.2.3, the name–name correction term becomes:

$$\sum_{i<j}^{n} \rho_{ij} u_{0,0,1}^{(ij)} = \sum_{i<j}^{n} R_{ij}^{(4)} w_{ij}^{(4)} \prod_{\substack{k=1 \\ k \neq i,j}}^{n} Q_k$$

$$= \frac{1}{2} n(n-1) R_{12}^{(4)} w_{12}^{(4)}(t,x,x) q^{n-2},$$

where from (13.68), $R_{12}^{(4)} = \rho \sigma^2(z)$.

In particular this implies that the approximation for the loss distribution given by (13.60) still holds if we replace the quantity B in (13.56) by $B + B_\rho$ where

$$B_\rho \equiv \frac{1}{2} R_{12}^{(4)} w_{12}^{(4)}(t,x,x), \tag{13.73}$$

so that (13.60) becomes

$$\mathbb{P}^*(L = k) \approx I_0 + A I_1 + (B + B_\rho) I_2. \tag{13.74}$$

Note that for a single maturity T the correlations generated by stochastic volatility and name–name correlation are of the same form as leading order. However, if one looks at the term structure of correlation across several maturities, an important aspect of CDO tranches, then the shape of the function $w_{12}^{(4)}$ is different from the shapes of $w_{12}^{(1)}$ and $w_{12}^{(3)}$ and therefore the nature of the correlation plays a role.

Table 13.3: *Joint survival probability for twenty five firms ($n = 25$) with stochastic volatility and name–name correlations*

ε	δ	ρ_{ij}	u_0	$\tilde{u}$	u_{MC}	Absolute (relative) error
1/50	1/20	0	0.471683	0.486892	0.4803	0.006 (1.4%)
1/50	1/20	0.05	0.471683	0.518151	0.5119	0.006 (1.2%)
1/50	1/20	0.1	0.471683	0.549409	0.5426	0.007 (1.3%)
1/50	1/20	0.2	0.471683	0.611926	0.5986	0.013 (2.2%)
1/50	1/20	0.4	0.471683	0.736961	0.6937	0.043 (6.2%)

Table 13.4: *Loss distribution for one hundred firms ($n = 100$) with stochastic volatility and name–name correlations. All omitted probabilities are less than 10^{-5}. The expected number of defaults is 2.88 when stochastic volatility is present, while it is 2.96 in the independent constant volatility case*

0	1	2	3	4
0.16	0.26	0.17	0.062	0.047
5	6	7	8	9
0.078	0.086	0.065	0.037	0.017
10	11	12	13	14
6.5×10^{-3}	2.2×10^{-3}	6.2×10^{-4}	1.6×10^{-4}	3.7×10^{-5}

Numerical Illustration In order to illustrate the quality of the approximation of the joint survival probability given by (13.72), we have conducted the following numerical experiments. For $n = 25$ names (Table 13.3), we compute the zero-order approximation u_0 given by (13.23), and the first-order approximation $\tilde{u}$ given by (13.72) and the explicit formulas derived in Fouque *et al.* (2008). We present the results for $\varepsilon = 0.02$, $\delta = 0.5$, and five cases of $\rho_{ij} = \rho$. The true value proxy u_{MC} is obtained by Monte Carlo simulations with 10^5 realizations and using an Euler scheme with time step of 10^{-4}, as in Section 13.2.2. The other parameters are as in Section 13.2.2.

The presence of a small name–name correlation ρ enhances the importance of the correction $(\tilde{u} - u_0)$ derived here, as can be seen by comparing u_0 and u_{MC} in Table 13.2 and in Table 13.3.

Table 13.4 gives a loss distribution generated with stochastic volatility and name–name correlation, as computed by (13.74). The results are plotted in Figure 13.10. The parameters are taken from Section 13.2.2, with $n = 100$, $\varepsilon = 1/50$, $\delta = 1/20$, $\rho_{ij} = 1/10$. With these parameters, the

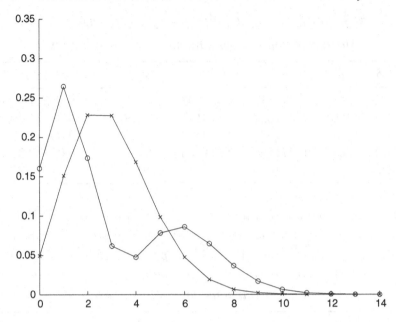

Figure 13.10   Plot of the loss distribution (○) given in Table 13.4. The mass function of the corresponding binomial distribution (×) is superimposed for comparison.

coefficients in (13.74) are: $A = 6.607 \times 10^{-4}$, $B = -0.014 \times 10^{-4}$, and $B_\rho = 2.08 \times 10^{-4}$. The binomial distribution plotted in Figure 13.10 is obtained by setting $A = B = B_\rho = 0$ in (13.74).

The loss distribution shown in Figure 13.10 clearly has a bimodal structure. With the given parameters, one can see that the event of having two to five defaults is significantly less likely than if the stocks were independent.

14

Credit Risk II: Multiscale Intensity-Based Models

As discussed in Section 13.1, the Black–Cox structural model with constant volatility predicts essentially zero-yield spreads at short maturities, because defaults are "predictable" in the sense that we can see them coming. One way to remedy this, as detailed in Chapter 13, was with multiscale stochastic volatility. More traditionally, in the credit risk literature, this observation serves as a motivation for intensity-based (or reduced-form) models in which the default occurs at the jump of an exogenous Poisson-type process. The element of surprise means that, even at short maturities, credit spreads are significant since the default time is no longer announced as in structural models. In addition to the references for background on credit modeling given in the Notes at the end of Chapter 1, we also refer to Bielecki and Rutkowski (2004) for more details on reduced-form models.

The multiname material presented in Section 14.3 is from Fouque *et al.* (2009) and the material in Section 14.4 with additional grouping structure, is from Papageorgiou and Sircar (2009).

14.1 Background on Stochastic Intensity Models

We start with a brief review of the basics of Poisson and Cox processes.

14.1.1 Poisson Process

We begin by recalling the defining properties of a Poisson process $(N_t)_{t \geq 0}$ with parameter $\lambda > 0$:

(i) It starts at zero: $N_0 = 0$.
(ii) It is a counting process, meaning that it takes successively the values $0, 1, 2, \ldots$
(iii) It has independent increments.

(iv) Its increments are Poisson distributed with parameter λ times the length of the increment:

$$N_t - N_s \sim \text{Poisson}(\lambda(t-s)), \quad t > s.$$

As a consequence of the distribution of the increments, that is

$$\mathbb{P}\{N_t - N_s = k\} = e^{-\lambda(t-s)}\frac{(\lambda(t-s))^k}{k},$$

for non-negative integers k, we obtain (with $k = s = 0$) the distribution of the first jump time

$$\tau = \inf\{t > 0 \mid N_t > 0\},$$

by

$$\mathbb{P}\{\tau > t\} = \mathbb{P}\{N_t = 0\} = e^{-\lambda t}.$$

In other words, the holding (or inter-arrival) times are exponentially distributed with mean

$$\mathbb{E}\{\tau\} = 1/\lambda.$$

The parameter λ (which is in units of time^{-1}) is called the *rate* or *intensity* of the Poisson process.

We denote by $(\mathcal{H}_t)_{t \geq 0}$ the filtration generated by N. Then we have for $t > s$:

$$\mathbb{E}\{N_t \mid \mathcal{H}_s\} = \mathbb{E}\{(N_t - N_s) + N_s \mid \mathcal{H}_s\}$$
$$= N_s + \lambda(t-s),$$

using the independence and Poisson distribution of the increments. If we define

$$M_t = N_t - \lambda t,$$

then

$$\mathbb{E}\{M_t \mid \mathcal{H}_s\} = M_s,$$

so that M is a martingale with respect to $(\mathcal{H}_t)$. The "process" λt (that which we need to subtract to create a martingale) is called the compensator of the Poisson process N.

The Poisson process is a continuous-time Markov chain on the state space $\{0, 1, 2, \ldots\}$ whose infinitesimal generator is a bidiagonal matrix $\mathcal{L}$ with diagonal entries $\mathcal{L}_{i,i} = -\lambda$ and above diagonals $\mathcal{L}_{i,i+1} = \lambda$. Therefore $\lambda \Delta t$ plays the role of the local jump probability over the small time period of length Δt. Here, it is a constant (that is, it does not depend on time t).

Later we will make it a deterministic function of time and then a stochastic process.

14.1.2 Inhomogeneous Poisson Process

Based on the last observation, the formal definition of an inhomogeneous Poisson process $(H_t)_{t \geq 0}$ with intensity $\lambda(t)$, a non-negative deterministic function of time with $\int_0^t \lambda(s)\, ds < \infty$ for all $t > 0$, is:

(i) It starts at zero: $H_0 = 0$.
(ii) It is a counting process.
(iii) It has independent increments.
(iv) Local jump probabilities are given by

$$\mathbb{P}\{H_{t+\Delta t} - H_t = 1 \mid H_t\} = \lambda(t)\Delta t + o(\Delta t),$$
$$\mathbb{P}\{H_{t+\Delta t} - H_t \geq 2 \mid H_t\} = \mathcal{O}(\Delta t).$$

In this formulation, we have simply generalized the Poisson process by allowing the intensity to vary with time.

Alternatively, we can replace the last (local) property with the property that the increments are Poisson distributed with parameter equal to the integral of λ over the time interval:

$$H_t - H_s \sim \text{Poisson}\left(\int_s^t \lambda(u)\, du\right), \qquad t > s.$$

Indeed, it is natural to think of obtaining an inhomogeneous Poisson process by time-changing a standard Poisson process. In other words, let N be a standard Poisson process with parameter 1, and let

$$\Lambda(t) = \int_0^t \lambda(s)\, ds$$

define a time change. Then we could define $H_t = N_{\Lambda(t)}$, from which it is easy to verify the properties above. In particular,

$$H_t - H_s = N_{\Lambda(t)} - N_{\Lambda(s)}$$
$$\sim \text{Poisson}(\Lambda(t) - \Lambda(s))$$
$$= \text{Poisson}\left(\int_s^t \lambda(u)\, du\right).$$

As with the standard case, we can define the compensator of H. Let $(\mathcal{H}_t)$ be the filtration generated by (H_t). Then we have

$$\mathbb{E}\{H_t - H_s \mid \mathcal{H}_s\} = \Lambda(t) - \Lambda(s),$$

so that (M_t) defined by

$$M_t = H_t - \Lambda(t) = H_t - \int_0^t \lambda(u)\,du$$

satisfies the martingale property with respect to $(\mathscr{H}_t)$. So, the compensator of the inhomogeneous Poisson process is $\int_0^t \lambda(u)\,du$.

14.1.3 Doubly Stochastic or Cox Processes

We generalize next to a stochastic intensity process (λ_t), so that default times are generated by a two-step randomization process described by λ and N. We start with a probability space $(\Omega, \mathscr{G}, \mathbb{P})$ and a filtration $(\mathscr{G}_t)$. Let (H_t) be a counting process with jump times $\tau_1, \tau_2, \ldots$ We will assume that H is nonexplosive, meaning that

$$\lim_{n \to \infty} \tau_n = +\infty \quad \text{a.s.}$$

Then, (λ_t) is called the intensity of H if:

(i) λ is a non-negative predictable process.
(ii) $\mathbb{E}\{\int_0^t \lambda_s\,ds\} < \infty$ for all $t > 0$.
(iii) The process

$$M_t = H_t - \int_0^t \lambda_s\,ds$$

is a $\mathscr{G}_t$-martingale.

Now, suppose that $(\mathscr{F}_t)$ is a subfiltration, that is $\mathscr{F}_t \subset \mathscr{G}_t$ for all t. Then H is called a doubly stochastic process driven by $(\mathscr{F}_t)$ (or a $(\mathscr{F}_t)$-conditional Poisson process, or a Poisson process conditioned by $(\mathscr{F}_t)$) if λ_t is $\mathscr{F}_t$-predictable and, for $t > s$, conditional on $\mathscr{G}_s \vee \mathscr{F}_t$, the increment $H_t - H_s$ is Poisson distributed with parameter $\int_s^t \lambda_u\,du$.

In practice, the filtration $(\mathscr{F}_t)$ will be generated by some factor processes $(\mathbf{Y}_t)$ and $\lambda_t = f(\mathbf{Y}_t)$, for some (non-negative and bounded) function f. The jump times can be generated in the following way. Let $Z_0, Z_1, \ldots$ be i.i.d. exponentially distributed random variables with mean 1, independent of λ. Define $\tau_0 = 0$. Then the jump times are given by

$$\tau_n = \inf\left\{ t \geq \tau_{n-1} : \int_{\tau_{n-1}}^t \lambda_s\,ds = Z_n \right\}.$$

From these, we define the counting process (H_t) and the filtration $(\mathscr{H}_t)$ generated by it. Finally, the full filtration is $\mathscr{G}_t = \mathscr{F}_t \vee \mathscr{H}_t$.

The jump times τ_n of H are *not* predictable stopping times, where predictable was defined in the context of structural models in Section 1.8.2. Even stronger, they are "totally inaccessible" stopping times, meaning that, given any predictable stopping time S, $\mathbb{P}\{\tau_n = S < \infty\} = 0$. This is the element of surprise we have mentioned many times before, in contrast to first-passage times in diffusion structural models.

The construction described above forms the basis of a simulation method in which we draw unit exponential independent random variables, simulate the path of λ, and register a jump in H when the accumulated time since the last jump $\int_{\tau_{n-1}}^{t} \lambda_s \, ds$ reaches the drawn Z_n. This also reveals that, since the standard Poisson process N has unit exponential holding times between jumps, H can be viewed as a time-changed Poisson process, where the deformed time is given by $\Lambda_t = \int_0^t \lambda_s \, ds$. In other words, $H_t = N_{\Lambda_t}$.

14.1.4 Connection with Interest Rate Derivatives

One reason for the rapid adoption of intensity-based models is that various standard calculations central to pricing defaultable securities are similar to interest rate derivative calculations, so that models which are effective in fixed income (for example, because they give closed-form solutions) can be adapted for credit derivatives.

To see this, consider a defaultable bond, which pays \$1 on date T as long as there is no default by then and nothing if there has been. We assume there is a market-chosen risk-neutral probability $\mathbb{P}^\star$, under which H is a Cox process, and the default time τ is the first jump time of H:

$$\tau = \inf\{t > 0 \mid H_t > 0\}.$$

The arbitrage-free price of the defaultable bond under a constant interest rate r is given by the expectation of its discounted payoff:

$$P_D(0,T) = \mathbb{E}^\star\{e^{-rT}\mathbf{1}_{\{\tau>T\}}\} = \mathbb{E}^\star\{e^{-rT}\mathbb{E}^\star\{\mathbf{1}_{\{\tau>T\}} \mid \mathscr{F}_T\}\}$$
$$= \mathbb{E}^\star\left\{e^{-\int_0^T (r+\lambda_s)\,ds}\right\}, \qquad (14.1)$$

where we have used the property that, conditioned on the information $\mathscr{F}_T$ (in which case the path $(\lambda_s)_{0 \le s \le T}$ is given), H is an inhomogeneous Poisson process with intensity given by that path. Therefore, we reduce to computing (14.1), which is like pricing a bond with stochastic interest rate $r + \lambda$. The *yield spread* is given by

$$Y(0,T) = -\frac{1}{T}\log P_D(0,T) - r.$$

Note that if $\lambda_t \equiv \lambda$ is constant, in which case H is a Poisson process, then the yield spread is simply $Y(0,T) = \lambda$ for all maturities T. However, market data shows that yield spread curves are not flat. The next step would be to consider deterministically time-varying $\lambda(t)$, in which case H is an imhomogeneous Poisson process, then the yield spread is given by $Y(0,T) = \frac{1}{T}\int_0^T \lambda(s)\,ds$. This can capture the initial yield spread from data, but not the observed random evolution of the curve. This leads to introducing stochastic intensities, in which case H is a Cox process.

For instance, if λ is chosen as a CIR process, the calculation is given in Section 12.4. Suppose we have

$$d\lambda_t = a(\bar{\lambda} - \lambda_t)\,dt + \sigma\sqrt{\lambda_t}\,dW_t^\star, \tag{14.2}$$

with $W^\star$ a $\mathbb{P}^\star$-Brownian motion. This is a mean-reverting process (with long-run mean level $\bar{\lambda} > 0$) that stays positive with probability one as long as $\sigma^2 \leq 2a\bar{\lambda}$. The time-zero price of a defaultable bond is given by

$$e^{-rT}\mathbb{E}^\star\{e^{-\int_0^T \lambda_s\,ds}\} = A(T)e^{-B(T)\lambda_0 - rT},$$

where

$$A(s) = \left(\frac{2\theta e^{(\theta+a)s/2}}{(\theta+a)(e^{\theta s}-1)+2\theta}\right)^{2a\bar{\lambda}/\sigma^2}, \tag{14.3}$$

$$B(s) = \frac{2(e^{\theta s}-1)}{(\theta+a)(e^{\theta s}-1)+2\theta},$$

$$\theta = \sqrt{a^2+2\sigma^2}.$$

The yield spread is given by

$$Y(0,T) = -\frac{1}{T}(\log A(T) - B(T)\lambda_0).$$

It is easy to show that

$$Y(0,T) \to \frac{1}{2}(\theta - a) = \frac{1}{2}(\sqrt{a^2+2\sigma^2} - a) \quad \text{as } T \to \infty,$$

a positive constant. In other words, long-term spreads do not converge to zero as observed in the structural Black–Cox model. Additionally, it is straightforward to see that $\lim_{T\downarrow 0} Y(0,T) = \lambda_0$, and so we can match the nonzero observed short-term yield spread.

However, it is well documented that one diffusion factor driving the intensity is not enough to capture the rich dynamics of observed yield spread curves. Incorporating multiscale stochastic volatility into the intensity is one way to add flexibility while keeping computational tractability. As the analysis is identical to that of Chapter 12, we do not repeat it here. Instead, we move on to the far more challenging problem of multiname credit derivatives, after introducing in the next section the single-name credit default swap contract.

14.1.5 Credit Default Swap

A credit default swap (CDS) contract is a derivative that provides insurance against the default of a *reference entity*, which is usually a corporation or a sovereign (i.e., a national government). The results given above on defaultable bonds generalize easily to this important contract. We provide below a brief description following Papageorgiou and Sircar (2008), to which we refer for further details and the implementation of the stochastic volatility corrections.

In a single-name CDS the *protection buyer*, that is, the counterparty that receives a payoff if the reference entity defaults, pays a periodic premium to the *protection seller*. In return, the protection seller has to compensate the protection buyer in the case where the reference entity defaults prior to a predetermined maturity time. The premiums the protection buyer pays to the protection seller are usually paid quarterly or semiannually until the maturity of the credit default swap contract, or until the time of the default event, whichever comes first.

To fix ideas, suppose that the CDS is written on a bond issued by the reference entity. In the case where the reference entity defaults before the maturity of the CDS contract, we assume that the protection seller compensates the protection buyer with a *cash settlement*. In particular, the protection seller will make a cash payment to the protection buyer equal to the difference of the notional amount of the bond and its post-default market value, which is usually determined by polling several dealers. Here we assume that in default the bond recovers $1 - q$ of its face value – which is known as *recovery of face value* – and the protection seller provides the remaining proportion q of the face value to the protection buyer.

While there are other types of settlement used in practice, we will not go into details here. For a detailed specification of the repayment methods at default, more information on what constitutes a default event, usual practices for settlements in the financial industry, and other legal issues we refer to the books by Duffie and Singleton (2003), Schönbucher (2003), and Lando (2004).

The pricing of a credit default swap amounts to determining the *CDS spread*, which determines the amount paid by the protection buyer to the protection seller on each payment date. Assume that there are M such scheduled periodic payments and let $\mathbf{T}$ be the *payment tenor*, used to denote the sequence of these payment dates T_m, $m = 1, \ldots, M$ or in other words $\mathbf{T} = (T_1, \ldots, T_M)$, with $T_1 < \cdots < T_M$. We refer to T_M as the *maturity* of the CDS contract.

The CDS spread is quoted in basis points, or one-hundredths of one percent (i.e., $1/10000$), and thus the payment of the protection buyer to the protection seller at each payment date is the product of the CDS spread and the face value of the bond which is designated in the CDS contract. We will also make the simplifying assumptions that

 (i) the bond coupon dates match the payment dates of the CDS, and

 (ii) if a default occurs, the settlement takes place at a coupon date following the default but we do not consider the accrued interest of the intermediate period.

Instead of determining the CDS spread for a credit default swap that would be active immediately, we will price a *forward CDS* (forward contract on a CDS). A forward CDS is a CDS contract between a protection buyer and a protection seller with payment tenor $\mathbf{T}$ that is active after some initial time T_0 (the *effective date*, where $0 \le T_0 < T_1$). In the case where the credit event happens prior to the effective date, the forward CDS is worthless and no payments are made from either counterparty.

Since the CDS contract is made up of two distinct counterparties' payoffs, its pricing follows from no-arbitrage arguments based on the pricing of each counterparty's position. We will determine the payment of the protection buyer, c^{pb}, and that of the protection seller, c^{ps}, at some time t with $t \le T_0$. We denote the forward CDS spread at time t with effective date T_0 and payment tenor $\mathbf{T}$ as $c^{ds}(t, T_0; \mathbf{T})$, and so the spread of a CDS is given by the spread of a forward CDS with immediate effective date, i.e., $c^{ds}(t, t; \mathbf{T})$.

Consider the case of the protection buyer first. Such an individual will pay a premium at each payment date T_m ($m = 1, \ldots, M$) until maturity or until default, whichever comes first, for every dollar of the face value of the bond. Recalling that τ is the time of the first jump of the doubly stochastic Poisson process H introduced in Section 14.1.3, we can express this premium as

$$c^{pb}(t, T_0; \mathbf{T}) =$$

$$\mathbb{E}^\star \left\{ \sum_{m=1}^M \exp\left(-\int_t^{T_m} r_s \, ds \right) \mathbf{1}_{\{\tau > T_m\}} c^{ds}(t, T_0; \mathbf{T}) \mid \mathscr{F}_t \vee \sigma\{H_s; \, 0 \leq s \leq t\} \right\}$$

$$= c^{ds}(t, T_0; \mathbf{T}) \sum_{m=1}^M \mathbb{E}^\star \left\{ \exp\left(-\int_t^{T_m} (r_s + \lambda_s) \, ds \right) \mid \mathscr{F}_t \right\}, \quad t < \tau, \quad (14.4)$$

where here we allow for varying interest rate. We arrived at the expression (14.4) as in the defaultable bond pricing expression (14.1). Indeed, the conditional expectation that appears in (14.4) is simply the price of a zero-coupon bond at time t with zero recovery of market value (or zero recovery of face value) and maturity T_m, which we denote by $p(t; T_m)$, and is given by

$$p(t; T_m) = \mathbf{1}_{\{t < \tau\}} \mathbb{E}^\star \left\{ \exp\left(-\int_t^{T_m} (r_s + \lambda_s) \, ds \right) \mid \mathscr{F}_t \right\}.$$

The payment of the protection buyer is zero if $t \geq \tau$.

Under the cash settlement assumption we made, the present (time t) payment by the protection seller at the default time τ is

$$c^{ps}(t, T_0; \mathbf{T}) = \mathbb{E}^\star \left\{ \exp\left(-\int_t^\tau r_s \, ds \right) \mathbf{1}_{\{T_0 \leq \tau \leq T_M\}} q \mid \mathscr{F}_t \vee \sigma\{H_s; \, 0 \leq s \leq t\} \right\}$$

$$= \mathbb{E}^\star \left\{ \int_{T_0}^{T_M} \exp\left(-\int_t^u (r_s + \lambda_s) \, ds \right) q \lambda_u \, du \mid \mathscr{F}_t \right\}, \quad t < \tau.$$

Similarly, the protection seller payment is zero on $\{t \geq \tau\}$.

The forward CDS spread is, by definition, the spread that equates the payments of the two counterparties, thus

$$c^{ds}(t, T_0; \mathbf{T}) = \frac{\mathbb{E}^\star \left\{ \int_{T_0}^{T_M} \exp\left(-\int_t^u (r_s + \lambda_s) \, ds \right) q \lambda_u \, du \mid \mathscr{F}_t \right\}}{\sum_{m=1}^M p(t; T_m)}, \quad t < \tau.$$

A similar expression can be derived in the context of structural models described in the Section 13.1.

14.2 Multiname Credit Derivatives

The much harder problem in credit risk is to value exotic credit derivatives whose payoffs are contingent on defaults of many (sometimes hundreds) of corporate bonds of different firms. These defaults are certainly not independent: difficult times, for example in the airline industry, increase the risk of default by many airlines; default by a supplier may impact the credit-worthiness of the firms they do business with; it may also improve the situation of their competitors. The dramatic losses in the credit derivatives market in 2007 and subsequently illustrate that the problem of appropriate modeling and pricing of large portfolios of debt obligations is challenging and also a largely open question.

The mathematical challenge is to model the default times of the firms (also referred to as names or underlyings) and, most importantly, the correlation between them. Part of the challenge is that incorporating heterogeneity and correlations may appear as intractable due to the curse of combinatorial complexity. Here we consider pricing of collateralized debt obligations (CDOs) using intensity-based models with multiscale stochastic volatility. The CDO contract is explained in more detail below. Copulas have been the standard tool in the industry for creating correlation structures (Li, 2000) in the last few years. This is even the case at present after the recent credit crisis. The main drawback of this approach is the fact that these are static models which do not take into account the time evolution of joint default risks. This has been recognized in the academic literature on dynamic models, with recent developments in the multiname structural approach (Fouque *et al.*, 2008; Hurd, 2009), reduced-form models (Duffie and Gârleanu, 2001; Papageorgiou and Sircar, 2009; Mortensen, 2006; Eckner, 2007), and top-down models (Errais *et al.*, 2010; Lopatin and Misirpashaev, 2007; Schönbucher, 2006). In Chapter 13, we already discussed CDOs from the point of view of structural modeling.

CDO payoffs depend on the default events of a basket portfolio of many firms over periods from three to ten years in length, with five years being most common. As long as there are no defaults, investors in CDO tranches enjoy high yields, but as defaults start occurring they affect first the high-yield equity tranche, then the mezzanine tranches, and perhaps on to the senior and super-senior tranches. CDOs have been attractive because of the high yields they offer relative to poorly performing stocks and bonds in recent years. They also allowed certain pension funds, mutual funds, and endowments which are prohibited from investing directly in debt below a certain rating grade to indirectly participate in the higher yields such bonds offer by repackaging the basket into (reinsurance) tranches which had higher credit ratings than some of the components, even though such ratings remain questionable and controversial.

In the rest of the chapter, we will approach the CDO valuation problem first with the convenient symmetric Vasicek models with stochastic volatility, and then with a CIR-based model that pools firms into several homogeneous groups.

14.2.1 The CDO Contract

CDOs are designed to securitize portfolios of defaultable assets. Their main feature is that the total nominal associated with the names or obligors is sliced into tranches. Each tranche is then insured against default. The first default events apply to the first tranche, and so on. The protection seller for the first tranche, the equity tranche, is therefore strongly exposed to credit risk relative to the protection sellers for the subsequent mezzanine and senior/super-senior tranches, and the CDO provides a prioritization of credit risk. Two credit derivative indexes are the US-based CDX and the European iTraxx. Each tranche is described by a lower and an upper *attachment point*. In the CDX case the decomposition into tranches corresponds to {0–3, 3–7, 7–10, 10–15, 15–30}% of the total nominal and this is the decomposition we shall use below in our computational examples.

Let α_ℓ be the yield associated with tranche ℓ, that is, the rate at which the insurance buyer pays for protection of tranche ℓ. In the event of default in this tranche, the protection seller pays a fraction $1 - R$ of the loss, with R being the recovery, to the tranche holder (the buyer). We stress here that the recovery is chosen as constant, while in general it could also be modeled as being random. We shall also assume that the payments are made at a set of predetermined times T_k, $k \in \{1, 2, \ldots, K\}$.

For valuation purposes, all expectations are taken under the pricing measure $\mathbb{P}^\star$. We have that α_ℓ is determined by equating expected cashflows from the insurance buyer with expected cashflows from the insurer. This gives

$$\sum_k e^{-rT_k} f_\ell(T_{k-1}) \alpha_\ell (T_k - T_{k-1}) = \sum_k e^{-rT_k} \left(f_\ell(T_{k-1}) - f_\ell(T_k) \right) (1 - R),$$

(14.5)

where r is the constant short rate and $f_\ell(T_k)$ is the expected fraction of tranche ℓ left at time T_k, given as follows. Let $p_n(t)$ be the probability of n names defaulting by time t, and let a_ℓ, b_ℓ denote respectively the lower and upper attachment points associated with tranche ℓ. We then have

$$f_\ell(t) = \sum_{n=0}^{(a_\ell-1)N} p_n(t) + \sum_{n=a_\ell N}^{(b_\ell-1)N} p_n(t)(b_\ell - n/N)/(b_\ell - a_\ell + 1),$$

with N the number of names. We stress that the tranche yields are determined by the loss distribution, $p_n(t), n = 0, \ldots, N, t = T_1, \ldots, T_K$. We refer to Section 14.4.4 for additional details on CDO mechanics.

This is therefore the central technical question that we address, how to consistently model and compute the loss distribution over N names at a set of times T_k. The expression (14.5) is an approximation corresponding to the defaults occurring in the time interval from T_{k-1} to T_k being accounted at the end of this time interval. This is the simple model contract we use below when the tranches are those associated with the CDX. Further details can be found in Elizalde (2005), for instance. In the following, we shall compute the tranche prices α_ℓ when the defaults of the obligors are modeled in terms of the Vasicek reduced-form model, and we next describe this doubly stochastic modeling.

14.3 Symmetric Vasicek Model

We start with a simple reduced-form modeling approach for multiname defaults. The model is based on the Vasicek or Ornstein–Uhlenbeck model for the intensity of default of the underlying names. We analyze the impact of volatility time scales on the default distribution and CDO spreads. We demonstrate how correlated fluctuations in the parameters of the name intensities affect the loss distribution. In particular, we show that the senior tranches of the CDO are quite strongly affected by the presence of parameter time scales. The effect of stochastic parameter fluctuations is to change the shape of the loss distribution.

We shall consider the case when we use the Vasicek model in the context of multiname reduced-form modeling of credit risk. In this doubly stochastic framework there are N underlying names that each may default, and the conditional default times are modeled as the first arrival times of risk-neutral Poisson processes. Here, we let the intensities of the N names be specified as correlated Ornstein–Uhlenbeck processes. Clearly, with this modeling, the intensity might become negative. However, as stated in Duffie and Singleton (2003), "the computational advantage with explicit solutions may be worth the approximation error associated with this Gaussian formulation". This is particularly the case when it comes to calibration, which typically involves a large number of evaluations of the default probabilities as part of an iterative procedure.

Here, we are in the context of bottom-up models of credit risk. By bottom-up, we mean an approach where we explicitly model the default events of the different firms involved, rather than for instance modeling the default distribution directly (top-down). The bottom-up framework has the advantage of modeling the different firms and their default events explicitly. Thus, information about marginal default probabilities and commonality of names between different credit derivatives can be incorporated and the model parameters have more explicit financial interpretation. Our focus will be on correlation effects that is caused by multiscale stochastic parameter variation. However, there may also be sector-wide contagion effects in the bottom-up framework, and we remark that such phenomena can be modeled by a regime-based framework or grouping of names, as we discuss in Section 14.4.

We consider N obligors or underlying names. The event that a particular obligor i defaults is modeled in terms of the first arrival τ_i of a Cox process with stochastic intensity or hazard rate $X^{(i)}$, the conditional hazard rate. Conditioned on the paths of the hazard rates, the default times τ_i of the firms are independent, and the probability that obligor i has survived till time T is given by $\exp(-\int_0^T X_s^{(i)} ds)$. Therefore, the unconditional survival probability is

$$\mathbb{P}^\star\{\tau_i > T\} = \mathbb{E}^\star\left\{e^{-\int_0^T X_s^{(i)} ds}\right\},$$

with the expectation taken with respect to the risk-neutral pricing measure $\mathbb{P}^\star$.

Consider a subset $\{i_1, \ldots, i_n\}$ of obligors. The probability of the joint survival of this set till time T is then, under the doubly stochastic framework,

$$\mathbb{P}^\star\{\tau_{i_1} > T, \ldots, \tau_{i_n} > T\} = \mathbb{E}^\star\left\{e^{-\sum_{j=1}^n \int_0^T X_s^{(i_j)} ds}\right\}.$$

14.3.1 Vasicek Intensities

The intensities $(X_t^{(i)}), 1 \leq i \leq N$, are given by correlated Ornstein–Uhlenbeck processes:

$$dX_t^{(i)} = \kappa_i\left(\theta_i - X_t^{(i)}\right) dt + \sigma_i dW_t^{(i)}, \tag{14.6}$$

where the $(W_t^{(i)})$ are correlated Brownian motions, with the correlation matrix $\mathbf{c}$ given by:

$$d\left\langle W^{(i)}, W^{(j)}\right\rangle_t = c_{ij} dt. \tag{14.7}$$

We denote the survival probability for name i by

$$S_i(T; x_i) = \mathbb{P}^\star(\tau_i > T \mid X_0^{(i)} = x_i) = \mathbb{E}^\star\left\{e^{-\int_0^T X_s^{(i)} ds} \mid X_0^{(i)} = x_i\right\}.$$

We also denote the joint survival probability of all N names by

$$S(T;\mathbf{x},N) = \mathbb{E}^{\star}\left\{ e^{-\int_0^T \sum_{i=1}^N X_s^{(i)} ds} \mid X_0^{(1)} = x_1, \ldots, X_0^{(N)} = x_N \right\},$$

with $\mathbf{x} = (x_1, \ldots, x_N) \in \mathbb{R}^N$. From the Feynman–Kac formula it follows that the joint survival probability from time t till time T:

$$u(t,\mathbf{x}) = \mathbb{E}^{\star}\left\{ e^{-\int_t^T \sum_{i=1}^N X_s^{(i)} ds} \mid \mathbf{X}_t = \mathbf{x} \right\},$$

solves the partial differential equation

$$\frac{\partial u}{\partial t} + \frac{1}{2}\sum_{i,j=1}^N (\sigma_i \sigma_j c_{ij}) \frac{\partial^2 u}{\partial x_i \partial x_j} + \sum_{i=1}^N \kappa_i(\theta_i - x_i)\frac{\partial u}{\partial x_i} - \left(\sum_{i=1}^N x_i\right) u = 0,$$

(14.8)

with terminal condition $u(T,\mathbf{x}) = 1$.

Assume first that the covariance matrix $\mathbf{c}$ is the identity matrix, corresponding to the components of the intensity process $\mathbf{X}$ being independent. Then, as is well known, or can be readily checked, the solution is given by

$$u(t,\mathbf{x}) = \prod_{i=1}^N A_i(T-t) e^{-B_i(T-t)x_i},$$

where we introduce

$$B_i(s) = \int_0^s e^{-\kappa_i \xi}\, d\xi = \frac{1 - e^{-\kappa_i \tau}}{\kappa_i}, \qquad (14.9)$$

$$A_i(s) = e^{-\theta_i \int_0^s \kappa_i B_i(\xi)\, d\xi + \frac{1}{2}\sigma_i^2 \int_0^s B_i^2(\xi)\, d\xi} = e^{-\left(\theta_i^d (s - B_i(s)) + \frac{\sigma_i^2}{4\kappa_i} B_i^2(s)\right)}, \qquad (14.10)$$

$$\theta_i^d = \theta_i - \frac{\sigma_i^2}{2\kappa_i^2}.$$

In the general correlated case, we can write

$$u(t,\mathbf{x}) = A^c(T-t) \prod_{i=1}^N A_i(T-t) e^{-B_i(T-t)x_i}, \qquad (14.11)$$

with

$$A^c(s) = e^{\frac{1}{2}\sum_{i=1}^N \sum_{j\neq i=1}^N (\sigma_i \sigma_j c_{ij}) \int_0^s B_i(\xi) B_j(\xi)\, d\xi}. \qquad (14.12)$$

The last integral is given explicitly by

$$\int_0^s B_i(\xi) B_j(\xi)\, d\xi = \frac{s}{\kappa_i \kappa_j} - \frac{B_i(s)}{\kappa_i(\kappa_i + \kappa_j)} - \frac{B_j(s)}{\kappa_j(\kappa_i + \kappa_j)} - \frac{B_i(s)B_j(s)}{(\kappa_i + \kappa_j)}.$$

(14.13)

14.3.2 Symmetric Names Case

In this section, we analyze the symmetric names case where the dynamics and the starting points of the intensities are the same for all the names. This is convenient to understand the effects of the correlation and the size of the portfolio. We return to the heterogeneous case in Section 14.4.3.

Specifically, we have

$$dX_t^{(i)} = \kappa \left(\theta - X_t^{(i)} \right) dt + \sigma \, dW_t^{(i)}, \qquad X_0^{(i)} = x,$$

with the parameters κ, θ, σ, and x assumed constant and positive. Moreover, we assume that the correlation matrix is defined by $c_{ij} = \rho_X$, for $i \neq j$, with $\rho_X \geq 0$, and ones on the diagonal. We remark that such a correlation structure can be obtained by letting

$$W_t^{(i)} = \sqrt{1 - \rho_X} \, \tilde{W}_t^{(i)} + \sqrt{\rho_X} \, \tilde{W}_t^{(0)}, \qquad (14.14)$$

where $\tilde{W}^{(i)}$, $i = 0, 1, \ldots, N$, are independent standard Brownian motions.

It follows from (14.11) that the joint survival probability for n given names, say the first n names, is

$$S(T; (x, \ldots, x), n) = \mathbb{E}^\star \left\{ e^{- \int_0^T (X_s^{(1)} + \cdots + X_s^{(n)}) \, ds} \mid X_0^{(1)} = x, \ldots, X_0^{(n)} = x \right\}$$

$$= e^{-n \left[\theta_\infty (T - B(T)) + [1 + (n-1)\rho_X] \sigma^2 B^2(T)/(4\kappa) + x B(T) \right]}, \qquad (14.15)$$

with

$$B(T) = \frac{1 - e^{-\kappa T}}{\kappa},$$

$$\theta_\infty = \theta - [1 + (n-1)\rho_X] \frac{\sigma^2}{2\kappa^2}. \qquad (14.16)$$

This expression shows explicitly how the joint survival probability depends on the correlation ρ_X and the "basket" size n. Note in particular how the basket size enhances the correlation effect. We consider next how a characterization of the loss distribution follows from (14.15).

14.3.3 The Loss Distribution

The loss distribution at time T of a basket of size N is given by its mass function

$$p_n = \mathbb{P}^\star \{ (\#\text{names defaulted at time } T) = n \}, \qquad n = 0, 1, \ldots, N, \qquad (14.17)$$

and explicitly written as

$$p_n = \binom{N}{n} \sum_{j=0}^{n} \binom{n}{j} S_{N+j-n}(-1)^j, \tag{14.18}$$

using the shorthand notation $S_n = S(T;(x,\ldots,x),n)$ for the joint survival probability of n names (see Section 13.2.3 for a derivation of this classical formula). Note that a direct implementation of this formula is not numerically stable due to catastrophic cancellation errors in finite precision arithmetics. This challenge will be addressed in Section 14.4.1 by using a recursion algorithm. We use here an alternative implementation of (14.18) based on a conditioning argument. Note first that, from the formula (14.15) for the survival probability, we can write

$$S_n = e^{-d_1 n + d_2 n^2}, \tag{14.19}$$

with d_i explicitly given as

$$d_1 = d_1(T,x) = \theta T + (x-\theta)B(T) - \frac{1}{2}\sigma^2(1-\rho_X)B^{(2)}(T), \tag{14.20}$$

$$d_2 = d_2(T) = \frac{1}{2}\sigma^2\rho_X B^{(2)}(T), \tag{14.21}$$

$$B^{(2)}(T) = \int_0^T B^2(s)\,ds = \frac{(T-B(T))}{\kappa^2} - \frac{B(T)^2}{2\kappa},$$

and we assume that the model parameters are chosen so that $d_1 > 0$. In the independent case $\rho_X = 0$, we get the binomial distribution

$$p_n = \binom{N}{n}(1 - e^{-d_1})^n e^{-(N-n)d_1} =: \tilde{p}_n(d_1).$$

In the general case, we can write

$$S_n = \mathbb{E}\left\{ e^{-d_1 n + n\sqrt{2d_2}Z} \right\},$$

for Z a zero-mean, unit-variance Gaussian random variable. Therefore, in the general case we find

$$p_n = \mathbb{E}\left\{ \tilde{p}_n(d_1 + \sqrt{2d_2}\,Z) \right\}. \tag{14.22}$$

Thus, we get the loss distribution stably and fast by integrating (nonnegative) binomial distributions with respect to the Gaussian density. We remark that this essentially corresponds to conditioning with respect to the correlating Brownian motion, $W^{(0)}$, in (14.14). The argument $d_1 + \sqrt{2d_2}\,Z$ will be negative for Z negative and with large magnitude. This reflects the fact that we are using a Vasicek model where the intensity may be negative. Below, we condition the Gaussian density to $Z > d_1/\sqrt{2d_2}$ and choose parameters such that the complementary event has probability less than 10^{-3}.

Example with Constant Parameters and Strong Correlations

In the model (14.6) we choose the parameters

$$\theta = 0.02, \quad \kappa = 0.5, \quad \sigma = 0.015, \quad x = 0.02,$$

and we let the time to maturity $T = 5$ and the number of names $N = 125$, corresponding to the most common CDO contracts on the CDX and iTraxx. Here and below, we use 20 equally spaced payment dates over the period of the contract. The loss distributions with $\rho_X = 0$ and $\rho_X = 0.75$, respectively, are shown in Figure 14.1 (left). Note how the strong correlation widens the loss distribution. Hence, it will strongly affect tranche prices. We consider the tranche prices for the CDX, defined in Section 14.2. The short rate is chosen to be fixed at 3%, and the recovery is 40%. In Figure 14.1 (right) we show the tranche prices plotted against the upper attachment point of each tranche, associated with the loss distributions on the left. The top plot is on a linear scale and the bottom on a log scale to visually resolve the senior tranches. Note how the strong correlation affects all tranches, and that its relative effects are strongest for the senior tranches. The equity tranche is also strongly affected by the correlation with a negative correction.

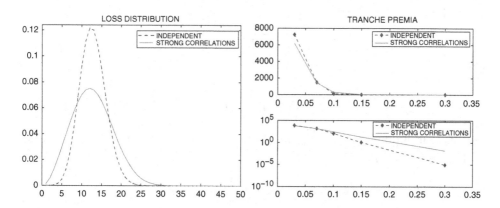

Figure 14.1 The left figure shows the loss distribution, p_n, the dashed line without correlation, $\rho_X = 0$, and the solid line for strong name–name correlation, $\rho_X = 0.75$. The right figure shows the tranche premia, the CDX tranche prices α_ℓ, against upper attachment point, the diamond–dashed line for constant parameters and the solid line for strong name–name correlation (top plot is on a linear scale and bottom plot on a log scale).

14.3.4 Stochastic Volatility Effects

We consider next the extension to the case where the volatility is modeled in terms of a stochastic volatility evolving on a fast and slow scale. Under the risk-neutral probability measure we assume the model

$$dX_{it} = \kappa_i(\theta_i - X_t^{(i)})dt + \sigma_i\,dW_t^{(i)}, \qquad (14.23)$$

for $1 \le i \le N$, where the $W^{(j)}$'s are correlated Brownian motions as in (14.7). Following the modeling point of view of Chapter 4, the volatilities are stochastic and depend on a fast evolving factor Y and a slowly evolving factor Z:

$$\sigma_i = \sigma_i(Y,Z). \qquad (14.24)$$

The fast process is modeled by

$$dY_t = \frac{1}{\varepsilon}(m - Y_t)dt + \frac{v\sqrt{2}}{\sqrt{\varepsilon}}\,dW_t^{(y)}, \qquad (14.25)$$

with the small parameter ε corresponding to the short time scale of the process Y. As before, it is not important which particular model we choose for the fast scale, the important aspects of the process Y are that it is ergodic and that it evolves on a fast time scale. We assume the correlations

$$d\left\langle W^{(i)}, W^{(y)} \right\rangle_t = \rho_Y\,dt, \quad \text{for} \quad 1 \le i \le N. \qquad (14.26)$$

We remark that, in view of (14.14), such correlations can be realized by the modeling

$$W_t^{(y)} = \sqrt{1 - \rho_y}\,\tilde{W}_t^{(y)} + \sqrt{\rho_y}\,\tilde{W}_t^{(0)},$$

for

$$\rho_y = \rho_Y/\sqrt{\rho_X},$$

and $\tilde{W}_t^{(y)}$ independent of $\tilde{W}_t^{(j)}$, $j \in \{0,\ldots,N\}$.

The slow factor evolves as

$$dZ_t = \delta c(Z)dt + \sqrt{\delta}g(Z)\,dW_t^{(z)}, \qquad (14.27)$$

with the large parameter $1/\delta$ corresponding to the long time scale of the process Z.

The functions c and g are assumed to be smooth, and we assume the correlations

$$d\left\langle W^{(i)}, W^{(z)} \right\rangle_t = \rho_Z\,dt, \qquad (14.28)$$

realized by the modeling

$$W_t^{(z)} = \sqrt{1 - \rho_z}\, \tilde{W}_t^{(z)} + \sqrt{\rho_z}\, \tilde{W}_t^{(0)},$$

for

$$\rho_z = \rho z / \sqrt{\rho x},$$

and $\tilde{W}_t^{(z)}$ independent of $\tilde{W}_t^{(j)}$ and $\tilde{W}_t^{(y)}$.

The joint survival probabilities now become

$$q(T; \mathbf{x}, y, z, N) = \mathbb{E}^\star \left\{ e^{-\sum_{i=1}^N \int_0^T \lambda_s^{(i)} ds} \mid \mathbf{X}_0 = \mathbf{x}, Y_0 = y, Z_0 = z \right\},$$

with the intensity processes given by

$$\lambda_t^{(i)} = x_i e^{-\kappa_i t} + \theta_i (1 - e^{-\kappa_i t}) + \int_0^t \sigma_i (Y_s, Z_s) e^{-\kappa_i (t-s)}\, dW_s^{(i)}. \quad (14.29)$$

In this case, the joint survival probability from time t:

$$u^{\varepsilon,\delta}(t, \mathbf{x}; y, z, N) = q(T - t, \ldots, T - t; \mathbf{x}, y, z)$$

$$= \mathbb{E}^\star \left\{ e^{-\int_t^T \sum_{i=1}^N \lambda_s^{(i)} ds} \mid \mathbf{X}_t = \mathbf{x}, Y_t = y, Z_t = z \right\},$$

solves the partial differential equation

$$\frac{\partial u^{\varepsilon,\delta}}{\partial t} + \mathscr{L}_{(\mathbf{x},y,z)} u^{\varepsilon,\delta} - \sum_{i=1}^N x_i u^{\varepsilon,\delta} = 0, \quad (14.30)$$

with the terminal condition $u^{\varepsilon,\delta}(T, \mathbf{x}, y, z) = 1$, and where $\mathscr{L}_{(\mathbf{x},y,z)}$ denotes the infinitesimal generator of the Markov process $(\mathbf{X}_t, Y_t, Z_t)$ under the risk-neutral measure. We define the operator $\mathscr{L}^{\varepsilon,\delta}$ by

$$\mathscr{L}^{\varepsilon,\delta} = \frac{\partial}{\partial t} + \mathscr{L}_{(\mathbf{x},y,z)} - \sum_{i=1}^N x_i \cdot,$$

so that equation (14.30) and its terminal condition can be written as

$$\mathscr{L}^{\varepsilon,\delta} u^{\varepsilon,\delta} = 0,$$
$$u^{\varepsilon,\delta}(T, \mathbf{x}, y, z) = 1.$$

We expand the operator $\mathscr{L}^{\varepsilon,\delta}$ in terms of the small parameters (ε, δ) as

$$\mathscr{L}^{\varepsilon,\delta} = \frac{1}{\varepsilon} \mathscr{L}_0 + \frac{1}{\sqrt{\varepsilon}} \mathscr{L}_1 + \mathscr{L}_2 + \sqrt{\delta}\, \mathscr{M}_1 + \delta \mathscr{M}_2 + \sqrt{\frac{\delta}{\varepsilon}}\, \mathscr{M}_3,$$

where the operators $\mathscr{L}_i$ and $\mathscr{M}_i$ are defined by:

$$\mathscr{L}_0 = v^2 \frac{\partial^2}{\partial y^2} + (m-y)\frac{\partial}{\partial y}, \tag{14.31}$$

$$\mathscr{L}_1 = \sqrt{2}v\rho_Y \sum_{i=1}^{N} \sigma_i(y,z)\frac{\partial^2}{\partial x_i \partial y}, \tag{14.32}$$

$$\mathscr{L}_2 = \frac{\partial}{\partial t} + \frac{1}{2}\sum_{i,j=1}^{N} c_{ij}\sigma_i(y,z)\sigma_j(y,z)\frac{\partial^2}{\partial x_i \partial x_j} \tag{14.33}$$

$$+ \sum_{i=1}^{N} \kappa_i(\theta_i - x_i)\frac{\partial}{\partial x_i} - \sum_{i=1}^{N} x_i\cdot,$$

$$\mathscr{M}_1 = \rho_Z g(z)\sum_{i=1}^{N} \sigma_i(y,z)\frac{\partial^2}{\partial x_i \partial z}, \tag{14.34}$$

$$\mathscr{M}_2 = \frac{1}{2}g^2(z)\frac{\partial^2}{\partial z^2} + c(z)\frac{\partial}{\partial z},$$

$$\mathscr{M}_3 = \sqrt{2}vg(z)(\rho_Y\rho_Z/\rho_X)\frac{\partial^2}{\partial y\partial z}. \tag{14.35}$$

Note that:

- $\varepsilon^{-1}\mathscr{L}_0$ is the infinitesimal generator of the Ornstein–Uhlenbeck process Y.

- $\mathscr{L}_1$ contains the mixed derivatives due to the correlation between $\mathbf{X}$ and Y.

- $\mathscr{L}_2$ is the differential operator corresponding to the unperturbed problem in (14.8), but evaluated at the volatilities $\sigma_i(y,z)$.

- $\mathscr{M}_1$ contains the mixed derivatives due to the correlation between $\mathbf{X}$ and Z.

- $\delta\mathscr{M}_2$ is the infinitesimal generator of the process Z.

- $\mathscr{M}_3$ contains the mixed derivatives due to the correlation between Y and Z.

Next, we use the results of the singular and regular perturbation techniques framework introduced in Chapter 4 and adapted to the credit risk problem above to obtain an accurate characterization of the price in the regime where ε and δ are small. This will enable us to characterize how the fluctuations in the volatility affect the loss distribution and tranche prices.

Time-Scale Perturbations We expand the survival probability in the small parameters ε and δ as

$$u^{\varepsilon,\delta} \sim \tilde{u}^{\varepsilon,\delta} = u_0 + \sqrt{\varepsilon}u_{1,0} + \sqrt{\delta}u_{0,1}, \tag{14.36}$$

where $u_{1,0}$ and $u_{0,1}$ are the first corrections due to fast and slow volatility scales, respectively.

The leading-order term u_0 of the joint survival probability is obtained by solving the problem (14.8) with the effective diffusion matrix

$$\{\mathbf{d}(z)\}_{ij} = d_{ij}(z) := \int c_{ij}\sigma_i(y,z)\sigma_j(y,z)\Phi(y)\,dy, \tag{14.37}$$

with Φ being the invariant distribution for the Y process. Since the process Y evolves on the fast scale, its leading-order effect is obtained by integration with respect to Φ. The process Z evolves on a relatively slow scale and at this level of approximation its effect corresponds to just evaluating this process at its current "frozen" level z. We introduce the effective operator

$$\mathscr{L}^e(\mathbf{d}(z)) = \frac{\partial}{\partial t} + \frac{1}{2}\sum_{i,j=1}^{N}\{d_{ij}(z)\}\frac{\partial^2}{\partial x_i \partial x_j} + \sum_{i=1}^{N}\kappa_i(\theta_i - x_i)\frac{\partial}{\partial x_i} - \left(\sum_{i=1}^{N}x_i\right)\cdot, \tag{14.38}$$

then we have

Definition 14.1 The leading-order term u_0 is the survival probability which solves

$$\mathscr{L}^e(\mathbf{d}(z))u_0 = 0, \tag{14.39}$$
$$u_0(T,\mathbf{x};z) = 1. \tag{14.40}$$

Next, we obtain $u_{1,0}$, the correction to the survival probability due to the fast volatility factor Y. We start by introducing the operator

$$\mathscr{A}_{1,0} = \langle \mathscr{L}_1 \mathscr{L}_0^{-1}(\mathscr{L}_2 - \langle \mathscr{L}_2 \rangle)\rangle, \tag{14.41}$$

which will be given explicitly below and where the triangular brackets represent integration with respect to the invariant distribution for the Y process.

Definition 14.2 The function $u_{1,0}(t,\mathbf{x},z)$ solves the inhomogeneous problem

$$\mathscr{L}^e(\mathbf{d}(z))u_{1,0} = \mathscr{A}_{1,0}u_0, \tag{14.42}$$
$$u_{1,0}(T,\mathbf{x};z) = 0. \tag{14.43}$$

Thus, $u_{1,0}$ is obtained by solving with respect to the effective operator $\mathcal{L}^e(\mathbf{d}(z))$, but now the problem involves a source term $\mathcal{A}_{1,0}u_0$, defined in terms of the leading-order survival probability u_0, and with a zero terminal condition.

We consider next the correction $u_{0,1}$ due to the slow volatility factor. In this case we introduce the operator

$$\mathcal{A}_{0,1} = - \langle \mathcal{M}_1 \rangle, \tag{14.44}$$

and obtain

Definition 14.3   The function $u_{0,1}(t,\mathbf{x},z)$ is the solution of the problem

$$\mathcal{L}^e(\mathbf{d}(z))u_{0,1} = \mathcal{A}_{0,1}u_0, \tag{14.45}$$
$$u_{0,1}(T,\mathbf{x};z) = 0, \tag{14.46}$$

which is again a source problem with respect to the operator $\mathcal{L}^e(\mathbf{d}(z))$ and with a zero terminal condition. Next, we obtain an expression for $\tilde{u}^{\varepsilon,\delta}$ and start by introducing the symmetric matrix $\Phi(y,z)$ satisfying

$$\mathcal{L}_0\Phi_{i_1,i_2} = c_{i_1,i_2}\sigma_{i_1}(y,z)\sigma_{i_2}(y,z) - d_{i_1,i_2}(z),$$

and the coefficients

$$V_3^\varepsilon(z,i_1,i_2,i_3) = -\sqrt{\varepsilon}\frac{\rho_Y\nu}{\sqrt{2}}\left\langle \sigma_{i_3}\frac{\partial\Phi_{i_1i_2}}{\partial y} \right\rangle. \tag{14.47}$$

Using the definitions in (14.31)–(14.33), one then obtains that the scaled operator $\sqrt{\varepsilon}\mathcal{A}_{1,0}$ can be written

$$\sqrt{\varepsilon}\mathcal{A}_{1,0} = -\sum_{i_1,i_2,i_3=1}^{N} V_3^\varepsilon(z,i_1,i_2,i_3)\frac{\partial^3}{\partial x_{i_1}\partial x_{i_2}\partial x_{i_3}}. \tag{14.48}$$

Using this and the expressions (14.10) and (14.12) for the survival probability in the constant volatility case, we compute explicitly the form for the correction due to fast volatility fluctuations:

$$\sqrt{\varepsilon}u_{1,0} = \left(\sum_{i_1,i_2,i_3=1}^{N} V_3^\varepsilon(z,i_1,i_2,i_3)\int_0^T B_{i_1}(s)B_{i_2}(s)B_{i_3}(s)\,ds \right)u_0. \tag{14.49}$$

Therefore, they depend on the underlying model structure in a complicated way, but only the effective *market group parameters* $V_3^\varepsilon(\cdot)$ are needed to compute the fast time scale correction $u_{1,0}$.

Using next the definition (14.34), we find that the (scaled) operator $\sqrt{\delta}\mathscr{A}_{0,1}$ in the source term of the $u_{0,1}$ problem (14.45) can be written

$$\sqrt{\delta}\mathscr{A}_{0,1} = -\sqrt{\delta}\langle\mathscr{M}_1\rangle = -\sum_{i=1}^{N} V_1^{\delta}(z,i)\frac{\partial^2}{\partial z\partial x_i}, \tag{14.50}$$

where we introduced

$$V_1^{\delta}(z,i) = g(z)\rho_Z\langle\sigma_i\rangle. \tag{14.51}$$

It then follows, again using the expressions (14.10) and (14.12), that

$$\sqrt{\delta}u_{0,1}$$

$$= \left(\frac{1}{2}\sum_{i_1=1}^{N} V_1^{\delta}(z,i_1)\sum_{i_2=1}^{N}\sum_{i_3=1}^{N}\frac{\partial}{\partial z}(d_{i_1 i_2})\int_0^T B_{i_1}(v)\int_0^v B_{i_1}(s)B_{i_2}(s)\,ds\,dv\right)u_0. \tag{14.52}$$

Note that the effective market parameters V_3^{ε} and V_1^{δ} depend on the underlying model in a complicated way, as explained above. However, this particular dependence will not be needed in applying the asymptotic theory since these market group parameters – rather than the full underlying model – will be calibrated to market data.

Stochastic Volatility Effects in the Symmetric Case We consider the simplified form for the asymptotic survival probabilities in the symmetric case (14.6). One readily computes that in this case

$$q(T;x,z,n) = u^{\varepsilon,\delta}(0;x,z,n) \sim \tilde{u}^{\varepsilon,\delta}(0;x,z,n)$$

$$= \left(1 + D^{\varepsilon}(T;z,n) + D^{\delta}(T;z,n)\right)\exp\left(-n[\theta_{\infty}(z)(T-B(T))\right.$$
$$\left. + [1+(n-1)\rho_X]\bar{\sigma}^2(z)B^2(T)/(4\kappa)+xB(T)]\right), \tag{14.53}$$

where

$$\bar{\sigma}^2(z) = \langle\sigma(\cdot,z)^2\rangle, \tag{14.54}$$

$$D^{\varepsilon}(T;z,n) = v_3(z)n^2(1+(n-1)\rho_X)B^{(3)}(T), \tag{14.55}$$

$$D^{\delta}(T;z,n) = v_1(z)n^2(1+(n-1)\rho_X)\tilde{B}^{(3)}(T), \tag{14.56}$$

$$B^{(3)}(T) = \int_0^T B^3(s)\,ds, \tag{14.57}$$

$$\tilde{B}^{(3)}(T) = \int_0^T B(s)B^{(2)}(s)\,ds, \tag{14.58}$$

$$v_1(z) = \frac{\sqrt{\delta}}{2} g(z) \rho_Z \left\langle \sigma(\cdot, z) \right\rangle \frac{\partial}{\partial z} \left\langle \sigma^2(\cdot, z) \right\rangle, \qquad (14.59)$$

$$v_3(z) = -\sqrt{\varepsilon} \frac{\rho_Y v}{\sqrt{2}} \left\langle \sigma(\cdot, z) \frac{\partial \Phi(\cdot, z)}{\partial y} \right\rangle, \qquad (14.60)$$

with here

$$\mathscr{L}_0 \Phi = \sigma^2(y, z) - \left\langle \sigma^2(\cdot, z) \right\rangle.$$

Note that θ_∞ is computed as in (14.16), but evaluated as $\sigma = \bar{\sigma}(z)$:

$$\theta_\infty(z) = \theta - [1 + (n-1)\rho_X] \frac{\bar{\sigma}^2(z)}{2\kappa^2}.$$

We therefore have, using the notation introduced in (14.19),

$$q(T; x, z, n) \sim \left(1 + n^3 \rho_X \left(v_3(z) B^{(3)}(T) + v_1(z) \tilde{B}^{(3)}(T)\right)\right)$$
$$\times \exp\left(-n \tilde{d}_1(T, x, z) + n^2 d_2(T, z)\right), \qquad (14.61)$$

with

$$\tilde{d}_1(T, x, z) = d_1(T, x, z) + (1 - \rho_X)\left(v_3(z) B^{(3)}(T) + v_1(z) \tilde{B}^{(3)}(T)\right), \qquad (14.62)$$

where d_1 and d_2 are computed as in (14.20) and (14.21):

$$d_1(T, x, z) = \theta T + (x - \theta) B(T) - (1 - \rho_X)(\bar{\sigma}(z)^2/2) B^{(2)}(T),$$
$$d_2(T, z) = \rho_X(\bar{\sigma}(z)^2/2) B^{(2)}(T),$$
$$B^{(2)}(T) = \int_0^T B^2(s)\, ds.$$

By the remarks below equation (14.19), we easily compute the loss distribution that follows from (14.61) in the case $\rho_X = 0$. That is, we compute it by (14.22) with

$$\tilde{p}_n(x') = \binom{N}{n} \sum_{j=0}^{n} \binom{n}{j} \exp(-x'(N + j - n))(-1)^j \qquad (14.63)$$

$$= \binom{N}{n} (1 - \exp(-x'))^n \exp(-x')^{N-n}, \qquad (14.64)$$

and d_1 replaced by $\tilde{d}_1$. In the general case with $\rho_X \neq 0$, we obtain the loss distribution by the generalization of (14.22):

$$p_n = \mathbb{E}\{\tilde{p}_n(\tilde{d}_1 + \sqrt{2d_2}\,X) + \left(\rho_X(v_3 B^{(3)} + v_1 \tilde{B}^{(3)})\right)\tilde{p}_n'''(\tilde{d}_1 + \sqrt{2d_2}\,X)\}.$$

$$(14.65)$$

We use this procedure to calculate the loss distribution in the following numerical examples. Note that

$$\sum_{n=0}^{N} \tilde{p}_n'''(x) = \frac{d^3}{dx^3}\sum_{n=0}^{N} \tilde{p}_n(x) = 0,$$

so that indeed $\sum_{n=0}^{N} p_n = 1$. We note however that outside the domain of validity of the approximation we may have $p_n < 0$. Thus, when applying the approximation the v_i must be chosen small enough so that the computed p_n define a distribution. In the above model, the v_i are small of order $\mathcal{O}(\sqrt{\delta} + \sqrt{\varepsilon})$. From the representation (14.65) we see that the effect of the stochastic volatility in the uncorrelated case with $\rho_X = 0$ is a modification of the hazard rate, to the order we consider. While the combined effect of correlation and stochastic volatility is qualitatively different and gives a correction to the binomial shape. We can also observe that the effects of the slow and fast volatility scales are qualitatively similar, giving the computed correction a canonical character; it gives the structure of the correction under a large class of underlying models.

We continue the numerical example introduced in Section 14.3.3. We choose parameters

$$\theta = 0.03, \quad \kappa = 0.5, \quad \sigma = 0.02, \quad x = 0.03, \qquad (14.66)$$

and we let the time to maturity $T = 5$ and the number of names $N = 125$. The short rate is chosen to be fixed at 3% and the recovery is 40% as before. Here and below, when we show numerical examples, they are based on the asymptotic approximations of the type (14.65). Our analysis has shown that name correlation can be generated in various ways. Either by directly correlating the innovations or Brownian motions driving the hazard rates of the names or alternatively, by introducing time scale effects in the volatility. We remark though that the time evolution in the loss distribution depends somewhat on how the correlation is generated. In Figure 14.2, we illustrate the relatively strong effect of combined name correlation and stochastic volatility, with $\rho_X = 0.01$ and $v_3 = 3 \times 10^{-4}$. We let $v_1 = 0$; the influence of this parameter is similar to that of v_3. Note in particular the relatively strong effect on the senior tranches, and that in this case the shape of the loss distribution is affected, however, the equity tranches are relatively less affected.

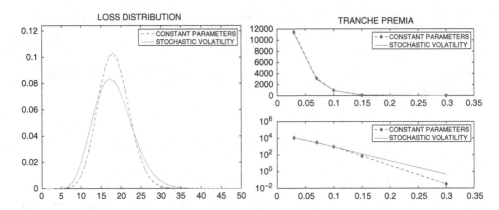

Figure 14.2   The left figure shows the loss distribution with and without stochastic volatility. The right figure shows the CDX tranche premia with and without stochastic volatility (top plot is on a linear scale and bottom on a log-scale).

14.4 Homogeneous Group Structure

There is already a large literature on CDOs concerning their financial purpose and the mathematical models used to determine their value. We present here an extension to the current methodology for correlating default intensities in an attempt to price CDOs and variants of this type. The pricing of multiname credit derivatives requires (i) realistic modeling of the firms' default times and the correlation between them, and (ii) efficient computational methods for computing the portfolio loss distribution. *Dynamic factor models*, a widely used class of pricing models in the literature, can be computationally tractable despite the large dimension of the pricing problem, thus satisfying issue (ii), but to have any hope of calibrating CDO data, numerically intense versions of these models have to be implemented. In the approach we analyze here, we start from intensity-based models for default risk, and with the aforementioned issues in mind, we propose improvements (a) via incorporating fast mean-reverting stochastic volatility in the default intensity processes, and (b) by breaking the original set of firms into homogeneous groups. Let us also note that the models considered so far are part of the *bottom-up* framework of multiname credit derivatives pricing models, since we build the loss distribution from the distributions of the default times of the individual firms. On the contrary, *top-down* models describe directly the dynamics of the loss distribution of the portfolio without explicit specification of the constituent single-name behaviors. The model we propose here can be viewed as intermediate between these approaches.

We begin with a brief guided tour of factor models, introduce our homogeneous group structure, and explain the computational benefits.

14.4.1 Factor Models

Let $(\Omega, \mathcal{H}, \mathbb{P}^\star)$ be a probability space, and suppose there are n Cox processes $N^i = (N^i_t)_{\geq 0}$ with non-negative (and nontrivial) intensity processes $\lambda^i = (\lambda^i_t)_{t \geq 0}$, $i = 1, \ldots, n$. Here, $\mathbb{P}^\star$ is the risk-neutral pricing measure reflected by market prices of credit derivatives. We are not interested in the historical measure for our valuation problem. The process N^i is dubbed the *default process* to denote that the first arrival from it signifies the *default* of firm i, for every $i = 1, \ldots, n$. We define τ_i to be the *time of default* of firm i, or $\tau_i := \inf\{t > 0 : N^i_t > 0\}$. Then, if we define the filtration $\mathscr{F}^i = (\mathscr{F}^i_t)_{t \geq 0}$ via

$$\mathscr{F}^i_t := \sigma\{\lambda^i_s; 0 \leq s \leq t\},$$

which describes the history of the default intensity process of the ith firm, the default time τ_i satisfies

$$\mathbb{P}^\star\{\tau_i > t \mid \mathscr{F}^i_s\} = \mathbf{1}_{\{\tau_i > s\}} \mathbb{E}^\star\left[\exp\left\{-\int_s^t \lambda^i_u\, du\right\} \mid \mathscr{F}^i_s\right], \quad 0 \leq s \leq t.$$

We define the *loss process* $L = (L_t)_{t \geq 0}$ by

$$L_t := \sum_{i=1}^{n} \mathbf{1}_{\{\tau_i \leq t\}}, \quad t \geq 0,$$

which counts the number of defaulted firms until time t. The path $t \mapsto L_t$ starts from zero, is right-continuous, and it increases only by jumps of unit size (because there cannot be simultaneous defaults). As already seen in Sections 13.2.3 and 14.3.3, and revisited below in Section 14.4.4, the pricing of portfolio derivatives amounts to specifying the probability law of the process L, which we refer to as the *loss distribution*. We give some examples below of choices for the intensity processes and the resulting loss distributions to enhance our intuition about the loss process.

Example 14.4 (Deterministic and identical) If every default intensity process is deterministic (but time-dependent) and equal to each other, i.e., we have $\lambda^1_t \equiv \cdots \equiv \lambda^n_t \equiv \lambda(t)$, the loss distribution is obviously binomial, with mass function $m \mapsto \mathbb{P}^\star\{L_t = m\}$ given by

$$\mathbb{P}^\star\{L_t = m\} = \binom{n}{m} \left(1 - e^{-\int_0^t \lambda(s)\, ds}\right)^m \left(e^{-\int_0^t \lambda(s)\, ds}\right)^{n-m}, \quad m = 0, \ldots, n.$$

$$(14.67)$$

Example 14.5 (Independent and identically distributed)   If the default intensity processes are independent and identically distributed, the loss distribution remains binomial, but the mass function now becomes

$$\mathbb{P}^\star\{L_t = m\} = \binom{n}{m}\left(1 - \mathbb{E}^\star\left[e^{-\int_0^t \lambda_s\, ds}\right]\right)^m \mathbb{E}^\star\left[e^{-\int_0^t \lambda_s\, ds}\right]^{n-m}, \quad m = 0,\ldots,n,$$

where the process $\lambda = (\lambda_t)_{t\geq 0}$ is equal in distribution to the independent and identically distributed intensity processes $\lambda^1,\ldots,\lambda^n$.

Example 14.6 (Independent)   If the default intensity processes are independent but not necessarily identically distributed, the time-t loss distribution $(P^n(m))_{m=0,\ldots,n}$ can be built up by first selecting an arbitrary ordering of the firms. Then, let $P^k(m)$ be the probability of m defaults from the pool of k firms ($k = 1,\ldots,n$), which can be calculated recursively via

$$P^{k+1}(m) = P^k(m)(1 - p_{k+1}) + P^k(m-1)p_{k+1}, \quad m = 1,\ldots,k+1,$$
$$P^{k+1}(0) = P^k(0)(1 - p_{k+1}),$$

for $k = 1,\ldots,n-1$, and

$$P^0(m) = \mathbf{1}_{\{m=0\}}, \quad m = 0,\ldots,n,$$
$$P^k(m) = 0, \quad m > k,$$

where p_k is the probability the kth firm defaults by time t, or

$$p_k = 1 - \mathbb{E}^\star\left[e^{-\int_0^t \lambda_s^k\, ds}\right], \quad k = 1,\ldots,n.$$

This is the recursion algorithm suggested by Andersen *et al.* (2003) and Hull and White (2004) for the computation of the loss distribution. Notice that the distribution $(P^n(m))_{m=0,\ldots,n}$ is always the same, regardless of the ordering of firms that is used for this recursion algorithm (but $(P^k(m))_{m=0,\ldots,k}$, $k < n$ do not have to be).

Example 14.7 (Identical)   If there is a common intensity process driving *all* Cox processes, i.e., there is $\lambda = (\lambda_t)_{t\geq 0}$ with $\lambda_t^1 = \cdots = \lambda_t^n = \lambda_t$ a.s., the *conditional loss distribution* given the path of the common intensity $(\lambda_s)_{s\leq t}$ is binomial as argued in (14.67). This implies then that

$$\mathbb{P}^\star\{L_t = m\} = \mathbb{E}^\star\left[\binom{n}{m}\left(1 - e^{-\int_0^t \lambda_s\, ds}\right)^m \left(e^{-\int_0^t \lambda_s\, ds}\right)^{n-m}\right], \quad m = 0,\ldots,n. \tag{14.68}$$

Of course, the loss distribution is no longer binomial in general.

The assumption of independent default intensities as illustrated in Example 14.5 simplifies the computation of the loss distribution to the computation of a Laplace transform of the integrated intensity process. On the other hand, the situation described in Example 14.7 is simplistic as well, but, as a starting point for a more complicated approach, it is far from the unrealistic binomial case of Example 14.5. The computation of the loss distribution in the common intensity process of Example 14.7 may also be reduced to the computation of the Laplace transform of the integrated intensity process if we rewrite the probability in (14.68) as

$$\mathbb{P}^{\star}\{L_t = m\} = \binom{n}{m} \sum_{j=0}^{m} \binom{m}{j} (-1)^{m-j} \mathbb{E}^{\star}\left[e^{-(n-j)\int_0^t \lambda_s ds}\right], \quad m = 0,\ldots,n.$$

Summations of this type, where the terms have alternating signs, are known as *Euler–Maclaurin sums*. As n increases, the evaluation of these sums becomes computationally intensive due to round-off numerical errors committed by subtracting large numbers that differ only in trailing decimals. Even though there exist efficient algorithms for accurate numerical computation of these summations, it is necessary to limit the number of firms in the portfolio, n, by 30. (Also, there is a numerical error generated by the multiplication of large binomial coefficients with the small Laplace transform terms.) Obviously, this is a disadvantage considering that the typical size of the name pool is 125 (for the CDX and iTraxx indices) and up to 600 for bespoke CDOs.

Example 14.8 (Common factor) The default intensity process λ^i of firm i, $i = 1,\ldots,n$, is decomposed into an idiosyncratic and a systematic component:

$$\lambda^i = X^i + c_i Z,$$

where X^i and Z are positive and independent processes, and $c_1,\ldots,c_n$ are positive numbers. The processes $X^1,\ldots,X^n$ are independent, as well. With these assumptions, the market-wide behavior of default risk influences all firms increasingly with its c-coefficient, while there is also a firm-specific component to every firm's default time. The name–name correlation is imposed only through the dependence of each λ^i on the market factor Z.

Let us define $U = (U_t)_{t \geq 0}$ to be the *integrated systematic* process, i.e., $U_t = \int_0^t Z_s \, ds$. When the processes $X^1,\ldots,X^n$ have the same law, the loss process is computed with the binomial distribution of Example 14.5, after conditioning on the path of the common process Z. This leads to

$$\mathbb{P}^\star\{L_t = m\} = \mathbb{E}^\star\left[\binom{n}{m}\left(1 - \mathbb{E}^\star\left[e^{-\int_0^t \lambda_s\, ds} \mid U_t\right]\right)^m \mathbb{E}^\star\left[e^{-\int_0^t \lambda_s\, ds} \mid U_t\right]^{n-m}\right].$$

When the processes $X^1,\ldots,X^n$ do not, necessarily, have the same law (but are still independent), the conditional loss distribution is computed according to Example 14.6. Usual stochastic processes considered for the dynamics of $X^1,\ldots,X^n$ and Z are affine diffusions or affine jump-diffusions so that their integrated versions have closed-form, or nearly closed-form, expressions for the Laplace transforms required.

In the following subsections, we detail our framework which comprises the partitioning of the firms into groups, and allowing for stochastic volatility in the firms' default intensity processes.

14.4.2 Homogeneous Group Structure

The two modeling situations that are described in Examples 14.5 and 14.7 are intuitively at opposite extremes. We take advantage of their computational efficiencies by breaking the large number of firms into a smaller number of groups and by linking the default intensities across the groups by a common factor.

Assumption 14.9 *There are k homogeneous groups of firms ($k < n$), and each of the n firms belongs in one group only. We denote by n_i the number of firms in the ith group, $i = 1,\ldots,k$.*

Assumption 14.10 *Within each group, the firms share a common intensity of default, therefore given a default from the group each (remaining) firm within that group is equally likely to be the defaulted one.*

Let $L^i = (L^i_t)_{t\geq 0}$ be the loss process corresponding to the firms of group i, $i = 1,\ldots,k$, and obviously the total loss process $L = L^1 + \cdots + L^k$. For $i = 1,\ldots,n$, we redefine $\lambda^i = (\lambda^i_t)_{t\geq 0}$ to denote the default intensity process shared by the firms of group i.

Assumption 14.11 *The default intensity process shared by all firms of the homogeneous group i, $\lambda^i = (\lambda^i_t)_{t\geq 0}$, is given by*

$$\lambda^i_t = X^i_t + c_i Z_t, \quad i = 1,\ldots,k. \tag{14.69}$$

The idiosyncratic *factors $X^1,\ldots,X^k$ are independent from each other and independent of the* systematic *factor Z. The processes $X^1,\ldots,X^k$ and Z are non-negative almost surely, and the parameters $c_1,\ldots,c_k$ are positive constants.*

In our setup, the default times of the firms within a given group are heavily correlated since they share identical intensities, and the loss distributions of different groups are correlated through the common factor Z. As we shall see in the calibration section, the presence of the systematic factor enhances enormously the capability of the model to capture the data.

We define the integrated common factor U by

$$U_t = \int_0^t Z_s \, ds, \quad t > 0.$$

Then the conditional mass function of the loss in the ith group is given by:

$$\mathbb{P}^\star\{L_t^i = m \mid U_t = v\}$$
$$= \binom{n_i}{m} \sum_{j=0}^{m} \binom{m}{j} (-1)^{m-j} e^{-c_i(n_i-j)v} \mathbb{E}^\star \left[e^{-(n_i-j)\int_0^t X_s^i \, ds} \right], \quad (14.70)$$

for $m = 0, \ldots, n_i$. The conditional distribution of the entire portfolio is then given by the convolution

$$\mathbb{P}^\star\{L_t = m \mid U_t = v\} = \sum_{i_1=0}^{m} \cdots \sum_{i_{k-1}=0}^{m-i_1-\cdots-i_{k-2}} \mathbb{P}^\star\{L_t^1 = i_1 \mid U_t = v\} \cdots$$
$$\cdot \mathbb{P}^\star\{L_t^{k-1} = i_{k-1} \mid U_t = v\} \mathbb{P}^\star\{L_t^k = m - i_1 - \cdots - i_{k-1} \mid U_t = v\}, \quad (14.71)$$

where $m = 0, \ldots, n$. (Recall also that for every $i = 1, \ldots, k$, $\mathbb{P}^\star\{L_t^i = m \mid U_t = v\} = 0$, for $m \notin \{0, \ldots, n_i\}$.) The numerical computation of the nested sums is very intensive due to the large number of firms, n, and number of groups, k. Instead, we compute the conditional distribution of L_t given U_t from the distributions of $L_t^1, \ldots, L_t^k$ by inverting the conditional generating function:

$$\mathbb{E}^\star \left[z^{L_t} \mid U_t = v \right] = \mathbb{E}^\star \left[z^{L_t^1} \mid U_t = v \right] \cdots \mathbb{E}^\star \left[z^{L_t^k} \mid U_t = v \right]. \quad (14.72)$$

A popular inversion method of the left-hand side of (14.72) is the fast Fourier transform (FFT) method, which we use here.

From (14.70) we see that we need to compute the Laplace transform of the integrated idiosyncratic component X^i of the intensity process λ^i. The class of affine continuous stochastic processes yields closed-form expressions for such expectations. We discuss such models and important extensions in the next section. Finally, the portfolio loss distribution is computed from

$$\mathbb{P}^{\star}\{L_t = m\} = \int_0^{+\infty} \mathbb{P}^{\star}\{L_t = m \mid U_t = v\}\, \mathbb{P}^{\star}\{U_t \in dv\}, \quad m = 0, \ldots, n.$$

$$(14.73)$$

14.4.3 Stochastic Volatility in the Default Intensity Process

Modeling the default intensity processes in a Cox process setup with a single-scale diffusion process was found to be inadequate in producing appropriate loss distributions to capture real data. In particular, the market-implied loss distribution exhibits longer tails than the ones corresponding to diffusion processes. Adding jumps yields satisfactory results. As we argue below, the introduction of stochastic volatility in the asymptotic approximation is enough to allow for a *heavier tail* in the loss distribution, which offsets the need for jump characteristics, and maintains closed-form expressions (up to multiscale asymptotic approximations) for the conditional loss distribution.

Fast Mean-Reverting Stochastic Volatility We fix a group i ($i = 1, \ldots, k$). The model for the two components of the default intensity process in (14.69) is described by the system of stochastic differential equations:

$$dX_t^i = \alpha_i(\bar{x}_i - X_t^i)\, dt + f_i(Y_t^i)\sqrt{X_t^i}\, dW_t^i, \tag{14.74}$$

$$dY_t^i = \frac{1}{\varepsilon} X_t^i(\bar{y}_i - Y_t^i)\, dt + \frac{v_i\sqrt{2}}{\sqrt{\varepsilon}}\sqrt{X_t^i}\, dW_t^{Y^i}, \tag{14.75}$$

$$dZ_t = \alpha_z(\bar{z} - Z_t)\, dt + \sigma_z\sqrt{Z_t}\, dW_t^Z.$$

Here, W^i and W^{Y^i} are Wiener processes such that

$$d\langle W_{\cdot}^i, W_{\cdot}^{Y^i}\rangle_t = \rho_i\, dt, \quad d\langle W_{\cdot}^i, W_{\cdot}^{Y^j}\rangle_t = 0,$$

$$d\langle W_{\cdot}^i, W_{\cdot}^j\rangle_t = 0, \quad d\langle W_{\cdot}^{Y^i}, W_{\cdot}^{Y^j}\rangle_t = 0,$$

for $i = 1, \ldots, k$, $j = 1, \ldots, k$, $i \neq j$, and ρ_i in $[-1, 1]$. Due to the factor model Assumption 14.11, the Wiener process W^Z is independent of W^i and W^{Y^i}.

The idiosyncratic intensity process X^i is a square-root diffusion process with stochastic volatility; it is a mean-reverting process with rate of mean-reversion α_i, and level of mean-reversion $\bar{x}_i$. The diffusion coefficient is driven by another mean-reverting process Y^i via a function f_i which we assume to be positive and bounded in order to (i) have strong solutions to (14.74), and (ii) satisfy certain growth conditions needed for the proof of the approximation theorem later. The condition $2\alpha_i \bar{x}_i > f_i(\cdot)^2$ is sufficient for X^i to remain strictly positive, and the justification follows that from the constant volatility case.

The presence of the small positive number ε in (14.75) makes Y^i a *fast mean-reverting process*. Furthermore, we choose to correlate X^i and Y^i not only through their Wiener processes, but also from the presence of X^i in (14.75). We do so in this specific way to construct closed-form approximations for the Laplace transforms of the integrated X^i process. The systematic component, Z, is modeled as a square-root diffusion with constant volatility σ_z.

Asymptotic Approximation We start with the Laplace transform of $\int_t^T X_s \, ds$ that is used to compute the conditional (on the common factor Z) group loss distribution and then the total loss portfolio loss distribution in Section 14.4.3. In what follows we simplify the notation for the processes involved and instead of X^i and Y^i we write X and Y, respectively. The stochastic differential equations (14.74) and (14.75) remain the same with the appropriate drop of the index i.

We construct an approximation to the type of expectation in expression (14.70):

$$u(t,T,x,y;m) := \mathbb{E}^* \left[e^{-m \int_t^T X_s \, ds} \mid X_t = x, Y_t = y \right], \quad t \le T. \qquad (14.76)$$

The function u satisfies the partial differential equation (PDE)

$$\mathscr{L}u(t,T,x,y;m) = 0, \quad t < T, \quad x \in \mathbb{R}_+, y \in \mathbb{R}, \qquad (14.77)$$
$$u(T,T,x,y;m) = 1,$$

with

$$\mathscr{L} := \frac{1}{\varepsilon}\mathscr{L}_0 + \frac{1}{\sqrt{\varepsilon}}\mathscr{L}_1 + \mathscr{L}_2,$$

$$\mathscr{L}_0 := x\left(v^2\frac{\partial^2}{\partial y^2} + (\bar{y}-y)\frac{\partial}{\partial y}\right),$$

$$\mathscr{L}_1 := \sqrt{2}\rho v f(y)x\frac{\partial^2}{\partial x \partial y},$$

$$\mathscr{L}_2 := \frac{\partial}{\partial t} + \frac{1}{2}f(y)^2 x\frac{\partial^2}{\partial x^2} + \alpha(\bar{x}-x)\frac{\partial}{\partial x} - mx\cdot.$$

Recall that in the absence of stochastic volatility, that is, $f(y) \equiv \sigma > 0$, the expression for u in (14.76) is given by

$$u(t,T,x,y;m) = u(t,T,x;m) = A(T-t)e^{-B(T-t)x}, \qquad (14.78)$$

where A and B are given by:

$$A(s) = \left[\frac{2\gamma e^{(\alpha+\gamma)s/2}}{(\alpha+\gamma)(e^{\gamma s}-1)+2\gamma} \right]^{\frac{2\alpha\bar{x}}{\sigma^2}}, \qquad (14.79)$$

$$B(s) = \frac{2m(e^{\gamma s}-1)}{(\alpha+\gamma)(e^{\gamma s}-1)+2\gamma}, \qquad (14.80)$$

$$\gamma = \sqrt{\alpha^2 + 2m\sigma^2}. \qquad (14.81)$$

We construct an expansion of the form

$$u = u_0 + \sqrt{\varepsilon}\, u_1 + \varepsilon\, u_2 + \varepsilon^{3/2} u_3 + \cdots, \qquad (14.82)$$

where each term $u_0, u_1, \ldots$ is independent of ε. By now standard calculations lead to

$$u_0(t,T,x;m) = A(T-t)e^{-B(T-t)x}, \quad t \le T, \qquad (14.83)$$

where A and B were defined in (14.79) and (14.80) after replacing the term σ^2 with $\bar{\sigma}^2$, defined by

$$\bar{\sigma}^2 := \langle f(y)^2 \rangle = \int_{\mathbb{R}} f(y)^2 \frac{1}{\sqrt{2\pi v^2}} \exp\left\{ -\frac{(y-\bar{y})^2}{2v^2} \right\} dy.$$

By introducing the notation $\tilde{u}_1 := \sqrt{\varepsilon}\, u_1$, letting $V_1 := \sqrt{\varepsilon}\rho v\langle f\varphi'\rangle/\sqrt{2}$, and observing from (14.83) that $\partial^3 u_0/\partial x^3 = -B^3 u_0$, we write the following PDE problem for $\tilde{u}_1$:

$$\langle \mathscr{L}_2 \rangle \tilde{u}_1(t,T,x;m) = -V_1 B(T-t)^3 x u_0(t,T,x;m), \quad t < T, \qquad (14.84)$$
$$\tilde{u}_1(T,T,x;m) = 0.$$

We make the ansatz

$$\tilde{u}_1(t,T,x;m) = (D_1(T-t)x + D_2(T-t))u_0(t,T,x;m), \qquad (14.85)$$

and we get the ODEs for D_1 and D_2 by separating in (14.84) the x-terms and the terms independent of x:

$$D_1' + (\bar{\sigma}^2 B + \alpha) D_1 - V_1 B^3 = 0, \qquad D_1(0) = 0, \qquad (14.86)$$
$$D_2' - \alpha\bar{x}D_1 = 0, \qquad D_2(0) = 0. \qquad (14.87)$$

By setting $\theta := \bar{\sigma}^2/(\alpha\bar{x})$, the solution of D_1 in (14.86) is

$$D_1(s) = \frac{m^2 V_1}{\gamma^2 \bar{\sigma}^2} e^{-\alpha s} A(s)^{\theta} \left[\frac{2\gamma^2}{m\alpha\bar{x}} \log\left(A(s)\right) + \frac{\alpha + \gamma - 2\gamma e^{\gamma s}}{\gamma e^{\gamma s}} \right.$$
$$\left. + \frac{(\gamma - \alpha) e^{\gamma s}}{\gamma} + 2\alpha s \right], \qquad (14.88)$$

where A and γ were defined in (14.79) and (14.81) (replacing σ^2 by $\bar{\sigma}^2$). The solution of (14.87) for D_2 is given by

$$D_2(s) = \frac{2V_1}{(\bar{\sigma}^2)^2} \left[\frac{8\alpha\gamma}{h(s)} + \frac{(\alpha + \gamma)(\gamma + 3\alpha)B(s)}{2m} \right] \log(A(s))$$
$$+ \frac{2mV_1 \alpha\bar{x}}{\bar{\sigma}^2 \gamma(\gamma - \alpha)} \left[\frac{4m(3\gamma^2 + \alpha^2)}{(\alpha + \gamma)h(s)} + (3\gamma + \alpha)B(s) \right] s$$
$$- \frac{2V_1 \alpha\bar{x}(\alpha^2 + 3\gamma^2)B(s)}{(\bar{\sigma}^2)^2 \gamma^2}, \qquad (14.89)$$

where B was defined in (14.80) and

$$h(s) := (\alpha + \gamma)(e^{\gamma s} - 1) + 2\gamma.$$

In conclusion, the approximation of u as defined in (14.76) up to terms of order ε is given by

$$u_0(t, T, x; m) + \tilde{u}_1(t, T, x; m)$$
$$= [1 + D_1(T - t)x + D_2(T - t)]A(T - t)e^{-B(T-t)x}, \qquad (14.90)$$

with A and B defined in (14.79) and (14.80) with $\bar{\sigma}^2$ replacing σ^2. From previous analyses, the accuracy of approximation is of order $\mathcal{O}(\varepsilon)$. The asymptotic approximation in (14.90) may become negative for certain values of the aggregate parameter and large maturity T. Of course, this is expected due to the type of the approximation, but it is a disadvantage nonetheless. To account for this problem, we introduce the alternative approximation

$$\bar{u}^{\varepsilon}(t, T, x; m) = u_0(1 + \tanh(\tilde{u}_1/u_0))$$
$$= u_0[1 + \tanh(D_1(T - t)x + D_2(T - t))], \qquad (14.91)$$

which is non-negative and preserves the order of accuracy of the approximation (see Fouque and Zhou (2008) for more details).

Summary and Effects of Stochastic Volatility on Loss Distributions

Using the asymptotic approximation $\bar{u}^{\varepsilon}$ as given in (14.91) for u (defined in (14.76)), we provide an approximation for the conditional probability loss density of a group via (14.70), or

$$\mathbb{P}^\star\{L_t^i = m \mid U_t = v\}$$

$$= \binom{n_i}{m} \sum_{j=0}^{m} \binom{m}{j} (-1)^{m-j} e^{-c_i(n_i-j)v} u^i(0,t,x_i;n_i-j)$$

$$\approx \binom{n_i}{m} \sum_{j=0}^{m} \binom{m}{j} (-1)^{m-j} e^{-c_i(n_i-j)v}$$

$$\times [1 + \tanh(D_1^i(t)x_i + D_2^i(t))] u_0^i(0,t,x_i;n_i-j). \qquad (14.92)$$

The functions D_1^i and D_2^i are defined in (14.88) and (14.89), respectively, modified appropriately to reflect the use of the parameters from the processes X^i and Y^i (in place of the generic parameters of X and Y). Note also that $X_0^i = x_i$.

The total portfolio conditional loss distribution is obtained from the numerical inversion of the conditional generating function (14.72), while the distribution of U_t is taken from the numerical inversion of its characteristic function. Both numerical inversions are done using the fast Fourier transform algorithm. Even though the generating function inversion procedure seems more complicated than the convolution computation (14.71), it is dramatically faster, on the order of 50 times for five equally sized homogeneous groups of 25 firms per group. Finally, the loss distribution is obtained from the numerical evaluation of the integral (14.73).

Remark 14.12 We remark that the recursion algorithm illustrated in Example 14.6 is not applicable here because the grouping does not simply reduce the problem to the sum of conditionally independent Bernoulli random variables. Therefore, we need to use inversion of the generating function to obtain the conditional loss distribution of the sum of the random variables $L_t^1, \ldots, L_t^k$, given the integrated systematic factor U_t.

Figure 14.3 exhibits the effect of the stochastic volatility approximation on the probability loss distribution of a group of $n_i = 20$ firms with identical default intensity processes X^i as given in (14.74) and stochastic volatility process $(f_i(Y_t^i))_{t \geq 0}$ and systematic intensity process Z. The parameters used are listed in the caption of the figure. Notice that the stochastic volatility correction allows for the mass of the distribution to be shifted to the right, which is vital for the calibration of the model to market data.

The presence of stochastic volatility in our approximation is also present in the loss process for the entire portfolio as shown in Figure 14.4, which shows the probability mass function $m \mapsto \mathbb{P}^\star\{L_t = m\}$ from the formula

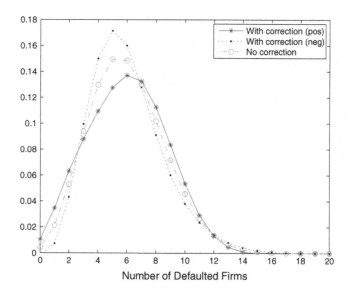

Figure 14.3 The probability loss density function within a single homogeneous group $m \mapsto \mathbb{P}^{\star}\{L_t^i = m\}$ with and without the correction due to stochastic volatility, as shown in (14.73). The solid line corresponds to the density function with positive parameter $V_{1,i} = 0.00015$, while the dotted line has negative $V_{1,i}$ parameter ($V_{1,i} = -0.00015$). They can be compared to the dashed line, which corresponds to the constant volatility density function (i.e., $V_{1,i} = 0$). The other parameters used are $n_i = 20$, $\alpha_i = 0.15$, $\sigma_i^2 = 0.1^2$, $\bar{x}_i = 0.015$, $x_i = 0.04$, $\alpha_z = 0.1$, $\bar{z} = 0.005$, $z = 0.004$, $\sigma_z = 0.01$, $t = 5$ years.

(14.73) with five equal-sized groups ($k = 5$) of 25 firms ($n_1 = \cdots = n_5 = 25$) for a total of 125. Notice again the mass shift in the density function compared to the constant volatility case (no correction). The presence of five independent groups results in a lower peak and thicker tail of the density function. This feature allows the model to fit real data very adequately, as we will see in the fitting results of Section 14.4.5.

14.4.4 CDO Mechanics

We are interested in pricing the *tranches* of a CDO. Let K_1 and K_2 be the *attachment* and *detachment* points of a tranche, or in other words, the endpoints of the range of percentage losses the contract is providing insurance for. The Dow Jones standardized indices CDX and iTraxx have five benchmark tranches, the attachment and detachment points of which are given in Table 14.1. The first tranche is referred to as *equity*, since it is the riskiest and requires the highest premium. The last tranche is called

Table 14.1: *The attachment (K_1) and detachment (K_2) points for the tranches of the Dow Jones CDX and iTraxx indices*

	Equity		Mezz 1		Mezz 2		Mezz 3		Senior	
	K_1	K_2	K_1	K_2	K_1	K_2	K_1	K_2	K_1	K_2
CDX	0%	3%	3%	7%	7%	10%	10%	15%	15%	30%
iTraxx	0%	3%	3%	6%	6%	9%	9%	12%	12%	22%

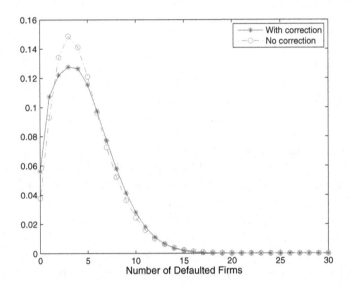

Figure 14.4   The effect of stochastic volatility on the mass function $m \mapsto \mathbb{P}^*\{L_t = m\}$ of the number of defaulted firms at time t for a basket of 125 firms separated into five equal-sized groups of 25 firms per group. The parameters used are $\alpha_1 = \alpha_2 = \alpha_3 = 0.15$, $\alpha_4 = \alpha_5 = 0.1$, $\sigma_1^2 = \cdots = \sigma_5^2 = 0.1^2$, $\bar{x}_1 = \cdots = \bar{x}_5 = 0.015$, $x_1 = \cdots = x_5 = 0.005$, $V_{1,1} = -0.0001$, $V_{1,2} = -0.00005$, $V_{1,3} = 0.0001$, $V_{1,4} = 0.00015$, $V_{1,5} = 0.0002$, $\alpha_z = 0.1$, $\bar{z} = 0.005$, $z = 0.004$, $\sigma_z = 0.01$, $t = 5$ years. The dashed line illustrates the absence of stochastic volatility approximation by setting $V_{1,1} = \cdots = V_{1,5} = 0$. (We graph the mass function over $\{0, 1, \ldots, 30\}$ instead of $\{0, 1, \ldots, 125\}$.)

senior, and in almost all cases it is of very high credit quality. Finally, the three intermediate tranches are called *mezzanine* tranches. Obviously, the insurance premia are decreasing across the tranches, with the equity tranche having the largest premium, and the senior tranche the smallest.

The tranche loss function $\ell_{[K_1, K_2]}$ with attachment and detachment points K_1 and K_2 is defined as

$$\ell_{[K_1, K_2]}(x) = (x - K_1)^+ - (x - K_2)^+, \quad 0 \le x \le 1.$$

We assume that each firm has the same notional amount in the portfolio, which without loss of generality we take to be $\$1/n$. We also assume a deterministic recovery $\delta \in [0,1)$, that is, the loss to the portfolio from each default is $\$(1-\delta)/n$. Then the portfolio loss at time t is given by

$$\Lambda_t = \frac{(1-\delta)L_t}{n}, \quad t \geq 0.$$

There are two counterparties in a CDO agreement. The *protection buyer* (usually institutional investors) of the $[K_1, K_2]$ tranche enters into the contract and purchases insurance against any losses between $K_1\%$ and $K_2\%$ of the total notional of the portfolio over the time period $[0, T]$. The protection buyer makes M periodic payments at times $t_1 < t_2 < \cdots < t_M = T$, which for the case of CDX and iTraxx are quarterly, i.e., $t_1 = 0.25, t_2 = 0.5$, and so on. The *protection seller* (usually an investment bank or the originator of the original pool of securitized assets) receives the payments, and in return will compensate the protection buyer as agreed. Hence, the *pricing* of the tranches of a CDO is the specification of the constant coupon rate the protection buyer will pay to the protection seller on every payment due date. This coupon for the mezzanine and senior tranches (but not equity) is given by

$$s = \frac{\sum_{j=1}^{M} e^{-r(t_j+t_{j-1})/2} \left(\mathbb{E}^{\star}[\ell_{[K_1,K_2]}(\Lambda_{t_j})] - \mathbb{E}^{\star}[\ell_{[K_1,K_2]}(\Lambda_{t_{j-1}})] \right)}{\sum_{j=1}^{M} (t_j - t_{j-1}) e^{-rt_j} \left[K_2 - K_1 - \frac{1}{2} \left(\mathbb{E}^{\star}[\ell_{[K_1,K_2]}(\Lambda_{t_j})] + \mathbb{E}^{\star}[\ell_{[K_1,K_2]}(\Lambda_{t_{j-1}})] \right) \right]}.$$

Details can be found in Lando (2004), for instance.

For the equity tranche it is standard to assume that there is an *up-front* fee in the inception of the contract paid to the protection seller by the protection buyer plus a fixed running premium s of 500 basis points (bps) or 5% of the tranche notional $K_2 - K_1 = K_2$. Therefore, if we define the up-front fee as the fraction s_{eq} of K_2, we have

$$s_{eq} = \frac{1}{K_2} \left[\sum_{j=1}^{M} e^{-r(t_j+t_{j-1})/2} \left(\mathbb{E}^{\star}[\ell_{[0,K_2]}(\Lambda_{t_j})] - \mathbb{E}^{\star}[\ell_{[0,K_2]}(\Lambda_{t_{j-1}})] \right) \right.$$

$$- 0.05 \sum_{j=1}^{M} (t_j - t_{j-1}) e^{-rt_j} \left[K_2 - \frac{1}{2} \left(\mathbb{E}^{\star}[\ell_{[0,K_2]}(\Lambda_{t_j})] \right. \right.$$

$$\left. \left. \left. + \mathbb{E}^{\star}[\ell_{[0,K_2]}(\Lambda_{t_{j-1}})] \right) \right] \right].$$

14.4.5 Dow Jones CDX Fitting

To exhibit the implementation of the combination of stochastic volatility default intensity models with the homogeneous group structure, we fit the five-year Dow Jones CDX index tranche spreads.

The Dow Jones CDX is a credit default swap index introduced in 2003, and is composed of 125 equally weighted North American corporations of investment grade credit rating. The firm components are chosen from seven financial market sectors, and the index is interpreted as the average insurance premium required for protection against the default of an investment-grade company within five years. (There are also seven- and ten-year versions of the index, but the five-year CDX quotes are the most liquid.) In order for the index to reflect the most up-to-date corporate representation, and reflect accurately the credit risk environment, the composition of the index is re-examined frequently and firms are added or removed according to certain eligibility criteria. In particular, a new CDX series is introduced every six months on March 20 and September 20 of each year (or the next business day if the prespecified day is on a weekend, or a holiday as designated by the New York Stock Exchange).

We perform a calibration and give analytic details on the tranche prices of the Series 7 Dow Jones CDX index (ticker symbol DJ.CDX.NA.IG.7) for October 31, 2006 as shown in the first two rows of Table 14.3. Also, for comparison purposes, we provide the fitting results on the Series 4 CDX index for August 23, 2004 (which are not calibrated to real market data, though) and compare it against the results of Mortensen (2006).

Let us note that the credit quality of the CDX index has been improving with every new series that is introduced, as shown from the spread data in the first two rows of Tables 14.3 and 14.5. As a result, both the overall level of the tranche spreads and their bid/ask spreads are decreasing, and for this reason we choose to illustrate the calibration results for both credit environments that are roughly two years apart.

Calibration Method We apply the homogeneous group structure framework to assign the 125 components of the CDX index into separate groups that will describe the credit quality of the portfolio in more detail. We choose seven groups of firms with their sizes listed on the titles of the plots of Figure 14.5. The first group is comprised of the eight firms with the smallest 5-year CDS spreads, the second group has the ten firms with the next smallest spreads, and so on.

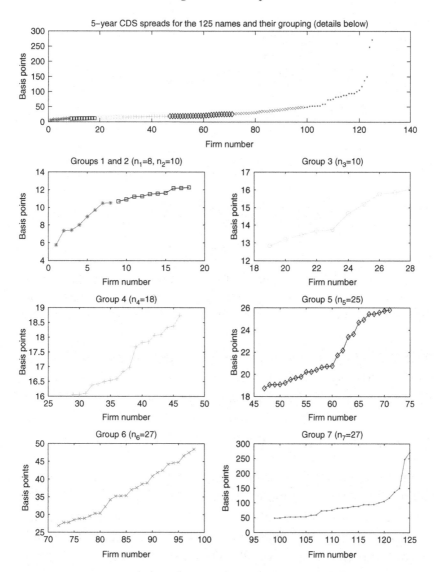

Figure 14.5 The 5-year CDS spreads of the 125 names of the Dow Jones CDX index on October 31, 2006 and their grouping.

Figure 14.5 gives the breakdown of the 125 firms into the seven groups according to their 5-year CDS spread. The last two groups (groups 6 and 7) represent the above-average credit risk firms (when compared to the overall average of the CDX index), whereas the remaining firms are classified into five groups with relatively small default risk. The default intensity processes $\lambda^1, \ldots, \lambda^7$ as given in (14.69) for the seven homogeneous groups will have term structure given by the average 1-, 3-, 5-, 7-, and 10-year CDS spreads of the firms within each group. These term structures on October 31, 2006 are given in Figure 14.6.

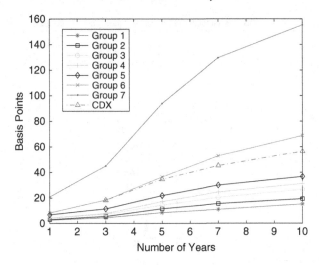

Figure 14.6   CDS term structure of the CDX index (dashed line) and the average CDS term structure of the firms of the seven homogeneous groups on October 31, 2006. *Source*: Bloomberg Generic Database (CBGN).

Since the default intensity process λ^i is decomposed into an idiosyncratic, X^i, and a systematic risk factor, Z, we will calibrate the common market risk factor against the CDX index spread, i.e., the coupon that is paid for protection against the entire index (or the 0–100% tranche). The term structure is plotted in Figure 14.6 (dashed line). Notice that it is available only in 3-, 5-, 7-, and 10-year quotes.

Assuming a piecewise constant hazard rate, we construct the survival term structure for each of the groups and the CDX index. The term structure is given in Table 14.2, and notice that groups 6 and 7 are considerably riskier than the rest, consistent with the average CDS term structure in Figure 14.6. Furthermore, the overall CDX index on October 31, 2006 has a term structure between those of the fifth and sixth homogeneous groups.

There are 46 free parameters in the loss distribution expressions:

$$\alpha_1,\ldots,\alpha_7,\overline{\sigma_1^2},\ldots,\overline{\sigma_7^2},x_{0,1},\ldots,x_{0,7},\bar{x}_1,\ldots,\bar{x}_7,V_{1,1},\ldots,$$
$$V_{1,7},c_1,\ldots,c_7,\alpha_z,\sigma_z^2,z_0,\bar{z},$$

and only five tranche spreads, so our model is, obviously, over-parameterized. As ill-conditioned as this might look, it is rather a typical situation when a rich modeling framework is used for the pricing of basket credit derivatives. For instance, the total number of parameters to be estimated in the model proposed by Mortensen (2006) is $6+7n$ (before making the dimension reduction by considering equal parameters among the processes for different firms).

Table 14.2: *The survival curve for the seven groups and the Dow Jones CDX Series 7 index on October 31, 2006, assuming piecewise constant intensities*

	Size	Year 1	3	5	7	10
Group 1	8	0.0017	0.0031	0.0065	0.0089	0.0139
Group 2	10	0.0020	0.0039	0.0088	0.0129	0.0177
Group 3	10	0.0028	0.0050	0.0111	0.0172	0.0252
Group 4	18	0.0028	0.0055	0.0133	0.0210	0.0297
Group 5	25	0.0045	0.0080	0.0169	0.0257	0.0350
Group 6	27	0.0054	0.0129	0.0291	0.0488	0.0746
Group 7	27	0.0135	0.0325	0.0846	0.1617	0.2561
CDX	125	—	0.0127	0.0272	0.0400	0.0570

Table 14.3: *Dow Jones CDX tranche Series 7 (mid) spreads and bid/ask spreads on October 31, 2006, and homogeneous groups with stochastic volatility model (HGSV) fit. The equity tranche is quoted in percentage points of 3% (the detachment point) plus 500 bps running spread, while the other tranches are quoted in bps. The parameters are given in Table 14.4*

	Tranche 0–3%	3–7%	7–10%	10–15%	15–30%	RMSE
Mid-market spread	23.66%	88.26	28.75	7.25	3.43	
Bid/ask spread	0.31%	1.25	1.11	0.64	0.97	
HGSV	23.76%	87.35	31.40	7.17	0.08	1.91

As such, this is not a disadvantage of the proposed setup because we explicitly allow for separate dynamics within each homogeneous group. Fundamentally, this is an improvement over the case where each underlying firm in the pool of the 125 names would have its own dynamics. By separating firms of *similar* credit risk (as measured by their 5-year CDS spread) and modeling their aggregate credit risk within their homogeneous group, we obtain more accurate estimates of their default probabilities rather than assuming the existence of a unique specification of each firm's credit risk.

We fit the tranches with the remaining parameters, after setting the recovery rate, δ, equal to 0.35, and the risk-free interest rate, r, equal to 0.05. The results of the fitting are shown in Table 14.3 and the parameters used

Table 14.4: *Parameters for the fit of Table 14.3. Note also that* $\delta = 0.35$
and $r = 0.05$

	Group							
	1	2	3	4	5	6	7	Common
n	8	10	10	18	25	27	27	—
α	0.06	0.1	0.11	0.12	0.13	0.15	0.2	0.05
$\bar{\sigma}$	0.06	0.07	0.07	0.07	0.07	0.09	0.23	0.01
$V_1/10^{-4}$	−0.7	−1.07	−1.2	−1.43	1.43	1.35	1.65	—
$\bar{x}/10^{-3}$	2.1	2.2	2.9	2.1	1.8	3.2	9.9	2.72
$x_0/10^{-3}$	1.6	1.5	2.6	2.3	1.9	1.8	5.6	1.27
c	0.65	0.69	0.96	0.65	0.54	1.07	3.64	—

are listed in Table 14.4. The root-mean-squared error (RMSE) is listed in the last column.

Effect of the Homogeneous Group Setup From the fitting results of Table 14.3 we see that the proposed setup with seven homogeneous groups and the calibrated parameters shown in Table 14.4 fits the equity and the three mezzanine tranches very well, each within one bid/ask spread deviation or less. The senior tranche (15–30%) is underpriced; the model computes a premium of less than one-tenth of a basis point for protection against losses in that range, while the market quote was 3.43 basis points. Considering the bid/ask spread of one basis point, this deviation is rather large.

Past results regarding CDO pricing with intensity-based models have indicated that pure diffusion models for the default intensity process are not able to generate loss distributions with the market-implied characteristics. This characteristic is simply the longer tails required for the probability mass function of L_t resulting from high-correlation situations. However, if that was the case, the other tranches would have been considerably mispriced as well, since this is exactly the behavior of a pure diffusion model as Mortensen (2006) notes in the calibration example (early peaks and thin tails of the loss distribution). Furthermore, the presence of the stochastic volatility correction term in the Laplace transform, required for the computation of the conditional probability mass function, introduces additional control over the probability density of the mid-range number of losses, but does not affect the probability of a small number of defaults.

The most important feature of the proposed framework is the categorization of the portfolio of firms into groups with homogeneous characteristics. As aggregate as it seems, modeling the firms' default intensities as a single process as long as they belong in the same homogeneous group is *more accurate*, in terms of analyzing the overall credit risk of the portfolio, compared to some other pricing methodologies that are used in practice. In applications of factor models to the pricing of CDOs, the inevitable simplifying assumption (or, compromise) is to consider that all firms' idiosyncratic components of the default intensity processes are, statistically, the same. This is done by setting all, or almost all, the parameters equal to each other and thus using the common ones to fit the tranche spreads to the market observed. The reason for this is to reduce the number of free parameters available for calibration, and hopefully use an optimization program for the minimization of an error measure.

The credit quality of the 125 firm constituents of the Dow Jones CDX index, as measured by their 5-year CDS spread, can vary considerably (see top panel of Figure 14.5 for an example on October 31, 2006) and they should not be considered homogeneous. Dealer quotes for the tranches of CDX or iTraxx are, almost exclusively, computed by models having the assumption of a single idiosyncratic component among all the 125 processes. This mollifies the effect very high-risk firms have on the portfolio loss distribution, thus attributing the overall default risk evenly among all the firms, or, worse, to the systematic factor component of the default intensity. In turn, this attributes a larger-than-real weight of the default probabilities to the common market factor, which overestimates the probability of many defaults, which consequently prices the senior tranche very highly.

Therefore, the overly simplistic assumption of extreme homogeneity of the idiosyncratic components of the default intensity explains, partially, the large premia of senior and super-senior tranches (i.e., the 15–30% and 30–100% tranches for CDX and the 12–22% and 22–100% tranches for iTraxx) we see in data. This is in accordance with Duffie and Gârleanu (2001) where, even though the asymmetrical information of the underlying assets is alleviated by the tranching of losses, there is still a considerable *liquidity premium* for the virtually improbable losses. This premium associated with the senior and super-senior tranches is exogenous to the default premium dictated by pricing models. Sircar and Zariphopoulou (2010) reach the same conclusion via an alternative pricing approach using utility valuation.

We stress the fact that the breakdown of the 125 names into seven groups of variant credit risk explains the tranche spreads very closely, and is offered for the pricing of CDOs, and particularly bespoke CDOs, where a more inhomogeneous pool of firms (including speculative-grade firms) is usually used.

Table 14.5: *Market spreads for the Dow Jones CDX Series 4 index on August 23, 2004, and fit comparisons between the HGSV and five other popular models. The equity tranche is quoted in percentage points and all the other tranches in bps. The parameters are given in Table 14.6*

| | \multicolumn{6}{c}{Tranche} |
	0–3%	3–7%	7–10%	10–15%	15–30%	RMSE
Mid-market spread	40%	312.5	122.5	42.5	12.5	
Bid/ask spread	2%	15	7	7	3	
HGSV	**40.8%**	**311.5**	**123.4**	**40.4**	**1.6**	**1.64**
Jump-diffusion intensities	46.9%	340.2	119.7	61.9	14.3	2.17
Pure-diffusion intensities	49.3%	442.9	94.9	16.8	0.4	5.34
Gaussian copula	46.8%	474.4	131.8	36.9	2.9	5.3
RFL Gaussian copula	48.6%	334.9	125.5	66.5	9.2	2.59
Double-*t* copula	45.1%	367.0	114.9	54.9	20	2.44

Source: Last five rows Mortensen (2006).

Table 14.6: *Parameters for the fit of Table 14.5. Note also that $\delta = 0.35$ and $r = 0.05$*

| | \multicolumn{7}{c}{Group} | |
	1	2	3	4	5	6	7	Common
n	8	10	10	19	25	26	27	—
α	0.08	0.09	0.1	0.21	0.23	0.165	0.17	0.05
$\bar{\sigma}$	0.06	0.07	0.07	0.18	0.2	0.41	0.45	0.01
$V_1/10^{-4}$	−0.07	−1.07	−1.2	14.5	14.43	14.5	20.0	—
$\bar{x}/10^{-3}$	4.5	4.7	5.9	6.2	8.0	10.5	10.6	2.72
$x_0/10^{-3}$	3.42	5.5	5.7	5.5	6.5	8.0	14.9	1.27
c	0.65	0.69	0.96	0.65	0.54	1.07	3.49	—

We also perform a fitting example on the Dow Jones CDX tranches from August 23, 2004, and compare against the fitting of Mortensen (2006). The results are given in Table 14.5 and the parameters are listed in Table 14.6. Let us note that Mortensen (2006) uses the US swap curve for the default-free interest rate, instead of our constant $r = 0.05$, but keeps $\delta = 0.35$.

It is not surprising that the model of homogeneous groups with stochas-tic volatility fits the equity and mezzanine tranches considerably better than all the other models, since it has many more free parameters that govern the dynamics. However, even after an exhaustive survey of the model's free parameters, the senior tranche remains underpriced, which attests to the aforementioned argument of the existence of an additional premium included in the senior tranche spread.

15

Epilogue

The Credit Crisis that erupted in Summer 2007, leading eventually in September 2008 to a full-blown Financial Crisis, has led to many questions about the role of quantitative models in the financial industry and whether they were (and still are) being used to provide an illusory crutch to highly risky trading activities. It is worth noting the origin of the crisis is in mortgage-backed securities (MBS), which is the least quantitatively modeled and academically studied of all derivatives markets. Nonetheless, one can suspect that the niceties of risk-neutral pricing, the arbitrage-free equivalent martingale measure, particularly as a formal justification for *mark-to-market*, were over-extended from liquid options and fixed income markets into unregulated and poorly understood high-dimensional credit markets.

We saw in the previous chapter that, in the CDO arena, before the crisis, the market valued tranches as if there was a high probability of a small number of losses, a small probability of around 10% being lost over the five years, and a *relatively* high probability of a larger number of losses around 20%. The market's distribution seems to be double-humped. It seems natural that some of the prices or spreads seen in credit markets today are due to "crash-o-phobia" in a relatively illiquid market, with the effect enhanced in large baskets. For a number of years up till 2007, the few basis points offered to the tranche holder of a CDX senior tranche, for protection against the risk of 15–30% of investment grade US firms defaulting over the next five years, was considered anomalous as the market was ascribing a seemingly large probability to what would almost be "the end of the world as we know it." After recent events, spreads in the senior tranche (as well as the others) have jumped, for example from 3.43 bps on October 31, 2006 to 79.53 bps on October 16, 2008. There is even heightened interest in the super-senior 30–100% tranche, where previously there was very little. It is also interesting to note that in early November 2008, the first mezzanine tranche for the CDX (3–7%) joined the equity tranche in being quoted with upfront points plus a running spread, instead of the usual spread convention, illustrating the extreme counterparty risk aversion among primary dealers.

In May 2005, when Ford and GM were simultaneously downgraded, single-name default probabilities jumped in reaction across the credit market, but spreads on the mezzanine tranches actually fell. The reason was that many funds and banks were worried that their exposure to default risk by holding mezzanine tranches was suddenly very high, having previously felt there was little likelihood of those being hit, and they desperately tried to unwind those positions. There were many very large losses among well-established hedge funds and investment banks. Moreover, the one-factor Gaussian copula model couldn't even fit the data around this time (i.e., no correlation parameter $\rho \in [0, 1]$ worked), and there were considerable efforts inside and outside industry to come up with workable alternatives, particularly one that is *dynamic* (as opposed to the static copula framework). The front-page *Wall Street Journal* (September 12, 2005) article "How a Formula Ignited Market That Burned Some Big Investors," while not mentioning mortgage-backed securities, remains remarkably prescient about events that occurred later, particularly the effect of CDOs on investors and banks.

The overuse of credit derivatives, particularly in the mortgage arena, contributed massively to the ongoing financial crisis. While the May 2005 ripple in the corporate CDO market served as a warning of worse things to come, its lessons were largely ignored as the risks abated. The attraction of unfunded returns on default protection proved too great, and the culture of unbounded bonus-based compensation for traders led to excessive risk-taking that jeopardized numerous long-standing and once fiscally conservative financial institutions. No one can argue with a trader bringing in tens of millions each year, and he himself is relatively indifferent to his job security and the stability of the firm while making millions in yearly bonuses.

A key chicken–egg question then is the role that quantitative models played in motivating or justifying these trades. The *Wall Street Journal* article from 2005 highlights the practice of playing CDO tranches off against each other to form a dubious hedge ("correlation trading"). In some sense this was "quantified" by calculations based on the Gaussian copula model (particularly implied correlation and its successor, the so-called base correlation) which were developed in-house by quants hired by and answering to traders. These models were developed for simplicity (one parameter dictating a complex instrument dependent on hundreds of underlyings and correlated risks) and speed. Many (surviving) banks have since restructured their quant/modeling teams so that they now report directly to management instead of to traders.

Numerous media outlets and commentators have blamed the crisis on financial models, and called for *less* quantification of risk. This spirit is distilled in Warren Buffet's comment from 2003: "Beware of geeks bearing formulas." Indeed, the caveat about the use of models to *justify a posteriori* "foolproof" hedges is justified. But this crisis, as with past crises, only highlights the need for more mathematics, and quantitatively trained people at the highest level, not least at ratings and regulatory agencies. On the morning of the *Wall Street Journal* article from 2005, the CEO of a major (still surviving) bank reputedly called his head credit derivatives trader to make sure that "*we* are not using the Gaussian copula." (Of course they were, and the trader's punchline in this anecdote is the CEO's mis-pronunciation of the words Gaussian and copula). The case for sanity about mathematical models is made in the excellent March 21, 2009 *Financial Times* editorial "Maths and markets."

But the real damage was done in the highly unquantified market for mortgage backed securities. Here the notion of independence and diversification through tranching was taken to ludicrous extremes. An MBS is itself a CDO on a pool of mortgages whose tranches were rated by the agencies as AAA for the senior and, for example, BBB for lower tranches. Pooling of risks and tranching them as reinsurance contracts to pass on to investors with differing risk profiles is natural, and the basis for insurance markets for dozens of years at least.

Of course the AAA tranches sold well, while there was less interest in the BBB tranches, and many institutions were prohibited from investing in such low-grade products. Here is where the financial alchemy or wizardry took hold. An MBS CDO is a CDO on MBS tranches. The BBB-rated tranches were pooled together and that new pool was tranched. The senior tranche of this CDO was accorded AAA status by the ratings agencies, because it was senior in its class, the cream of the crap, as it were. In fact, the MBS CDO is a CDO^2, a CDO on CDOs. Such products were introduced in the corporate credit world around 2005, but quickly abandoned as they were deemed impossible to value due to their insane complexity. There is a simple back-of-the-envelope calculation in the paper "Credit Crunch" by Hull (2008) that demonstrates how default risks are magnified enormously in CDO^2s. All this was hidden under the AAA rating that allowed restricted institutions now to invest in junk mortgages. The telling title of one of Paul

Krugman's *New York Times* columns from 2008 is "Just say AAA." The
level of interest in the new MBS CDO tranches was so great in the period
up to 2007, that it created pressure to write more (subprime) mortgages on
which to write the derivatives.

The MBS desk at major banks was relatively free of quants and quantitative analysis compared with the corporate CDO desk, even though the
MBS book was many many times larger. The excuse given was that these
products were AAA and therefore like US Government bonds, did not need
any risk analysis. As it turned out, this was a mass delusion willingly played
into by banks, hedge funds, and the like. The level of quantitative analysis
performed by the ratings agencies was to test outcomes on a handful of scenarios, in which, typically, the worst-case scenario was that US house prices
would appreciate at the rate 0.0%. The rest is history.

Following is a *polemic* (written by Peter Cotton and RS) on some of the
cultural mistakes on Wall Street in recent years.

Sideways

*Peter Cotton (Benchmark Solutions) & Ronnie Sircar (Princeton Univer-
sity)*

October 2008

```
Flying across the Atlantic in dense fog, Lindbergh
descended several times to an altitude of just ten feet.
He judged wind direction from the spray of the waves, and
maneuvered his plane sideways to catch a glance of what
was ahead. The Spirit of St Louis had no windshield;
looking sideways was Lindbergh's only option.

     On Wall Street, looking sideways at the prices of
similar securities is called marking the book. Without
sideways-looking models, business cannot be transacted and
bonuses cannot be paid. Wall Street models enabled pilots
to see that everyone else was more or less in formation...
as they all ploughed into the mountain.

     Ratings agencies, meanwhile, looked backwards. Past is
prologue, they said, and many believe them to this day.
Based on past events, they judged events that have now
occurred to be one in a thousand, ignoring the one in two
chance their basic premise was wrong. For example, recent
Libor movements are twelve standard deviation events,
historically speaking. That probability is so small you
```

can't compute or display it in Excel. Actual corporate
spread movements in the last twelve months would have been
deemed impossible, according to rating agency models still
in use today.

Engineers cannot predict the weather, but they can
design planes which fly through storms. Wall Street could
have taken a scientific approach to financial product
design based on forward-looking simulation. Most
institutions chose not to. On Wall Street, in the heyday
of credit default swaps and collateralized debt
obligations (the notorious CDOs), the trader bringing in
tens of millions in risky bets was king, and the quants,
the mathematicians, if they wanted to keep their jobs,
served up models for one purpose: looking sideways to mark
the books.

Many traders and trading managers have broad skill
sets, essential to driving revenue. Most were not
scientists by training however, and their ability to
manage the scientific task of model creation was limited.
This, and other demands of a frenetic business led to the
predominance of trader-friendly sideways-looking financial
models. The models computed numbers quickly, but gave
scant insight into what might lie ahead.

On the credit derivatives desks of major financial
institutions, the windshield was boarded up. A proper
financial model weighs risk by generating hypothetical
simulations of the future. On many of these simulations
investors make a nice return; on some they lose their
shirts. In 2004, a team at J.P. Morgan proposed a model
for corporate CDOs that served the objective of marking
the books, but could not be used to simulate the future.
Nobody had thought to do that before, and it was a bold
success. Known as the ''base correlation'', it was the
quintessential trader model. Forward-looking models can
alert management to contingent hedging costs, and might
prevent a deal being done in the first place. Nobody
wanted that while easy money was rolling in.

In May 2005, CDO prices slipped out of the models'
grasp, hedging positions straddling the fault line were
torn apart, and investors lost money. The glaring

over-simplification of industry-standard models for CDOs
was exposed, and the ''Gaussian Copula'', a crucial
ingredient, made the front page of the *Wall Street
Journal* in September of that year. The public and
Government were warned of the slender underpinnings of a
multi-trillion dollar market, but quickly forgot. The only
thing considered a problem by some traders was failure to
calibrate to the market, that is to be able to see
sideways. As conditions became calmer, the industry
standard model became ever more popular. Most traders
loved it. Most academics rolled their eyes. Most quants
did what they were told.

Looking sideways took Lindbergh to Paris, and CDOs to
Sydney. But like aircraft, complex financial products
ultimately require careful design. Careful design requires
simulation. Simulation requires models that make sense,
and making models that make sense takes time.
Sideways-looking pseudoscience can be created in a week,
but models which look forward require research,
collaboration, reflection, experimentation, and
mathematical advances. The current crisis must spur the
remaining financial institutions to invest seriously in
such R&D. The problem on Wall Street has hardly been the
over-influence of quantitative modeling, as some have
claimed, but rather, in some derivatives markets, its near
absence.

References

Achdou, Y. and Pironneau, O. (2005). *Computational Methods for Option Pricing*, SIAM Frontiers in Applied Mathematics.

Alòs, E. (2006). A generalization of Hull and White formula and applications to option pricing approximation. *Finance & Stochastics* **10**(3): 353–65.

Alòs, E. (2009). A decomposition formula for option prices in the Heston model and applications to option pricing approximation. *Economics working papers*, Department of Economics and Business, Universitat Pompeu Fabra.

Andersen, L. and Piterbarg, V. (2007). Moment explosions in stochastic volatility models. *Finance & Stochastics* **11**(1): 29–50.

Andersen, L., Sidenius, J. and Basu, S. (2003). All your hedges in one basket. *RISK* pp. 67–72.

Avellaneda, M., Boyer-Olson, D., Busca, J. and Friz, P. (2002). Reconstructing the smile. *RISK*.

Avellaneda, M., Friedman, C., Holmes, R. and Samperi, D. (1997). Calibrating volatility surfaces via relative-entropy minimization. *Applied Mathematical Finance* **4**(1): 37–64.

Ball, C. and Roma, A. (1994). Stochastic volatility option pricing. *Journal of Financial and Quantitative Analysis* **29**(4): 589–607.

Batalova, N., Maroussov, V. and Viens, F. (2006). Selection of an optimal portfolio with stochastic volatility and discrete observations. *Transactions of the Wessex Institute on Modelling and Simulation* **43**: 371–80.

Bayraktar, E. and Yang, B. (2011). A unified framework for pricing credit and equity derivatives. *Mathematical Finance*.

Benabid, A., Bensusan, H. and El Karoui, N. (2009). Wishart stochastic volatility: Asymptotic smile and numerical framework, *Preprint*.

Benaim, S., Friz, P. and Lee, R. (2009). On the Black–Scholes implied volatility at extreme strikes, in R. Cont (ed.), *Frontiers in Quantitative Finance Volatility and Credit Modeling*. New York: Wiley Finance, pp. 19–45.

Bensoussan, A. and Lions, J.-L. (1982). *Applications of Variational Inequalities in Stochastic Control*, Vol. 12 of *Studies in Mathematics and its Applications*. Amsterdam: North-Holland.

Berestycki, H., Busca, J. and Florent, I. (2004). Computing the implied volatility in stochastic volatility models. *Communications on Pure and Applied Mathematics* **57**(10): 1352–73.

Bergomi, L. (2009). Smile dynamics IV, *Preprint*.

Bielecki, T. and Rutkowski, M. (2004). *Credit Risk: Modeling, Valuation and Hedging*, Springer Finance. Berlin: Springer.

Björk, T. (2004). *Arbitrage Theory in Continuous Time*, 2nd edn. Oxford: Oxford University Press.

Black, F. and Cox, J. (1976). Valuing corporate securities: Some effects of bond indenture provisions. *Journal of Finance* **31**: 351–67.

Black, F. and Scholes, M. (1973). The pricing of options and corporate liabilities. *Journal of Political Economy* **81**: 637–59.

Blankenship, G. and Papanicolaou, G. (1978). Stability and control of stochastic systems with wide-band noise disturbances. *SIAM Journal on Applied Mathematics* **34**(3): 437–76.

Boyarchenko, N. and Levendorskii, S. (2007). The eigenfunction expansion method in multifactor quadratic term structure models. *Mathematical Finance* **17**(4): 503–39.

Breiman, L. (1992). *Probability*, Vol. 7 of *Classics in Applied Mathematics*, SIAM.

Carmona, R. and Tehranchi, M. (2006). *Interest Rate Models: an Infinite Dimensional Stochastic Analysis Perspective*, Springer Finance. Berlin: Springer.

Carmona, R., Fouque, J.-P. and Vestal, D. (2009). Interacting particle systems for the computation of rare credit portfolio losses. *Finance & Stochastics* **13**(4): 613–33.

Carr, P. and Madan, D. (2001). Optimal positioning in derivative securities. *Quantitative Finance* **1**: 19–37.

Chen, X., Chadam, J., Jiang, L. and Zheng, W. (2008). Convexity of the exercise boundary of the American put option on a zero dividend asset. *Mathematical Finance* **18**(1): 185–97.

Chernov, M., Gallant, R., Ghysels, E. and Tauchen, G. (2003). Alternative models for stock price dynamics. *Journal of Econometrics* **116**: 225–57.

Chi, Y., Jaimungal, S. and Lin, S. (2010). An insurance risk model with stochastic volatility. *Insurance: Mathematics and Economics* **46**(1): 52–66.

Choi, S.-Y., Fouque, J.-P. and Kim, J.-H. (2010). Option pricing under hybrid stochastic and local volatility, *Preprint*.

Christoffersen, P., Jacobs, K. and Vainberg, G. (2008a). Forward looking betas, *Manuscript, McGill University*.

Christoffersen, P., Jacobs, K., Ornthanalai, C. and Wang, Y. (2008b). Option valuation with long-run and short-run volatility components, *CREATES Research Papers*.

Collin-Dufresne, P. and Goldstein, R. (2001). Do credit spreads reflect stationary leverage ratios? Reconciling structural and reduced form frameworks. *Journal of Finance* **56**: 1929–58.

Conlon, J. and Sullivan, M. (2005). Convergence to Black-Scholes for ergodic volatility models. *European Journal of Applied Mathematics* **16**(3): 385–409.

Cont, R. and Tankov, P. (2003). *Financial Modelling with Jump Processes*. London: Chapman and Hall/CRC.

Cotton, P., Fouque, J.-P., Papanicolaou, G. and Sircar, R. (2004). Stochastic volatility corrections for interest rate derivatives. *Mathematical Finance* **14**(2): 173–200.

Cox, J. (1975). Notes on option pricing I: Constant elasticity of variance diffusion, *Technical report*, Stanford University, Graduate School of Business. Reprinted in: *Journal of Portfolio Management*, Vol. 22, 1996.

Cvitanic, J. and Rozovskii, B. (2006). A filtering approach to tracking volatility from prices observed at random times. *Annals of Applied Probability* **16**(3): 1633–52.

Dembo, A. and Zeitouni, O. (1998). *Large Deviation Techniques and Applications*, 2nd edn. Berlin: Springer.

Derman, E. and Kani, I. (1994). Riding on a smile. *RISK* **7**: 32–9.

DeSantiago, R., Fouque, J.-P. and Sølna, K. (2008). Bond markets with stochastic volatility, in J.-P. Fouque and K. Sølna (eds), *Econometrics and Risk Management*, Vol. 22 of *Advances in Econometrics*. Bingley, UK: Emerald, pp. 215–42.

Duffie, D. (2001). *Dynamic Asset Pricing Theory*, 3rd edn. Princeton, NJ: Princeton University Press.

Duffie, D. and Gârleanu, N. (2001). Risk and valuation of collateralized debt obligations. *Financial Analysts Journal* **57**(1): 41–59.

Duffie, D. and Lando, D. (2001). Term structures of credit spreads with incomplete accounting information. *Econometrica* **69**: 633–64.

Duffie, D. and Singleton, K. (2003). *Credit Risk*. Princeton, NJ: Princeton University Press.

Duffie, D., Pan, J. and Singleton, K. (2000). Transform analysis and asset pricing for affine jump-diffusions. *Econometrica* **68**(6): 1343–76.

Dumas, B., Fleming, J. and Whaley, R. (1998). Implied volatility functions: Empirical tests, *Journal of Finance* **53**(6): 2059–106.

Dupire, B. (1994). Pricing with a smile. *RISK* **7**: 18–20.

Dupire, B. (2010). Historical moments, Presentation at Research in Options Conference, Rio.

Durrett, R. (2004). *Probability: Theory and Examples*, 3rd edn. Pacific Grove, CA: Duxbury Press.

Eckner, A. (2007). Computational techniques for basic affine models of portfolio credit risk. *Journal of Computational Finance*.

El Karoui, N. and Quenez, M. (1995). Dynamic programming and pricing of contingent claims in an incomplete market. *SIAM Journal on Control and Optimization* **33**: 29–66.

Elizalde, A. (2005). Credit risk models IV: Understanding and pricing CDOs. www.abelelizalde.com.

Engle, R. and Patton, A. (2001). What good is a volatility model? *Quantitative Finance* **1**: 237–45.

Eom, Y., Helwege, J. and Huang, J. (2004). Structural models of corporate bond pricing: An empirical analysis. *Review of Financial Studies* **17**: 499–544.

Errais, E., Giesecke, K. and Goldberg, L. (2010). Affine point processes and portfolio credit risk. *SIAM Journal on Financial Mathematics* **1**: 642–65.

Etheridge, A. (2002). *A Course in Financial Calculus*. Cambridge: Cambridge University Press.

Feng, J., Forde, M. and Fouque, J.-P. (2010). Short maturity asymptotics for a fast mean-reverting Heston stochastic volatility model. *SIAM Journal on Financial Mathematics* **1**: 126–41.

Fengler, M. (2005). *Semiparametric Modeling of Implied Volatility*. Berlin: Springer.

Figlewski, S. (2010). Estimating the implied risk neutral density for the U.S. market portfolio, in M. Watson, T. Bollerslev and J. Russell (eds), *Volatility and Time Series Econometrics: Essays in Honor of Robert Engle*. Oxford: Oxford University Press.

Fiorentini, G., Leon, A. and Rubio, G. (2002). Estimation and empirical performance of Heston's stochastic volatility model: the case of a thinly traded market. *Journal of Empirical Finance* **9**(2): 225–55.

Fleming, W.H. and Soner, H.M. (1993). *Controlled Markov Processes and Viscosity Solutions*. Berlin: Springer-Verlag.

Forde, M. and Jacquier, A. (2010). Robust approximations for pricing Asian options and volatility swaps under stochastic volatility. *Applied Mathematical Finance* **17**(3): 241–59.

Fouque, J.-P. and Han, C. (2004a). Asian options under multiscale stochastic volatility, in G. Yin and Q. Zhang (eds), *Proceedings of the AMS-IMS-SIAM Summer Conference on Mathematics of Finance*, Vol. 351 of *Contemporary Mathematics*, AMS.

Fouque, J.-P. and Han, C. (2004b). Variance reduction for Monte Carlo methods to evaluate option prices under multi-factor stochastic volatility models. *Quantitative Finance* **4**(5): 597–606.

Fouque, J.-P. and Han, C. (2007). A martingale control variate method for option pricing with stochastic volatility. *ESAIM Probability & Statistics* **11**: 40–54.

Fouque, J.-P. and Kollman, E. (2011). Calibration of stock betas from skews of implied volatilities. *Applied Mathematical Finance*.

Fouque, J.-P. and Lorig, M. (2009). A fast mean-reverting correction to the Heston stochastic volatility model (submitted).

Fouque, J.-P. and Tashman, A. (2011a). Option pricing under a stressed-beta model. *Annals of Finance*.

Fouque, J.-P. and Tashman, A. (2011b). Portfolio optimization under a stressed-beta model. *Wilmott Journal*.

Fouque, J.-P. and Zhou, X. (2008). Perturbed Gaussian copula, in J.-P. Fouque and K. Sølna (eds), *Econometrics and Risk Management*, Vol. 22 of *Advances in Econometrics*. Bingley, UK: Emerald, pp. 103–21.

Fouque, J.-P., Garnier, J., Papanicolaou, G. and Sølna, K. (2007). *Wave Propagation and Time Reversal in Randomly Layered Media*, Stochastic Modelling and Applied Probability 56. Berlin: Springer.

Fouque, J.-P., Jaimungal, S. and Lorig, M. (2010). Spectral decomposition of option prices in fast mean-reverting stochastic volatility models (submitted).

Fouque, J.-P., Papanicolaou, G. and Sircar, R. (2000). *Derivatives in Financial Markets with Stochastic Volatility*. Cambridge: Cambridge University Press.

Fouque, J.-P., Papanicolaou, G. and Sircar, R. (2001a). From the implied volatility skew to a robust correction to Black–Scholes American option prices. *International Journal of Theoretical & Applied Finance* **4**(4): 651–75.

Fouque, J.-P., Papanicolaou, G. and Sircar, R. (2001b). Stochastic volatility and the epsilon-martingale decomposition, in M. Kohlmann and S. Tang (eds), *Mathematical Finance*. Berlin: Birkhauser, pp. 152–61.

Fouque, J.-P., Papanicolaou, G. and Sircar, R. (2004). Stochastic volatility and correction to the heat equation, in R. Dallang, M. Dozzi and F. Russo (eds), *Seminar on Stochastic Analysis, Random Fields and Applications IV*, Vol. 58 of *Progress in Probability*. Berlin: Birkhauser, pp. 267–76.

Fouque, J.-P., Papanicolaou, G., Sircar, R. and Sølna, K. (2003a). Multiscale stochastic volatility asymptotics. *SIAM Journal on Multiscale Modeling & Simulation* **2**(1): 22–42.

Fouque, J.-P., Papanicolaou, G., Sircar, R. and Sølna, K. (2003b). Short time-scale in S&P 500 volatility. *Journal of Computational Finance* **6**(4): 1–23.

Fouque, J.-P., Papanicolaou, G., Sircar, R. and Sølna, K. (2003c). Singular perturbations in option pricing. *SIAM Journal on Applied Mathematics* **63**(5): 1648–65.

Fouque, J.-P., Papanicolaou, G., Sircar, R. and Sølna, K. (2004a). Maturity cycles in implied volatility. *Finance & Stochastics* **8**(4): 451–77.

Fouque, J.-P., Papanicolaou, G., Sircar, R. and Sølna, K. (2004b). Timing the smile. *Wilmott Magazine*, pp. 59–65.

Fouque, J.-P., Sircar, R. and Sølna, K. (2006). Stochastic volatility effects on defaultable bonds. *Applied Mathematical Finance* **13**(3): 215–44.

Fouque, J.-P., Sircar, R. and Sølna, K. (2009). Multiname and multiscale default modeling. *SIAM Journal on Multiscale Modeling and Simulation* **7**(4): 1956–78.

Fouque, J.-P., Wignall, B. and Zhou, X. (2008). Modeling correlated defaults: first passage model under stochastic volatility. *Journal of Computational Finance* **11**(3): 43–78.

Fournie, E., Lebuchoux, J. and Touzi, N. (1997). Small noise expansion and importance sampling. *Asymptotic Analysis* **14**(4): 361–76.

Freidlin, M. (1985). *Functional Integration and Partial Differential Equations*. Princeton, NJ: Princeton University Press.

French, D., Goth, J. and Kolari, J. (1983). Current investor expectations and better betas. *Journal of Portfolio Management*, pp. 12–17.

Frey, R. (1996). Derivative asset analysis in models with level-dependent and stochastic volatility. *CWI Quarterly* **10**(1): 1–34.

Friedman, A. (2006). *Stochastic Differential Equations and Applications*. New York: Dover.

Fukasawa, M. (2011). Asymptotic analysis for stochastic volatility: martingale expansion. *Finance & Stochastics*.

Gatheral, J. (2006). *The Volatility Surface: a Practitioner's Guide*. New York: John Wiley and Sons, Inc.

Geske, R. (1979). The valuation of compound options. *Journal of Financial Economics* **7**: 63–81.

Giesecke, K. (2004). Credit risk modeling and valuation: An introduction, in D. Shimko (ed.), *Credit Risk: Models and Management*, Vol. 2. New York: RISK Books.

Glasserman, P. (2003). *Monte Carlo Methods in Financial Engineering*, Applications of Mathematics 53. Berlin: Springer.

Gobet, E. and Miri, M. (2010). Time-dependent Heston model. *SIAM Journal on Financial Mathematics* **1**: 289–325.

Goldman, M., Sosin, H. and Gatto, M. (1979). Path-dependent options: Buy at the low, sell at the high. *Journal of Finance*.

Hagan, P., Kumar, D., Lesniewski, A. and Woodward, D. (2002). Managing smile risk. *Wilmott Magazine*, pp. 84–108.

Henry-Labordère, P. (2005). A general asymptotic implied volatility for stochastic volatility models. *Proceedings "Petit déjeuner de la Finance."*

Heston, S. (1993). A closed-form solution for options with stochastic volatility with applications to bond and currency options. *Review of Financial Studies* **6**(2): 327–43.

Hikspoors, S. and Jaimungal, S. (2008). Asymptotic pricing of commodity derivatives for stochastic volatility spot models. *Applied Mathematical Finance* **15**(5): 449–77.

Hilberink, B. and Rogers, L.C.G. (2002). Optimal capital structure and endogenous default. *Finance and Stochastics* **6**(2): 237–63.

Hobson, D. (1996). Stochastic volatility, *Technical report*, School of Mathematical Sciences, University of Bath.

Howison, S. (2005). Matched asymptotic expansions in financial engineering. *Journal of Engineering Mathematics* **53**: 385–406.

Hu, F. and Knessl, C. (2010). Asymptotics of barrier option pricing under the CEV process. *Applied Mathematical Finance* **17**(3): 261–300.

Hull, J. (2008). *Options, Futures and Other Derivative Securities*, 7th edn. New Jersey: Prentice Hall.

Hull, J. and White, A. (1987). The pricing of options on assets with stochastic volatilities. *Journal of Finance* **42**(2): 281–300.

Hull, J. and White, A. (1988). An analysis of the bias in option pricing caused by a stochastic volatility. *Advances in Futures and Options Research* **3**: 29–61.

Hull, J. and White, A. (2004). Valuation of a CDO and an *n*-th to default CDS without Monte Carlo simulation. *Journal of Derivatives* **12**(2): 8–23.

Hurd, T. (2009). Credit risk modeling using time-changed Brownian motion. *International Journal of Theoretical and Applied Finance* **12**(8): 1213–30.

Hurd, T. and Kuznetsov, A. (2008). Explicit formulas for Laplace transforms of stochastic integrals. *Markov Processes and Related Fields* **14**: 277–90.

Ilhan, A., Jonsson, M. and Sircar, R. (2004). Singular perturbations for boundary value problems arising from exotic options. *SIAM Journal on Applied Mathematics* **64**(4): 1268–93.

Jackel, P. and Kahl, C. (2005). Not-so-complex logarithms in the Heston model. *Wilmott Magazine*.

Jackwerth, J. and Rubinstein, M. (1996). Recovering probability distributions from contemporaneous security prices. *Journal of Finance* **51**(5): 1611–31.

Jeanblanc, M., Yor, M. and Chesney, M. (2009). *Mathematical Methods for Financial Markets*, Springer Finance. Berlin: Springer.

Jonsson, M. and Sircar, R. (2002a). Optimal investment problems and volatility homogenization approximations, in A. Bourlioux, M. Gander and G. Sabidussi (eds), *Modern Methods in Scientific Computing and Applications*, Vol. 75 of *NATO Science Series II*. Dordrecht: Kluwer, pp. 255–81.

Jonsson, M. and Sircar, R. (2002b). Partial hedging in a stochastic volatility environment. *Mathematical Finance* **12**(4): 375–409.

Karatzas, I. and Shreve, S. (1991). *Brownian Motion and Stochastic Calculus*, Vol. 113 of *GTM*, 2nd edn. Berlin: Springer.

Karatzas, I. and Shreve, S. (1998). *Methods of Mathematical Finance*. Berlin: Springer-Verlag.

Karlin, S. and Taylor, H. (1981). *A Second Course in Stochastic Processes*. New York: Academic Press Inc.

Kushner, H. (1984). *Approximation and Weak Convergence Methods for Random Processes, with Applications to Stochastic Systems Theory*, MIT Press Series in Signal Processing, Optimization, and Control, 6. Boston: MIT Press.

Kuske, R. and Keller, J. (1998). Optimal exercise boundary for an American put option. *Applied Mathematical Finance* **5**(2): 107–16.

Lamberton, D. and Lapeyre, B. (1996). *Introduction to Stochastic Calculus Applied to Finance*. New York: Chapman & Hall.

Lando, D. (2004). *Credit Risk Modeling: Theory and Applications*, Princeton Series in Finance. Princeton, NJ: Princeton University Press.

LeBaron, B. (2001). Stochastic volatility as a simple generator of apparent financial power laws and long memory. *Quantitative Finance* **1**(6): 621–31.

Lee, R. (1999). Local volatilities under stochastic volatility. *International Journal of Theoretical and Applied Finance* **4**(1): 45–89.

Lewis, A. (2000). *Option Valuation under Stochastic Volatility*. London: Finance Press.

Li, D. (2000). On default correlation: a copula approach. *Journal of Fixed Income* **9**: 43–54.

Lipton, A. (2001). *Mathematical Methods For Foreign Exchange: A Financial Engineer's Approach*. Singapore: World Scientific.

Longstaff, F. and Schwartz, E. (1995). Valuing risky debt: A new approach. *Journal of Finance* **50**: 789–821.

Lopatin, A. and Misirpashaev, T. (2007). Two-dimensional Markovian model for dynamics of aggregate credit loss. *Technical report*, Numerix.

Lord, R. and Kahl, C. (2006). Why the rotation count algorithm works. *SSRN eLibrary*.

Markowitz, H. (1952). Portfolio selection. *Journal of Finance* **7**: 77–99.

Merton, R.C. (1969). Lifetime portfolio selection under uncertainty: the continous-time case. *Review of Economics and Statistics* **51**: 247–57.

Merton, R.C. (1971). Optimum consumption and portfolio rules in a continuous-time model. *Journal of Economic Theory* **3**(1/2): 373–413.

Merton, R.C. (1973). Theory of rational option pricing. *Bell Journal of Economics* **4**(1): 141–83.

Mikosch, T. (1999). *Elementary Stochastic Calculus With Finance in View*. Singapore: World Scientific.

Mortensen, A. (2006). Semi-analytical valuation of basket credit derivatives in intensity-based models. *Journal of Derivatives* **13**(4): 8–26.

Musiela, M. and Rutkowski, M. (2002). *Martingale Methods in Financial Modelling*, 2nd edn. Berlin: Springer-Verlag.

Nayak, S. and Papanicolaou, G. (2007). Stochastic volatility surface estimation, *Preprint*.

Oksendal, B. (2007). *Stochastic Differential Equations: An Introduction with Applications*, Universitext, 6th edn. Berlin: Springer.

Papageorgiou, E. and Sircar, R. (2008). Multiscale intensity models for single name credit derivatives. *Applied Mathematical Finance* **15**(1): 73–105.

Papageorgiou, E. and Sircar, R. (2009). Multiscale intensity models and name grouping for valuation of multi-name credit derivatives. *Applied Mathematical Finance* **16**(4): 353–83.

Platen, E. and Heath, D. (2006). *A Benchmark Approach to Quantitative Finance*, Springer Finance. Berlin: Springer.

Renault, E. and Touzi, N. (1996). Option hedging and implied volatilities in a stochastic volatility model. *Mathematical Finance* **6**(3): 279–302.

Rogers, L.C.G. (1995). Which model for the term-structure of interest rates should one use?, *Mathematical Finance*, Vol. 65 of *IMA*. New York: Springer NY, pp. 93–116.

Rubinstein, M. (1985). Nonparametric tests of alternative option pricing models. *Journal of Finance* **40**(2): 455–80.

Rubinstein, M. (1994). Implied binomial trees. *Journal of Finance* **69**: 771–818.

Samuelson, P. (1973). Mathematics of speculative prices. *SIAM Review* **15**: 1–39.

Scholes, M. and Williams, J. (1977). Estimating betas from nonsynchronous data. *Journal of Financial Economics* **5**: 309–27.

Schönbucher, P. (2003). *Credit Derivatives Pricing Models*. New York: John Wiley.

Schönbucher, P. (2006). Portfolio losses and the term structure of loss transition rates: A new methodology for the pricing of portfolio credit derivatives. *Technical report*, ETH Zurich.

Schweizer, M. (1999). A guided tour through quadratic hedging approaches. *Preprint,* Technische Universitat, Berlin.

Scott, L. (1987). Option pricing when the variance changes randomly: theory, estimation, and an application. *Journal of Financial and Quantitative Analysis* **22**(4): 419–38.

Sharpe, W. (1996). Mutual fund performance. *Journal of Business* **39**: 119–38.

Shaw, W. (2006). Stochastic volatility, models of Heston type.

Shephard, N. (ed.) (2005). *Stochastic Volatility: Selected Readings*. Oxford: Oxford University Press.

Shreve, S. (2004). *Stochastic Calculus for Finance II: Continuous-Time Models*. Berlin: Springer.

Siegel, A. (1995). Measuring systematic risk using implicit beta. *Management Science* **41**: 124–8.

Sircar, R. and Papanicolaou, G. (1999). Stochastic volatility, smile and asymptotics. *Applied Mathematical Finance* **6**(2): 107–45.

Sircar, R. and Zariphopoulou, T. (2005). Bounds & asymptotic approximations for utility prices when volatility is random. *SIAM Journal on Control & Optimization* **43**(4): 1328–53.

Sircar, R. and Zariphopoulou, T. (2010). Utility valuation of credit derivatives and application to CDOs. *Quantitative Finance* **10**(2): 195–208.

Souza, M. and Zubelli, J. (2007). On the asymptotics of fast mean-reversion stochastic volatility models. *International Journal of Theoretical and Applied Finance* **10**(5): 817–35.

Stein, E. and Stein, J. (1991). Stock price distributions with stochastic volatility: an analytic approach. *Review of Financial Studies* **4**(4): 727–52.

Touzi, N. (1999). American option exercise boundary when the volatility changes randomly. *Applied Mathematics and Optimization* **39**: 411–22.

Vecer, J. (2002). Unified pricing of Asian options. *RISK*.

Wiggins, J. (1987). Option values under stochastic volatility. *Journal of Financial Economics* **19**(2): 351–72.

Willard, G. (1996). *Calculating prices and sensitivities for path-independent derivative securities in multifactor models*. PhD thesis, Washington University in St. Louis.

Wilmott, P., Howison, S. and Dewynne, J. (1996). *Mathematics of Financial Derivatives: A Student Introduction*. Cambridge: Cambridge University Press.

Yin, G. and Zhang, Q. (1998). *Continuous-Time Markov Chains and Applications. A Singular Perturbation Approach*, Applications of Mathematics 37. Berlin: Springer.

Zariphopoulou, T. (2001). A solution approach to valuation with unhedgeable risks. *Finance & Stochastics* **5**(1): 61–82.

Zhang, J. and Shu, J. (2003). Pricing Standard & Poor's 500 index options with Heston's model, *Computational Intelligence for Financial Engineering, 2003. Proceedings 2003 IEEE International Conference*, pp. 85–92.

Zhou, C. (2001a). An analysis of default correlations and multiple defaults. *Review of Financial Studies* **14**(2): 555–76.

Zhou, C. (2001b). The term structure of credit spreads with jump risk. *Journal of Banking and Finance* **25**: 2015–40.

Zhu, J. and Avellaneda, M. (1997). A risk-neutral stochastic volatility model. *International Journal of Theoretical and Applied Finance* **1**(2): 289–310.

Index

1987 crash, 53, 84

adapted, 2
adjoint operator, 91, 100
affine
    LMMR, 151, 269, 304
    models, 310
    term structure, 311
American options, 11, 19, 32, 50
Asian options, 42, 185
asymptotic analysis, 190, 199
    non-Markovian case, 228
averaging effect, 204

barrier options, 12, 36, 50, 181, 347
Bellman principle, 288
Black–Cox model, 343, 345
Black–Scholes
    formula, 17, 52, 76, 149
    model, 51
    operator, 16, 128, 233
    partial differential equation, 15, 16, 31, 57
    theory, 1, 56
bond, 307, 342
bond options, 335
bottom-up, 388, 402
Brownian motion, 2, 50

calibration, 56, 73, 84, 148, 153, 154, 209, 304,
    351, 416
call option, 145, 149
call options, 10, 19, 263
Cameron–Martin formula, 22
Capital Asset Pricing Model (CAPM), 296
Cauchy–Schwarz, 5
CDO tranches, 387, 413
centering condition, 93, 128, 250, 292
characteristic function, 2

Collateralized Debt Obligation (CDO), 371,
    386, 393, 402, 413, 424
complete market, 49, 51, 57
compound options, 41
conditional shift, 229
contingent claims, 10
control policy, 288
copula, 386, 425
correction
    hedging, 206
    optimal asset allocation, 292, 295
    pricing, 187
correlation, 59, 62–64, 68, 83, 206, 296
Cox process, 380, 389
Cox–Ingersoll–Ross (CIR), 62, 64, 104, 109,
    123, 260, 307, 330
Credit Default Swap (CDS), 342, 383, 416
credit derivatives
    multiname, 353, 385
    single-name, 342
credit risk, 43, 50, 342, 377

data, 56, 71, 110, 114, 154, 304, 351, 416
day effect, 117, 249
default, 43, 342, 403
derivative
    contracts, 10
    price, 11
    pricing, 11, 65
digital (binary) options, 179
distortion power, 293
dividends, 5, 122, 232
Dupire's formula, 61
dynamic programming, 288

equivalent martingale measure, 22, 32, 62, 69,
    71
ergodic

process, 64, 95, 252
  property, 88, 111, 120
exotic options, 12, 179

fast mean-reversion, 113
fat tails, 64
Feynman–Kac formula, 29, 32, 48, 124, 143,
    390
filtration, 2, 66, 71, 380
financial crisis, 424
first passage, 43, 343
flow property, 27
forward measure, 336
forward rates, 313
forward-start options, 41
free boundary value problems, 19, 33, 189

Gamma distribution, 105, 109, 123
Gaussian distribution, 101
geometric Brownian motion, 8, 58, 200
Girsanov's theorem, 23, 47, 50, 69, 140, 285
Greeks, 19, 137, 184, 207, 216, 236, 346
Gronwall's lemma, 5
group market parameters, 123, 130, 134, 138,
    149, 235, 323

Hamilton–Jacobi–Bellman equation, 288, 290
HARA utility function, 288
hedging, 14, 16, 17, 51, 66, 83, 199
  ratio, 18, 57, 71, 207, 218
  strategy, 199
    cumulative cost, 210
    mean self-financing, 206
    P&L, 202
hedging strategy total cost, 202, 210
Hermite polynomials, 102
Heston model, 64, 74, 259
Hull–White formula, 76, 78

implied volatility, 52, 74, 148, 217
  surface, 53, 72, 84, 152, 206
incomplete market, 49, 51, 62, 199
infinitesimal generator, 28, 68, 98, 100, 105,
    124, 223, 250
intensity-based models, 377
interest rate, 122, 234, 307
invariant distribution, 88, 91, 111, 113, 122, 223
inverse problem, 51
Itô's formula, 6, 7, 43, 46, 50

jumps, 85, 95, 97, 111, 120, 251

Kolmogorov equation, 30, 60, 90

Laguerre polynomials, 106
large deviation, 276
leverage effect, 63

lognormal, 8, 51
lookback options, 39
loss distribution, 365, 376, 391

market price of volatility risk, 68, 70, 71, 123,
    139
Markov property, 11, 25, 27, 29, 49, 89, 95
martingale, 7, 22, 23, 25, 202
  approach, 220
  decomposition result, 222
  property, 4, 29, 221
  representation theorem, 25, 49, 50, 71
mean-reversion, 63, 88, 100, 116, 122, 308
  rate of, 93, 103, 111
Merton
  model, 43
  problem
    constant volatility, 287
    stochastic volatility, 289
  solution, 287, 291, 295
method of images, 50
Monte Carlo
  control variates, 286
  importance sampling, 283
  variance reduction, 283
Mortgage-Backed Securities (MBS), 424
multidimensional volatility models, 254
multiscale, 86, 118, 121, 271, 388

no-arbitrage principle, 11, 14, 15, 24, 33, 189
non-Markovian volatility models, 226

optimal asset allocation problem, 287
optimal stopping, 33, 50, 190
options, 10
Ornstein–Uhlenbeck process, 64, 100, 107, 112,
    120, 123, 255, 308

path-dependent derivatives, 36, 181
penalization methods, 198
perturbation, 121, 125, 146, 164, 199
perturbed binomial formula, 369
Poisson
  equation, 93, 106, 107, 109, 129, 209, 250,
    253, 257, 292, 294, 319
  process, 98, 377
probability space, 2, 66, 70
put options, 10, 20, 189
put–call parity, 18

quadratic model, 327
quadratic variation, 4, 7

Radon–Nikodym derivative, 23, 79, 284, 336
recovery, 383
Renault–Touzi theorem, 76
replicating portfolios, 13, 14, 70

returns, 110, 111
reversibility, 91, 96, 99, 101, 105
risk management, 199
risk-neutral
  pricing, 21
  probability, 23, 57, 122
  valuation, 24
  world, 22, 26, 206

S&P 500, 53, 88, 114, 155, 172, 306
self-financing, 14, 16, 23, 24, 71, 200
semigroup, 90
Sharpe ratio, 68, 261
skew, 55, 80, 83, 84, 113, 159, 176, 245, 269
small noise effect, 205, 211
smile, 53, 76
  curve, 84
spectral gap, 91, 102, 106
spectrum, 117
stability of parameters, 84
stochastic
  differential equation, 1, 5, 50, 57, 62, 122
  integral, 3, 50
stochastic volatility, 94
  models, 62, 74
  multifactor, 353
  two-factor, 86

structural models, 43, 342
super-replication, 72, 84

time
  homogeneity, 27
  scales, 86, 90, 120
  series, 57, 72, 120
top-down, 386, 388, 402
transaction costs, 71

utility of wealth, 288

variogram, 115
Vasicek model, 307, 388
VIX, 88
volatility, 4
  effective, 128, 257
  integrated square, 89, 94
  local, 57
  process, 62
  time-dependent, 58
  uncorrelated, 75

Wiener measure, 2

yield curve, 311
yield spread curve, 343, 348

Printed in the United States
by Baker & Taylor Publisher Services